The Minerals, Metals & Materials Series

Lei Zhang · Jaroslaw W. Drelich
Neale R. Neelameggham
Donna Post Guillen · Nawshad Haque
Jingxi Zhu · Ziqi Sun · Tao Wang
John A. Howarter · Fiseha Tesfaye
Shadia Ikhmayies · Elsa Olivetti
Mark William Kennedy
Editors

Energy Technology 2017

Carbon Dioxide Management and Other Technologies

TMS

Springer

Editors
Lei Zhang
University of Alaska Fairbanks
Fairbanks, AK
USA

Tao Wang
Nucor Steel
Manila, AR
USA

Jaroslaw W. Drelich
Michigan Technological University
Houghton, MI
USA

John A. Howarter
Purdue University
West Lafayette, IN
USA

Neale R. Neelameggham
IND LLC
South Jordan, UT
USA

Fiseha Tesfaye
Åbo Akademi University
Turku
Finland

Donna Post Guillen
Idaho National Laboratory
Idaho Falls, ID
USA

Shadia Ikhmayies
Al Isra University
Amman
Jordan

Nawshad Haque
CSIRO
Canberra
Australia

Elsa Olivetti
Massachusetts Institute of Technology
Cambridge, MA
USA

Jingxi Zhu
SYSU-CMU Joint Institute of Engineering, Sun Yat-sen University
Guangzhou
China

Mark William Kennedy
Proval Partners SA
Lausanne
Switzerland

Ziqi Sun
Queensland University of Technology
Brisbane
Australia

ISSN 2367-1181 ISSN 2367-1696 (electronic)
The Minerals, Metals & Materials Series
ISBN 978-3-319-52191-6 ISBN 978-3-319-52192-3 (eBook)
DOI 10.1007/978-3-319-52192-3

Library of Congress Control Number: 2016963179

Printed on acid-free paper

This Springer imprint is published by Springer Nature
The registered company is Springer International Publishing AG
The registered company address is: Gewerbestrasse 11, 6330 Cham, Switzerland

Preface

This volume contains selected papers presented at the Energy Technologies Symposium organized in conjunction with the TMS 2017 Annual Meeting & Exhibition in San Diego, CA, USA, and organized by the TMS Energy Committee. The papers in this volume intend to address the issues, intricacies, and challenges related to energy and environmental science. This volume also contains selected papers from the Solar Cell Silicon Symposium, Deriving Value from Challenging Waste Materials: Recycling and Sustainability Joint Session Symposium, and Advances in Environmental Technologies: Recycling and Sustainability Joint Session Symposium.

The Energy Technologies Symposium was open to the participants from both industry and academia and focused on energy-efficient technologies, including innovative ore beneficiation, smelting technologies, recycling, and waste heat recovery. This volume also covers various technological aspects of sustainable energy ecosystems, processes that improve energy efficiency, reduce thermal emissions, and reduce carbon dioxide and other greenhouse emissions. Papers addressing renewable energy resources for metals and materials production, waste heat recovery and other industrial energy efficient technologies, new concepts or devices for energy generation and conversion, energy efficiency improvement in process engineering, sustainability and life cycle assessment of energy systems, as well as the thermodynamics and modeling for sustainable metallurgical processes are included. This volume also includes topics on CO_2 sequestration and reduction in greenhouse gas emissions from process engineering, sustainable technologies in extractive metallurgy, as well as the materials processing and manufacturing industries with reduced energy consumption and CO_2 emission. Contributions from all areas of non-nuclear and non-traditional energy sources, such as solar, wind, and biomass, are also included in this volume.

We hope this volume will provide a reference for the materials scientists and engineers as well as metallurgists for exploring innovative energy technologies and novel energy materials processing.

We would like to acknowledge the contributions from the authors of the papers in this volume, the efforts of the reviewers dedicated to the manuscripts review process, and the help received from the publisher. We appreciate the efforts of Energy Committee Chair Cong Wang for enhancing the *Energy Technology 2017* proceedings. We also acknowledge the organizers of the three symposia (Solar Cell Silicon Symposium, Deriving Value from Challenging Waste Materials: Recycling and Sustainability Joint Session Symposium, and Advances in Environmental Technologies: Recycling and Sustainability Joint Session Symposium) contributing the papers to this volume.

Energy Technologies Symposium Organizers

Lei Zhang
Jaroslaw W. Drelich
Neale R. Neelameggham
Donna Post Guillen
Nawshad Haque
Jingxi Zhu
Ziqi Sun
Tao Wang
John A. Howarter
Fiseha Tesfaye
Shadia Ikhmayies
Elsa Olivetti
Mark William Kennedy

Contents

Part I Energy Technologies: Energy Technologies

Continuous Optimization of the Energy Input—The Success Story of AOS 3
Felix Wolters and Michael Schütt

Part II Energy Technologies: CO_2 Management and Sustainable Metallurgical Processes

Transforming the Way Electricity is Consumed During the Aluminium Smelting Process 15
Mark Dorreen, Linda Wright, Geoff Matthews, Pretesh Patel and David S. Wong

The Thermodynamics of Slag Forming Inorganic Phases in Biomass Combustion Processes 27
Daniel K. Lindberg and Fiseha Tesfaye

Leaching of Sb from TROF Furnace Doré Slag 43
P. Halli, S. Jolivet, A. Klöfverskjöld, P. Latostenmaa, B.P. Wilson and M. Lundström

Potential CO_2 Emission Reduction and H_2 Production Using Industrial Slag Wastes Originating from Different Industrial Sectors 51
Jinichiro Nakano, James Bennett and Anna Nakano

Part III Energy Technologies: Novel Technologies

Modeling Anthropogenic Heat Flux in Climate Models 63
Ganesan Subramanian and Neale R. Neelameggham

In-Situ Microscopic Study of Morphology Changes in Natural Hematite and Cu-Spinel Particles During Cyclic Redox Gas Exposures for Chemical Looping Applications 69
Anna Nakano, Jinichiro Nakano and James Bennett

Thermodynamic Stability of Condensed Phases in the Ternary System CaO–Cu–O by the EMF Method 79
Joseph Hamuyuni, Dmitry Sukhomlinov, Mari Lundström and Pekka Taskinen

Experimental Study on Electro-Spraying of Ethanol Based on PDA Measurement 87
Haige Li, Yunhua Gan, Xiaowen Chen, Yang Tong and Meilong Hu

Part IV Energy Technologies: Heat Recovery

Integrated Utilization of Sewage Sludge for the Cement Clinker Production 95
Zhenzhou Yang and Zuotai Zhang

Valuable Metals and Energy Recovery from Electronic Waste Streams 103
Fiseha Tesfaye, Daniel Lindberg and Joseph Hamuyuni

Dry Granulation of Hot Metal and Heat Recovery from Off-Gas 117
Wenchao He, Xueqin Li, Xuewei Lv, Jie Qiu and Jie Dang

The Energy Recovery of Livestock Waste in Taiwan 123
Esher Hsu and Chen-Ming Kuo

Life Cycle Assessments of Incineration Treatment for Sharp Medical Waste 131
Maryam Ghodrat, Maria Rashidi and Bijan Samali

Part V Energy Technologies: Poster Session

Effect of Granularity on Pretreatment of Coke With Microwave Irradiation 147
Qing-hai Pang, Zhi-jun He, Jun-hong Zhang, Wen-long Zhan, Teng-fei Song and Zhe Ning

Effect of Microwave and Ultrasonic Coupling Treatment on Granularity and Microstructure of Pulverized Coal 155
Zhi-jun He, Ji-hui Liu, Qing-hai Pang, Jun-hong Zhang, Wen-long Zhan and Zhe Ning

Influence of Sodium on Coke Microstructure in Different Reaction Atmosphere 161
Zhijun He, Wenlong Zhan, Junhong Zhang, Qinghai Pang, Sen Zhang and Chen Tian

Part VI Deriving Value from Challenging Waste Materials: Recycling and Sustainability Joint Session: Deriving Value from Challenging Waste I

Maximizing the Values of Steelmaking Slags 173
Naiyang Ma

Recycling of Zinc from the Steelmaking Dust in the Sintering Process 181
Piotr Palimaka, Stanislaw Pietrzyk and Michal Stepien

Direct Preparation of Metal Doping Ni–Zn Ferrite from Zn-Containing Electric Arc Furnace Dust by Calcination Method 191
Hui-gang Wang, Min Guo and Mei Zhang

Part VII Deriving Value from Challenging Waste Materials: Recycling and Sustainability Joint Session: Deriving Value from Challenging Waste II

Hydrometallurgical Processing of Copper Smelter Dust for Copper Recovery as Nano-particles: A Review 205
D.O. Okanigbe, A.P.I. Popoola and A.A. Adeleke

Part VIII Deriving Value from Challenging Waste Materials: Recycling and Sustainability Joint Session: Deriving Value from Challenging Waste III

Chromium Removal from Iron-Rich Waste Generated During Processing Lateritic Nickel Ores 229
Hong Vu, Tomas Frydl, Petr Dvorak, Jana Selucka and Petra Starkova

Removal of Magnesium from Liquor Produced by Nickel Mining by Crystallization 239
Kristine Bruce Wanderley, Denise Crocce Romano Espinosa and Jorge A. Soares Tenório

Synthesis of Magnesium Oxide from Ferronickel Smelting Slag Through Hydrochloric Acid Leaching-Precipitation and Calcination 247
M.Z. Mubarok and A. Yudiarto

Thermodynamic Analysis of the Recycling of Aircraft AL Alloys 259
Senlin Cui and In-Ho Jung

Lithium-Ion Battery Recycling Through Secondary Aluminum Production 267
Reza Beheshti, Ali Tabeshian and Ragnhild E. Aune

Part IX Deriving Value from Challenging Waste Materials: Recycling and Sustainability Joint Session: Poster Session

Alternative Method for Materials Separation from Crystalline Silicon Photovoltaic Modules 277
Pedro F.A. Prado, Jorge A.S. Tenório and Denise C.R. Espinosa

Bioleaching Process for Metal Recovery from Waste Materials 283
Solange Kazue Utimura, Carlos Gonzalo Alvarez Rosario, Amilton Barbosa Botelho, Jorge Alberto Soares Tenório and Denise Crocce Romano Espinosa

Chemical Analysis of Sludge Originating from Industrial Painting Performed in Brazil 291
Rita de Cássia S.S. Alvarenga, Henrique de Paula Santos, Beatryz C. Mendes, Maurício Paulo F. Fontes, Eduardo Antônio G. Marques and Kléos M.L. Cesar

Preparing Ferrosilicon Alloy with Copper Slag 301
Ruirui Wei, Mingrui Yang and Xuewei Lv

Research on Optimization of Sintering Mixture with Low-Grade Complex Ore 311
Yuchuan Ding, Zizong Zhu, Zhiqiang Zhou, Hao Xiong and Libin Zhu

Part X Solar Cell Silicon: Silicon Production, Crystallization, and Properties

Study on Producing Solar Grade Silicon by Carbothermal Reduction of Andalusite Ore 325
Shilai Yuan, Huimin Lu and Panpan Wang

Phase Analysis of the Si-O_2 System 333
Shadia J. Ikhmayies

Characterization of Composition, Morphology, and Structure of Disi Raw Sandstones in Jordan 343
Shadia J. Ikhmayies, Bothina M. Hamad, Abdulkader M. Abed, Belal S. Amireh and Yulia Valery Meteleva

Part XI Solar Cell Silicon: Silicon Impurity Removal and Refining

Effect of Magnesium Addition on Removal of Impurities from Silicon by Hydrometallurgical Treatment 355
Stine Espelien, Gabriella Tranell and Jafar Safarian

Evaporation Removal of Boron in Molten Silicon Using Reactive Fluxes 367
Ye Wang and Kazuki Morita

Part XII Solar Cell Silicon: Silicon Photovoltaics

Electrodynamic Eddy Current Separation of End-of-Life PV Materials 379
York R. Smith, James R. Nagel and Raj K. Rajamani

Investigation on Quartz Crucibles for Monocrystalline Silicon Ingots for Solar Cells 387
M. Di Sabatino, F.W. Thorsen, A. Lanterne, Y. Hu, J.A. Bones and E. Øvrelid

Influence of Oxygen Content on the Wettability of Silicon on Graphite 395
Zineb Benouahmane, Lifeng Zhang and Yaqiong Li

Particle Separation in Silicon Ingot Casting Using AC Magnetic Field 403
V. Bojarevics, G. Djambazov and K. Pericleous

Part XIII Advances in Environmental Technologies: Recycling and Sustainability Joint Session: Advances in Environmental Technologies: Characterization and Uncertainty

Characteristics of Municipal Solid Waste Incineration Bottom Ash With Particulate Matters PM 2.5–PM 10 413
T. Thriveni, Ch. Ramakrishna and Ahn Ji Whan

Part XIV Advances in Environmental Technologies: Recycling and Sustainability Joint Session: New Areas of Value Recovery

Accelerating Life-Cycle Management Protocols for New Generation Batteries 423
Timothy W. Ellis and John A. Howes

Jaroslaw W. Drelich received his B.S. degree in chemistry and M.S. degree in chemical technology from the Technical University of Gdansk (TUG), Poland, in 1983, and earned his Ph.D. degree in metallurgical engineering from the University of Utah in 1993. He came to Michigan Technological University (Michigan Tech) in 1997 and currently is a professor of materials science and engineering. His main research interests are in applied surface chemistry and interfacial engineering for ore dressing and materials processing, materials characterization, formulation, modification, and testing of biomaterials and antimicrobial materials. In addition to teaching several courses on characterization and processing of materials at Michigan Tech, Dr. Drelich has edited six books, published nearly 200 technical papers (cited 4600 times as of September 2016), holds nine patents, has made over 40 invited talks including keynote addresses, and has delivered or co-authored nearly 100 conference presentations.

Dr. Drelich is the editor in chief for the *Surface Innovations* journal. He is an active member of The Minerals, Metals & Materials Society (TMS), Society for Mining, Metallurgy and Exploration (SME), and American Chemical Society (ACS), and has served on a number of different committees, most recently he served as chair of the TMS Energy Committee in 2012–2014. Dr. Drelich is also an advocate of interdisciplinary capstone senior design projects and promotes interdisciplinary projects between engineering and business to encourage the students to think about both technical and business viability and interconnections. In recognition of his entrepreneurial activities, Dr. Drelich has been awarded with the *2012 Food Safety Innovation Award* from Great Lakes Entrepreneur's Quest in Michigan.

Neale R. Neelameggham is "The Guru" at IND LLC, involved in international consulting in the field of metals and associated chemicals (boron, magnesium, titanium, and lithium and rare earth elements), thiometallurgy, energy technologies, soil biochemical reactor design, etc. He was a visiting expert at Beihang University of Aeronautics and Astronautics, Beijing, China. He was a plenary speaker at the Light Metal Symposium in South Africa—on low carbon dioxide emission processes for magnesium.

Dr. Neelameggham has more than 38 years of expertise in magnesium production and was involved in the process development of its start-up company NL Magnesium through to the present US Magnesium LLC, UT until 2011. Neelameggham and Brian Davis authored the ICE-JNME award winning (2016) paper—"21st Century Global Anthropogenic Warming Convective Model"—which notes that constrained air mass warming is independent of the energy conversion source—fossil or renewable energy. He is presently developing Agricoal™ and agricoalture to improve arid soils.

Dr. Neelameggham holds 16 patents and patent applications and has published several technical papers. He has served in the Magnesium Committee of Light Metals Division (LMD) of TMS since its inception in 2000, chaired it in 2005, and in 2007 he was made a permanent co-organizer for the Magnesium Symposium. He has been a member of the Reactive Metals Committee, Recycling Committee, Titanium Committee, and Programming Committee Representative of LMD and LMD council.

Dr. Neelameggham was the inaugural chair, when in 2008, LMD and EPD (Extraction and Processing Division) created the Energy Committee, and has been a coeditor of the energy technology symposium proceedings through the present. He received the LMD Distinguished Service Award in 2010. While he was the chair of Hydro and Electrometallurgy Committee, he initiated the rare metal technology symposium in 2014. He is coeditor for the 2017 proceedings for the symposia on magnesium technology, energy technology, rare metal technology, and solar cell silicon.

Donna Post Guillen holds the position of distinguished research engineer in the Materials Science and Engineering Department at the Idaho National Laboratory. Dr. Guillen earned a B.S. in mechanical engineering from Rutgers University, an M.S. in aeronautics from Caltech, and a Ph.D. in engineering and applied science from Idaho State University. She has more than 30 years of research experience and has served as principal investigator for numerous multidisciplinary research projects on the topics of waste heat recovery, synthetic fuels production, nuclear reactor fuels and materials experiments, and waste glass vitrification. The focus of her research is on multiphase computational fluid dynamics and thermal hydraulics for sustainable energy technologies. She applies numerical modeling techniques to provide understanding of a wide variety of complex systems such as greenhouse gas generation/sequestration for dairies, working with industry (General Electric) to develop a direct evaporator for an organic rankine cycle, and modeling the waste vitrification process for the Hanford Waste Treatment Plant and the fluidized bed for the Idaho Waste Treatment Unit. She is experienced with irradiation testing and thermal hydraulic analysis for nuclear reactor experiments and serves as principal investigator/technical lead for the Nuclear Science User Facility Program. She is the lead inventor on two patents related to the development of a new composite material to produce a fast reactor environment within a pressurized water reactor. She actively mentors students, routinely chairs and organizes technical meetings for professional societies, serves in leadership capacity for the American Nuclear Society (Thermal Hydraulics Executive and Program committees), The Minerals, Metals & Materials Society (TMS—Energy Committee Past-Chair) and the American Society of Mechanical Engineers (Thermal Hydraulics and Computational Fluid Dynamic Studies Track Co-Chair), provides subject matter reviews for proposals and technical manuscripts, has published more than 100 conference papers, reports, and journal articles, and written/edited three books.

Nawshad Haque is a senior scientist and team leader at the Australian national research agency CSIRO. He is leading a range of projects that evaluates technology for mining industries for saving energy, water, and operating costs. Dr. Haque joined CSIRO Mineral Resources as a research scientist (Process Modelling) in 2007. His current research focuses on process, project, and technology evaluation applying life cycle assessment (LCA) methodology and techno-economic capabilities using various tools, software, and databases. He has contributed to develop a number of novel technologies and flow sheets for "Mine to Metal" production at CSIRO. His publications and industry reports are widely used internally and externally and assist in decision-making both in Australia and internationally. Dr. Haque completed his doctorate with the Department of Chemical Engineering at the University of Sydney on process modelling, simulation, and optimization in 2002. He commenced work as a materials scientist at New Zealand Forest Research Institute (Scion) and later seconded to CSIRO at Clayton to conduct research on drying process simulation and technology evaluation for industries. He is an active leader in professional societies such as The Minerals, Metals & Materials Society (TMS), Australian Life Cycle Assessment Society, and AusIMM for events such as conferences and meetings. Dr. Haque supervised undergraduate and Ph.D. students. He has coordinated and offered mineral processing and life cycle assessment courses and workshops at universities. He has a number of international collaborations with the universities and publicly funded research laboratories on mineral and metal processing and sustainability. Kindly refer to his weblink for more information: http://people.csiro.au/H/N/Nawshad-Haque.aspx or www.users.tpg.com.au/nawshadul.

Jingxi Zhu is currently an assistant professor of the SYSU-CMU Joint Institute of Engineering of Sun Yat-sen University located in Guangzhou, China. She received her B.E. degree in metallurgical engineering from University of Science and Technology Beijing, China, in 2003 and a M.E. degree in metallurgical physical chemistry from Central Iron & Steel Research Institute, Beijing, China in 2006. She earned a M.Sc. degree and then Ph.D. degree in materials science and engineering from Carnegie Mellon University (CMU) in 2009 and 2011, respectively.

She has had research collaboration with CMU's Center for Iron and Steel Research and the Data Storage System Center as well as National Energy Technology Laboratory of the US Department of Energy and the Defense Advanced Research Projects Agency (DARPA), US Army. Her current research interests include novel energy application-related electronic devices enabled by a wide range of functional materials, including electronic, magnetic, and multiferroic metals and ceramics. She also specializes in advanced characterization with electron microscopy and micro/nanofabrication processing techniques.

Ziqi Sun is a senior lecturer at the School of Chemistry, Physics and Mechanical Engineering, Queensland University of Technology, Australia. He received his Ph.D. degree in 2009 from Institute of Metal Research, Chinese Academy of Sciences and his B.E. degree in 1999 from Central South University China. He was awarded with some prestigious awards and fellowships, including the TMS Young Leaders Development Award from The Minerals, Metals & Materials Society (TMS, 2015), Discovery Early Career Research Award from Australian Research Council (DECRA, 2014), Alexander von Humboldt Fellowship from AvH Foundation Germany (2009), Australian Postdoctoral Fellowship from Australian Research Council (APD, 2010), and the Vice-Chancellor's Research Fellowship from University of Wollongong (2013). He is also serving as vice chair of the Energy Committee within TMS, an editorial board member of *Scientific Reports* (Nature Publishing Group), an editorial board member of *Journal of Materials Science and Technology*

(Elsevier), a guest professor in Shenzhen Institute, Peking University, and an honorary fellow in University of Wollongong. Ziqi is the program leader for two ongoing Australian Research Council Projects. He held the roles as symposium organizers and session chairs in some prestigious conferences such as TMS conferences and ACerS annual conferences. His major research interest is the rational design of bio-inspired metal oxide nanomaterials for sustainable energy harvesting, conversion, and storage.

Tao Wang is currently Castrip Research Lab Supervisor at Nucor Steel. He is also the lead engineer in the process and product research and development areas. Dr. Wang's current focus is to develop and modify a novel thin strip casting technology which uses up to 90% less energy to process liquid steel into hot rolled steel sheets than conventional casting methods. Dr. Wang has rich experience in metallurgical thermodynamics, thermal energy storage and transfer, steelmaking, metal solidification and casting, and metal corrosion. Dr. Wang obtained his Ph.D. and M.S. from the University of Alabama, and he received his B.S. from Xi'an Jiao Tong University in China. In his areas of research, Dr. Wang has published multiple papers and patents which led to breakthroughs in thermodynamic modeling, high efficiency thermal energy transfer medium development, and thin strip metal casting technology.

Dr. Wang received the 2013 Light Metals Division (LMD) Best Energy Paper Award from TMS and is the 2016 TMS Young Leader Award winner. Also he was selected to become a member of TMS Emerging Leader Alliance (ELA) in 2015. Dr. Wang serves technical committees including the Energy Committee and Pyrometallurgy Committee in TMS, and Metallurgy-Steelmaking & Casting Technology Committee, Continuous Casting Technology Committee, Southeast Chapter in AIST.

John A. Howarter is an assistant professor in Materials Engineering at Purdue University with a joint appointment in Environmental & Ecological Engineering. His research interests are centered on synthesis, processing, and characterization of sustainable polymers and nanocomposites. His research impacts water treatment, thermal management in electronic devices, and material design for recycling and value recovery. John has been involved in the Public and Government Affairs (P&GA) Committee of The Minerals, Metals & Materials Society (TMS), where he currently serves as incoming chair. In addition to working with P&GA, Prof. Howarter is a regular contributor to the technical programming for both TMS and Materials Science & Technology (MS&T) conferences in areas of energy, recycling, and materials sustainability. John earned a B.Sc. from The Ohio State University in 2003 and Ph.D. from Purdue University in 2008, both in materials engineering. From 2009 to 2011, he was a National Research Council postdoctoral scholar in the Polymers Division of the National Institute of Standards and Technology in Gaithersburg, Maryland.

Fiseha Tesfaye received his M.Sc. in Materials Processing Technology from Helsinki University of Technology (2009) and Ph.D. in Metallurgy from Aalto University (2014). He worked at the Helsinki University of Technology from 2008 to 2010 in the Department of Materials Science and Engineering. In 2010, he joined the Metallurgical Thermodynamics and Modeling research group at Aalto University, School of Chemical Technology. He is currently a senior researcher at Åbo Akademi University in the Johan Gadolin Process Chemistry Centre. His achievements include significant improvement of research in solid-state electrochemistry for thermodynamic studies of chalcogenide materials, and he contributed new experimental thermodynamic data for several alloys and compounds. Dr. Tesfaye's research interests are in the areas of extractive/process metallurgy, recycling technology, and materials science. He has published more than 30 papers, 12 of which are in high impact journals, and a book in these areas. He is an active member of professional associations American Chemical Society (ACS) and The Minerals, Metals & Materials Society (TMS). He has been serving as a reviewer for journals and research project proposals.

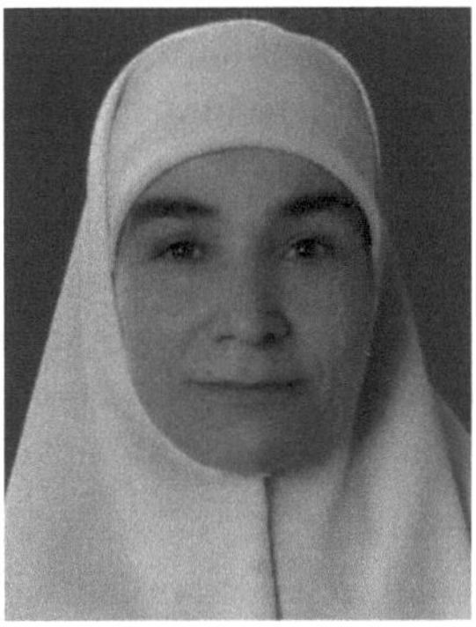

Shadia Ikhmayies had received the B.Sc. from the physics department in the University of Jordan in 1983, the M.Sc. in molecular physics from the same university in 1987, and the Ph.D. in producing CdS/CdTe thin film solar cells from the same university in 2002. Now she works in Al Isra University in Jordan as an associate professor. Her research is focused on producing and characterizing semiconductor thin films and thin-film CdS/CdTe solar cells. Dr. Ikhmayies also works in characterizing quartz in Jordan for the extraction of silicon for solar cells and characterizing different materials by computation. She published 40 research papers in international scientific journals, three chapters in books, and 62 research papers in conference proceedings. She is the author of two books for Springer, *Silicon for Solar Cell Applications*, and *Performance Optimization of CdS/CdTe Solar Cells*, which are in production, and is editor in chief of the book *Advances in Silicon Solar Cells* for Springer and of the eBook series *Material Science: Current and Future Developments* for Bentham Science Publishers.

Shadia is a member of the The Minerals, Metals & Materials Society (TMS) and the World Renewable Energy Network (WREN). She is a member of the International Organizing Committee and the International Scientific Committee in the European Conference on Renewable Energy Systems (ECRES2015, ECRES2015). She is a member of the editorial board of the *International Journal of Materials and Chemistry* for Scientific & Academic Publishing, and the editor in chief of a book for the Research Signpost. She was a technical advisor/ subject editor for *JOM*. She is a guest editor for a special section in the *Journal of Electronic Materials*: European Conference on Renewable Energy Systems (2015). Shadia is a reviewer in 22 international journals, and she is chair of the TMS Materials Characterization Committee (2016–2017).

Elsa Olivetti is the Thomas Lord Assistant Professor in the Department of Materials Science and Engineering at Massachusetts Institute of Technology (MIT). She received her B.S. in engineering science from the University of Virginia and her Ph.D. in materials science from MIT working on development of nanocomposite electrodes for lithium-ion rechargeable batteries. Olivetti joined MIT's faculty in 2014 where her current research focuses on improving the environmental and economic sustainability of materials in the context of rapid-expanding global demand. Olivetti leverages machine learning as well as data mining coupled with engineering and macroeconomic models to determine the scaled impact of novel materials and processes.

Mark William Kennedy is chief technology officer at Proval Partners SA of Lausanne Switzerland and a part-time associate professor at the Norwegian University of Science and Technology (NTNU), Department of Materials Science and Engineering (as of January 2017).

Part I
Energy Technologies: Energy Technologies

Continuous Optimization of the Energy Input—The Success Story of AOS

Felix Wolters and Michael Schütt

Abstract In over 40 years of operation, engineers at Aluminium Oxid Stade GmbH (AOS) have acquired a lot of experience whereby they have developed a series of optimizations regarding the specific primary energy consumption. The tube digester technology, invented by VAW in Germany and applied on an industrial scale for the first time in Stade, was brought to perfection by its innovative cleaning methods, which led to a constantly high heat transfer, low pressure loss and excellent availability of the digesters. Further improvements such as the heat recovery from the exhaust gas stream of the salt heater as well as the optimization of one of the fluid bed calciners and the installation of a combined heat and power generator (CHP) led AOS to where it is now—a leading alumina refinery with regard to the lowest energy consumption per ton of alumina.

Keywords Energy optimization · Tube digester · Alumina

Introduction

Aluminium Oxid Stade GmbH (AOS) is an alumina plant (Fig. 1) with an annual production capacity of approximately 1.000.000 mt Al_2O_3. From 1970 to 1973, the plant was built with a designed annual production capacity of 600.000 mt Al_2O_3. Due to continuous optimisation and smaller enhancements, the production has been increased to its present value. Apart from the ongoing increase in production, engineers at AOS developed a series of optimizations to minimize the specific primary energy consumption.

Basis for the plant's low primary energy consumption is for one thing the tube reactor, for another the fluid bed calciner. The tube reactor technology, developed by VAW and primarily used at AOS, makes digestion at temperatures as high as

F. Wolters (✉) · M. Schütt
Aluminium Oxid Stade GmbH, Johann-Rathje-Köser-Strasse,
21683 Stade, Germany
e-mail: f.wolters@aos-stade.de

L. Zhang et al. (eds.), *Energy Technology 2017*,
The Minerals, Metals & Materials Series, DOI 10.1007/978-3-319-52192-3_1

Fig. 1 Production area at AOS

270 °C possible. Therefore, practically all kinds of bauxite can be processed. Ever increasing energy costs, cleaning costs and environmental aspects led AOS to introduce their innovative cleaning methods, to install heat recovery systems and to build a combined heat and power generator.

Tube Digester at AOS—Overview

At AOS four high temperature tube digesters are operated independently at a digestion temperature of 270 °C (Fig. 2). Digestion feed liquor and bauxite slurry are mixed prior to the digestion process and are pumped with a piston diaphragm pump through eight vapour heat exchangers using regenerative flash vapour (Fig. 3). Final stage heating to 270 °C is then accomplished by using molten salt. After adding oxygen for the wet oxidation process and lime slurry to e.g. enhance the yield, the heated slurry runs through a retention time track to extract alumina, before it enters a cascade of eight flash tanks. In each flash tank, vapour is recovered to heat the digester feed slurry stepwise in countercurrent flow. In the last heat recovery step, the pregnant liquor leaving the cascade is fed to a heat exchanger transferring heat to the digestion feed slurry, which is entering. By using this heat recovery system, more than 75% of the necessary digestion energy is reached.

Throughout the operation of the digesters at AOS, optimizations regarding the flash tanks led to an excellent quality of the condensate leaving the digester heat exchangers. Typical conductivities of 80–100 μS/cm during the lifetime of the digesters allow minimal cleaning effort in the regenerative flash system as well as

Fig. 2 AOS tube digester

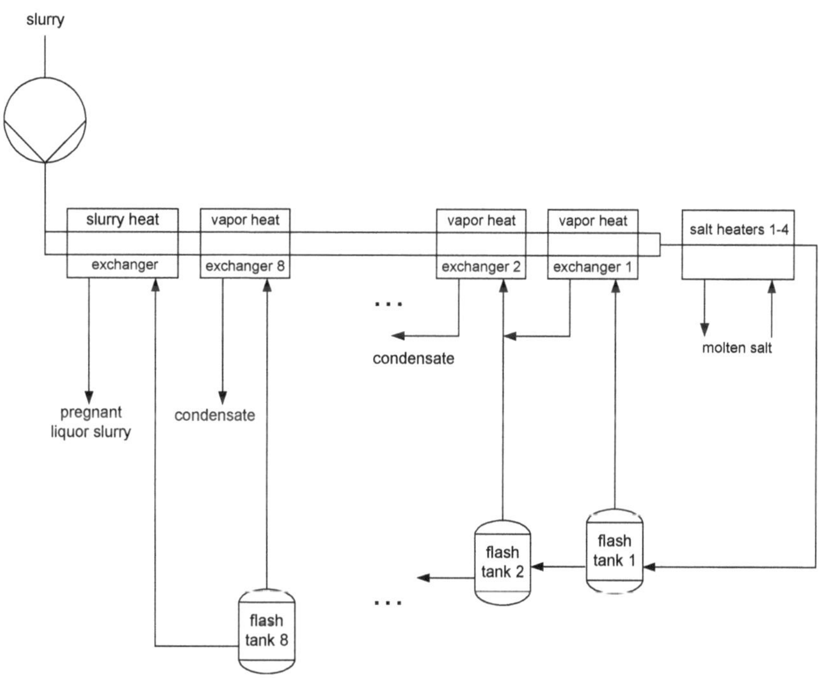

Fig. 3 Schematic of the tube digester at AOS

the washing of the product stream with clean condensate from the digestion process. Thus, energy-intensive evaporation plants are expendable.

Advantages and technical challenges with the digestion:

The tube digester technology developed in the 1960s by VAW in Germany and applied at AOS since 1973 includes several advantages such as [1, 2]:

- Approximately 40% less energy consumption compared to autoclaves under similar conditions due to the regenerative flash system and high turbulence in the digester pipes leading to an excellent heat and mass transfer and thus to a low need of caustic liquor for the digestion process.
- Using molten salt, which is heated with natural gas, the primary energy input is reduced.
- Clean condensate produced in the regenerative flash system of the digesters makes an evaporation plant expendable.
- Digestion temperatures up to 270 °C allow the use of practically all kinds of Bauxite. Bauxites such as Weipa, Gove and Boké have successfully been processed at AOS.
- High degree of automation due to continuous operation of the digesters leading to reduced labour intensity.

One challenge every Bauxite slurry plant has to face, though, is rapid scaling of reactor surfaces caused by silica and titanate. Typically, the rate of chemical scale deposition increases in zones of stagnation. Even though the turbulence in the tube digester pipes is high—Reynolds numbers higher than 1.4×10^5—silica scaling occurs in low temperature sections and hard titanate scaling at the high temperature end. Thus, periodical shutdowns of the digester and de-scaling of the pipes are necessary in order to increase the heat transfer capacity and reduce the pressure loss.

At AOS, the digesters are operated independently, so that a shutdown of only one digester at a time is possible. De-scaling is usually done by acid treatment followed by mechanical cleaning. Originally, mechanical cleaning of the tubes at AOS was conducted by high-pressure water jet cleaning. During a shutdown, every return bend of the digester pipes had to be removed to provide access to the digester. The impact of this cleaning method on production is significant since it is time-consuming and labour as well as cost intensive. Additionally, high-pressure water jet cleaning could cause serious damage to the digester pipes since a water pressure of more than 1000 bar is necessary to break the hard titanate scales.

Consequently, the cleaning methods were reviewed by AOS engineers in order to reduce the application of high-pressure water jet cleaning during digester shutdown. The method of choice was pigging. Cleaning tools were tested and optimized and pig launchers and receivers were installed. Acid treatment followed by pigging made high-pressure cleaning of the digester pipes expendable. The result was a significant reduction of downtime, maintenance and operating costs.

Nowadays, digesters at AOS are shut down approximately every 1450 operating hours resulting in four short (one day) and two long time (four to five days) shut downs per year for cleaning and maintenance processes. Every cascade of flash tanks has to be cleaned twice a year. During the lifetime of the digesters, heat transfer is high and pressure loss, caused by scaling, is low (Fig. 4), resulting in a minimum energy input, a full production rate and an excellent availability of the digesters of over 93%.

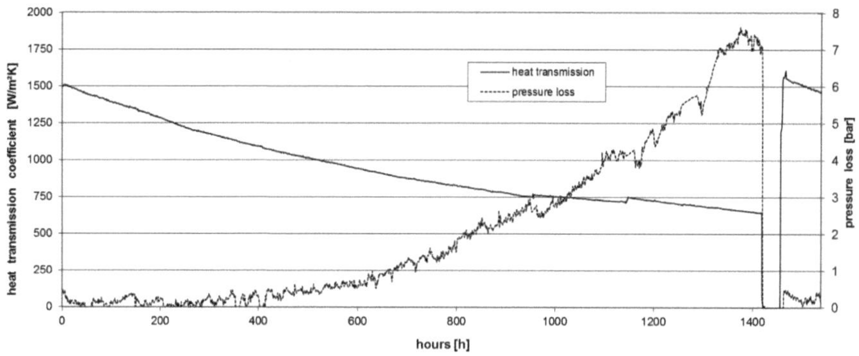

Fig. 4 Heat transmission of a salt heater and pressure loss caused by scaling over lifetime

Installation of a Combined Heat and Power Generator (CHP)

In several parts of the plant, steam is used as a heating medium. Before the installation of the CHP, steam was provided by the operation of three natural gas-fired steam boilers, and all the electrical energy needed was taken from the external electricity grid. Nowadays, two combined heat and power generators are installed at AOS to provide the plant with heat in the form of steam and electricity. With the two CHP it is possible to cover about 100% of the steam and 95% of the electricity demand at AOS. Due to the Renewable Energy Law, the expansion of renewable energies in Germany is being promoted, which may weaken the electricity grid. The CHP empowers AOS to work independently from the outer grid, particularly in cases of a sudden breakdown.

Environmental as well as economic reasons caused the installation of the CHP. With the use of natural gas instead of brown and pit coal as fuel for the electricity production, the CO_2 emission can be significantly reduced.

Generally, the total efficiency of processes, where heat and electricity are generated separately, is about 35%. In contrast to a high-efficiency CHP, as is installed at AOS, the total efficiency can be increased to more than 80%. To produce the same amount of electricity and heat in a separated process, about 35% more energy is needed, compared to a CHP-process [3, 4].

The comparison of a separated process (left part) and a combined process (right part) is illustrated in Fig. 5 where the same amount of thermal and electrical energy is generated. The sum of all individual losses during the generation of thermal and electrical energy in the separated process is much higher than in a combined process. With an optimal design and operation of a CHP plant, up to one third of the primary energy consumption can be saved in comparison to a separated generation of electrical and thermal energy. As shown in Fig. 5, instead of (60.5 + 75) % in a separated process, in a combined process only 100% of fuel is necessary to cover the demand of 54% thermal and 27% electrical energy.

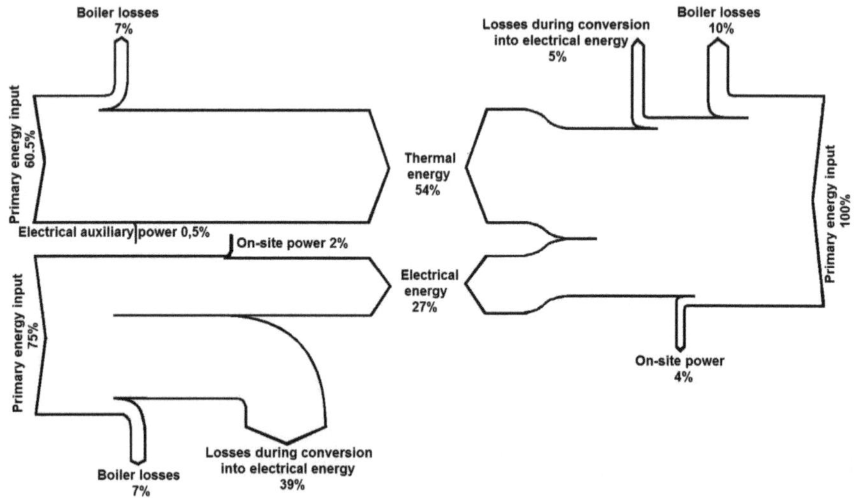

Fig. 5 Energy flow of a separated process compared to a combined process [4]

Modification of the Fluid Bed Calciner

The circulating fluid bed calciner (CFB) technique was developed by Lurgi-VAW with the aim of minimum energy consumption during the calcination process. In 1973, AOS became the first plant worldwide applying this technique to 100%. Three calciners with a designed capacity of 650 mt Al_2O_3/d each were installed.

Since the calcination process is still very energy consuming, optimizations of the calciners have been conducted in order to reduce the energy demand. In 2011, one calciner was modified by mainly adding two heat recovery steps [5]: Aluminium hydroxide entering the plant is preheated in an additional step which leads to a cooler exhaust gas stream. Furthermore, the secondary air stream is additionally preheated by alumina leaving the calciner. These changes resulted in a significant reduction of the primary energy demand of approximately 12%.

Heat Recovery System at the Salt Heater

To achieve the digestion temperature of 270 °C, AOS installed a salt heater, which is operated with natural gas. During the combustion and after the heat transfer to the molten salt, the exhaust gas has a temperature of about 500 °C. In the first step, the heat energy in the exhaust gas is used in a tube bundle heat exchanger to preheat the combustion air. The exhaust gas, which is cooled in the first step, still has a temperature of about 250 °C. Because of the utilization of natural gas, no sulphur content is generated; it is possible to cool the exhaust gas down to a temperature far below 100 °C, without producing condensate or even worse, sulphuric acid. The fact that the exhaust gas still has a high caloric energy and the possible cooling temperature of below 100 °C led to a second heat recovery step, where a new plant section was installed. With the installation of a recuperator in the exhaust gas stream it is now possible to further decrease the heat loss resulting from the hot exhaust gas and utilise the recovered energy for heating processes instead of steam.

The new plant section consists of a recuperator (Fig. 6) and a pressurized water system to transport the heat energy where it is needed in the plant (Fig. 7). For example, at AOS the recovered heat energy is applied to the diluter. Energy savings of more than 23.500 MWh/y are not uncommon, which is equivalent to about V_n = 2.13 mm^3 of natural gas.

Fig. 6 Heat recovery system at AOS

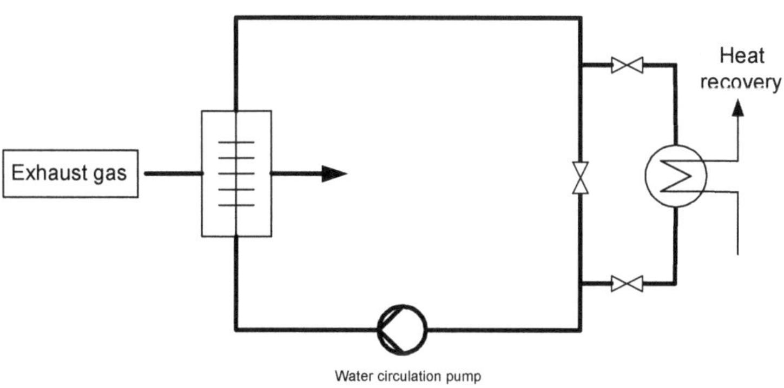

Fig. 7 Schematic of the heat recovery system at the salt heater

Total Energy Consumption

With the optimisations mentioned above, the total primary energy consumption has been significantly reduced over the last 10 years (Fig. 8). Compared to 2004, savings of over 16% have been achieved. Further improvements such as the replacement of control valves by speed-controlled engines or the substitution of energy intense lamps by LED lamps added to the savings.

The only increase in the energy input was observed in 2009 during the financial crisis. Due to difficulties on the alumina market, AOS was forced to reduce the production drastically. Based on the data provided by The International Aluminium institute [6], the same effect can be applied on Europe's refineries (Fig. 8).

Nevertheless, AOS still has the potential for further optimizations such as the modification of two calciners. Even though AOS would not reach the minimum total energy demand of a Greenfield 6 GJ/mt Al_2O_3 high temperature refinery postulated by Mach [7], AOS is still in the lead in comparison to the average energy input in Europe and worldwide. By 2015, the average energy consumption per tonne alumina was 36% (worldwide) and even 49% (Europe) higher than AOS's demand.

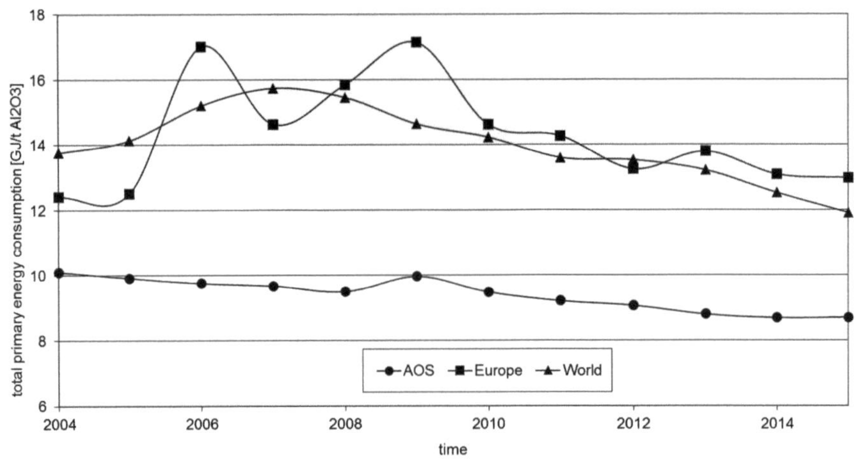

Fig. 8 Total energy consumption per tonne alumina [6]

Summary

The foundation for the success of AOS regarding the primary energy consumption per ton of alumina was laid by VAW engineers over 40 years ago. Then during operation, engineers at AOS found multiple ways to optimize the process and thus reduce the energy demand. For instance, the operation and cleaning methods of the digesters were improved, one calciner was modified and a heat recovery system recovering heat from the exhaust gas stream of the salt heater was installed. Additionally, CO_2 emissions were significantly reduced by the installation of a CHP. Even though every refinery aims to keep the energy demand of its plant to a minimum, AOS is still far ahead of the average in Europe as well as worldwide.

References

1. H. Peters, Rohraufschluss—später Erfolg einer VAW-Technologie. World Metallurgy—ERZMETALL **68**(2), 80–84 (2015)
2. F.W. Kämpf, *Der Bauxitaufschluss im Bayer-Verfahren. unveröffentlichte Ergebnisse* (Vereinigte Aluminium Werke AG, Lünen)
3. Bundesverband Kraft-Wärme-Kopplung e.V., Kraft-Wärme-Kopplung—Chance für Wirtschaft und Umwelt, URL: http://www.bkwk.de/fileadmin/users/bkwk/aktuelles/Broschur/BKWK_Chance_fuer_Wirtschaft_und%20Umwelt_Broschuere_A4_web.pdf (Update: 18.07.2016)
4. G. Schaumann, Kraft-Wärme-Kopplung, (Springer, 2010), pp. 6–19
5. L. Perander, A. Lalla, C. Klatt, B. Reeb, B. Petersen, M. Missalla, E. Guhl, Application of optimized energy efficient calcination configuration to AOS stade CFB calciners. in *Proceedings of the 9th International alumina quality workshop*, (2012)
6. The International Aluminium Institute, Metallurgical Alumina Refining Energy Intensity, Date of issue: 18 July 2016, URL: http://www.world-aluminium.org/statistics/metallurgical-alumina-refining-energy-intensity/#linegraph (Update: 18.07.2016)
7. T. Mach, Energy consumption in the bayer process, in *Proceedings of the 9th international alumina quality workshop*, (2012)
8. A. Lalla, R. Arpe, 12 years of experience with wet oxidation, in *TMS Light metals*, pp. 177–180 (2002)
9. E. Guhl, R. Arpe, Nearly 30 years of experience with lurgi calciners and influence concerning particle breakage, in *TMS light metals*, pp. 141–144 (2002)
10. A. Lalla, R. Arpe, 30 years of experience with tube digestion, in *Conference of Metallurgists*, pp. 105–113 (2006)

Part II
Energy Technologies: CO_2 Management and Sustainable Metallurgical Processes

Transforming the Way Electricity is Consumed During the Aluminium Smelting Process

Mark Dorreen, Linda Wright, Geoff Matthews, Pretesh Patel and David S. Wong

Abstract This paper examines the development of the new EnPot heat exchanger technology for aluminium smelters and the potential impact it could have on the sustainability and economics of primary aluminium production. The EnPot technology can be used to help the aluminium smelting industry be part of the solution to accommodate increased intermittency in our future renewable energy generation, post COP 21. The EnPot system provides for the first time, dynamic control of the heat balance of aluminium smelting pots across the potline, so that energy consumption and aluminium production can be increased or decreased by as much as plus or minus 30% almost instantaneously. This enables a new way of thinking to emerge when considering the relationship the aluminium smelter plays in connection to the power grid. The EnPot technology provides smelters with the means to free up power back to the grid, transforming the smelter from only an end user of power into a 'virtual battery' for the electricity network. This fundamentally changes the way electricity is consumed by the smelter and is a concept already being tested by several smelters.

Keywords Aluminium smelting · Energy flexibility · Virtual battery · Power grid · Enpot

M. Dorreen (✉) · D.S. Wong
Light Metals Research Centre, The University of Auckland,
Private Bag 92019, Auckland, New Zealand
e-mail: m.dorreen@auckland.ac.nz

L. Wright · G. Matthews
Energia Potior Ltd., Queenstown, New Zealand

P. Patel
Auckland UniServices Ltd., Private Bag 92019, Auckland, New Zealand

L. Zhang et al. (eds.), *Energy Technology 2017*,
The Minerals, Metals & Materials Series, DOI 10.1007/978-3-319-52192-3_2

Introduction

This paper examines the potential impact the newly developed 'EnPot' system [1] could have on the sustainability and economics of primary aluminium production. In particular, it tracks the development of the EnPot heat exchanger technology and explores how it could be used to help the aluminium smelting industry be part of the solution of accommodating increased intermittency in our future renewable energy generation, post COP 21 (2015 Paris Climate Conference).

The means by which our future growing energy needs will be met is one of the most globally discussed topics of our time. If the future of power generation is to come largely from renewable sources, replacing fossil fuel sources, then we must fundamentally change the way we consume energy.

Aluminium is one of the most important metals in our world today and is now the second most used metal globally after steel. Our per capita consumption of aluminium rises with our income as we purchase more consumer goods and seek to lightweight everything from phones and tablets, to cars. With the predicted rise in incomes in the developing world over the next 15 years, global demand for aluminium is expected to double over the same period and it is estimated that 40 to 50 new smelters will be needed to keep step with this increasing demand.

Aluminium smelting accounts for over 3% of the world's total electricity supply, which is equivalent to 16.5% of total global domestic consumption. In other words, the 200 smelters globally consume the equivalent amount of power as 1.2 billion people do domestically. The major challenge with aluminium smelting is that the process not only requires vast amounts of electricity, but it needs it continuously and at consistent levels to keep the electrolysis process running.

Current Model—The Energy-Production 'Straightjacket'

With the existing primary aluminium production process, the energy input to a smelter cannot be varied by more than roughly ±5% from its base operating point (Fig. 1). This means that a smelter essentially operates consistently at full capacity 24/7, 365 days a year, for its entire lifespan. The current maximum production 'straightjacket' that the aluminium industry is in has a significant and dramatic effect on both the cost of production for individual smelters, as well as the dynamics of the supply and demand curves of the industry as a whole. It also creates power supply issues for national grids, especially as nations seek to produce a higher percentage of their electricity from renewable sources (such as solar, wind, etc.), which inherently are more intermittent in their generation.

The electrochemical process used to make aluminium requires a powerful electric current (up to 600,000 A or 600 kA) to be driven through large carbon lined steel furnace pots filled with molten electrolytic bath (mostly cryolite, Na_3AlF_6) and aluminium oxide (Al_2O_3), kept at close to 950 °C. The electric

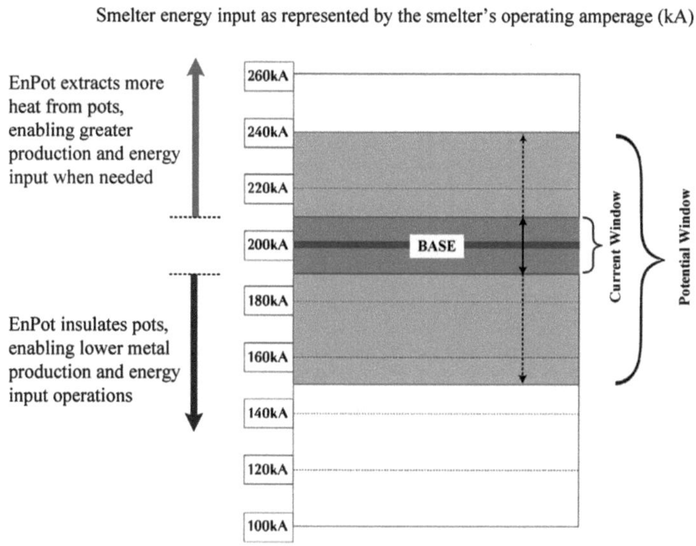

Fig. 1 Current window of flexibility in energy usage for smelters (±5%) compared to the potential operating window with EnPot technology (up to ±30%)

current breaks the molecular bonds of the oxide and the resulting denser metallic aluminium then collects at the bottom of the pot, ready for siphoning off and casting.

The inability of current smelters to be flexible in energy usage and production is due to the absolute importance of maintaining the aluminium reduction process within a very narrow "heat balance" temperature window, usually only within 30 °C, which must be maintained to keep the bath molten in pots and hence why smelters must consume electricity at a very consistent rate. The process produces a vast amount of heat as a by-product and how this heat is dispersed through the carbon lined steel pots is critical.

- If you want to *increase metal production*, you need to increase the amperage through the pot, increasing energy usage. This generates yet more heat, which then needs to be removed from the outside of the pot at a greater rate to prevent pot overheating and failure.
- Conversely if you want to *slow down production* or *reduce energy usage*, you need to reduce the amperage and then insulate the pot in order to keep it from cooling and solidifying (a worst case scenario resulting in loss of metal production entirely; reinstating full production is very costly, taking weeks if not months).

While smelters currently can carry out long-term increases and slow-downs in production and energy usage (through starting up/stopping individual pots from production, or small incremental increases in production after individual pots are

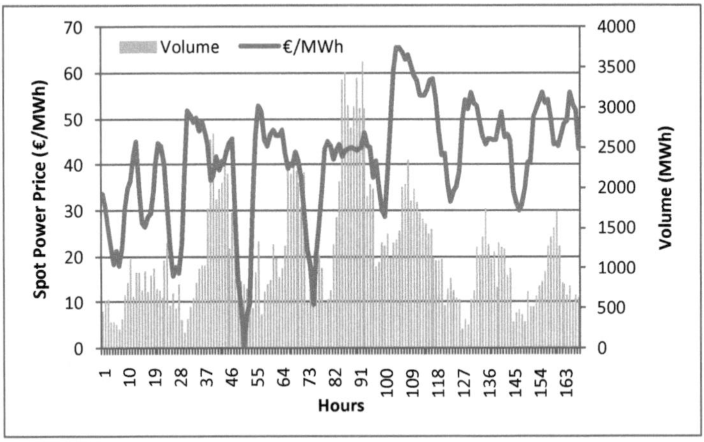

Fig. 2 European Energy Exchange (EEX) hourly power price and volume for the week of September 6, 2010. Reproduced from [2]

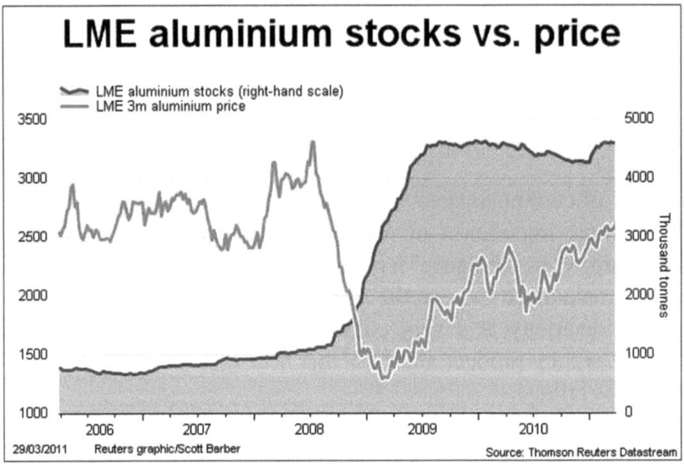

Fig. 3 Historic LME aluminium prices and how it can be affected by oversupplying and stockpiling in the aluminium market

upgraded to extract more heat), these require significant planning and capital expenditure. More importantly, the time scale for such changes (weeks/months) means that smelters are unable respond quickly to large daily/periodic shifts in either the energy market (spot power prices that smelters pay for electricity, for example Fig. 2), or in the metal commodity market (spot prices to sell metal produced, for example Fig. 3).

To operate flexibly and dynamically (on short notice) in both directions outside the current operating window therefore is practically impossible for most smelters.

With EnPot technology however, this fixed operating model has the potential to change fundamentally, providing an ability for smelters to increase/decrease production and energy usage almost instantaneously, whilst maintaining pots within a wider thermally balanced operating window.

EnPot Technology and Development

The EnPot heat exchanger technology was first developed and patented by a specialist team of researchers and engineers at the Light Metals Research Centre (LMRC) at The University of Auckland. Figure 4 highlights more than a decade of historical development, commencing with fundamental PhD research in 2003, lab to pilot scale development and commercialization over 2007–2008 [2, 3], and progressing onto smelter based industrial trials on individual and multiple reduction pots after 2009. Plans for a complete potline installation at a German smelter is now underway following a successful 12 pot trial in 2014–2016 [4].

As shown in the timeline, the EnPot technology was initially focused at providing smelters the ability to increase their metal production and amperage (particularly during the pre-2008 GFC era of high metal commodity prices), whilst maintaining the pot thermal balance within their operating window.

However, the ability to use the very same technology for *flexible amperage and production operations*, i.e. being able to scale production/amperage significantly both up and down, was soon realized. Teams at Energia Potior and the Light Metals Research Centre have since worked closely with Yunca Engineering to perfect the EnPot System to both cool or insulate aluminium reduction pots in order to act as an air-conditioner or a thermal blanket, depending on energy use and production requirements. The technology is now patented and has been developed significantly from the original research and prototypes into the now commercially available practical design and application package from Energia Potior Ltd.

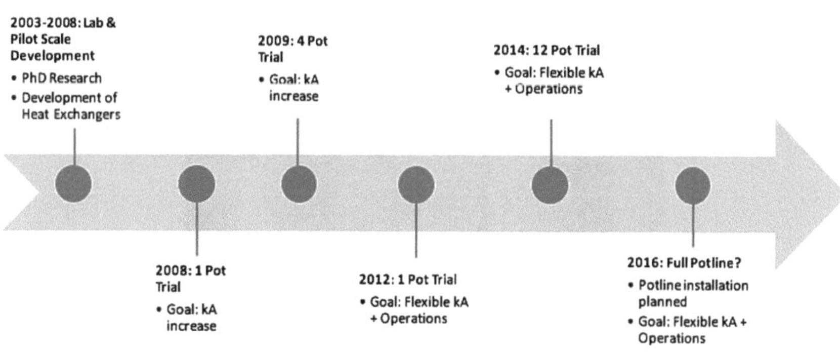

Fig. 4 Timeline showing historical development of EnPot technology at LMRC

The EnPot Heat Exchanger Concept

To maintain thermal balance on an aluminium reduction pot, roughly 50% of the energy input to a typical pot is lost as heat; 35% of this heat loss is through the sides of a pot (Fig. 5). Controlling this side-wall heat flow is critical, particularly during flexible amperage and production, as failure to do so results in either failure of pot integrity (when amperage and energy is increased) or solidification of the pot (when amperage and energy is ramped down). EnPot technology does this by regulating side-wall heat flow from pots, using a system of heat exchangers installed on the sides of each reduction pot. A schematic of individual exchangers is shown in Fig. 6, whereas an industrial installation is shown in Fig. 7; these units can be retrofitted non-invasively on existing pots in operation.

Fig. 5 Distribution of heat losses from a typical aluminium reduction cell

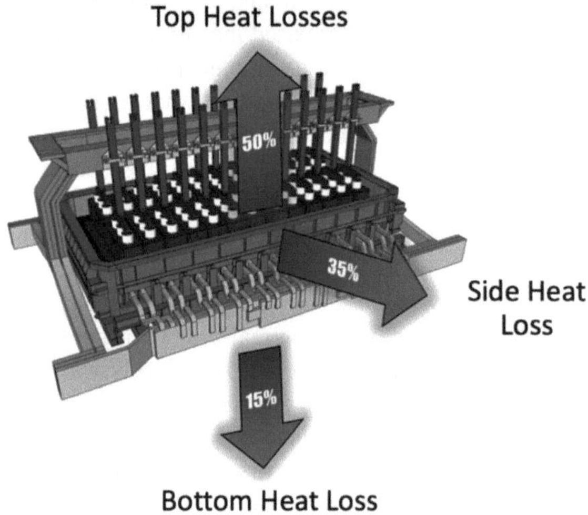

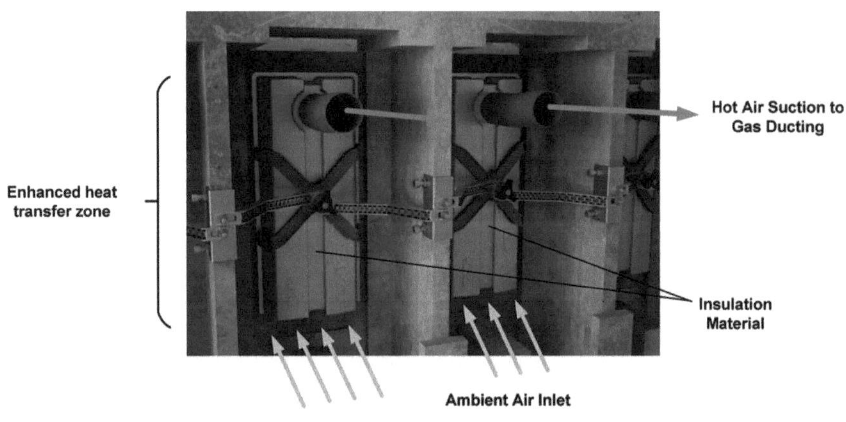

Fig. 6 Schematic showing the operation of two EnPot heat exchanger units installed on the sides of an aluminium reduction cell

Fig. 7 EnPot heat exchanger units fitted on one sidewall of an aluminium smelting pot, connected to ducting for hot gas extraction, figure reproduced from [4]

When *higher metal production* (increase in amperage/energy) is required, the exchangers work by extracting cool ambient air past the pot sidewalls to a system of gas ducting and fans. The transfer of heat from the pot to the EnPot exchangers is enhanced by patented technology. The rate of gas extraction can be easily dialled up or down in order to provide variable heat extraction at the desired production/energy level (Fig. 8).

Conversely, when *lower metal production* is desired (or *energy usage* needs to be reduced), gas flows through the exchangers are reduced or completely halted. Exchangers then act as a thermal blanket around the cell, insulating the side-walls and reducing heat loss.

Industrial Trials

The EnPot technology was initially tested on individual pots in a number of smelters before being installed on a 12-pot partitioned section of a smelter in Germany in June 2014. Since then the EnPot System has performed above expectations, providing the smelter operators the ability to turn energy consumption up and down by as much as +20% to −13% for 48 h periods [4], with even greater fluctuations planned after potline installation (ultimately a potential to ramp up the operating window to ±30%). This has enabled the smelter to 'modulate' or be flexible in its power consumption from the electrical grid, and essentially become an energy-buffer or 'virtual-battery'.

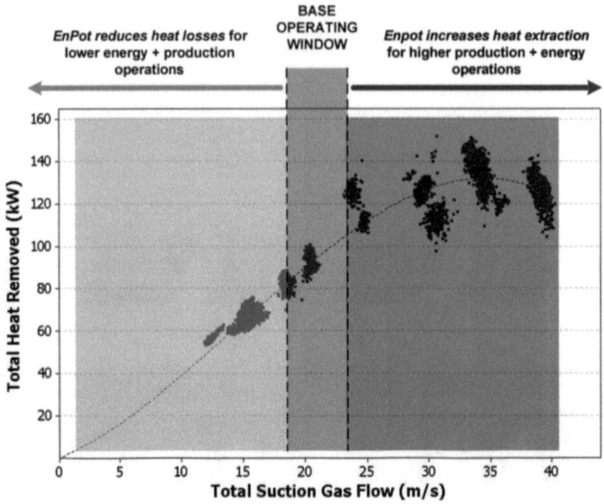

Fig. 8 Principle behind the EnPot system, with variable gas extraction flows providing dynamic control of the heat extracted from an aluminium smelting pot, figure adapted from [4]

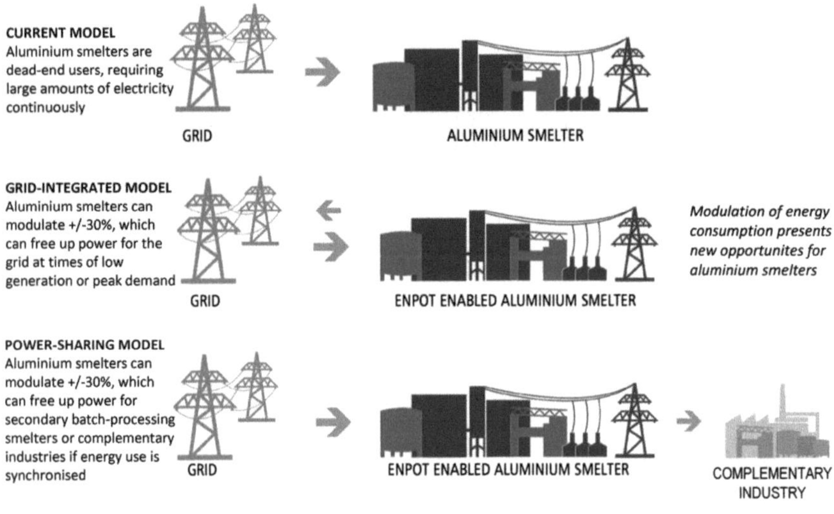

Fig. 9 Current end-user versus potential future models of smelter-power generator relationships with the use of EnPot technology, figure adapted from [5]

Smelters who apply this technology can sell or supply energy back to the grid during periods of peak power demand in the surrounding community; conversely, during off-peak periods, smelters can use the extra energy available to produce more metal. This provides a win-win situation for both the smelter and energy suppliers.

This model of operation (Fig. 9) fundamentally transforms the way electricity is consumed during the aluminium smelting process. This enables production output to be varied to match supply and demand and also to take advantage of off peak power prices and to accommodate the intermittency associated with renewable power supply.

Opportunities and a New Model for Smelters

Several new models of smelter-power generator relationships are now possible with the new EnPot technology (Fig. 9). In contrast to the current model, where smelters are only constant 'dead-end' users of electricity, EnPot provides smelters the potential to modulate their power usage by ±30%, enabling them to be either:

- 'Grid-integrated'—where smelters can free up power for the grid at times of low generation or peak demand, or
- 'Power-sharing'—where smelters free up power for other complementary industries or smelters with batch-energy requirements.

As such, the implications of this new technology are enormous, not only for the aluminium smelting industry, but also for the nations that host smelters and that struggle to generate enough clean power to meet domestic consumption at times of peak demand or low power generation, without the need to invest in new or back-up generation capacity. The EnPot technology could provide the first transformative change to the aluminium smelting industry since the 1950s; furthermore it is estimated that it could free up enough electricity regionally at peak consumption times to power 2.5 million homes in Europe, 1.1 million homes in North America and 25 million homes in China.

For many countries the installation of the EnPot technology in local aluminium smelters, could help alleviate problems with intermittency in their national grids, and could even enable the decommissioning of fossil fuel or nuclear power stations. It also frees up peak power for nations to use, accommodates intermittency in the gird, improves process efficiency, enables smelter operators to control production output, increases profits to smelter shareholders, reduces the need for subsidies, improves working conditions for employees and reduces insurance risk.

While not the focus of this paper, recovery of the 'waste' heat removed from smelters using EnPot technology (typically hot air extracted from pots in the range of 140–180 °C) presents another opportunity for the smelting industry. In addition, for the smelter to supplement its own energy needs, there is further potential to

Fig. 10 Potential future model of smelter-industry/community relationships recovering heat from EnPot technology to complement surrounding industries or heating needs in the community

complement the heating needs of surrounding industries or community facilities (e.g. swimming pools, domestic heating, etc.) as shown in Fig. 10. Breakthrough technologies to use or convert low-grade waste heat into power (e.g. modified organic rankine cycle [6], multiple effect distillation) will enable further opportunities for the industry.

Conclusions

The EnPot technology was originally designed to provide smelters the ability to increase their amperage and therefore metal production, whilst maintaining the pot thermal balance within their operating window. However, due to changes in the economic environment, development of the EnPot technology has since moved in a transformative new direction. The technology now has the potential to provide smelters with *±30% flexibility in metal production and energy usage*, in a production process that is well known for its inflexibility. Smelters utilizing this technology have the potential to fundamentally change their entire model of operations, from a constant dead-end user of electricity, to a 'grid-integrated' user who can free up power for the grid at times of low generation or peak demand, or a 'power-sharing' user, offloading power to nearby industries with complementary energy needs. Rather than just being a consumer of energy, the EnPot technology enables smelters to be a vital contributor to energy security of the community—a need that will arise with the intermittency of renewable energy sources.

Who We Are

Energia Potior is a partnership between Auckland UniServices Limited (the commercialisation arm of the University of Auckland) and Yunca Holdings Ltd. The EnPot technology was first developed and patented by the Light Metals Research Centre (part of Auckland UniServices Ltd.). During this development, Yunca provided the experience and expertise to retrofit the technology in operating aluminium smelters.

References

1. Energia Potior Limited. *Energia Potior*. 2016 [cited 2016, 23rd September]; Available from: http://www.energiapotior.com/
2. P. Lavoie et al., Increasing the power modulation window of aluminium smelter pots with shell heat exchanger technology, in *Light Metals 2011*, ed. by S.J. Lindsay (Wiley, 2011), pp. 367–374
3. S. Namboothiri et al., Controlled cooling of aluminium smelting cell sidewalls using heat exchangers supplied with air, in *TMS 2009 Annual Meeting and Exhibition,* San Francisco, CA (2009)

4. N. Depree et al., The 'Virtual Battery'—operating an aluminium smelter with flexible energy input, in *Light Metals 2016*, ed. by E. Williams (Wiley, 2016), pp. 571–576
5. L. Wright, Opening the window of opportunity. Aluminium Int. Today (Sept/Oct 2016)
6. B.F. Tchanche et al., Low-grade heat conversion into power using organic Rankine cycles—a review of various applications. Renew. Sustain. Energy Rev. **15**(8), 3963–3979 (2011)

The Thermodynamics of Slag Forming Inorganic Phases in Biomass Combustion Processes

Daniel K. Lindberg and Fiseha Tesfaye

Abstract To reduce the use of fossil fuels and increase self-sufficiency in energy, nowadays, there is an increasing interest to produce heat, power and transportation fuels from renewable resources. Solid biomass is one of the most important renewable energy sources for meeting this target. However, fouling, slagging, and corrosion threaten long-term operation availability and costs of biomass power plants. Slags accumulated on the surfaces of superheaters, which decrease thermal efficiency, often constitute a considerable percentage of complex inorganic phases. However, thermodynamic properties of the complex inorganic phases and their combined effect, which will help to deal with the slag related problems during high-temperature biomass combustion processes, are not well known. In the present paper, thermodynamic properties of K-, Ca-, and Na-based inorganic phases and their mixtures under different gas conditions are both critically reviewed and experimentally studied. The obtained results are presented and discussed.

Keywords Thermodynamic modeling · Ash chemistry · Slag · Molten salt · Biomass combustion

Introduction

The use of low-grade fuels such as various biomasses is increasing rapidly as an alternative to conventional fossil fuels. The overall share of energy consumption from renewable sources was about 16% within the European Union in 2014, and biomass-derived fuels are among the largest sources for renewable energy. Biomasses are often inexpensive energy sources, especially regionally, for example connected to the forest/pulp and paper industry or the agricultural industry. Biomass fuels also have a low greenhouse gas footprint and are considered CO_2-neutral.

D.K. Lindberg (✉) · F. Tesfaye
Johan Gadolin Process Chemistry Centre, Åbo Akademi University,
Piispankatu 8, 20500 Åbo/Turku, Finland
e-mail: Daniel.lindberg@abo.fi

L. Zhang et al. (eds.), *Energy Technology 2017*,
The Minerals, Metals & Materials Series, DOI 10.1007/978-3-319-52192-3_3

However, their utilization as fuels for thermal conversion, mainly combustion, but also gasification and pyrolysis, is connected with a number of technical challenges.

Many challenges are related to the minor constituents or impurities contained in these fuels, especially ash-forming metals and elements, such as chlorine, sulfur and phosphorus. The ash species that are formed in the thermal conversion can lead to several operational challenges, such as corrosion of the furnace walls or heat exchanger surfaces, formation of deposits that hinder the heat transfer to the heat exchangers, or agglomeration of bed materials in fluidized bed combustion. Figure 1 summarizes some of the chemistry related challenges in biomass combustion. These challenges are similar to the ones in combustion systems using conventional fuels such as coal. However, in biomass combustion the phenomena and solutions needed may be very different. The chemical compositions of the biomasses influence the extent and severity of the problems illustrated in Fig. 1.

To control the challenges shown in Fig. 1 requires a good understanding of the chemical behavior of the minor constituents and impurities of the fuel during the combustion process and in the flue gases. Both detailed experimental data on fuel and ash behavior, as well as various modeling tools, are needed to predict the behavior of the impurities. Thermodynamic modeling has become a useful tool to predict the interactions between multiple elements during combustion, for example

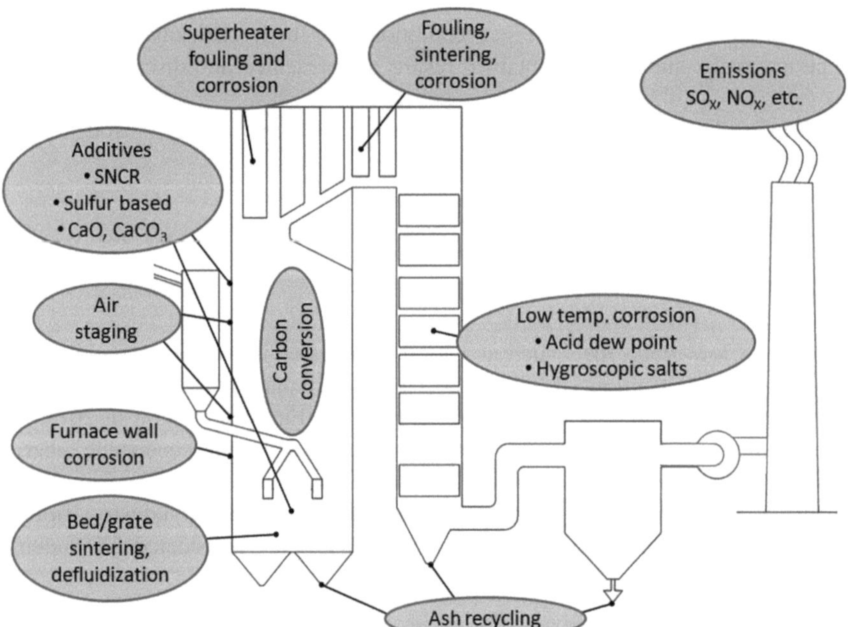

Fig. 1 Chemistry related challenges in biomass combustion

which phases and species are stable at certain conditions, how the ash melts and deposits. Together with detailed fuel analysis, chemical kinetics data, the use of thermodynamic modeling is a key tool in designing new boilers firing various types of biomass fuels. This paper discusses different aspects of the use of thermodynamic modeling to predict the chemical behavior of ash-forming elements in biomass boilers. The paper is a follow-up of previous reviews on the thermodynamic modeling of ash chemistry [1].

Overview of Ash Chemistry

The ash-forming elements in biomass and waste-derived fuels vary greatly. This is also reflected in the chemical composition and physical properties of the ash matter that is formed during combustion and gasification. Zevenhoven et al. [2] and Nordin [3] gave an overview of the chemical composition of the ash-forming elements in biomass fuels. Werkelin et al. [4, 5] also showed the variation of ash-forming elements in different parts of four tree species (spruce, pine, aspen, birch), as well as identifying the organic and inorganic ionic species that are present in wood-based fuels prior to combustion. The main ash-forming elements for biomass fuels are K, Na, Ca, Mg, Fe, Al, Si, P, S, Cl together with C, H, O. Vassilev et al. [6–8] have published several review papers on the chemical composition of the ash-forming elements in different types of biomass, as well as the composition of the ash phases that are formed during combustion. They identified three main elemental associations: (1) Si–Al–Fe–Na–Ti (mostly glass, silicates and oxyhydroxides); (2) Ca–Mg–Mn (commonly carbonates, oxyhydroxides, glass, silicates and some phosphates and sulphates); and (3) K–P–S–Cl (normally phosphates, sulphates, chlorides, glass and some silicates and carbonates). In addition, waste-derived fuels can also contain considerable amounts of Pb and Zn of anthropogenic origin. The chemistry of these elements is important for deposition and corrosion in boilers firing waste fuels due to the volatility of certain Pb and Zn compounds and their affinity for chlorine, which can lead to the formation of low-melting chloride-rich melts. The ash-forming matter in biomass fuels occurs in different modes. According to Zevenhoven et al. [2], the ash-forming matter can be found as:

1. Organically-bound matter
2. Dissolved salts
3. Included minerals
4. Foreign matter/excluded minerals

The release and the chemical reactions of the ash-forming matter vary depending on several factors, such as association of the elements, combustion conditions etc.

The reactions of the element may vary considerably during combustion of the actual fuel depending on its association in the fuel. For example, K or Na in certain silicate minerals will not be volatilized as readily as organically-bound or dissolved alkali metals due to the different bonding of the elements. However, these differences cannot be resolved by thermodynamic modeling without additional experimental input.

Zevenhoven et al. [9] combined advanced fuel analysis with advanced thermodynamic equilibrium calculations to separately predict the fly ash and furnace bottom ash reactions. Chemical fractionation by stepwise leaching was used to distinguish between organically-associated or dissolved ash-forming elements and included or excluded mineral matter. The first group was considered to be more reactive and form the fine particles fly ash and the second group was assumed to form the less reactive bottom ash.

Certain distinctions have been made in dividing the ash into two chemically distinct categories [10]: (a) An oxide/silicate system, and (b) a salt mixture system. However, this is a simplified distinction, as most condensed phases in the ash are mainly ionic compounds. The simplifications may be connected to other categorizations of the ash behavior. Silicates form polymeric silicate anions. These tend to have high-melting points, low volatility and be highly viscous in the liquid state. Silicates are often found in the furnace wall slag or bottom ash.

The so-called salt mixtures consist of different ionic compounds of relatively simple anions, such as Cl^-, Br^-, SO_4^{2-}, CO_3^{2-}, O^{2-}, S^{2-}, PO_4^{3-}, together with different metal cations. The melting points of these compounds vary from relatively low temperatures ($ZnCl_2$: 318 °C) to high temperatures (CaO: 2597 °C). The volatility is often higher than for the silicates but highly variable, and the viscosity of the liquid phase is generally rather low. These salt-like compounds are often found in the fly ash. However, exceptions exist, which renders the silicate/salt classification a simplification. Phosphates also form polymeric anions that resemble the anionic silicate species. The compound $ZnCl_2$ also forms polymeric species and resembles pure SiO_2 in that the interatomic bonds are more covalent than ionic. The term 'molten salt' is used for non-silicate ionic liquid phases, whereas the term 'slag' is reserved for an oxide liquid phase that can contain SiO_2.

Thermodynamic Software and Databases

Thermodynamic modeling is a commonly used tool for predicting the chemical behavior of complex systems. It is often based on minimization of the Gibbs energy of the system, which can be calculated using modern software if thermodynamic data exist for all the phases considered. Thermodynamic data for multicomponent solution phases are based on the thermodynamic data of the end-member components; mixing of the solution components, which is related to the structure of the phase; and, on the interaction parameters of the solution model describing the Gibbs energy of the solution phase. The computational methods involved in calculating multiphase multicomponent thermodynamic chemical equilibrium revolve around Gibbs energy minimization [11]. The true chemical equilibrium can be calculated by considering the Gibbs energy of all phases and minimizing the total Gibbs energy of the system (G). Here, G can be calculated either from the knowledge of the chemical potential $\left(\overline{G_i}, \mu_i\right)$ of component i, by

$$G = \sum_i n_i \overline{G_i}, \tag{1}$$

where n_i is the amount of component i, or alternatively by

$$G = \sum_\phi N^\phi G_m^\phi, \tag{2}$$

where N^ϕ is the amount of the phase and G_m^ϕ is the Gibbs energy of the phase. For a given set of constraints, such as fixed P, T, and overall composition, the Gibbs energy minimization algorithms find the amounts of the various phases and the composition of the solution phases which give a global minimum in the total Gibbs energy of the system.

The Gibbs energy is related to enthalpy (H), entropy (S), and heat capacity (C_p) in the following manner:

$$G(T) = \Delta H_{298} + \int_{298}^{T} C_p(T)dT - T \cdot (S_{298} + \int_{298}^{T} \frac{C_p(T)}{T} dT) \tag{3}$$

The values for ΔH_{298}, S_{298}, and $C_p(T)$ are often stored in thermodynamic data compilations.

There are several different types of software for thermodynamic equilibrium calculations. Recently, two volumes of the Calphad Journal have been dedicated to describe the available thermodynamic equilibrium modeling software [12, 13]. Most of these software packages focus on phase diagram calculations for metallurgical applications. Examples of different software that are currently used for thermodynamic equilibrium modeling are Factsage [14–16], HSC Chemistry [17], MTDATA [18] and Thermo-Calc [19]. Some of the main features of the software, as well as examples of other software are given in Table 1. For additional details of the software, see the review of Lindberg et al. [1].

Table 1 Examples of thermodynamic software used for modeling ash chemistry

Thermodynamic software	Own database	Complex solutions	Special features	References
ChemApp	No	Yes	User-defined programming library for custom-made applications	[20–22]
ChemSheet	No	Yes	Spreadsheet application for customized process modeling	[23]
Factsage	Yes	Yes	Contains extensive databases for molten salts and slags	[14, 15]
HSC	Yes	No	Contains extensive thermodynamic databases for stoichiometric compounds	[17]
MTDATA	Yes	Yes	Contains extensive databases for molten salts and slags	[18]
Thermo-Calc/DICTRA	Yes	Yes	Contains extensive databases for metallurgical applications and can be used for simulating diffusion-controlled phase transformations	[19]

Thermodynamic Models for Liquid and Solid Solutions

Pelton [24] published a review on the thermodynamic models for slags, fluxes and salts. Many of these models have been used to develop large thermodynamic databases for multicomponent salt and oxide systems. These databases are developed by what is known as the 'Calphad method' (**CAL**culations of **PHA**se **D**iagrams) [25]. Thermodynamic and phase equilibrium data for unary, binary, and ternary subsystems are critically evaluated and model equations for each phase are written. These equations give the Gibbs energy, G, as a function of temperature, T, and composition. For all solution phases, the Gibbs energy is given by the general formula

$$G = G^{\circ} + \Delta G_{mix}^{ideal} + G_{mix}^{excess}, \tag{4}$$

where G° is the contribution of the pure components of the phase to the Gibbs energy, ΔG_{mix}^{ideal} is the ideal mixing contribution, and G_{mix}^{excess} is the excess Gibbs energy of mixing, which is the contribution involving the non-ideal interactions between the components. The optimized model parameters of these equations should reproduce the experimental data within the experimental error limits.

The models are then used to predict the properties of multicomponent solutions from the optimized binary and ternary parameters, which are stored in databases. The databases are accessed by Gibbs energy minimization software to calculate thermodynamic properties, phase diagrams, etc. for multiphase, multicomponent systems. The review of Pelton [24] was restricted to models that are being used in the development of multicomponent databases. In the following section, models that are relevant to ash chemistry modeling will be shortly described. These models are for solid and liquid salts and oxide/silicate phases. For the models to be useful for ash chemistry applications, they must be sufficiently realistic to have good predictive ability for compositions and temperature conditions that have not been studied experimentally.

Alkali Salt Database for Combustion Applications

Fly ash particles and ash deposits in biomass combustion are often rich in potassium salts, such as chlorides or sulfates. On the other hand, particle residues in black liquor combustion are generally mixtures of sodium sulfates, carbonates and chlorides. The black liquor recovery boiler smelt also contains considerable amount of Na_2S at the reducing conditions in the smelt bed. The development by Lindberg et al. [26–31] of the alkali salt database for combustion applications considered all binary, ternary, and higher order systems of salt mixtures of the Na–K–C–S–O–H–Cl chemical system. The liquid model considers two cations, Na^+ and K^+. The anions are CO_3^{2-}, SO_4^{2-}, $S_2O_7^{2-}$, S^{2-}, S_n^{2-} (n = 2–8), O^{2-}, Cl^-, OH^- and Va^- (charged vacancies to model metal solubility). The solid solutions were modeled using the Compound Energy Formalism [32]. Lindberg et al. developed the thermodynamic database for alkali sulfide mixtures [28], alkali polysulfides [30], and acidic alkali sulfates [27], which are components of interest in some special combustion cases in black liquor recovery boilers. The thermodynamic evaluation for alkali chloride mixtures [29] is the most relevant for all biomass firing fuels. Melting properties of mixtures containing all of the ions sodium, potassium, sulfate, carbonate and chloride can be predicted with reasonable accuracy. The lowest calculated melting temperature of the complete $NaCl–Na_2SO_4–Na_2CO_3–KCl–K_2SO_4–K_2CO_3$ system is 501 °C. The calculated liquidus temperatures correspond well with the measured values [33, 34] for the whole compositional range. Predictions for sulfide-containing mixtures show larger uncertainties—mainly due to difficulties in obtaining reliable experimental data. The solidus temperature (first-melting temperature) was calculated for deposits and fly ash samples with known compositions from black liquor combustion systems. The calculated values correspond well with the measured solidus temperatures [35–37], in most cases with an accuracy of ±10 °C.

The alkali salt database is being expanded to also consider salts of Ca, Mg, Zn, and Pb, as well as chromates (CrO_4^{2-}, $Cr_2O_7^{2-}$) and phosphates. No thermodynamic database has previously been published or is currently commercially available for the Na^+, K^+, Ca^{2+}, Mg^{2+}, Zn^{2+}, Pb^{2+}/CO_3^{2-}, SO_4^{2-}, S^{2-}, O^{2-}, Cl^- system. The subsystem $NaCl–KCl–CaCl_2–MgCl_2–ZnCl_2–PbCl_2$ is available in the FTsalt database [14] and has been partly published [38, 39], and the database of Lindberg et al. [26–31] for the Na^+, K^+/CO_3^{2-}, SO_4^{2-}, S^{2-}, Cl^- system has been published. The evaluation of Ca-mixtures has also been published [40]. However, a combination of the databases taking into account all relevant binary and ternary interactions has not been presented previously.

Silicate-Oxide Slags

Silicon is one of the most common ash-forming elements in biomass fuels and coals. Molten silicates consist of polymeric anionic silicate units, which make them highly viscous. In the review of Pelton [24], several thermodynamic models for oxide/silicate melts were described. The main models are associate models, ionic models, and quasichemical models. The associate model considers different solutions associates or molecules, such as K_2O, SiO_2, K_4SiO_4, and K_2SiO_3 for the K_2O–SiO_2 system. However, these associates are not necessarily real species in the liquid phase, but sometimes chosen for fitting the experimental data. This means that the associate models may have low predictive ability if the choice of species is unrealistic. Ionic models considers the distribution of certain species, such as metal cations (K^+, Ca^{2+}), on a cationic sublattice, and anionic species (O^{2-}, $SiO_4{}^{2-}$) on an anionic sublattice. The quasichemical models, such as the model of Pelton and Blander [41, 42] and Pelton et al. [43], consider short-range ordering in the liquid phase. This is taken into account by the quasichemical reaction among second-nearest neighbor (SNN) pair:

$$(\mathrm{K}-[\mathrm{O}]-\mathrm{K})_{\mathrm{pair}} + (\mathrm{Si}-[\mathrm{O}]-\mathrm{Si})_{\mathrm{pair}} = 2(\mathrm{K}-[\mathrm{O}]-\mathrm{Si})_{\mathrm{pair}},$$

The model can be used to model systems with very strong non-ideal interactions, whereas sublattice models, such as the model of Pelton [44], inherently assume weak interactions for proper mathematical description. Kang and Pelton [45] utilized the Quasichemical Quadruplet Model [43] to model also salt dissolution in the oxide/silicate melt. All metallic cations and Si^{4+} mix on the cationic sublattice, while O^{2-} and the salt anions, such as S^{2-}, mix on the anionic sublattice.

Possibly the most comprehensive thermodynamic database for solid and liquid silicates and oxides is the FToxid database by Pelton et al. [14, 41], which covers the following oxide components:

$$\begin{gathered}\mathrm{Al_2O_3,\ As_2O_3,\ B_2O_3,\ CaO,\ CoO,\ CrO,\ Cr_2O_3,\ Cu_2O,}\\ \mathrm{FeO,\ Fe_2O_3,\ GeO_2,\ K_2O,\ MgO,\ MnO,\ Na_2O,\ NiO,}\\ \mathrm{PbO,\ SiO_2,\ SnO,\ TiO_2,\ Ti_2O_3,\ ZnO,\ ZrO_2}\end{gathered}$$

The thermodynamic database contains the thermodynamic properties for several complex solid solutions and for a molten phase, which has been modeled using the modified quasichemical model [42]. The database also considers the solubility of salt components and gases in the silicate/oxide melt [45–48]. The dissolved salt species include alkali carbonates, sulfates, sulfides, halides, and phosphates, while the dissolved gas species include H_2O, CO_2, and N_2.

The Al_2O_3–CaO–FeO–Fe_2O_3–MgO–SiO_2 subsystem has been fully optimized over the whole compositional range and gives accurate predictions of the solid-liquid phase equilibria [49–52]. The chemical system covers metallurgical slags and coal ash slags. However, addition of Na_2O and K_2O to the above-mentioned system has not been optimized over the whole compositional range in the FToxid databases. Therefore, uncertainties are to be expected in the predictions of melting temperatures and phase equilibria. The main subsystem of silicate/oxides in biomass ashes is the K_2O–CaO–SiO_2 system. The binary systems K_2O–SiO_2 [53–55] and CaO–SiO_2 [56] have been modeled and optimized based on phase equilibrium data and thermodynamic measurements. The phase diagram of the ternary K_2O–CaO–SiO_2 system has been measured by Morey et al. [57]. Several solid K–Ca silicates exist, but their thermodynamic properties have not been measured. No thermodynamic evaluation of the ternary K_2O–CaO–SiO_2 has been reported. Therefore, thermodynamic predictions of the phase equilibria of K_2O–CaO–SiO_2 show considerable uncertainties. Berjonneau et al. [58] compared thermodynamic predictions of the liquidus of synthetic K_2O–CaO–SiO_2 mixtures and miscanthus and straw ashes using the FToxid database from FactSage [14] with experimental determinations of their own. They showed that calculated liquidus temperatures of K_2O-rich mixtures differ with up to 200 °C compared to measured values. Berjonneu et al. [58] recommended additional experimental determinations and thermodynamic assessments of the K_2O–CaO–SiO_2 and K_2O–MgO–SiO_2 subsystems in order to improve the accuracy for phase equilibrium predictions for silicate ashes that have high contents of K_2O.

A recent development in the oxide/silicate databases [14, 41] is the introduction of the charge compensation effect [59, 60], whereby Na^+ or K^+ ions pair with Al^{3+} ions in Na_2O–K_2O–Al_2O_3–SiO_2 melts. This has considerably improved the accuracy of predictions for the liquid properties of alkali-aluminum silicate melts. Solid and liquid phases of unary, binary and many ternary P_2O_5-containing systems in the P_2O_5–CaO–MgO–Al_2O_3–SiO_2–BaO–FeO–Fe_2O_3–MnO–Na_2O system have been recently updated. Yazhenskikh et al. [55, 61–63] have recently also developed a thermodynamic database for the Na_2O–K_2O–Al_2O_3–SiO_2 system using an associate model.

Thermodynamic Modeling of Molten Ash-Related Issues

Applications of thermodynamic modeling to study molten ash-related issues are numerous. Thermodynamic modeling can be used as a standalone tool to study the overall chemistry of the ash-forming elements or to study specific phenomena occurring locally during combustion or on heat exchanger surfaces. Multicomponent/multiphase thermodynamic modeling can also be combined with other modeling tools, such as CFD. Some examples of the use of advanced thermodynamic modeling for predicting ash-related processes in biomass combustion are given below.

Pejryd and Hupa [64] were among the first to use advanced thermodynamic modeling to predict combustion and ash chemistry in boilers. They calculated the composition of the lower furnace gas and the smelt bed in black liquor recovery boilers. The phase equilibrium, as well as the smelt and gas composition were calculated as a function of temperature and air/fuel ratio to study the recovery boiler chemistry. The results were compared with observations from the recovery boiler and Pejryd and Hupa [64] were able to identify furnace processes that can be explained by chemical thermodynamics (e.g. smelt composition), but they also pinpointed phenomena that cannot be explained by only chemical thermodynamics and instead are controlled by chemical kinetics and mass transfer.

The melting of ash particles is of interest for several phenomena in biomass combustion and gasification. The presence of molten ash in deposits tends to increase corrosion of superheater tubes [65]. This is due to the ionic nature of molten silicates and salts, which increases the electric or ionic conductivity of ash. Backman et al. [66] also observed increased corrosion of the furnace floor tubes of black liquor recovery boilers, due to contact of the tubes with molten alkali polysulfides. This was shown with both conductivity measurements, chemical analysis showing sulfur enrichment in the smelt bed close to the furnace floor, and with thermodynamic predictions showing a melt formation at around 200–300 °C.

Backman et al. [67] introduced the terms T_0, T_{15}, T_{70} and T_{100} to describe the characteristic temperatures of gradually melting salt mixtures. T_0 gives the first melting temperature, which is of great interest as already the presence of the first molten phase may enhance corrosion. T_0 corresponds to the solidus temperature. The term T_{100} corresponds to the liquidus temperature.

Completely or partly molten ash particles tend to stick to surfaces, such as superheater tubes, which can lead to growth of ash deposits. Backman et al. [67], and Tran et al. [68] have shown that a minimum liquid content is needed for ash particles in recovery boilers to stick to superheater surfaces at impact. The liquid fraction of synthetic alkali salt mixtures (Na^+, K^+//Cl^-, SO_4^{2-}, CO_3^{2-}) was determined through advanced multicomponent phase equilibrium calculations and the deposition rate of the synthetic mixtures were experimentally observed in an entrained flow reactor with simulated superheater tubes and conditions. It was experimentally shown that 15–20% liquid on mass basis is needed to make the particles sticky. The temperature at which a particle has a liquid fraction of 15% is called the **sticky temperature, T_{15}**. It has also been shown that if the ash particles or the outer surface of a deposit contain more than 70% liquid, a steady-state deposit thickness is reached on vertical surfaces, as new material will flow down from the surface. This characteristic temperature for deposits is called the **flow temperature, T_{70}**. Multicomponent phase equilibrium modeling is the only relatively simple method to determine the T_{15} and T_{70} temperatures of the salt mixtures. The criterion of 15 and 70% liquid for the sticky temperature and flow temperature is only valid for alkali salt and alkaline earth salt mixtures (chlorides, sulfates, carbonates) according to Backman et al. [69]. Silicate liquids have much higher viscosities and the sticky and flow properties of ash particles containing molten silicates have therefore much more complicated relations to the liquid fraction and should not be calculated using the T_{15} and T_{70} concepts.

Mueller et al. [70] utilized a combination of advanced fuel analysis, advanced multicomponent thermodynamic modeling and CFD simulations to predict deposit formation in a 105 MW biomass-fired BFB boiler. The advanced fuel analysis used chemical fractionation methods to identify a reactive ash fraction that forms fly ash particles, which may form deposits in the boiler. The melting curve of the reactive ash fraction was calculated with thermodynamic modeling and the sticky temperature (T_{15}) was predicted for the particles. The CFD calculations predicted the flight trajectories of the ash particles and their impacts on boiler walls. Calculating the wall temperatures/ash particle temperature and using the stickiness criterion (liquid fraction >15%), deposition predictions were made. A comparison of the advanced deposition predictions with observations made in the boiler during a shutdown period showed very good qualitative agreement.

Conclusions

Advanced multicomponent/multiphase thermodynamic modeling is an important tool in predicting ash-related processes in biomass and waste combustion and gasification. Thermodynamic modeling is used to predict ash deposition, corrosion of heat exchangers, bed agglomeration of fluidized beds, smelt bed behavior in recovery boilers, and trace element speciation. Consistent and accurate thermodynamic data are needed for all the ash phases that may form in furnace and boiler conditions. These phases include complex liquid and solid ionic phases, such as molten salts and silicate slags, as well as a multicomponent gas phase.

Advanced thermodynamic modeling can be used as a stand-alone tool or be combined with other modeling tools to predict ash-related issues. Thermodynamic modeling has been successfully used in combination with advanced fuel analysis and computational fluid dynamics to predict ash deposition in real biomass boilers. Many other processes, such as the behavior of volatile heavy metals (e.g. Pb, Zn) in waste combustion and gasification has also been predicted. Thermodynamic modeling is a powerful predictive tool of the ash chemistry in combustion processes. Together with accurate thermodynamic data, innovative approaches for using experimental data and observations in the predictions, and by combining the tool with other modeling tools, the modeling capabilities to study new processes will expand further.

Acknowledgements The authors are grateful to the Academy of Finland for financial support. This work was made under the project "Behavior and properties of molten ash in biomass and waste combustion" as part of the activities of the Johan Gadolin Process Chemistry Center at Åbo Akademi University.

References

1. D. Lindberg et al., Towards a comprehensive thermodynamic database for ash-forming elements in biomass and waste combustion—current situation and future developments. Fuel Process. Technol. **105**, 129–141 (2013)
2. M. Zevenhoven, P. Yrjas, M. Hupa, in *Ash-forming Matter and Ash-related Problems*, ed. By M. Lackner, F. Winter, A.K. Agarwal. Handbook of Combustion Vol. 4: Solid Fuels (WILEY-VCH Verlag GmbH & Co. KGaA, Weinheim, 2010), pp. 495–531
3. A. Nordin, Chemical elemental characteristics of biomass fuels. Biomass Bioenergy **6**(5), 339–347 (1994)
4. J. Werkelin, B.-J. Skrifvars, M. Hupa, Ash-forming elements in four Scandinavian wood species. Part 1: summer harvest. Biomass Bioenergy **29**(6), 451–466 (2005)
5. J. Werkelin et al., Chemical forms of ash-forming elements in woody biomass fuels. Fuel **89** (2), 481–493 (2010)
6. S.V. Vassilev et al., An overview of the composition and application of biomass ash. Part 1. Phase-mineral and chemical composition and classification. Fuel **105**, 40–76 (2013) (Copyright (C) 2016 American Chemical Society (ACS). All Rights Reserved)
7. S.V. Vassilev et al., An overview of the organic and inorganic phase composition of biomass. Fuel, 2012. **94**, 1–33 (2012) (Copyright (C) 2016 American Chemical Society (ACS). All Rights Reserved)
8. S.V. Vassilev, D. Baxter, C.G. Vassileva, An overview of the behaviour of biomass during combustion: Part I. Phase-mineral transformations of organic and inorganic matter. Fuel **112**, 391–449 (2013) (Copyright (C) 2016 American Chemical Society (ACS). All Rights Reserved)
9. M. Zevenhoven-Onderwater et al., The prediction of behaviour of ashes from five different solid fuels in fluidised bed combustion. Fuel **79**(11), 1353–1361 (2000)
10. R. Backman, A. Nordin, in *High Temperature Equilibrium Calculations of Ash Forming Elements in Biomass Combustion/Gasification Systems-State of the Art, Possibilities and Applications*. International Biomass Ash Workshop (Graz, Austria, 1998)
11. N. Saunders, A.P. Miodownik, *CALPHAD (Calculation of Phase Diagrams): A Comprehensive Guide*, ed. By R.W. Cahn Pergamon Materials Series, vol. 1 (Pergamon, 1998), p. 479
12. L. Kaufman, Foreword. Calphad **26**(2), 141–141 (2002)
13. B. Hallstedt, Z.-K. Liu, Software for thermodynamic and kinetic calculation and modelling. Calphad **33**(2), 265–265 (2009)
14. C.W. Bale et al., FactSage thermochemical software and databases—recent developments. Calphad **33**(2), 295–311 (2009)
15. C.W. Bale et al., FactSage thermochemical software and databases. Calphad **26**(2), 189–228 (2002)
16. C.W. Bale et al., FactSage thermochemical software and databases, 2010–2016. Calphad **54**, 35–53 (2016)
17. http://www.hsc-chemistry.com/
18. R.H. Davies et al., MTDATA—thermodynamic and phase equilibrium software from the national physical laboratory. Calphad **26**(2), 229–271 (2002)
19. J.-O. Andersson et al., Thermo-Calc & DICTRA, computational tools for materials science. Calphad **26**(2), 273–312 (2002)
20. G. Eriksson, E. Koenigsberger, FactSage and ChemApp: two tools for the prediction of multiphase chemical equilibria in solutions. Pure Appl. Chem. **80**(6), 1293–1302 (2008)
21. G. Eriksson, P. Spencer, H. Sippola, in *A general thermodynamic software interface* (Lehrstuhl Theoretische Huttenkunde, RWTH Aachen, Germany, 1995), pp. 115–126
22. S. Petersen, K. Hack, The thermochemistry library ChemApp and its applications. Int. J. Mater. Res. **98**(10), 935–945 (2007)

23. P. Koukkari et al., ChemSheet—an efficient worksheet tool for thermodynamic process simulation. EUROMAT 99, Biannu. Meet. Fed. Eur. Mater. Soc. (FEMS) **3**, 323–330 (2000)
24. A.D. Pelton, Thermodynamic models and databases for slags, fluxes and salts. Trans. Inst. Min. Metall. Sect. C Min. Process. Extr. Metall. **114**(3), 172–180 (2005)
25. R. Schmid-Fetzer et al., Assessment techniques, database design and software facilities for thermodynamics and diffusion. Calphad **31**(1), 38–52 (2007)
26. D. Lindberg, Thermochemistry and melting properties of alkali salt mixtures in black liquor conversion processes. Doctoral thesis, in Laboratory of Inorganic Chemistry, Åbo Akademi University: Turku, Finland, 2007, p. 109
27. D. Lindberg, R. Backman, P. Chartrand, Thermodynamic evaluation and optimization of the (Na_2SO_4 + K_2SO_4 + $Na_2S_2O_7$ + $K_2S_2O_7$) system. J. Chem. Thermodyn. **38**(12), 1568–1583 (2006)
28. D. Lindberg, R. Backman, P. Chartrand, Thermodynamic evaluation and optimization of the (Na_2CO_3 + Na_2SO_4 + Na_2S + K_2CO_3 + K_2SO_4 + K_2S) system. J. Chem. Thermodyn. **39** (6), 942–960 (2007)
29. D. Lindberg, R. Backman, P. Chartrand, Thermodynamic evaluation and optimization of the (NaCl + Na_2SO_4 + Na_2CO_3 + KCl + K_2SO_4 + K_2CO_3) system. J. Chem. Thermodyn. **39** (7), 1001–1021 (2007)
30. D. Lindberg et al., Thermodynamic evaluation and optimization of the (Na + K + S) system. J. Chem. Thermodyn. **38**(7), 900–915 (2006)
31. D. Lindberg et al., A thermodynamic and phase equilibrium model for the smelt in recovery boilers. Annual Meeting Preprints—Pulp and Paper Technical Association of Canada, 91st, Montreal, QC, Canada, 8–10 Feb 2005. vol. **B**, pp. B31–B35 (2005)
32. B. Sundman, J. Ågren, A regular solution model for phases with several components and sublattices, suitable for computer applications. J. Phys. Chem. Solids **42**(4), 297–301 (1981)
33. A.G. Bergman, A.K. Sementsova, The ternary systems K_2Cl_2–Na_2SO_4–Na_2CO_3 and Na_2Cl_2–K_2SO_4–K_2CO_3. Zh. Neorg. Khim. **3**(2), 393–402 (1958)
34. A.K. Sementsova, A.G. Bergman, The ternary system of five ions, Na_2CO_3–K_2Cl_2–K_2SO_4. Zh. Obshch. Khim. **26**, 992–996 (1956)
35. N.J. Stead et al., in *Formation of Low Melting Deposits in a Modern Kraft Recovery Boiler*. 1995 International Chemical Recovery Conference (CPPA, Toronto, Ontario, Canada, 1995)
36. J. Frederick, Wm. James et al., Mechanisms of sintering of alkali metal salt aerosol deposits in recovery boilers. Fuel **83**(11–12), 1659–1664 (2004)
37. A.R. Walsh, A. Verloop, J.F. La Fond, Thermal analysis of recovery boiler deposits. Tappi J. **76**(6), 208–209 (1993)
38. P. Chartrand, A.D. Pelton, Thermodynamic evaluation and optimization of the LiCl–NaCl–KCl–RbCl–CsCl–$MgCl_2$–$CaCl_2$ system using the modified quasi-chemical model. Metall. Mater. Trans. A **32A**(6), 1361–1383 (2001)
39. C. Robelin, P. Chartrand, Thermodynamic evaluation and optimization of the (NaCl + KCl + $MgCl_2$ + $CaCl_2$ + $ZnCl_2$) system. J. Chem. Thermodyn. **43**(3), 377–391 (2011)
40. D. Lindberg, P. Chartrand, Thermodynamic evaluation and optimization of the (Ca + C + O + S) system. J. Chem. Thermodyn. **41**(10), 1111–1124 (2009)
41. M. Blander, A.D. Pelton, Thermodynamic analysis of binary liquid silicates and prediction of ternary solution properties by modified quasichemical equations. Geochim. Cosmochim. Acta **51**, 85–95 (1987)
42. A.D. Pelton, M. Blander, Thermodynamic analysis of ordered liquid solutions by a modified quasichemical approach-application to silicate slags. Metall. Mater. Trans. B **17B**, 805–815 (1986)
43. A.D. Pelton, P. Chartrand, G. Eriksson, The modified quasi-chemical model: Part IV. Two-sublattice quadruplet approximation. Metall. Mater. Trans. A **32**(6), 1409–1416 (2001)
44. A.D. Pelton, A database and sublattice model for molten salts. Calphad **12**(2), 127–142 (1988)

45. Y.-B. Kang, A. Pelton, Thermodynamic model and database for sulfides dissolved in molten oxide slags. Metall. Mater. Trans. B **40**(6), 979–994 (2009)
46. A.D. Pelton, Thermodynamic calculations of chemical solubilities of gases in oxide melts and glasses. Glass Sci. Technol. **72**(7), 214–226 (1999)
47. I.-H. Jung, Thermodynamic modeling of gas solubility in molten slags (II)-water. ISIJ Int. **46** (11), 1587–1593 (2006)
48. I.-H. Jung, Thermodynamic modeling of gas solubility in molten slags (I)-carbon and nitrogen. ISIJ Int. **46**(11), 1577–1586 (2006)
49. I.-H. Jung, S. Decterov, A. Pelton, Critical thermodynamic evaluation and optimization of the MgO–Al_2O_3, CaO–MgO–Al_2O_3, and MgO–Al_2O_3–SiO_2 systems. J. Phase Equilib. Diffus. **25** (4), 329–345 (2004)
50. I.-H. Jung, S. Decterov, A. Pelton, Critical thermodynamic evaluation and optimization of the FeO–Fe_2O_3–MgO–SiO_2 system. Metall. Mater. Trans. B **35**(5), 877–889 (2004)
51. I.-H. Jung, S.A. Decterov, A.D. Pelton, Critical thermodynamic evaluation and optimization of the CaO–MgO–SiO_2 system. J. Eur. Ceram. Soc. **25**(4), 313–333 (2005)
52. E. Jak et al., Coupled experimental and thermodynamic modeling studies for metallurgical smelting and coal combustion slag systems. Metall. Mater. Trans. B **31B**(4), 621–630 (2000)
53. P. Wu, G. Eriksson, A.D. Pelton, Optimization of the thermodynamic properties and phase diagrams of the sodium oxide-silica and potassium oxide-silica systems. J. Am. Ceram. Soc. **76**(8), 2059–2064 (1993)
54. S. Forsberg, Optimization of thermodynamic properties of the K_2O–SiO_2 system at high temperatures. J. Phase Equilib. **23**(3), 211–217 (2002)
55. E. Yazhenskikh, K. Hack, M. Müller, Critical thermodynamic evaluation of oxide systems relevant to fuel ashes and slags. Part 1: Alkali oxide-silica systems. Calphad **30**(3), 270–276 (2006)
56. G. Eriksson et al., Critical evaluation and optimization of the thermodynamic properties and phase diagrams of the MnO–SiO_2 and CaO–SiO_2 systems. Can. Metall. Q. **33**(1), 13–21 (1994)
57. G.W. Morey, F.C. Kracek, N.L. Bowen, The ternary system potassium oxide, calcium oxide, silica. J. Soc. Glass Technol. **14**, 149–87T (1930)
58. J. Berjonneau et al., Determination of the liquidus temperatures of ashes from the biomass gazification for fuel production by thermodynamical and experimental approaches. Energy Fuels **23**(12), 6231–6241 (2009)
59. P. Chartrand, A.D. Pelton, Modeling the charge compensation effect in silica-rich Na_2O–K_2O–Al_2O_3–SiO_2 melts. Calphad **23**(2), 219–230 (1999)
60. P. Chartrand, A.D. Pelton, On the choice of "geometric" thermodynamic models. J. Phase Equilib. **21**(2), 141–147 (2000)
61. E. Yazhenskikh, K. Hack, M. Müller, Critical thermodynamic evaluation of oxide systems relevant to fuel ashes and slags Part 2: Alkali oxide-alumina systems. Calphad **30**(4), 397–404 (2006)
62. E. Yazhenskikh, K. Hack, M. Müller, Critical thermodynamic evaluation of oxide systems relevant to fuel ashes and slags, Part 4: Sodium oxide-potassium oxide-silica. Calphad **32**(3), 506–513 (2008)
63. E. Yazhenskikh, K. Hack, M. Müller, Critical thermodynamic evaluation of oxide systems relevant to fuel ashes and slags. Part 3: Silica-alumina system. Calphad **32**(1), 195–205 (2008)
64. L. Pejryd, M. Hupa, in *Bed and furnace gas composition in recovery boilers—advanced equilibrium calculations*. TAPPI Pulping conference (1984)
65. B.-J. Skrifvars et al., Corrosion of superheater steel materials under alkali salt deposits Part 1: The effect of salt deposit composition and temperature. Corros. Sci. **50**(5), 1274–1282 (2008)
66. R. Backman et al., Formation of sulfur-rich melt near the floor tubing of kraft recovery boilers. Tappi J. **1**(2), 15–20 (2002)
67. R. Backman, M. Hupa, E. Uppstu, Fouling and corrosion mechanisms in the recovery boiler superheater area. Tappi J. **70**, 123–127 (1987)

68. H.N. Tran et al., The sticky temperature of recovery boiler fireside deposits. Results confirm that a sticky particle must contain at least 15% liquid phase. Pulp Pap. Can. **103**(9), 29–33 (2002)
69. R. Backman, M. Hupa, B.-J. Skrifvars, in *Predicting Superheater Deposit Formation in Boilers Burning Biomasses*. Impact of Mineral Impurities in Solid Fuel Combustion (1999), 405–416
70. C. Mueller et al., Ash deposition prediction in biomass fired fluidised bed boilers—combination of CFD and advanced fuel analysis. Prog. Comput. Fluid Dyn. **3**(2/3/4), 112–120 (2003)

Leaching of Sb from TROF Furnace Doré Slag

P. Halli, S. Jolivet, A. Klöfverskjöld, P. Latostenmaa, B.P. Wilson and M. Lundström

Abstract Doré slag is secondary product of copper electrorefining anode slime treatment. The slag consists mainly of different slag formers, such as borax and silica based minerals together with soda and traces of sealing materials. However, Doré slag also contains metals present in the anode slime, such as Ag, Cu, Bi, Sb and Te. At some plants, slag cleaning has been conducted utilizing conventional grinding and flotation circuits in order to enrich the content of precious metals in the slag [1]. Nevertheless, the routine consideration of the slag as a secondary raw material has so far not been adopted on an industrial scale. The current study presents the preliminary results of the nitric acid and chloride leaching of antimony from Doré slag. Chemical and mineralogical analyses (SEM-EDS, AAS, XRD) were performed for the raw material, leaching residue, and leaching solution in order to ascertain the metal recovery and kinetics during leaching.

Keywords Antimony · TROF furnace · Doré slag · Leaching

Introduction

Doré slag is a secondary raw material produced in the extractive metallurgy field of copper [1]. The slag itself comprises mainly of the same elements that can be found in anode slimes and typical slag formers, like borax and silica based minerals. The anode slime is usually treated using several hydrometallurgical and/or pyrometallurgical unit processes in series, furnace such as TROF being used to retrieve the

P. Halli (✉) · A. Klöfverskjöld · B.P. Wilson · M. Lundström
Department of Materials Science and Engineering, School of Chemical Technology, Aalto University, Vuorimiehentie 2, P.O. Box 16200, 00076 Aalto, Finland
e-mail: petteri.halli@aalto.fi

S. Jolivet
Polytech Grenoble, 5 Rue de L'Oisans, 38400 Saint Martin D'Hères, France

P. Latostenmaa
Boliden Harjavalta, Teollisuuskatu 1, 29200 Harjavalta, Finland

L. Zhang et al. (eds.), *Energy Technology 2017*,
The Minerals, Metals & Materials Series, DOI 10.1007/978-3-319-52192-3_4

valuable metals and convert them into a more attractive form, such as pure elements. The selectivity of this kind of process depends on the slag formers used, their ratios to each other and to the matte. Nevertheless, converting processes are not ideal and in addition to impurities, small amounts of valuable elements always ends up in the slag. In this study, a comprehensive investigation of the leaching of Doré slag was performed in order to determine the leaching behavior of Sb in chloride and nitric media. Although the leaching of Sb by different solutions and under different conditions has been reported previously [2–4], there is no published data of a detailed investigation of Sb leaching from Doré slag.

Experimental Procedure

Characterization

The raw material used in this study was produced by Boliden Harjavalta Oy, Finland, and characterization was performed at Aalto University, applying similar characterization methods used earlier in anode slime investigations [5]. According to the XRD (X'Pert PRO Powder, NL), SEM-EDS (Zeiss LEO 1450 VP, Germany, and Oxford Instruments INCA software, UK) and AAS (Thermo Fisher Scientific iCE 3000, USA), the raw material contains a number of different slag formers like Ba, As, Mg, Pb, Fe, Si, and Na, together with precious metals and rare earth elements. The main phases of the Doré slag were barite ($BaSO_4$), quartz (SiO_2), lead oxide (Pb_xO_y) and a complex minerals formed between Sb, Bi, Te, Cu, Ag or O in different quantities. Sb was mainly in complex form with other elements mentioned. Particle size distribution analysis (Malvern Mastersizer 2000, UK) determined that the p90 was 244 μm and the mean size was 103 μm. The SEM-EDS, XRD and AAS results indicated, that Sb is present in a number of different forms in the raw material. According to the total leaching (aqua regia at room temperature) performed, the amount of Sb present in the raw material was 2290 ppm.

Leaching

The leaching experiments were carried out in batches using a constant solid concentration of 125 g/L. Total slurry volume was 800 mL and the leaching time was 300 min. Samples were taken at constant time intervals (t = 15, 30, 60, 120, 180, 240 and 300 min) to determine the metal recovery to the solution. Additionally, stirring and oxygen flow were kept constant throughout the experiments at 500 RPM and 1000 mL/min, respectively. In total 30 different leaching experiments (12 nitric and 18 chloride) were carried out in order to investigate the effect of leaching media on Sb dissolution from TROF Doré slag. The concentrations used in chloride leaching had [$FeCl_3$] = 3, 9 or 27 g/L (equivalent to 0.02, 0.06 and 0.17 M), [$NaCl$] = 2 or 4 M. During the experiments, pH was maintained to be below 1.0 with droplets of 12 M HCl and the temperature was varied between 40 and 80 °C. For the HNO_3 leaching, four different concentrations (0.5, 1.0, 1.5 and 2.0 M) and three temperatures (30, 40 and 50 °C) were investigated. All the chemicals used were technical grade.

The pH was measured at the same time when the solution samples were taken. Furthermore, a total of three different statistical analyses were performed on the raw material, leaching solution and leaching residue, respectively by SEM-EDS, XRD and AAS.

Results and Discussion

According to the AAS analysis, some of the Sb (5–24%) dissolved into the leaching solution in the early stage of leaching. Nevertheless, with increasing leaching time, soluble Sb concentration decreased regardless of the used media (HNO_3 or chloride leaching). Overall, the solution analysis suggested that chloride media leached more Sb compared to HNO_3. However, it should be noted that the HNO_3 concentrations were not equivalent to levels of Cl^- used in the chloride leaching experiments.

Chloride Leaching

The highest concentration of dissolved Sb achieved in chloride media was obtained after 30 min of leaching (550 mg/L, indicating a 24% yield) with 0.17 M $FeCl_3$, 4 M NaCl and T = 40 °C. According to solution analysis displayed in Fig. 1, the highest concentration of Sb in solution after five hours leaching was approximately 400 ppm at 40 °C, (17.2% recovery) and around 100 ppm at 60 and 80 °C (4.3% recovery). The observed decrease in the extraction of Sb with increasing chloride concentration has previously been investigated by Ukasik and Havlik [6], who suggested that this decrease could be the result of the formation of an antimony oxide found within the leaching residue.

As it can be seen in Fig. 1a, some of the Sb readily dissolved into the chloride solution at the early stage of the leaching, but the solution concentration decreased as a function of time. This indicates that precipitation may have occurred for Sb during chloride leaching at T = 40 °C. According to the results from the XRD and SEM-EDS, the precipitated Sb in the leaching residue was shown to be in the oxide form, Sb_2O_3, coordinated with the slag formers such as As, Pb, Ba and Sn.

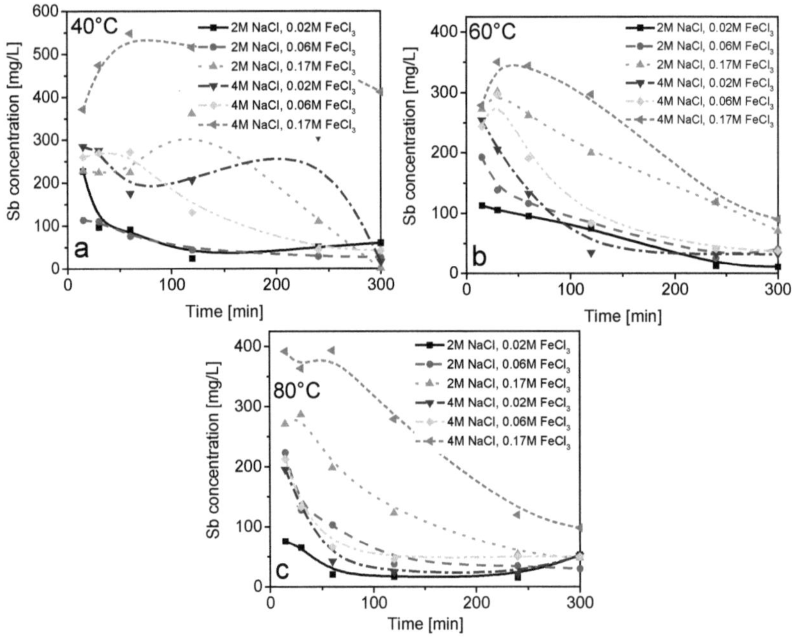

Fig. 1 Dissolution *curves* of chloride leached Sb from TROF Doré slag at **a** 40, **b** 60 and **c** 80 °C with different concentrations of NaCl and $FeCl_3$

According to Fig. 1b, the chloride leaching of Doré slag at 60 °C resulted in a similar precipitation behavior as at 40 °C at all chloride concentrations. Moreover, Fig. 1b shows that an increase in the oxidant ($FeCl_3$) concentration have had more effect on the Sb dissolution compared to NaCl concentration in the range investigated. However, the final extraction of Sb to the solution was below 5% at every chloride concentration after 5 h of leaching. At T = 60 °C the precipitated Sb in the residue was found in an oxide (Sb_2O_3) form from the leaching residue, according to XRD and SEM-EDS. At 80 °C (Fig. 1c) a similar level of dissolved Sb at the early stage of leaching and a similar precipitation behavior compared to those at 40 and 60 °C were obtained. With the highest chloride concentration, more Sb dissolved into the leaching solution, indicating that an increase in the chloride concentration enhanced Sb dissolution regardless of the experimental temperature used. The shape of the obtained dissolution curves also indicated a similar dissolution—precipitation phenomena at every experimental temperature. Figure 1a–c suggest that the leaching kinetics i.e. maximum Sb dissolution achieved was faster at higher temperatures, but more or less the leaching and precipitation phenomena were the same in the investigated parameter range. The back precipitation of Sb was shown to result in Sb_2O_3 precipitate at all temperatures, oxidant concentrations and chloride concentrations in chloride leaching.

Nitric Leaching

Figure 2 presents Sb concentration in the leaching solution as a function of leaching time at 30, 40 and 50 °C when different concentrations HNO_3 were employed. It can be seen that higher levels of HNO_3 provided higher Sb extraction to the solution, however, the extraction levels were low (1–6%). In addition, also during HNO_3 leaching, the precipitation of Sb was evident, precipitation rate being lower for HNO_3 when compared to chloride leaching at the same temperature. The highest Sb dissolution was achieved at the lowest temperature of 30 °C for HNO_3 concentrations above 1.0 M (Fig. 2a), however, this level of dissolved Sb was observed to decrease at higher dissolution times (>30 min). In contrast, the level of dissolved Sb at 30 °C was more or less constant in 0.5 M nitric solution over the whole duration of the experiment. Subsequent analysis of the leaching residue with SEM-EDS and XRD, showed that Sb was in the oxide form (Sb_2O_3) at T = 30 °C. According to Fig. 2b, the dissolution of Sb at 40 °C was lower than at 30 °C for HNO_3 concentrations of 1.5 and 2.0 M, whereas for 1.0 and 0.5 M, the initial amount of dissolved Sb was almost the same. According to SEM-EDS and XRD analyses, Sb was present both in the oxide form and in a more complex mineral together with Bi in the leaching residue. When the highest experimental

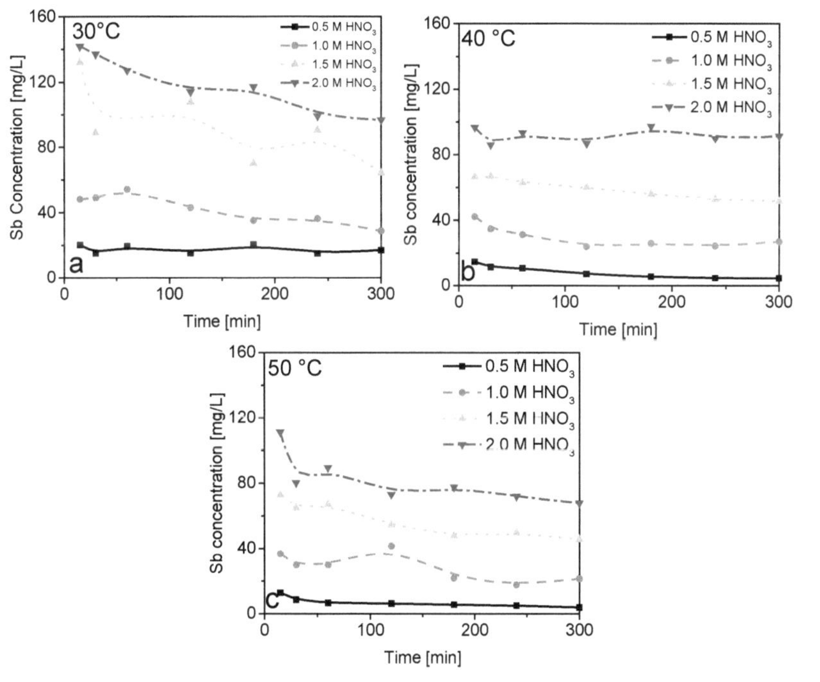

Fig. 2 Dissolution curves of HNO_3 leached Sb from TROF Doré slag at **a** 30, **b** 40 and **c** 50 °C with different concentrations

temperature for HNO_3 leaching was used (50 °C, Fig. 2c), the concentration of dissolved Sb was shown to decrease. Also at 50 °C, Sb was found both as oxide and a complex mineral, consisting of Sb, Bi and O, in the residue. XRD and SEM-EDS results showed that the precipitated Sb after nitric leaching was present not only in an oxide form within the leaching residue, but also as a complex mineral with Bi. Both of these phases most likely formed during the HNO_3 leaching, since none of them was present in the original raw material, according to SEM-EDS and XRD observations.

Conclusions

It was shown that during chloride and HNO_3 leaching of Doré slag, part of Sb (5–24% in chloride and <1–6% in nitric) present in the slag dissolved relatively rapidly in the initial stages into solution. However, after approximately 30 min, Sb begun to precipitate. Moreover, in both investigated solutions, the concentration used in the leach did not have a considerable impact on the precipitation of the Sb, whereas temperature was shown to have a more significant effect. The major difference observed between the chloride and HNO_3 leaching of Sb was that chloride containing solutions not only lead to a higher extraction of Sb to the solution but also, conversely, a faster precipitation rate of Sb was observed when compared to HNO_3 solutions. One of the reasons were that the initial concentration of the chlorides utilized were somewhat higher than the comparable HNO_3, which resulted in higher Sb dissolution and faster back precipitation rate. Similar back precipitation have been reported previously [6, 7] and the effect of increasing chloride concentration is also known to result in the increased solubility of Sb [8]. The back precipitation was due to the formation of Sb_2O_3, which is relatively insoluble mineral.

In the Doré slag, Sb was mainly in a complex form that contained Bi, Te, Cu, Ag and/or O. Results show that the back precipitation compound in the chloride solution was primarily antimony oxide (Sb_2O_3) whereas both Sb_2O_3 and a more complex compound consisting of Sb, Bi and O were observed in the HNO_3 solution. This finding could be the result of the fact that the HNO_3 concentrations were not equivalent to those of the chloride. Thus their dissolution properties are not similar and therefore, some of the original Sb–Bi–O—complex minerals did not fully dissolve.

Overall, the results of this study showed that the amount of Sb dissolved was approximately six times higher for chloride *cf.* HNO_3 media. This suggests that chloride based solutions are potentially more favorable for Sb leaching from TROF Furnace Doré Slag.

Acknowledgements This work has been financed by Metallinjalostajat ry and supported by "*System integrated metals processing*" (SIMP) project coordinated by the Finnish Metals and Engineering Competence Cluster FIMECC Ltd. and "*RawMatTERS Finland Infrastructure*" (RAMI). The provider of the raw material, Boliden Harjavalta, is greatly acknowledged by the authors.

References

1. P.W. Godbehere, US Patent, US4404022 A, 13 Sept 1983, 6 p
2. G. Cornelis, T.V. Gerven, C. Vandecasteele, Antimony leaching from uncarbonated and carbonated MSWI bottom ash. J. Hazard. Mater. **137**(3), 1284–1292 (2006)
3. G. Cornelis, T.V. Gerven, R. Snellings, B. Verbinnen, J. Elsen, C. Vandecasteele, Stability of pyrochlores in alkaline matrices: solubility of calcium antimonate. Appl. Geochem. **26**(5), 809–817 (2011)
4. G. Cornelis, T.V. Gerven, C. Vandecasteele, Antimony leaching from MSWI bottom ash: modelling of the effect of pH and carbonation. Waste Manag. **32**(2), 278–286 (2012)
5. M. Kiviluoma, M. Aaltonen, J. Aromaa, M. Lundström, O. Forsén, Development of characterization methods for adherent anode slimes in copper electrorefining. Physicochem. Probl. Miner. Process. **52**(1), 295–302 (2015)
6. M. Ukasik, T. Havlik, Effect of selected parameters on tetrahedrite leaching by ozone. Hydrometallurgy **77**(1–2), 139–145 (2005)
7. C.G. Anderson, The metallurgy of antimony. Chem. Erde Geochem. **72**(Suppl 4), 3–8 (2012)
8. S. Lafond, J.-F. Blais, R. Martel, G. Mercier, Chemical leaching of antimony and other metals from small arms shooting range soil. Water Air Soil Pollut. **224**(1, Article 1371), 15 p (2013)

Potential CO_2 Emission Reduction and H_2 Production Using Industrial Slag Wastes Originating from Different Industrial Sectors

Jinichiro Nakano, James Bennett and Anna Nakano

Abstract Previous empirical research demonstrated that by mixing gasification and metallurgical slags at a specific composition in a CO_2 enriched environment, enough exothermic heat would be generated to transform CO_2 to CO. The process capacity may be limited by the availability of the slags, however, it could be used to supplement existing energy or chemical production to reduce overall carbon emission. In this study, computational simulations were performed to explore optimal process conditions based on effects of individual slag constituents and thermal input for maximal gas conversion rate, heat production, and volume increase. In simulated conversions of CO_2 to CO and H_2O to H_2, comparisons were made by varying CaO, FeO, versus V_2O_3 slag constituents. The result indicated the optimal slag composition may be obtained by appropriate source selection and mixing of gasification and metallurgical slags in order to maximize carbon reduction and/or H_2 production.

Keywords Slag utilization · Iron and steelmaking · Gasification · CO_2 · CO · H_2O · H_2

Introduction

Waste byproducts such as H_2O, CO_2 and slag from iron and steelmaking and slagging gasification processes in the production of metal, power, and/or chemicals may be utilized to complement their production. 1.5 tons of CO_2 is emitted per ton of iron produced and an additional 0.9 tons of CO_2 per ton of steel produced is generated from subsequent steelmaking processes within the wastes [1]. Approximately 523 million tons/year of slag is generated from the iron and steelmaking industry

J. Nakano (✉) · J. Bennett · A. Nakano
U.S. Department of Energy National Energy Technology Laboratory,
1450 Queen Ave., Albany, OR 97321, USA
e-mail: jinichiro@gmail.com

J. Nakano · A. Nakano
AECOM, P.O. Box 1959, Albany, OR 97321, USA

L. Zhang et al. (eds.), *Energy Technology 2017*,
The Minerals, Metals & Materials Series, DOI 10.1007/978-3-319-52192-3_5

worldwide [2], with less than 65 wt% of it utilized for recovering valuable elements [3]. Currently, gasifiers generating 174 GW (thermal) are being built in the world for 2016 and planned through 2019 [4], including the Jamnagar Gasification Plant in India that would generate 9.8 GW (thermal) syngas as output [5]. When the existing worldwide gasifier facilities using petcoke as a carbon feedstock are combined with the new plants coming on-line by 2019 [4], a total of 310 GW (thermal) will be produced, generating up to 23,560 tons of petcoke slag every day.

Typically, slag generated from metallurgical plants are rich in calcium oxide, while petcoke slag produced in gasifiers is rich in vanadium oxide, as shown in Table 1. Upon mixing the two slags under certain conditions, the presence of calcium influences the valence of vanadium, forming calcium orthovanadate ($3CaO{\cdot}V_2O_5$). If the reaction takes place in the presence of CO_2 and/or H_2O, oxygen is removed from those gas molecules to form the vanadate in the system, which would follow the fundamental reaction:

$$3CaO + V_2O_3 + 2[(CO_2)_x + (H_2O)_{1-x}] \rightarrow (CaO)_3(V_2O_5) + 2\left[(CO)_x + (H_2)_{1-x}\right] + \text{heat} \quad (1)$$

Since the heat produced from the vanadate formation is larger than that required for decomposition of CO_2 and/or H_2O, excess heat is generated by the reaction, ranging from 104 to 163 kJ per mole of calcium vanadate at 1300 °C [6], depending on x (x = 0–1).

Previous research [6, 7] empirically demonstrated that by mixing gasification and metallurgical slags at a specific ratio in a CO_2 enriched environment, exothermic heat generated from the reaction would be enough to transform CO_2 to CO, which could be used to support the existing energy or chemical production,

Table 1 Compositions of metallurgical slags and petcoke ashes

	(wt%)	Al_2O_3	CaO	FeO_x	MgO	MnO	NiO	SiO_2	V_2O_5
Metallurgical slag	BF slag [2]	8–24	30–56	0.5–1	1–18	0.5–2	–	28–38	–
	BOF slag [2]	1–7	45–60	10–35	5–17	2–14	–	8–22	–
	EAF slag [2]	2–9	35–60	15–30	5–15	1–8	–	9–20	–
	LF slag [2]	5–35	30–60	0.1–15	1–10	0–5	–	2–35	–
Petcoke ash	Delayed coke ash [8]	6.9	2.2	5.3	0.3	–	12	10.1	58.2
	Shot coke ash [8]	5.9	3.6	4.5	0.6	–	10.2	13.8	57
	Fluid coke ash [8]	9.4	8.9	31.6	0.4	–	2.9	23.6	19.7
	Flexicoke ash [8]	0.5	2.4	2.5	0.2	–	11.4	1.6	74.5

Remaining materials not shown are minor constituents including K_2O, Na_2O, TiO_2, and SO_3
Note Vanadium content is typically reported as V_2O_5 but generally exists as V_2O_3 in metallurgical and gasification slags. BF = blast furnace, BOF = basic oxygen furnace, EAF = electric arc furnace, LF = ladle furnace

reducing overall carbon emissions. In this work, theoretical simulations were performed to guide the determination of optimal conditions to maximize the conversion phenomena. Effects of varying parameters on CO_2 conversion and H_2 production are discussed.

Simulation Approach

In order to understand general trends of the conversion reactions influenced by input parameters and conditions, thermodynamic simulations were conducted using FactSage™ 7.0 with databases of FactPS and FToxid [9]. In the thermodynamic calculations, compositions of metallurgical slags and gasifier slags (petcoke ash) were estimated by averaging values in Table 1 for simplicity - they do not represent slags from specific industrial processes or sources. Only Al_2O_3, CaO, FeO, SiO_2, and V_2O_3 were considered as slag constituents in the simulations. CaO, Fe_2O_3, and V_2O_3 were varied from 0–30 wt% as additions to the slag mixtures at different ratios to determine effects of individual slag constituents. Reaction temperatures were varied from 1000 to 1600 °C. Reactions were assumed to take place in pure CO_2 or pure H_2O in each calculation, where slag mixture:gas ratios equaled 1:1 by mass unless indicated otherwise. Note no interaction between slag and vanadium oxide was assumed due to a lack of such data in the current FactSage databases. Equilibrium calculations with calcium vanadate above 1380 °C were extrapolated from temperatures below that temperature. A total of approximately 4000 points were calculated and graphically summarized.

Results and Discussion

Effects of Temperature and Slag Constituents to the Previous Slag Mixture on CO_2 Conversion and H_2 Production

In the previous experimental work [6, 7], the CO_2 conversion phenomenon was demonstrated using a synthetic slag mixture at the proposed composition of $8.2Al_2O_3–24.7CaO–8.2Fe_2O_3–1.6SiO_2–57.3V_2O_3$ (by mass), where rapid conversion was observed upon heating in a CO_2 rich atmosphere. Theoretical changes in CO_2 conversion rates affected by reaction temperatures and V_2O_3 and CaO added to this mixture are shown in Fig. 1. In the two figures, a bottom axis where V_2O_3 addition = 0 wt% represents the slag mixture composition suggested in the previous work. It was found reaction temperature had minimal influence on the CO_2 conversion at temperatures between 1000 and 1600 °C for this slag chemistry. Higher

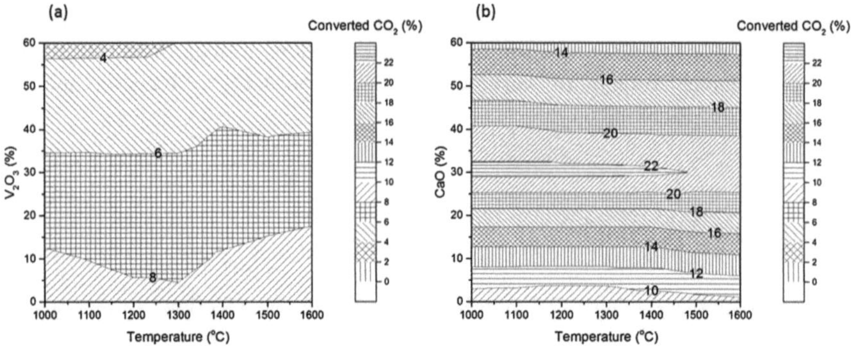

Fig. 1 Effect of **a** V_2O_3 and **b** CaO additions to the previous slag mixture [6, 7] on CO_2 conversion and H_2 production

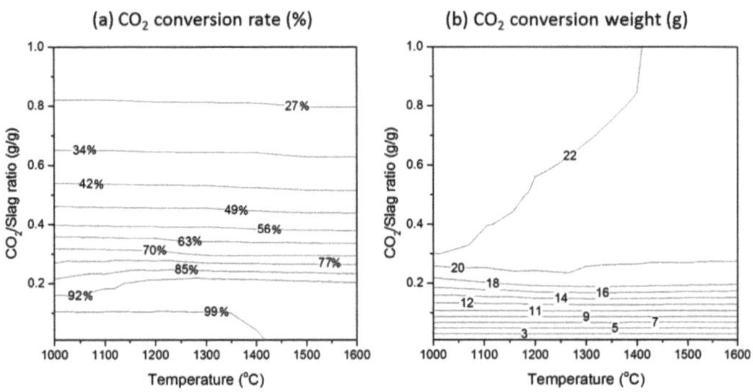

Fig. 2 Effects of CO_2/slag ratios and temperatures on **a** CO_2 conversion rate and **b** CO_2 conversion weight per unit slag mass (previous slag composition [6, 7] +30 wt% CaO addition)

V_2O_3 content with respect to this composition would have a negative impact on the conversion, which implies the previous composition had sufficient V_2O_3, while the CaO addition would improve the conversion at all the temperatures considered. This can be explained by the system already having high V_2O_3 content in the original slag mixture to react with CaO. The CaO addition, however, exhibited a peak around 30 wt% and the effect became negative, which may be related to destabilization of the vanadate over the formation of lime.

Effects of the CO_2 to slag ratio (by mass) on CO_2 conversion are shown in Fig. 2, using a slag composition proposed in the previous work [6, 7] plus 30 wt% CaO addition. For the calculations in Fig. 2, the slag weight was fixed at 100 g and only CO_2 weights were varied to determine the CO_2/slag ratios. Higher conversion rates can be achieved if smaller amounts of CO_2 are to be treated by a unit mass of slag as shown in Fig. 2a. However, smaller quantities of CO_2 can be converted per unit mass of slag at higher conversion rates as indicated in Fig. 2b. More CO_2 can be converted to CO using the same amount of slag if, in general, lower conversion rates are selected. These results imply the CO_2 input ratio must be considered in process designing in order to reduce more CO_2 emission (or more CO production), depending on the process goals (purity vs. mass). The simulated outcome still holds and indicates respective trends. In the previous work, 97% CO_2 conversion was experimentally measured, likely owing to the small amount of CO_2 with respect to the slag, while the present simulation using 1:1 ratio (gas to slag) shown in Fig. 1 indicated approximately 9% conversion for that composition. The previous CO_2/slag ratio was approximately 0.1 and the conversion occurred at temperatures between 1400 and 1460 °C, which may explain the 97% conversion as depicted in Fig. 2.

Effects of Temperature and Slag Constituents of Broad Mixed Compositions on CO_2 Conversion and H_2 Production

Theoretical CO_2 and H_2O conversion rates were determined with respect to slag ratios (metallurgical slag to petcoke ash), temperatures, and V_2O_3 are shown in Fig. 3. With no V_2O_3 addition, blending petcoke ash to metallurgical slag at 20–30 wt% would maximize the conversions as V_2O_3 is supplied from the petcoke ash. For both conversions, the higher slag V_2O_3 concentration exhibits stronger effects on the metallurgical slag rich side. V_2O_3 additions have minimal effects on the conversions on the petcoke ash richer side for both conversions. At 1000 °C, higher CO_2 conversion may be achieved if 10 wt% V_2O_3 is added to the 100 wt% metallurgical slag. The most effective addition of V_2O_3 was located around 20 wt% for 1200–1600 °C. The higher H_2O conversion rate can be achieved by around 10–20 wt% V_2O_3 additions to the metallurgical slag rich mixtures.

Figure 4 shows effects of the CaO addition to the averaged slag mixtures at different temperatures. Good conversion rates may be achieved without V_2O_3 additions if petcoke ash is added to metallurgical slag at 10–30 wt%. The CaO addition is more effective on the CO_2 and H_2O conversions using the petcoke ash richer side than the metallurgical slag side as more V_2O_3 is available to induce reaction (1). The CaO addition such as lime to the petcoke ash may contribute more to the conversion rates as compared to the V_2O_3 addition to the metallurgical slag, however more quantities are required. Excessive CaO addition seems to have negative impacts on the conversions.

Effects of the FeO addition to the averaged slag mixtures on CO_2 and H_2O conversions are shown in Fig. 5. The FeO addition exhibited more meaningful effects if petcoke ash is blended to the metallurgical slag at 25–35 wt%. The FeO effect was less on the petcoke ask richer side and the metallurgical slag richer side. Iron oxide, if present as FeO before reacting, would take oxygen from CO_2 and H_2O by transforming to Fe_3O_4 or Fe_2O_3, enhancing the conversion.

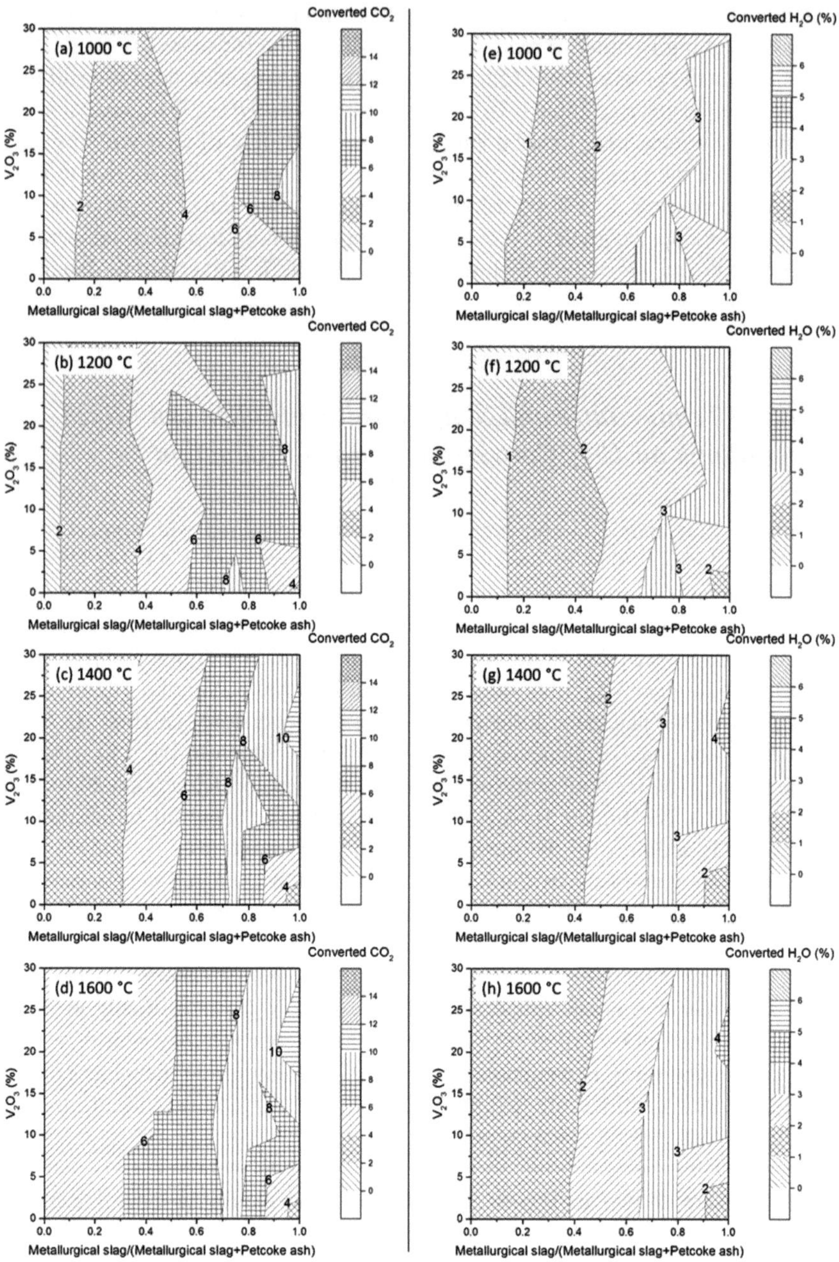

Fig. 3 Effect of additional V_2O_3 in slag on CO_2 conversion and H_2 production

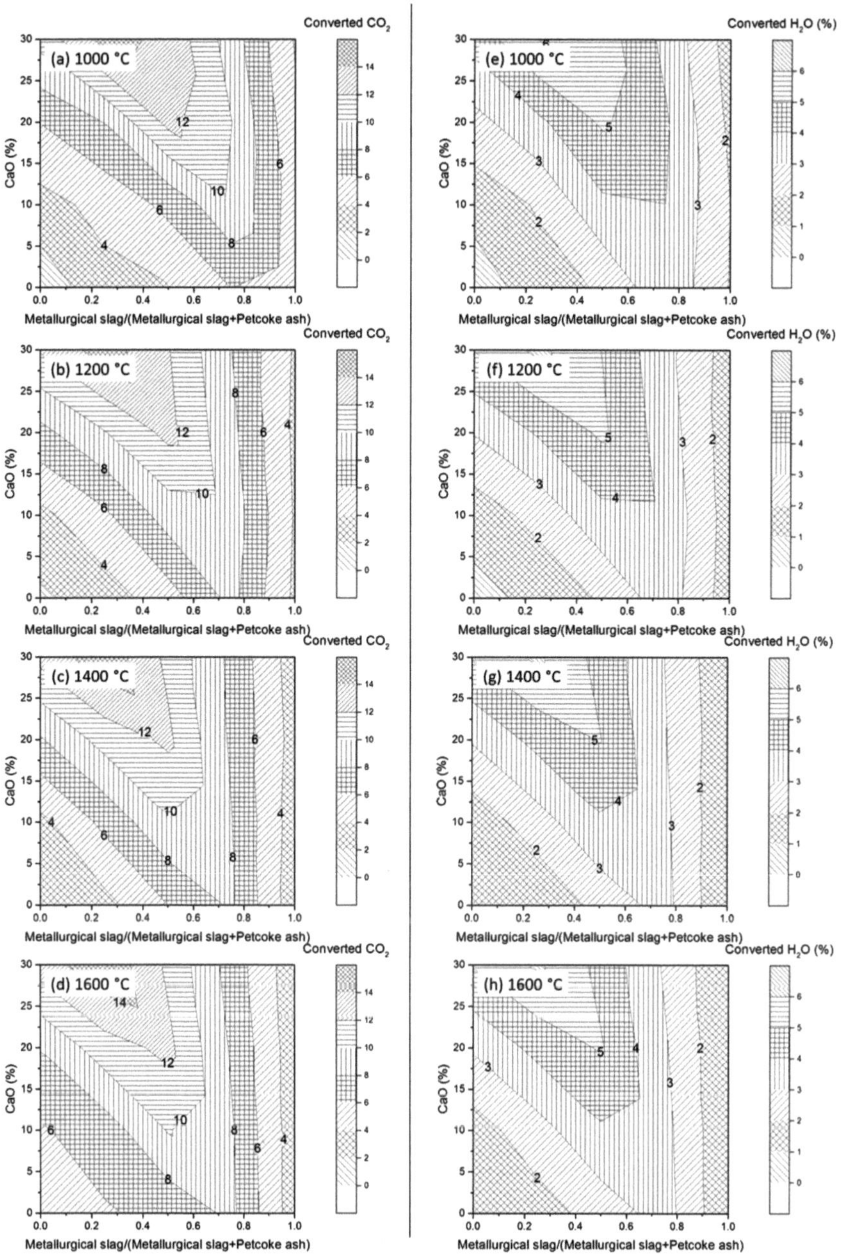

Fig. 4 Effect of additional CaO in slag on CO_2 conversion and H_2 production

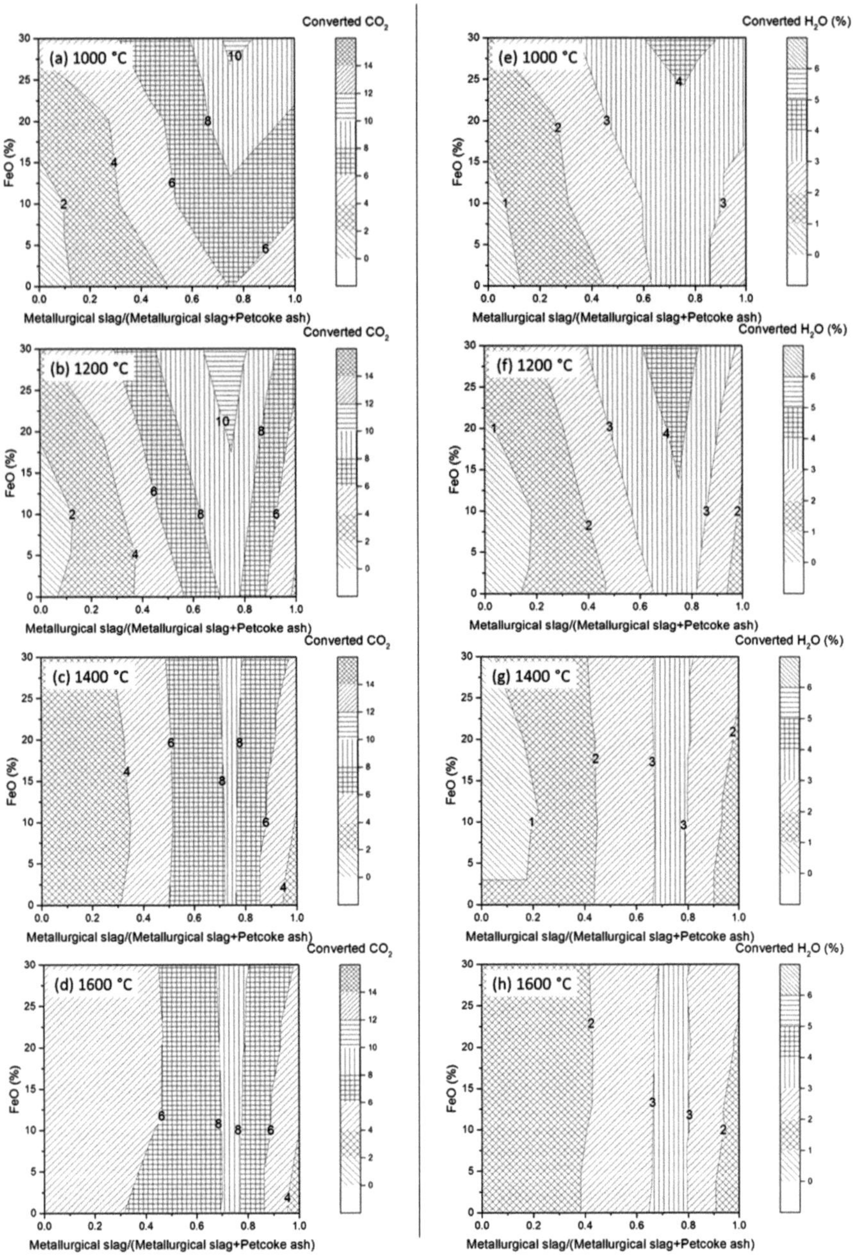

Fig. 5 Effect of additional FeO in slag on CO_2 conversion and H_2 production

Effects of Temperature and Slag Ratios on Heat Production and Volume Expansion

Theoretical heat evolution per 1 ton of slag and 1 ton of CO_2 (or H_2O), and volume expansion per 100 g slag and 100 g of CO_2 (or H_2O) for given slag ratios are shown in Fig. 6. Note enthalpy (H) given in the figures is absolute enthalpy [(a) and (b)]. More heat is generated from the slag mixtures with increasing metallurgical slag. Higher (negatively) enthalpy is expected if the slag mixtures are treated at lower temperatures. Higher concentrations of V_2O_3, CaO, and FeO are expected to increase heat generation as estimated from Figs. 3, 4 and 5. Volume, on the other hand, rapidly increased with temperature [(c) and (d)]. No effect by slag mixing was noted. At room temperature and 1 atm, a volume of 100 g CO_2 is approximately 56 l, which becomes over 320 l at 1500 °C (approx. 6 times). 100 g H_2O, if injected as vapor, has a volume of 170 l at 101 °C, which becomes around 800 l at 1500 °C (approx. 5 times). Volume expansion, along with heat and fuels generated from the conversion reactions, may be utilized to produce supplemental electric power. Note in the present process, energy required to heat the slag mixture is supplied by the metallurgical slag. Heat generation and volume expansion can be maximized through optimization of overall slag chemistry and process parameters.

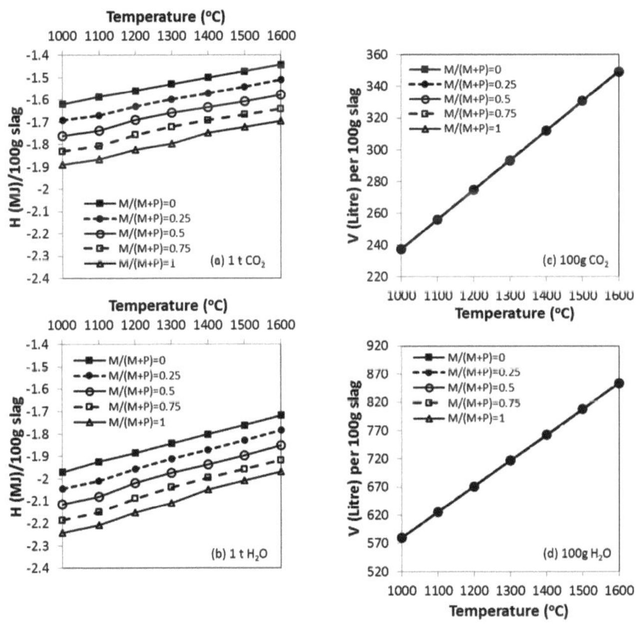

Fig. 6 Effect of temperature and slag ratios on heat production in **a** pure CO_2 and **b** pure H_2O, and volume expansion in **c** pure CO_2 and **d** pure H_2O. M = metallurgical slag and P = petcoke ash

Conclusions

Effects of V_2O_3, CaO, and Fe_2O_3 constituents in slag, reaction temperatures, and slag ratios were discussed using thermodynamic simulations. As proven by the empirical conversion value, a higher conversion may be achieved if a smaller amount of gas input compared to slag is used. The simulation indicated the possibility of reducing CO_2 emission and/or H_2 production by mixing slags from different industrial sectors. Sources and chemistry of the slags may be appropriately selected to achieve the optimal slag chemistry.

Acknowledgements This technical effort was performed in support of the National Energy Technology Laboratory's ongoing research under the RES contract DE-FE0004000.

References

1. R.J. Fruehan, Research on sustainable steelmaking. Metall. Mater. Trans. B **40**(2), 123–133 (2009)
2. M. Barati, S. Esfahani, T.A. Utigard, Energy recovery from high temperature slags. Energy **36** (9), 5440–5449 (2011)
3. A. Semykina et al., Kinetics of oxidation of divalent iron to trivalent state in liquid FeO–CaO–SiO_2 slags. Metall. Mater. Trans. B **41**(6), 1230–1239 (2010)
4. U.S. Department of Energy National Energy Technology Laboratory. Gasifipedia. 2015
5. Gasification Technology Council. World Gasification Database
6. J. Nakano, J. Bennett, CO_2 and H_2O gas conversion into CO and H_2 using highly exothermic reactions induced by mixed industrial slags. Int. J. Hydrogen Energ. **39**(10), 4954–4958 (2014)
7. J. Nakano, J. Bennett, A. Nakano, in *Energy Generation from Waste Slags: Beyond Heat Recovery*. Proceedings for TMS 2016, Nashville, TN, February 14–18, 2016 (2016), pp. 131–136
8. R.W. Bryers, Utilization of petroleum coke and petroleum coke/coal blends as a means of steam raising. Fuel Process. Technol. **44**(1–3), 121–141 (1995)
9. C.W. Bale et al., FactSage thermochemical software and databases—recent developments. Calphad **33**(2), 295–311 (2009)

Part III
Energy Technologies: Novel Technologies

Modeling Anthropogenic Heat Flux in Climate Models

Ganesan Subramanian and Neale R. Neelameggham

Abstract Anthropogenic Heat Flux (AHF) is defined as heat generated from use of non-renewable energy by humans in residential, commercial and industrial activity. State of the art Global Circulation Models (GCM) used to predict global impacts of climate change to help set related policy do not include this heat flux. AHF is considered to be a much smaller an insignificant forcing, solely based on the global mean values. Various Climate models have discussed the "Heat Island Effect" in high AHF areas. Through the use of simple climate models, it is shown that significant pockets of AHF can impact climate locally in ways significant enough to impact global climate. Inclusion of spatial AHF distribution, rather than a global mean value is important to accurate modeling either on a regional or global scale.

Keywords Climate change · Anthropogenic heating · Climate modeling · Simple models

Introduction

Climate modeling and prediction have advanced considerably in the last three decades on account of increased speed and availability of computing resources and the continuing global scale collection and analysis of climate data. Various kinds of paleo-climactic data have been used in reconstructing climate history.

Anthropogenic Heat Flux (AHF) is defined by mainstream modelers as the heat generated from use of non-renewable energy by humans in residential, commercial and industrial activity. This is not included in state of the art Global Circulation Models (GCM) used to predict global impacts of climate change and set related policy as the instantaneous and non-time based quantity is very small. Budyko and

G. Subramanian (✉)
Engineering Consultant, Houston, TX, USA
e-mail: gsubs@comcast.net

N.R. Neelameggham
The Guru, IND LLC, South Jordan, UT, USA

L. Zhang et al. (eds.), *Energy Technology 2017*,
The Minerals, Metals & Materials Series, DOI 10.1007/978-3-319-52192-3_6

Sellers who were pioneers in climate modeling [1, 2] attempted to include the AHF. This was done in simple climate models by noting the AHF to be uniformly distributed across the entire earth's surface as a global mean. This assumption in itself is subject to question. AHF relative to the solar forcing terms and energy transport by atmospheric and oceanic circulations is about 0.3% and thus neglected in climate models. A few recent climate models have discussed the "Heat Island effect" in high AHF areas along with the global impact using regional climate models (RCM) embedded in GCM's. Considerable AHF data is collected [3–5] on worldwide regions and can be used to predict climate locally. AHF will become significant relative to other terms in the energy balance, over a period of say 200 years or more so in the urban areas, due to population and industrial activity increases. Recently it is noted that large metropolitan areas with millions of residents show an energy consumption as large as 15.8 TW annually [6]. The Urban Heat Island effect such as this in concentrated small urban areas is over and above the effects simulated by GCMs [3, 4, 6]. The human energy consumption is only about 0.3% of the 5 PW total energy transported across the northern latitudes to the extra-tropics by atmospheric and oceanic circulations. This concentrated AHF could disrupt atmospheric circulation pattern and produce unforeseen effects on regional and global climate. Global climate models ignored these effects as the Global mean AHF is as small as the model's natural fluctuation, at least in short time simulations. However, long lasting effects could be significant. However, one of the regional model studies has found that the AHF could cause significant winter warming in Western Europe [5].

Neelameggham [7] and Davis studied the impact of AHF (generated by both renewable as well as non-renewable sources) through their convective model. They note that the effect of thermal waste storage into the atmosphere leads to climate change as excess precipitation and evapo-transpiration changes moderating the mean temperature of the atmospheric air mass that is constrained.

Using simple climate models, this paper demonstrates that significant pockets of AHF can impact climate locally in ways significant enough to impact global climate. Here we note that the inclusion of spatial AHF distribution, rather than a global mean value, is a key determinant to accurate modeling whether on a regional or a global scale.

The present work is an extension of the one dimensional energy model used by Brian Rose at the State University of New York at Albany, in one of his climate modeling classes [8]. This model is similar to the Budyko-Sellers model, where a diffusive mechanism describes the heat transport across latitudes. Over the tropics, the solar radiation absorbed exceeds the long wave emission to space. A comprehensive review of simple climate models have been made by North [9]. This causes a net gain of energy by radiation at the tropics. But in the mid-high latitudes, the earth loses energy by net radiation to space. The temperature gradient set up on this latitudinal basis is the driver for atmospheric and oceanic circulation poleward from the tropics. This temperature gradient is caused by the insolation and heat transport. The models of the global mean temperature, can ignore this issue because the transport just moves energy across latitude bands without creating or destroying energy.

Equations of motion and energy for the earth's surface, atmosphere and oceans are solved rigorously using grid based values in most Global climate models. We present the following set of equations used in the model discussed here as well as in other articles of this model. The net poleward heat transport that reduces the temperature gradient produced by radiant heat transfer has to be parametrized in our simple model similar to the Fourier law of heat transfer.

$$H = -K\,dT/dy$$

The diffusivity of the climate model system is denoted by K. The energy content, E(ϕ), of a column extending upwards from ocean/land is:

$$E(\phi) = C(\phi)T_s(\phi) \tag{1}$$

where

C(ϕ) is the effective heat capacity of the atmosphere-ocean column. The heat capacity isa function of the latitude and since most of the heat capacity is in the oceans, is proportional to the surface temperature. The energy balance for a thin zonal band centered at latitude Φ is

$$\frac{\partial E(\phi)}{\partial t} = \mathrm{ASR}(\phi) - \mathrm{OLR}(\phi) - \frac{1}{2\pi\alpha^2 \cos\phi}\frac{\partial H}{\partial \phi} \tag{2}$$

The heat transport is approximately represented as a down-gradient diffusion process as:

$$H(\phi) \approx -2\pi\alpha^2 \cos\phi\, D\frac{\partial T_s}{\partial \phi} \tag{3}$$

The thermal conductivity of the climate system is parameter D in W m^{-2} $°C^{-1}$. The value of D is typically tuned to match observations.

It is assumed that the solar term is temperature independent and

$$\mathrm{ASR} = (1-\alpha)Q \tag{4}$$

where α is the latitude dependent albedo and Q is the incoming radiation.

The outgoing Longwave radiation OLR is parametrized as

$$OLR = A + BT_s \tag{5}$$

where T_s is zonal average surface temperature in °C, A is constant in W m^{-2} and B is a constant in W m^{-2} $°C^{-1}$.

A and B are tuned to observed data as shown in Fig. 1.

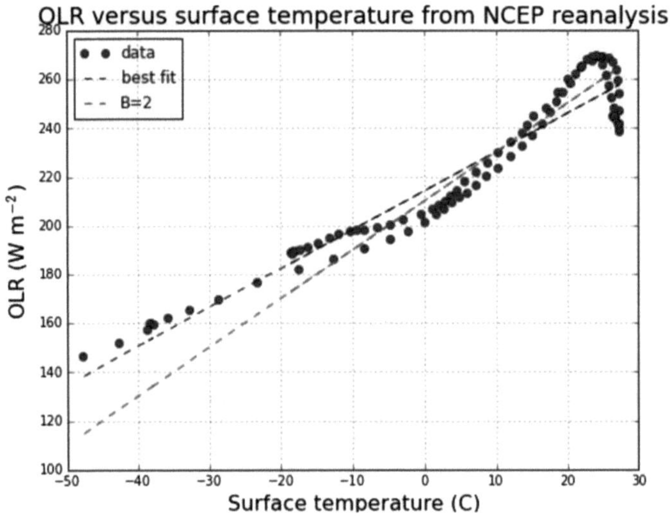

Fig. 1 QLR versus surface temperature from NCEP reanalysis

With Eqs. (1), (3)–(5), Eq. (2) for the surface temperature becomes

$$C\frac{\partial T_s}{\partial t} = (1-\alpha)Q - (A + BT_s) + \frac{D}{\cos\phi}\frac{\partial}{\partial\phi}\left(\cos\phi\frac{\partial T_s}{\partial\phi}\right) + \text{AHF}$$
$$\alpha(\phi) = \alpha_0 + \alpha_2 P_2(\sin\phi) \tag{6}$$

The land/ocean reflectivity, albedo α, is approximated by a smooth function that increases with latitude, where P_2 is the 2nd Legendre polynomial. $\boldsymbol{\alpha_0}$ and $\boldsymbol{\alpha_2}$ are fitted to observed albedo data. We have added, AHF, the anthropogenic heat flux to the energy balance equation as a function of time. This system that is no longer in equilibrium due to addition of the AHF term that reflects the heat added as the earth's population grows or changes in accordance with any specified climate policy, is then solved.. This AHF term can either be uniform across latitude or can be a function of latitude—representing areas with concentrated human activity.

We modeled a scenario where AHF starts at 0.015 watts/m^2 and increments by 3% annually as follows:

$$\text{AHF} = 0.015 * (1 + 0.03^{\text{n}})$$

where n is the number of years. This AHF depiction provides the same order of magnitude flux as correlated by Neelameggham [7] and is based on data constantly updated by British Petroleum statistics [10].

Equation (6) is then solved numerically by time stepping from the present through 200 years to obtain the temperature profile at any time with latitude. The global mean temperature can be derived from integrating Eq. (6) over all latitudes.

We can represent AHF as any other logical function depictive of human activities and solve the Eq. 6.

It is interesting to note that Arnold [13] extended a radiation based analysis by Dopel using a 2% annual increase in global energy consumption instead of the 3% annual increase in this model.

Results

The latitude ranges were chosen roughly based on the extent of human activity prevailing in the globe and noting that the world's total energy consumption in 2006 was equivalent to a constant-use rate of 16 terawatts (1 terawatt, or TW, equals 1 trillion watts). Of that, an average rate of 6.7 TW was consumed in 86 metropolitan areas in the Northern Hemisphere [5]. We find that the poles become warmer and the tropics cooler compared to the initial temperature as early as 10 years and then all latitudes warm up. There is significant additional warmup at all latitudes in spite of the AHF being active only in high human activity latitudes, compared to the case where AHF is distributed over all latitudes with a global mean value. One can note that the maximum heat islands have happened worldwide in the northern temperate latitudes with higher proportion of per capita energy consumptions. The detailed results of the model can be seen in our article submitted for to a climate science publication.

Conclusions

This paper demonstrates that significant pockets of AHF can impact climate locally in ways significant enough to impact global climate. Inclusion of spatial AHF distribution, rather than a global mean value is a key determinant to accurate modeling either on a regional or global scale.

Acknowledgements The authors are deeply indebted to Brian E. J. Rose, Department of Atmospheric and Environmental Sciences, University at Albany, Albany, New York for providing the course material and the Climlab package for modeling the simple climate model presented in this work.

References

1. M.I. Budyko, The effect of solar radiation variations on the climate of the earth. Tellus **21**(5), 611–619 (1969)
2. W. Sellers, A global climatic model based on the energy balance of the Earth-Atmosphere system. J. Appl. Meteor. **8**, 392–400 (1969)

3. A. Block, Impacts of anthropogenic heat on regional climate patterns. Geophys. Res. Lett. **31**, L12211 (2004). doi:10.1029/2004GL019852
4. M.G. Flanner, Integrating anthropogenic heat flux with global climate models. Geophys. Res. Lett. **36** doi:10.1029/2008gl036465. (2009) [Cross Ref]
5. G.J. Zhang, M. Cai,, A. Hu, Energy consumption and the unexplained winter warming over northern Asia and North America, Nat. Clim. Change, **3**, 466–470 (2013)
6. T. Ichinose, K. Shimodozono, K. Hanaki, Impact of anthropogenic heat on urban climate in Tokyo. Atmos. Environ. **33**, 3897–3909 (1999). doi:10.1016/S1352-2310(99)00132-6
7. N.R. Neelameggham, B.R. Davis, 21st century global anthropogenic warming, a convective model. J. Nanomater. Energy. **4**(2), 143–152 (December, 2015)
8. Brian Rose, Climate modeling package, http://dx.doi.org/10.5281/zenodo.48984
9. G.R. North, R.F. Cahalan, J.A. Coakley Jr., Energy balance climate models. Rev. Geophys. **19**(1), 91–121 (1981). doi:10.1029/RG019i001p00091
10. BP statistical review of world energy (June 2013), See http://www.bp.com/statistical_review for further details
11. J. Williams, G. Kromer, A. Gilchrist, The impact of waste heat release on climate: experiments with a general circulation model. J. Appl. Meteorol. **18**, 1501–1511 (1979)
12. Gerald R. North, Analytical solution to a simple climate model with diffusive heat transport. J. Atmos. Sci. **32**, 1301–1307 (1975)
13. H. Arnold, Global warming by authropogenic heat, a main problem of fusion (2016), https://www.db-thueringen.de/receive/dbt_mods_00029446. Accessed 8 July 2016

In-Situ Microscopic Study of Morphology Changes in Natural Hematite and Cu-Spinel Particles During Cyclic Redox Gas Exposures for Chemical Looping Applications

Anna Nakano, Jinichiro Nakano and James Bennett

Abstract Efficiency of the chemical looping combustion technology is adversely influenced by surface properties of oxygen carrier materials during oxidation–reduction process changes at elevated temperatures. In this work, natural hematite (Fe_2O_3) and alumina (Al_2O_3) supported Cu-spinel ($CuFe_2O_4$) oxygen carrier particles were isothermally exposed to redox gas cycles (air and 10 vol% CO–90 vol% Ar) at approximately 800 °C and studied using a high temperature confocal scanning laser microscope for real time alterations in surface morphology, roughness, and particle volume. 3D topological scans from each cyclic gas exposure indicated hematite particles exhibited noticeable volume expansion in reduction and shrinkage in oxidation, corresponding to respective phase transformations. Materials surface changes during the redox gas cycles are discussed.

Keywords Chemical looping · Oxidation · Reduction · In-situ study · Oxygen carrier

Introduction

Development of the oxygen carrier materials have been a critical area of research for chemical looping applications. Forutan et al. [1] reported the influences of alumina supported metals; including Fe, Cu, Co, and Mn; on oxygen exchange in the redox conditions studied at 850–1050 °C. Results of these experiments indicated that Fe based oxygen carriers had the highest resistance against sintering and

A. Nakano (✉) · J. Nakano · J. Bennett
U.S. Department of Energy National Energy Technology Laboratory,
1450 Queen Ave, 97321 Albany, OR, USA
e-mail: anna.nakano@netl.doe.gov

A. Nakano · J. Nakano
AECOM, P.O. Box 1959, 97321 Albany, OR, USA

L. Zhang et al. (eds.), *Energy Technology 2017*,
The Minerals, Metals & Materials Series, DOI 10.1007/978-3-319-52192-3_7

the maximum capacity for oxygen adsorption of those studied. Arjmand et al. [2] investigated the oxygen exchange characteristics of copper aluminate spinel ($CuAl_2O_4$) using a laboratory scale fluidized-bed reactor in the temperature range of 900–950 °C for 45 alternating redox cycles. The oxygen carrier exhibited reproducible and stable reactivity behavior at 900–925 °C; while, at the 950 °C tests, particles showed "soft" agglomeration and fragmentation. Fan et al. [3] analyzed the oxygen exchange kinetics using thermogravimetric analysis (TGA) of *M*-iron spinel (MFe_2O_4); with *M* being metals such as Co, Ni, Cu, Mg, Ca, Sr, and Ba. It was found that, the Group 2 metal ferrites had better reduction and oxidation rates than transition-metal ferrites. The use of natural hematite (Fe_2O_3) has also been targeted owing to its natural abundant resources [4–6]. Degradation of hematite during chemical looping processes would be related to the peculiar crystallographic nature of the hematite structure, such as twinning (single or contact twinned crystals) and fault formation [7, 8].

In this work, real time changes in surface morphology of both natural hematite (Fe_2O_3) and alumina (Al_2O_3) supported copper spinel ($CuFe_2O_4$) during redox gas cycles (air and 10 vol% CO–90 vol% Ar) at approximately 800 °C were investigated in-situ using a high temperature confocal scanning laser microscope (CSLM).

Experimental

Materials

Natural hematite (originating from the Wabush Mine in New Foundland and Labrador, Canada) was dried and screened by Nexceris Inc. Hematite samples were sieved to determine their particle size distributions, the majority of which were found to range between 150 and 250 μm in size.

The Al_2O_3 supported copper spinel ($CuFe_2O_4$) was synthesized from reagent grade powders of Al_2O_3, CuO, and Fe_2O_3 by spray drying. The majority of particles in the copper spinel samples ranged between 50 and 200 μm.

Test Procedures

The natural hematite and Al_2O_3 supported copper spinel ($CuFe_2O_4$) oxygen carrier samples were qualitatively and quantitatively characterized before isothermal redox gas cycle testing. Compositions of the samples were analyzed by X-ray Fluorescence (XRF, Rigaku ZSX Primus II) using EZScan L software and crystalline phases were identified by the X-ray diffraction analysis (XRD, Rigaku Ultima IV XRD spectrometer (Cu Kα, 40 kV, 40 mA)) and MDI Jade 9 XRD analysis software. The scanning electron microscopy with energy dispersive X-ray spectroscopy (SEM-EDX, FEI Inspect F and an Oxford INCA WAVE spectrometer) was used to investigate the surface morphology and chemistry of the particles.

To enable the in-situ observations at high temperatures under controlled atmospheres, a thin gold-plated heating chamber (Yonekura, 15TPS) was attached to a confocal scanning laser microscope (CSLM, Olympus, OLS3100), which was used to study behaviors of the hematite and Al_2O_3 supported copper spinel particles during redox gas cycles. A simplified schematic of the CSLM heating chamber is shown in Fig. 1. Before redox gas cycle tests, the temperature difference, created by the experimental configuration and the spatial deviation from the focal point in the gold-plated furnace, was estimated by repeated calibrations by melting high purity Au and Pd with known melting temperatures. The average overall deviation between the actual sample temperature and the thermocouple reading was determined and used to adjust each experimental temperature.

After the temperature calibration, the hematite or copper spinel particles were placed in a high density alumina crucible for redox gas cycle testing. The alumina crucible with the test sample was set on a sample holder that acted as a type R thermocouple positioned inside the heating chamber. The heating chamber consisted of two infrared sources (halogen lamps) on each side of the furnace, with the test sample sandwiched in the middle. The sample holder was set at the focal point where the two gold-plated elliptical cavities were joined, with each of the two halogen lamps effectively located at the other focal point. This configuration improved the furnace heating ability when compared to the traditional set up, where a single heating source is located at the furnace bottom. The test gas was introduced above the sample surface. The furnace system contained a laser with a wavelength of 408 ± 5 nm and a white LED light, which were used to investigate real-time changes occurring on the sample surface.

In-situ observations of surface changes on particles were conducted by heating a test material to approximately 800 °C at a rate of 800 °C/min using an oxidizing

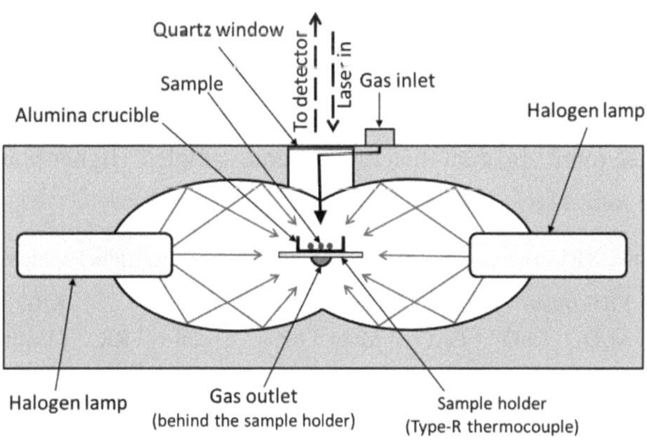

Fig. 1 A simplified schematic of the CSLM heating chamber

gas (air) flow rate of 50 ml/min. After 5 min of an isothermal hold at 800 °C, the oxidizing gas (air) was switched to a reducing gas mixture (10 vol% CO–90 vol% Ar) at the same flow rate. The particle surface morphology changes were visually recorded for 10 min in the reducing gas environment, then samples were evaluated for 5 min using a 3D topological laser scan equipped with the CSLM. During the 3D scanning, a sample surface was scanned over the particle height in a Z-direction with an increment of approximately 1.5 μm. After a total of 15 min exposure in a reducing environment, the test gas was switched back to oxidizing (air) and the same procedure was repeated. At the end of the last gas cycle, the sample was rapidly cooled by turning off the "heat" and blowing a He gas at 60 ml/min. Changes in surface morphology, roughness, and particle volume during the redox gas cycles were analyzed using 3D topological laser scan data and the LEXT-OLS software.

Results and Discussion

Sample Analysis Before Redox Gas Exposure

XRF and XRD analysis results on the original hematite and copper spinel samples (i.e., before redox gas exposure) are shown in Table 1. XRF indicated the bulk natural hematite was comprised mainly of Fe_2O_3, with small quantities of SiO_2 and MnO; while XRD indicated mainly Fe_2O_3 (hematite), with a small fraction of SiO_2 crystals. MnO could not be identified by XRD analysis due to its small concentration. The copper spinel was $CuFe_2O_4$ and Al_2O_3 crystals comprised of Al_2O_3, CuO, and Fe_2O_3 with CuO having the highest concentration by mass in the bulk.

SEM backscatter micrographs of surfaces of copper spinel and natural hematite are shown in Fig. 2. Hematite particles exhibited irregular angular morphology while the copper spinel particles were spherical in shape.

The particles of both samples seemed to be aggregates of multiple smaller particles or polycrystalline particles. The copper spinel aggregates were comprised of micron particles of $CuFe_2O_4$ with micron Al_2O_3 particles randomly distributed throughout (Fig. 2a). Grain boundaries (or particle boundaries) on the spinel aggregates were more apparent than the hematite samples. Higher porosity was also

Table 1 XRF and XRD analysis on original hematite and copper spinel samples

	XRF (mass%)							XRD
	Al_2O_3	CuO	Fe_2O_3	MnO	SiO_2	Total	ERR %	Identified crystalline phases
Natural hematite			86.27	3.51	10.23	100.0	0.50	Fe_2O_3, SiO_2
Copper spinel	31.19	37.49	31.32			100.0	0.50	$CuFe_2O_4$, Al_2O_3

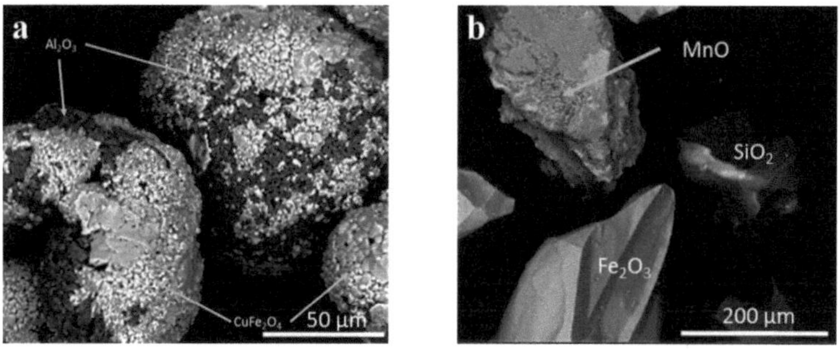

Fig. 2 **a** SEM of copper spinel samples and **b** SEM of natural hematite samples

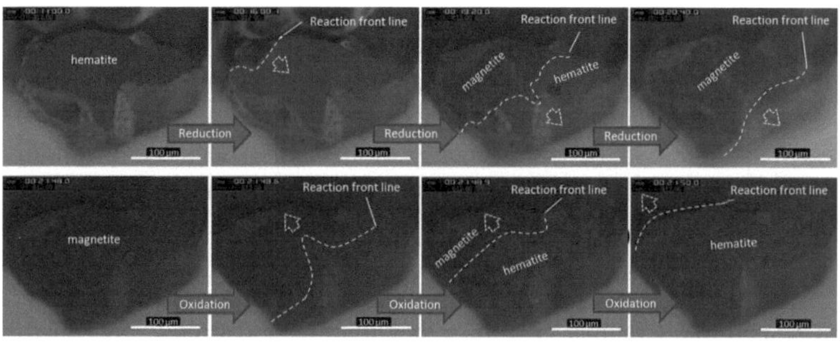

Fig. 3 Successive CSLM images (*white light*) showing hematite ↔ magnetite transition over an air ↔ 10 vol% CO–90 vol% Ar gas switch at 812 °C (1st cycle)

noted in the spinel samples. In the natural hematite sample, the SiO_2 was present as individual particles whereas MnO tended to be intermixed in the hematite particles (Fig. 2b).

In-Situ Microscopic Study During Redox Gas Cycles

Successive CSLM images (using white light) of natural hematite during gas changes (1st cycle) at 812 °C are shown in Fig. 3. When a gas was switched from air (oxidizing) to 10 vol% CO–90 vol% Ar (reducing), the hematite particle surface started to experience a lateral phase transformation from hematite to magnetite as indicated in Fig. 3. Note the reaction front moved at a faster rate along the fault (located perpendicular to the reaction front line in Fig. 3) to reduce the surface energy. After 5 min of incubation time (time before the transition started/was noted after the gas switch), the whole transition from hematite to magnetite at the top surface took approximately 10 min.

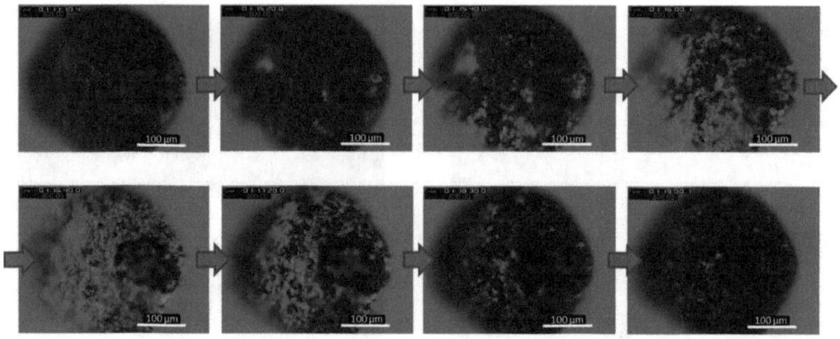

Fig. 4 Successive CSLM images (*white light*) showing scattered phase transition (*dark* → *bright* → *dark*) over a surface of copper spinel particle upon a gas switch from air to 10 vol% CO–90 vol% Ar at 800 °C (2nd cycle)

Note that the change occurred over an area spanning approximately 150 μm. Upon switching the gas back to air, the magnetite to hematite transition took approximately 2 s over the same area, with an incubation time of 48 s. Note the reversed transformation (from magnetite to hematite) occurred in an opposite direction. Both incubation and reaction times were shorter in oxidation compared to reduction in the hematite samples. After the 1st cycle, an increase in surface roughness was visually apparent. As more cycles were repeated, the incubation time and transitioning rate (frontline velocity on the surface) increased drastically, which may be related to alterations in surface including degradation.

Successive CSLM images showing phase transitions over a surface of the copper spinel particle during a gas change from air to 10 vol% CO–90 vol% Ar at 800 °C (2nd gas cycle) are shown in Fig. 4. It seems that a phase transformation in this sample occurred throughout the surface, likely corresponding to phase changes occurring in individual micron particles that make up the larger particle. Under white light, dark portions appeared to be bright (twinkling), then darkened again. The total reaction time (observable on surface) was approximately 5 min. It may be possible that the particle was experiencing two separate transformations or decompositions, which requires more investigations.

Volume Change and Roughness of the Hematite Surface

Changes in surface morphology, roughness, and particle volume during the redox gas cycles were analyzed using 3D topological laser scan data and are shown in Fig. 5.

As oxidation/reduction gas changes proceeded, particle surfaces tended to flatten and the overall height became extended as the cycle continued. Over an increasing number of transformation cycles, the surface became rougher and individual particles tended to agglomerate as seen in Fig. 5.

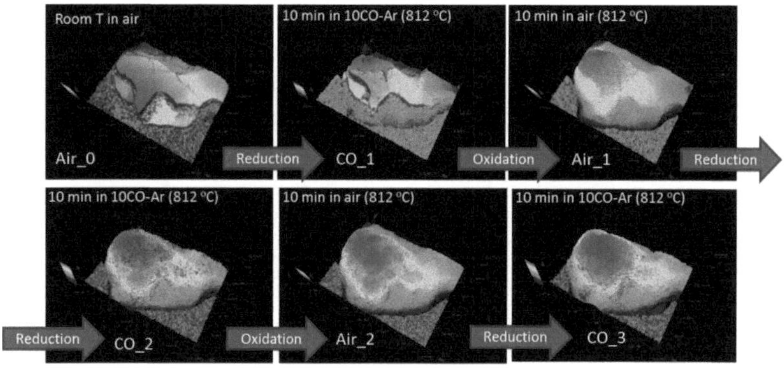

Fig. 5 3D scan images of natural hematite over reduction-oxidation gas cycles at 812 °C: 'Air_0'—Room temperature 3D scan; 'CO_1' and 'Air_1'—1st redox cycle; 'CO_2' and 'Air_2'—2nd redox cycle; 'CO_3'—3rd redox cycle followed by rapid cooling in He gas flow

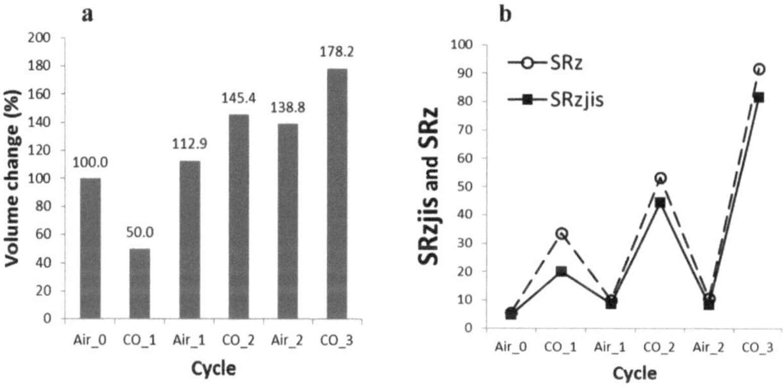

Fig. 6 **a** Changes in volume and **b** Changes in roughness of natural hematite over redox cycles. SRzjis: Ten-point mean roughness ("Rz" in JIS'94). SRz: Max. height of profile curved surface. 'Air_0'—Room temperature measurement; 'CO_1' and 'Air_1'—1st redox cycle; 'CO_2' and 'Air_2'—2nd redox cycle; 'CO_3'—3rd redox cycle followed by rapid cooling in He gas flow

The total volume of the particles over the cycles studied approached a theoretical limit (lattice expansion of 198% [9]) if one assumes a full transformation from pure hematite to pure magnetite without the formation of voids and cracks inside the particle (Fig. 6a). The increase in volume exhibited a zig-zag incline corresponding to the hematite ↔ magnetite phase transformations. Note volume measurements were performed under the assumption that a volume existed straight down below the top surfaces due to the CSLM limitation.

Roughness changes of natural hematite is shown in Fig. 6b. Overall roughness tended to increase over the redox cycles, implying potential formations of micro-voids and microbumps across the particle surfaces. However, it is not certain, from the surface observation, about their occurrence within the individual particle.

Roughness can be interpreted as an index of materials surface area changes. The particle surface tended to show higher roughness in the form of magnetite compared to that in hematite. However, overall roughness rapidly increased over the cycles, and may facilitate surface deterioration from constant cycling.

Conclusions

Oxygen carrier changes in surface morphology of natural hematite (Fe_2O_3) and alumina (Al_2O_3) supported copper spinel ($CuFe_2O_4$) caused by oxygen exchange during the chemical looping processes using redox gas cycles (between air and 10 vol% CO–90 vol% Ar) at approximately 800 °C were investigated using a high temperature confocal scanning laser microscope (CSLM). During the isothermal gas changes at the CSLM, the hematite particles exhibited lateral phase transitions on the surfaces corresponding to hematite transforming to magnetite when the environment was changed from oxidizing (air) to reducing (10 vol% CO–90 vol% Ar). The magnetite then experienced a reverse reaction to hematite with a shorter incubation time when the exposure environment was switched back to air. In general, particle volumes increased in transitioning to magnetite and decreased in transitioning back to hematite due to expansion/contraction induced by crystal structure phase changes while overall volume increased with cycles. The overall volume increase was consistent with the theoretical maximum. The copper spinel particle exhibited possibly two-step separate transformations/decompositions during the redox cycling. More study is required to understand the surface phenomena.

Acknowledgements This technical effort was performed in support of the National Energy Technology Laboratory's ongoing research under the RES contract DE-FE0004000.

Disclaimer This project was funded by the Department of Energy, National Energy Technology Laboratory, an agency of the United States Government, through a support contract with AECOM. Neither the United States Government nor any agency thereof, nor any of their employees, nor AECOM, nor any of their employees, makes any warranty, expressed or implied, or assumes any legal liability or responsibility for the accuracy, completeness, or usefulness of any information, apparatus, product, or process disclosed, or represents that its use would not infringe privately owned rights. Reference herein to any specific commercial product, process, or service by trade name, trademark, manufacturer, or otherwise, does not necessarily constitute or imply its endorsement, recommendation, or favoring by the United States Government or any agency thereof. The views and opinions of authors expressed herein do not necessarily state or reflect those of the United States Government or any agency thereof.

References

1. H.R. Forutan et al., Expert representation chemical looping reforming: a comparative study of Fe, Mn, Co and Cu as oxygen carriers supported on Al_2O_3. J. Ind. Eng. Chem. **21**, 900–911 (2015)
2. M. Arjmand et al., Evaluation of $CuAl_2O_4$ as an oxygen carrier in chemical-looping combustion. Ind. Eng. Chem. Res. **51**(43), 13924–13934 (2012)
3. Y.Y. Fan, R. Siriwardane, Novel new oxygen carriers for chemical looping combustion of solid fuels. Energ. Fuel **28**(3), 2248–2257 (2014)
4. E.R. Monazam, R.W. Breault, R. Siriwardane, Kinetics of magnetite (Fe_3O_4) oxidation to hematite (Fe_2O_3) in air for chemical looping combustion. Ind. Eng. Chem. Res. **53**(34), 13320–13328 (2014)
5. R.W. Breault, E.R. Monazam, Fixed bed reduction of hematite under alternating reduction and oxidation cycles. Appl. Energ. **145**, 180–190 (2015)
6. R.W. Breault, E.R. Monazam, J.T. Carpenter, Analysis of hematite re-oxidation in the chemical looping process. Appl. Energ. **157**, 174–182 (2015)
7. I. Sunagawa, Studies of crystal surfaces. Mineral. Soci. Am. Spec. paper **1**, 258–266 (1963)
8. L.E. Lagoeiro, Transformation of magnetite to hematite and its influence on the dissolution of iron oxide minerals. J. Metamorph. Geol. **16**(3), 415–423 (1998)
9. L.V. Bogdandy, H.J. Enngell (ed.), Fundamentals, in *The Reduction of Iron Ores—Scientific Bases and Technology*, (Springer, New York, 1971), p. 35

Thermodynamic Stability of Condensed Phases in the Ternary System CaO–Cu–O by the EMF Method

Joseph Hamuyuni, Dmitry Sukhomlinov, Mari Lundström and Pekka Taskinen

Abstract The standard thermodynamic properties of the compound Ca_2CuO_3 were determined electrochemically by the Electromotive Force (EMF) method. The ternary phase Ca_2CuO_3 was synthesized from the pure oxides CaO and CuO at CaO saturation using a ceramic route. The EMF measurements were performed over a temperature range $973 \leq T/K \leq 1124$. The standard Gibbs energy of formation of the ternary compound Ca_2CuO_3 was determined using the EMF technique as:

$$\Delta_{f(ox)}G^{\circ}_{Ca2CuO3} = 21.884 - 22.92 \cdot 10^{-3} \cdot T. \quad (1)$$

Keywords Gibbs energy · Electrochemistry · Solid electrolyte · Stabilized zirconia

Introduction

Phase relations and thermodynamic properties of the system Ca–Cu–O are central to understanding high temperature superconductor design [1–3] as well as processing of metals [4]. Sub solidus phase relations and thermochemical properties are important in the former while liquidus phase relations improve the understanding of the later [1–4].

Phase equilibria and thermochemical properties of the system Ca–Cu–O in literature are explained in six independent studies [5–10], mostly conducted in the late 1900s, between the years 1990 and 1997 [6–9]. The experimental study by Gadalla

J. Hamuyuni (✉) · M. Lundström
Department of Materials Science and Engineering Metallurgical Hydrometallurgy and Corrosion Research Group, Aalto University School of Chemical Technology, Aalto, Finland
e-mail: joseph.hamuyuni@aalto.fi

D. Sukhomlinov · P. Taskinen
Department of Materials Science and Engineering Metallurgical Thermodynamics and Modeling Research Group, Aalto University School of Chemical Technology, Vuorimiehentie 2K, PO Box 16200, FI-00076 Aalto, Finland

L. Zhang et al. (eds.), *Energy Technology 2017*,
The Minerals, Metals & Materials Series, DOI 10.1007/978-3-319-52192-3_8

and White dating back to 1963 [5] is an exception and has since been the main reference point for all recent studies. It should be mentioned that apart from Risold et al., [7] all the studies are experimental with each one covering only some of the phase relation aspects of the system Ca–Cu–O. Risold et al. reviewed the system Ca–Cu–O using binaries at 1 bar total pressure, and on this basis performed a thermodynamic assessment to achieve an optimized description of the system. Phase relations in the system Ca–Cu–O are depicted using the CaO–CuO pseudo binary in Fig. 1. The figure is calculated with MTDATA, a thermodynamic software and using the MTOX database version 8.1 [11].

In the experimental studies, a combined Four (4) methods have been applied namely, Thermogravimetric Analysis (TGA), X-Ray Powder Diffraction (XRPD), Scanning Electron Microscopy (SEM-EDS), and Electromotive Force (EMF), to investigate the phase relations and thermochemical properties.

By the thermobalance (TGA) method, Gadalla and White [5] in their phase equilibria study established that Ca_2CuO_3 was the only stable ternary compound. Stability of this compound was validated by Schulze et al. [3] and that this ternary compound was only stable below 1223 K.

Roth et al. [6] investigated the phase relations in the system CaO–CuO by XRD methods. Apart from confirming the ternary compound by an earlier study of Gadalla and white, two other ternary compounds $CaCu_2O_3$ and $Ca_{1-x}CuO_2$ were identified. Roth et al. further proposed that; Ca_2CuO_3 is only stable up to 1307 K where it melts incongruently to form CaO and oxide liquid, $CaCu_2O_3$ is stable between 1258 and 1291 K and that it decomposes to Ca_2CuO_3 and CuO below 1258 K. on the higher side, the compound melts incongruently to form Ca_2CuO_3

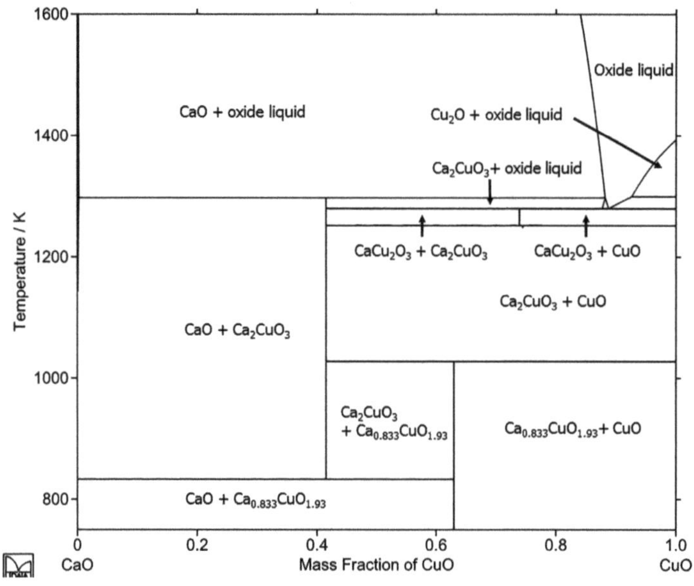

Fig. 1 Calculated phase relations in CaO–Cu–O system in O_2 at 1 atm. pressure

and liquid phase, $Ca_{1-x}CuO_2$ is only stable in air up to 1028 K and decomposes to Ca_2CuO_3 and CuO.

By combining XRD and SEM-EDS methods, Tsang et al. [9] and most recently Grivel [10], studied the phase relations in systems which include the CaO–CuO pseudo binary. After observing that earlier studies of the phase relations in the Ca–Cu–O system had been done in air, Tsang et al. investigated the phase equilibria of this system in oxygen at 1 atm and in a much wider temperature range between 1073 and 1573 K. They confirmed stability of Ca_2CuO_3 and $CaCu_2O_3$ although the stoichiometry of the later is given as $Ca_3Cu_7O_{10}$. Most recently, Grivel's study [10] of sub solidus phase relations of systems that include the CaO–CuO pseudo binary, these ternary compounds are investigated.

Using isothermal ternary sections of the system Cu–Cu–O at 1073 and 1223 K, Mathews et al. [8] determined the phase relations and thermochemical properties of the system Ca–Cu–O by the EMF, XRD and EPMA methods. At 1073 K both ternary compounds Ca_2CuO_3 and $Ca_{0.828}CuO_2$ were found to be stable but only the former was found at 1223 K. They then measured thermodynamic properties of these two ternary compounds. Thermodynamic properties of the ternary compound $CaCu_2O_3$ reported in by Roth et al. [6] were not measured. It is worth mentioning that this is the single most comprehensive experimental study of this system that combines both phase equilibria and thermochemical data to establish these phase relations.

From these studies, it is evident that experimentally measured thermochemical properties data is still scarce and that even in areas where studies have been conducted, validation is still necessary. Moreover, a complete thermodynamic description of any system is a function of both phase relations and thermochemical property data [12].

In this study thermochemical properties of the ternary compound Ca_2CuO_3 are determined experimentally by the EMF method using a solid electrolyte galvanic cell.

Experimental Section

Synthesis of Ca_2CuO_3 Saturated with CaO

Fine powders of pure CaO (99.9%, Sigma-Aldrich, USA) and CuO (99.995%, Alfa Aesar, Germany) were thoroughly mixed with a molar ratio 3:1 in an agate mortar and pressed into a pellet. Prior to use, CaO was dried at 1123 K for 4 h in air to improve its reactivity by removal of possible volatile impurities, i.e. H_2O and CO_2. The pelletized sample was heat-treated at 1223 K for 36 h in pure O_2 (99.999%, AGA, Finland) with an intermediate grinding and repelletizing (solid state reaction which was not complete, i.e. a certain fraction of CuO was remaining unreacted). It was further heat-treated at 1333 K for 14 h (temperature was slightly above the eutectic point in the CaO–CuO system generating a little fraction of a liquid phase which facilitated a rapid completion of the reaction) followed by annealing at 1223 K for 1 h.

The pelletized sample was placed in an alumina crucible during all described cycles of heat-treatment, but a direct contact between the sample and Al_2O_3 was avoided with the help of a Pt substrate placed at the bottom of the crucible. This was supported by the protective sacrificial pellet made of pure CaO placed in between the Pt substrate and the sample. Purity of the sample produced as well as its biphasic Ca_2CuO_3–CaO phase composition was confirmed by SEM-EDS and XRPD. Figure 2 depicts the Ca_2CuO_3–CaO phases. The XRPD pattern is shown in Fig. 3.

Fig. 2 Micrograph of the synthesized compound Ca_2CuO_3 saturated with CaO (Obtained with a LEO 1450 (Carl Zeiss Microscopy GmbH, Jena, Germany) scanning electron microscope with a Link Inca X-Sight 7366 Energy EDS analyzer (Oxford Instruments plc, Abingdon, Oxfordshire, UK))

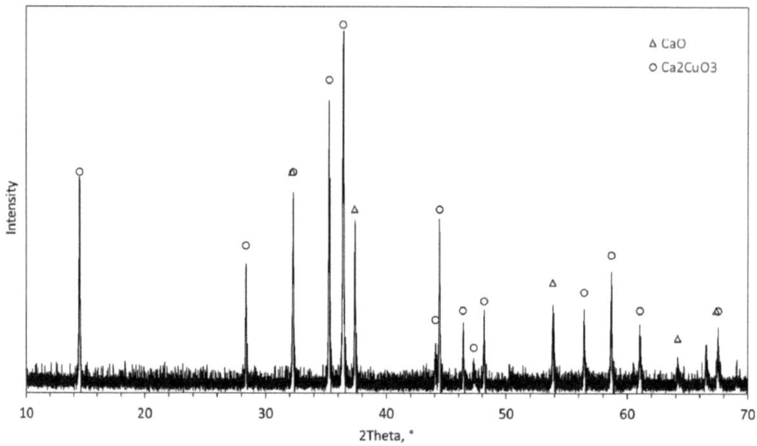

Fig. 3 XRPD pattern of the synthesized compound Ca_2CuO_3 saturated with CaO (Obtained with a PANalytical diffractometer (X'Pert Pro Powder, Almelo, Netherlands) equipped with CuKα radiation)

Apparatus and Galvanic Cells

EMF values of a galvanic cell of the following type were measured in isothermal conditions:

$$(-)\mathrm{Pt}|\ \mathrm{Ca_2CuO_3(s) + CaO(s) + Cu_2O(s)}\ |\mathrm{MSZ}|\ \mathrm{Cu_2O(s)\ CuO(s)}|\mathrm{Pt}(+), \quad (2)$$

where MSZ stands for magnesia-stabilized zirconia solid electrolyte.

Test electrode was formed by pressing into a pellet a mixture of synthesized Ca_2CuO_3 saturated with CaO and Cu_2O. Cuprous oxide (Cu_2O) was synthesized by oxidizing a cathode copper (99.996%, Boliden Harjavalta Oy, Finland). The pure copper was annealed in air above 1299 K for 120 h and quenched in air [13]. A mixture of CuO and Cu_2O was adopted as a reference electrode.

The electrochemical cell was housed in a horizontal furnace equipped with a fused silica gas-tight work tube. The arrangement of the cell is illustrated in Fig. 4. MSZ crucible was used as a solid electrolyte and two electrodes were pressed towards inner (reference) and other (test) surfaces of its bottom (see Fig. 4). Pure Pt wires (99.99%, 0.25 mm diameter, Johnson-Matthey Noble Metals, UK) were utilized as leads. In order to separate electrode compartments from each other and due to a rather high equilibrium oxygen partial pressure of the electrodes, an alumina cement was used as a sealing. A nearly static (5–6 ml min^{-1} flowrate) protective atmosphere of Ar (99.999%, AGA, Finland) was provided during the entire measurement.

The temperature measurement was performed from both side of the galvanic cells with two calibrated S-type Pt–Pt/Rh thermocouples (Johnson-Matthey Noble Metals, UK) connected to a Keithley 2010 DMM multimeters. A Pt100 resistance

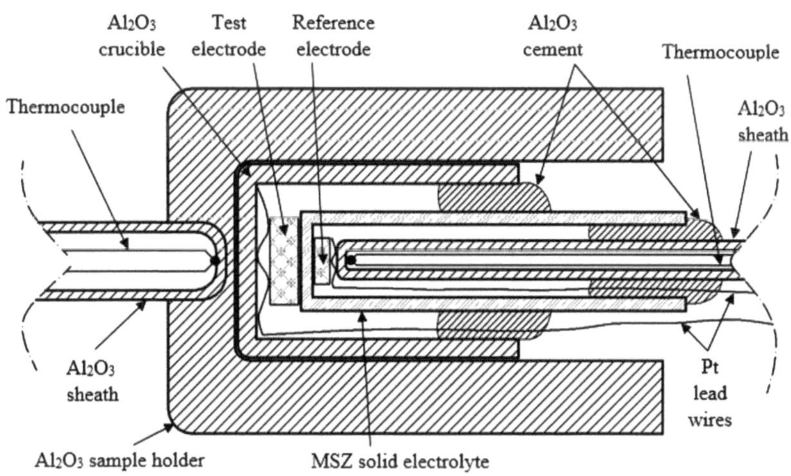

Fig. 4 Experimental arrangement of the galvanic cell

thermometer (SKS-Group, Finland, tolerance class B 1/10) connected to a Keithley 2000 DMM multimeter was utilized as a cold junction compensation. Measurements of the EMF were performed with Keithley 6517B electrometer with high input impedance of $2 \times 10^{14}\ \Omega$.

Results and Discussion

The reaction on the test (left hand side) electrode is:

$$2CaO(s) + 1/2Cu_2O(s) + 1/4O_2(g) = Ca_2CuO_3(s). \quad (3)$$

The reaction on the reference (right hand side) electrode is:

$$1/2Cu_2O(s) + 1/4O_2(g) = CuO(s). \quad (4)$$

Therefore, the overall (or virtual) cell reaction is:

$$2CaO(s) + CuO(s) = Ca_2CuO_3(s). \quad (5)$$

The Gibbs energy change of the virtual cell reaction can be calculated from the following expression:

$$\Delta_r G^\circ = -EF, \quad (6)$$

where E and F stand for measured EMF (V) and Faraday constant (96,485.3399 C mol^{-1}), respectively.

Linear regression calculated for the measured EMF as a function of temperature is expressed by:

$$E = -0.227 + 0.238 \cdot 10^{-3} \cdot T. \quad (7)$$

The determination coefficient is R^2 is 0.9894. The experimental temperature range is $973 \leq T/K \leq 1124$.

Therefore, the Gibbs energy change of the virtual cell reaction is given by the equation:

$$\Delta_r G^\circ = 21.884 - 22.92 \cdot 10^{-3} \cdot T. \quad (8)$$

Meanwhile, this expression represents the standard Gibbs energy of formation of Ca_2CuO_3 from its component oxides, $\Delta_{f(ox)}G^\circ{}_{Ca2CuO3}$. A comparison of the result obtained in this work with Mathews et al.'s [8] is given in Fig. 5.

The absolute values measured are in a reasonable agreement between this work and Mathews et al.'s results [8], however the slope is drastically different. By extrapolating the result of the present study to the lower temperature,

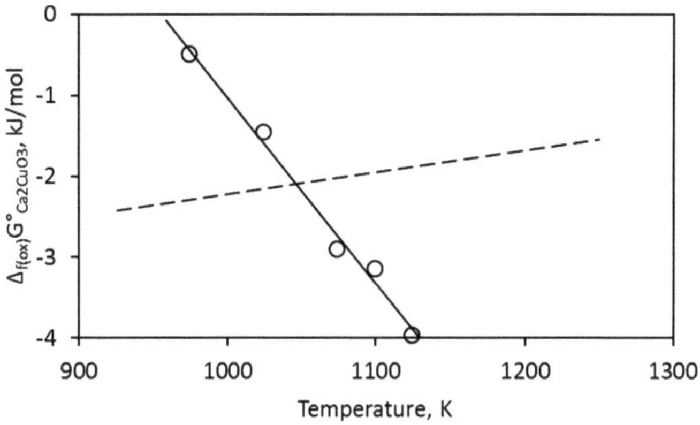

Fig. 5 Comparison of the standard Gibbs energy of formation of Ca_2CuO_3 from its component oxides acquired in this study (*bold line* with *round symbols* representing actual measurement points) with the previous work (*dashed line*) [8]

a decomposition of Ca_2CuO_3 can be expected at the temperature around 956 K, which is in a good agreement with the assessed CaO–CuO phase diagram reported by [11]. This is not in agreement with Mathews et al.'s results [8]. The difference could be attributed to the difference in the cell compartment designs. In this study, closed cell compartment compared to open isolated compartments employed in previous studies, have been used. This, however, still needs further experimental work to validate this claim.

Conclusions

The standard thermodynamic properties of solid phases were determined electrochemically by the EMF method. The ternary phase Ca_2CuO_3 was synthesized from the pure oxides CaO and CuO under CaO saturation using a ceramic route. The electromotive force measurements were performed over a wide temperature range which translates into more accurate extrapolations. The standard Gibbs energy of formation of the ternary compound Ca_2CuO_3 was found to be:

$$\Delta_{f(ox)}G^{\circ}_{Ca2CuO3} = 21.884 - 22.92 \cdot 10^{-3} \cdot T. \quad (9)$$

The results obtained in the present study are different from those in literature.

Acknowledgements The authors are grateful for financial support of this research by Association of Finnish Steel and Metal Producers and Systems Integrated Metal Processes (SIMP) project of FIMECC.

References

1. C. Sekar et al., Crystal growth of spin-ladder compound $La_2Cu_2O_5$ and ladder-like compound $CaCu_2O_3$. Physica C **378**, 678–682 (2002)
2. J.G. Bednorz, K.A. Müller, Possible High-T_c Superconductivity in the Ba-La-Cu-O System. Z. Phys. B **64**(2), 189–193 (1986)
3. K. Schulze et al., Phase equilibria in the system Bi_2O_3–SrO–CaO–CuO with emphasis on the high-Tc superconducting compounds. Z. Metallkund. **81**, 836–842 (1990)
4. J. Hamuyuni et al., Experimental determination of the liquidus of the binary system Cu_2O–CaO in air from 1050 to 1500 °C. J. Chem. Thermodyn. **77**(2014), 112–115 (2014)
5. A.M.M. Gadalla, J. White, Equilibrium relationships in the system CuO–Cu_2O–CaO. Trans. Br. Ceram. Soc. **65**, 181–190 (1966)
6. R.S. Roth et al., Phase equilibria in the systems CaO–CuO and CaO–Bi_2O_3. J. Am. Ceram. Soc. **74**, 2148–2151 (1991)
7. D. Risold, B. Hallstedt, L.J. Gauckler, Thermodynamic assessment of the Ca–Cu–O System. J. Am. Ceram. Soc. **78**, 2655–2661 (1995)
8. T. Mathews, J.P. Hajra, K.T. Jacob, Phase relations and thermodynamic properties of condensed phases in the system calcium-copper-oxygen. Chem. Mater. **5**, 1669–1675 (1993)
9. C.F. Tsang, J.K. Meen, D. Elthon, Phase equilibria of the calcium oxide-copper oxide system in oxygen at 1 atm. J. Am. Ceram. Soc. **78**, 1863–1868 (1995)
10. J.C. Grivel, Subsolidus Phase Relations of the CaO–REO_x–CuO Systems (RE = Eu, Tb, Dy, Ho, Er, Lu and Sc) at 900 °C in Air. J. Phase Equilib. Diffus. **37**, 601–610 (2016)
11. MTOX, *Release Notes for Version 8.1 of MTOX database.* (NPL, Teddington, U.K., 2015)
12. J. Hamuyuni, P. Taskinen, Experimental phase equilibria of the system Cu-O-CaO-Al_2O_3 in air. J. Eur. Ceram. Soc. **36**, 847–855 (2016)
13. J. Hamuyuni, P. Taskinen, Liquidus experimental data for the system Cu–O–Cr_2O_3 in air. Thermochim. Acta **638**, 96–102 (2016)

Experimental Study on Electro-Spraying of Ethanol Based on PDA Measurement

Haige Li, Yunhua Gan, Xiaowen Chen, Yang Tong and Meilong Hu

Abstract Ethanol is a kind of renewable fuel, which can be produced from biomass and applied in many energy areas, such as motorcar. The combustion of ethanol in meso-scale is facing great challenges, such as great heat loss, poor stability, and low combustion efficiency. Electro-spraying technique is an effective method to atomize liquid ethanol due to its great advantage of monodisperse and uniform characteristics. In present study, electro-spraying was applied in liquid meso-scale combustor to produce very small droplet diameter and uniform droplet distribution. Experiments were carried out based on nozzle-ring-mesh electrode structure using pure liquid ethanol as fuel. Electro-spraying characteristics were measured by Phase Doppler Anemometer (PDA) under different nozzle voltages, ring electrode voltages, flow rates, and steel mesh positions. The results may be helpful to the design of new meso-scale electro-spraying combustor using renewable liquid fuel.

Keywords Ethanol · Meso-scale combustor · Electro-spraying · PDA

Introduction

More and more attention has been paid to the micro-power systems using micro-scale combustors due to the higher specific energy of liquid hydrocarbons than that of batteries [1, 2]. The most important component in a micro energy system is the micro-combustor with stable flame [3]. But it is difficult to keep stable flame inside the micro-combustor, due to short residence time and high heat loss

H. Li · Y. Gan (✉) · X. Chen · Y. Tong
School of Electric Power, South China University of Technology,
Guangzhou 510640, China
e-mail: ganyh@scut.edu.cn

M. Hu
College of Materials Science and Engineering, Chongqing University,
Chongqing 400044, China

L. Zhang et al. (eds.), *Energy Technology 2017*,
The Minerals, Metals & Materials Series, DOI 10.1007/978-3-319-52192-3_9

rate [4, 5]. The electro-spraying method shares many irreplaceable advantages which can yield quasi-monodispersed droplets, control the droplet motion easily and avoid coalescence of droplets [6, 7]. Many researchers have been applying the electro-spraying technology to improve the performance and stability of micro-combustors. Kyritsis et al. [8] studied the performance of a combustor based on the liquid fuel electrospray injection which used the capillary-mesh system. Deng et al. [9] studied the liquid fuel micro-combustor using multiplexed capillary-ring-mesh electrospray system. Mikami et al. [10] studied the micro-scale combustor with capillary-ring-mesh structure, comparing the combustion characteristics with or without wire mesh. Gan et al. [11] studied the combustion characteristics with the two different combustors which used the capillary-mesh structure and capillary-ring-mesh structure respectively.

In this paper, experiments were carried out based on nozzle-ring-mesh electrode structure using pure liquid ethanol as fuel. Electro-spraying characteristics were studied under different nozzle voltages, ring electrode voltages, flow rates, and steel mesh positions. The results may give help for the design of new meso-scale electro-spraying combustor using renewable liquid fuel.

Experimental Description

Experimental Setup

The experimental setup is presented in Fig. 1. Ethanol was injected into the nozzle by a syringe pump (KDS100, KDS, USA). The droplet size was measured by a Phase Doppler anemometer (PDA) (classic PDA, Dantec Dynamics, DK).

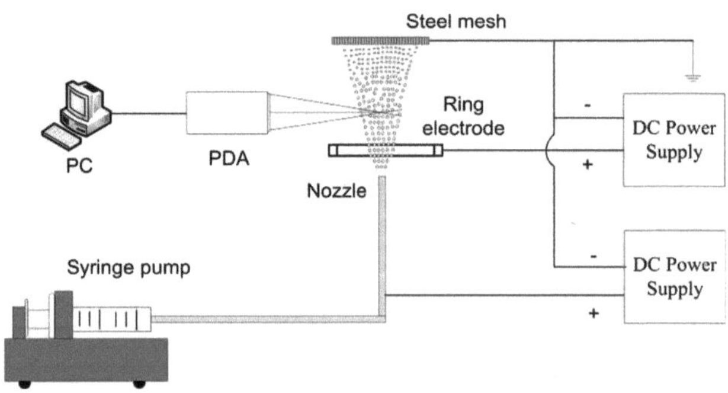

Fig. 1 Experimental setup

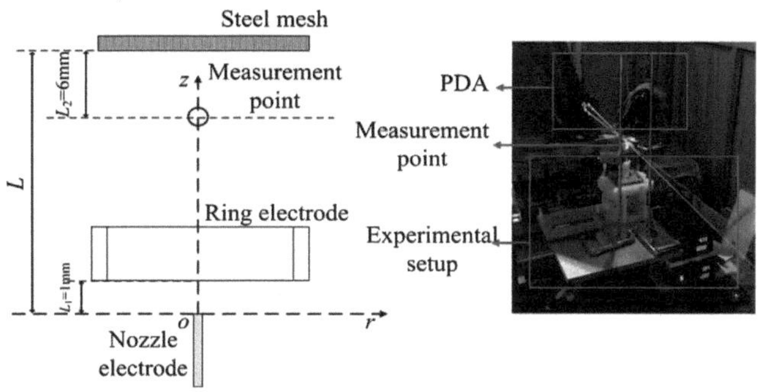

Fig. 2 The schematic of measurement points distribution

Test Section

The spraying system adopted a capillary-ring-mesh structure. Its nozzle electrode and ring electrode are connected to two different DC power sources respectively, and the steel mesh is grounded. The distance between nozzle outlet and the bottom of ring electrode is L_1 = 1 mm. The distance between nozzle outlet and steel mesh is L. The positions of nozzle and ring electrode maintains unchanged, and moving the position of steel mesh to change the L from 16 to 26 mm and 36 mm. The inner diameter and outer diameter of the nozzle are d_{in} = 0.9 mm and d_{out} = 1.2 mm respectively. The inner diameter and outer diameter of the ring electrode are D_{in} = 12 mm and D_{out} = 16 mm and its thickness is 5 mm. The diameter of steel mesh is 16 mm.

The experimental photo is shown in the right side of Fig. 2. The measurements section were performed at L_2 = 6 mm below the steel mesh. The coordinate of the measurement point is shown on the left side of Fig. 2.

Results

Effect of Volume Flow Rate on Electro-Spraying Characteristics

The droplet sizes were measured by PDA. The droplet sizes at different flow rates were compared. The droplet size decreases with an increasing nozzle voltage as presented in Fig. 3. The figure also shows that the droplet size decreases with a decrease in volume flow rate. This behavior may be attributed to the lower flow rate which caused a complete electrical breakdown [12]. Smaller droplet size is easy to be evaporated, which will reduce condition of incomplete combustion. So the combustion efficiency can be improved.

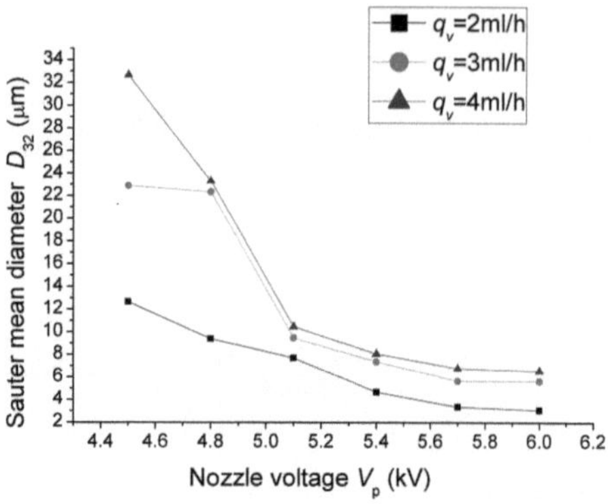

Fig. 3 Effect of volume flow rate on droplet size (q_v: volume flow rate; the distance between nozzle outlet and steel mesh L = 26 mm; ring electrode voltage V_r = 1.3 kV)

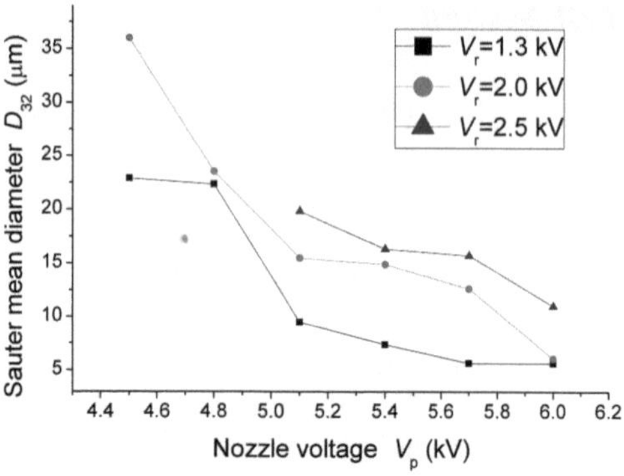

Fig. 4 Effect of ring electrode voltage on droplet size (volume flow rate q_v = 3 ml/h; the distance between nozzle outlet and steel mesh L = 26 mm)

Effect of Ring Electrode Voltage on Electro-Spraying Characteristics

The droplet sizes at different ring electrode voltages are compared. Figure 4 shows that the droplet size increases with an increase in ring electrode voltage. The reason for this phenomenon is that higher ring electrode voltage will decrease the droplet surface charge and increase droplet size [13].

Effect of the Position of Steel Mesh on Electro-Spraying Characteristics

In Fig. 5, it is observed that, the droplet size decreases as the distance between steel mesh and nozzle outlet increases. This behavior may be attributed to the effect of droplet evaporation.

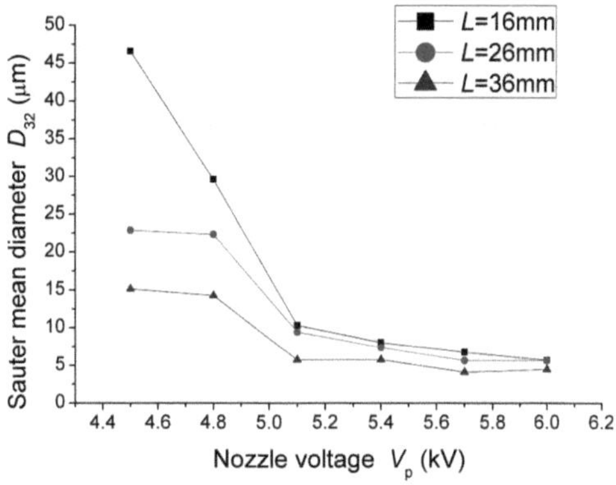

Fig. 5 Effect of the position of steel mesh on droplet size (volume flow rate q_v = 3 ml/h; ring electrode voltage V_r = 1.3 kV)

Conclusions

Experiment was carried out based on nozzle-ring-mesh electrode structure using ethanol as medium. Smaller droplet size is easy to be evaporated. Meanwhile, the combustion efficiency is improved due to reducing the incomplete combustion of ethanol with big size. Comparing the difference of droplet size in different conditions, some main conclusions can be drawn as following.

(1) Droplet size at low volume flow rate will be smaller than it at high volume flow rate.
(2) When the nozzle voltage is high enough, smaller droplet size will appear keeping the ring electrode voltage is at lower level.
(3) The distance between nozzle electrode and steel mesh has little influence on droplet size when nozzle voltage is high enough.

Acknowledgements The authors gratefully acknowledge the National Natural Science Foundation of China (51376066, 51611130194), State Key Laboratory of Engines, Tianjin University (K2016-01); Fundamental Research Funds for the Central Universities (201522083).

References

1. H.L. Cao, J.L. Xu, Thermal performance of a micro-combustor for micro-gas turbine system. Energ. Convers. Manag. **48**, 1569–1578 (2007)
2. Y.H. Gan et al., Effect of alternating electric fields on the behaviour of small-scale laminar diffusion flames. Appl. Therm. Eng. **89**, 306–315 (2015)
3. N.S. Kaisare, D.G. Vlachos, A review on microcombustion: fundamentals, devices and applications. Prog. Energ. Combust. Sci. **38**, 321–359 (2012)
4. G. Bagheri, S.E. Hosseini, M.A. Wahid, Effects of bluff body shape on the flame stability in premixed micro-combustion of hydrogen-air mixture. Appl. Therm. Eng. **67**, 266–272 (2014)
5. J. Li et al., Fundamental flame characteristics of premixed H_2–air combustion in a planar porous micro combustor. Chem. Eng. J. **283**, 1187–1196 (2016)

6. S.X. Zhao, G.S.P. Castle, K. Adamiak, Comparison of conduction and induction charging in liquid spraying. J. Electrostat. **63**, 871–876 (2005)
7. Y. Tian, P. Mehrani, Effect of particle size in fluidization of polyethylene particle mixtures on the extent of bed electrification and wall coating. J. Electrostat. **76**, 138–144 (2015)
8. D.C. Kyritsis et al., Optimization of a catalytic combustor using electrosprayed liquid hydrocarbons for mesoscale power generation. Combust. Flame **139**, 77–89 (2004)
9. W.W. Deng et al., Liquid fuel microcombustor using microfabricated multiplexed electrospray sources. Proc. Combust. Inst. **31**, 2239–2246 (2007)
10. M. Mikami et al., Combustion of Gaseous and Liquid Fuels in Meso-Scale Tubes with Wire Mesh. Proc. Combust. Inst. **34**, 3387–3394 (2013)
11. Y.H. Gan et al., The electro-spraying characteristics of ethanol for application in a small-scale combustor under combined electric field. Appl. Therm. Eng. **87**, 595–604 (2015)
12. Y.H. Gan et al., Effect of a ring electrode on the cone-jet characteristics of ethanol in small-scale electro-spraying combustors. J. Aerosol Sci. **98**, 15–29 (2016)
13. A.M. Ganan et al., The electrostatic spray emitted from an electrified conical meniscus. J. Aerosol Sci. **25**, 1121–1142 (1994)

Part IV
Energy Technologies: Heat Recovery

Integrated Utilization of Sewage Sludge for the Cement Clinker Production

Zhenzhou Yang and Zuotai Zhang

Abstract In the present study, we proposed a method that using sewage sludge (SS) as alternative fuel and raw material to produce cement clinker. It was found that the increasing amount of SS can favor for the formation of tricalcium silicate (C_3S) but the excessive amount can cause the impediment effect. Furthermore, SS contains high contents of trace elements, especially for zinc and manganese. The thermodynamic equilibrium calculation of Zn and Mn are also conducted to further understand their transformation behaviors and the results show that zine is predicted as a volatile element while manganese shows great condensed potential during the cement clinker calcination process.

Keywords Sewage sludge · Cement clinker · Trace element · Thermodynamic equilibrium

Introduction

With the rapid development of urbanization and industrialization, large quantity of waste is produced. Amongst these wastes, sewage sludge (SS) is a kind of waste that derives from sewage treatment plant [1, 2]. According to the statistics, more than 8 million tons of dried SS is produced in China every year with the increasing rate of 10% annually [3]. How to deal with such a large number of SS has become a severe environmental issue that needs to be timely resolved. In China, the common ways for the SS treatment are agriculture, landfill and incineration, but the shortage

Z. Yang
Beijing Key Laboratory for Solid Waste Utilization and Management and Department of Energy and Resource Engineering, College of Engineering, Peking University, Beijing 100871, China

Z. Zhang (✉)
School of Environmental Science and Engineering, South University of Science and Technology of China, Shenzhen 518055, China
e-mail: zgkdyzz@126.com

L. Zhang et al. (eds.), *Energy Technology 2017*,
The Minerals, Metals & Materials Series, DOI 10.1007/978-3-319-52192-3_10

of available of land and the increasing concern about of environment from public limit these methods application [4–6]. Utilization of SS in the cement kiln seems to be a promising method, which can completely destroy the hazardous substances due to the high temperature and long residual time that cement kiln involves [7]. Besides, due to SS contains significant heat value and the main components of its ash are SiO_2, Al_2O_3, CaO and Fe_2O_3, which are the main component of cement raw meals [8]. Therefore, SS can be used as alternative fuel and raw materials to produce cement clinker.

The utilization of SS in the cement kiln presents multiple advantages but also can exist some problems, such as trace element problem. Trace element has attracted more and more attention due to their hazardous effect on human health and ecosystem [9]. A better understanding of the trace elements transformation behaviors during cement clinker calcinations process can efficiently prevent its secondary pollution. However, this is still unclear. Besides, the transformation behaviors of trace elements can be influenced by various conditions, such as their thermodynamic properties, the combustion temperature and the characteristics of raw materials. Thermodynamic equilibrium calculation is widely used to study the trace elements behaviors under different conditions. To the best knowledge of author, no thermodynamic equilibrium study has been conducted on the trace elements behaviors during cement clinker calcination process.

The object of this study is to understand the effect on the cement clinker with different amounts of SS addition. Additionally, the trace elements transformation behaviors during cement clinker calcinations process are also studied using FactSage 6.4.

Materials and Method

The dried sewage sludge (SS) pellet was collected from a municipal wastewater treatment plant located in Beijing, China. The chemical composition of SS was determined by X-ray fluorescence spectrometer (XRF, S4-Explorer, Bruker), as shown in Table 1. It can be seen that major component of SS are SiO_2, Al_2O_3, Fe_2O_3, CaO, which are the essential component for the cement clinker manufacture

Table 1 The chemical composition of sewage sludge

Chemical composition	Content (wt%)
SiO_2	25.19
Al_2O_3	5.63
Fe_2O_3	4.15
CaO	6.35
MgO	1.54
K_2O	1.18
P_2O	4.85

and the mineralogical composition of SS was determined by X-ray diffraction (XRD, D/Max 2500, Rigaku).

The raw materials were prepared by the reagent of $CaCO_3$, SiO_2, Al_2O_3, Fe_2O_3 and various amounts of SS. The raw materials were pressed to $\varphi 20$ mm × 5 mm slices with the pressure of 10 Mpa and then put into the electrically heated tube furnace. The temperature was set from room temperature to 1450 °C at the rate of 10 °C/min.

The trace elements in the SS were measured by inductively coupled plasma-atomic emission spectroscopy (ICP-AES, Prodigy XP, Leeman) and the thermodynamic equilibrium calculation is conducted using FactSage 6.4.

Results and Discussion

Mineralogical Characterization of the Eco-Cement Clinker

The XRD pattern of eco-cement clinker with different amounts of SS addition is shown in the Fig. 1. It can be seen that the mainly crystalline composition in the blank sample are C_3S (Ca_3SiO_5), C_2S (Ca_2SiO_4), C_3A [$Ca_3Al_2O_6$ and $Ca_3(Al,Fe)_2O_6$] and C_4AF ($Ca_4Fe_2Al_2O_{10}$), which is similar with that in the commercial cement clinker. The most important characteristic peak of C_3S appeared at 2θ about 32°, which is the major component of Portland cement clinker that determines its quality, as shown in Fig. 2. The intensity of it can indicate the

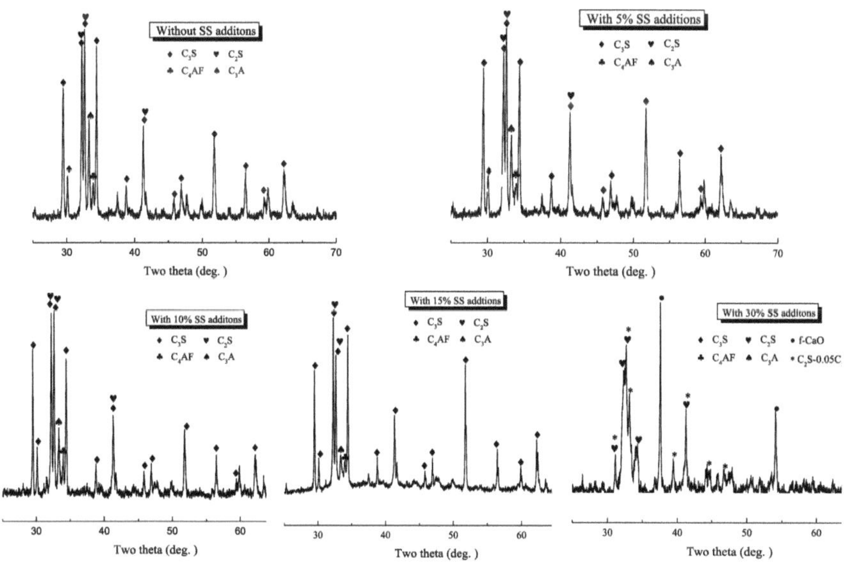

Fig. 1 The XRD pattern of cement clinker with different amount of SS addition

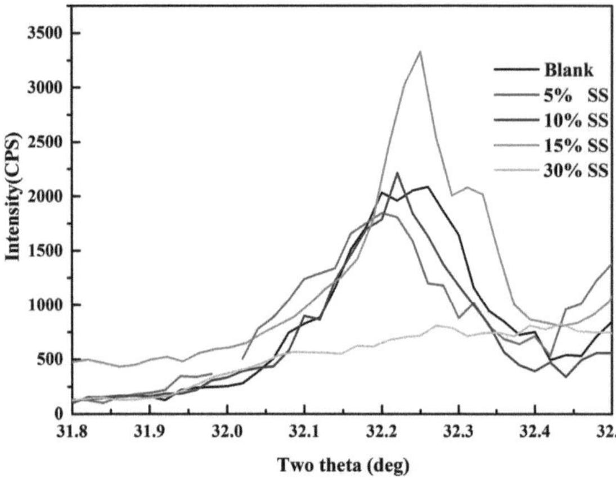

Fig. 2 The characteristic peak of C_3S

relative content of C_3S in the clinker to some extent [10, 11]. It can be seen that the characteristic peak of C_3S shows no obvious change as SS addition up to 10wt%. However, it obviously increases with 15wt% SS addition. With further increase of SS, the phase of C_3S disappears and the phase of f-CaO, C_2S–$0.05C_3P$ are formed, which might be attributed to the high content of phosphorus in SS, as shown in Table 1. During the cement clinker calcinations, the excessive phosphorus can decompose the C_3S into C_2S and f-CaO, and then causing the formation of phosphorus in C_2S solid solution (C_2S–$0.05C_3P$). This indicates that the high intake of SS can cause the undesirable effect on the C_3S formation. Therefore, the addition amounts of SS should be strictly controlled.

Micrographs Observation

The C_3S in the cement clinker is commonly produced by the reaction between C_2S and CaO in liquid phase. The liquid phase in the clinker can be observed by SEM, which is shown in Fig. 3. It can be seen that the big crystal grains are close to each other and are stacked together with distinct contours and boundary in the blank sample. With the SS addition, the outline of the crystal grains became more blurred. As the addition of SS reaches 15wt%, the boundary lines of the crystalline grains get mixed with the interphase and cannot be clearly identified. Furthermore, the crystal grains exhibit independent existence and are connected directly to the liquid phase as the amount of SS addition increases to 15 and 30wt%. Therefore, the addition of SS can raise the amount of liquid phase. As for the crystal size, the present results show that SS addition has no obvious influence on the growth of the crystalline grains until the amount of SS reached 15wt%. In addition, a large amount of pores were observed to be distributed in the blank sample. The pores were found to decrease with the SS addition and the crystalline grains were connected to each other. These results differ from those reported by Lin et al. [12],

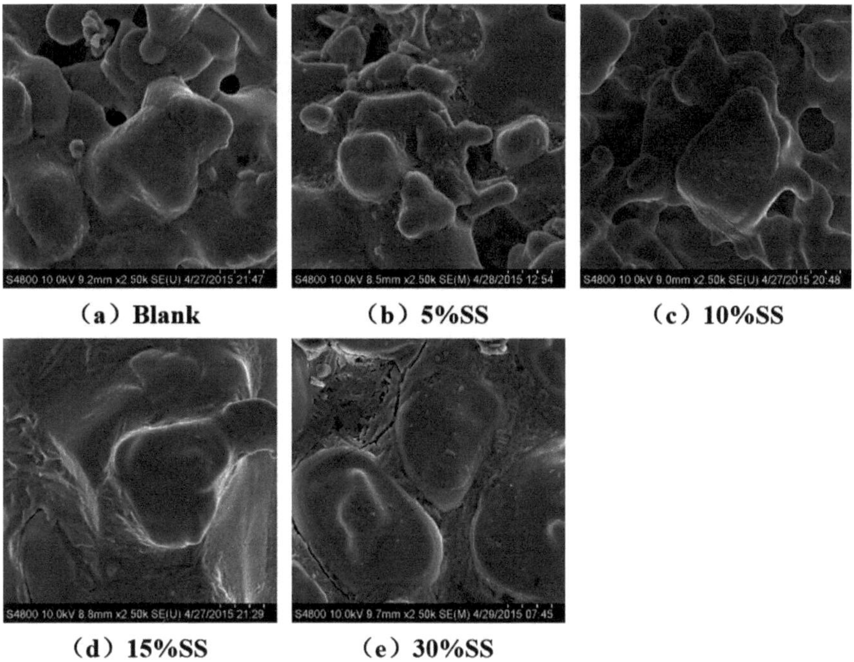

Fig. 3 SEM micrographs of the eco-cement clinkers with different amounts SS

which might be due to the increasing liquid phase with SS addition, resulting in a more compact cement clinker with decreased porosity.

The Transformation of Trace Element During Cement Clinker Calcinations

It is known that equilibrium calculations have limitations due to its neglect on kinetic effects, nonuniform distribution as well as the mode of occurrence of the elements [13]. However, from the current study, it can be seen that equilibrium calculation helps to understand the general disciplines of the elements behavior under different variables. To further understand the influence of different process variables and chemical compositions on trace elements behavior, thermodynamic equilibrium calculations were conducted using FactSage 6.4 assuming a closed environment and 1 kg of raw materials is used as input data. Several calculation conditions are considered: The system only containing C, H, N and S is calculation as basic system. The introduction of Cl in the system is considered to identify the possible impact. The mineral contents are added into the system to examine the possible interaction. The calculations are performed at the range of temperature 1000–2000 K and this range of temperature includes the clinker calcinations at 1723 K.

Table 2 The trace elements content in SS (mg/kg)

	As	Ba	Cd	Co	Cr	Cu	Mn	Ni	Pb	Zn
SS	80	256	2	11	124	358	543	41	48	1084

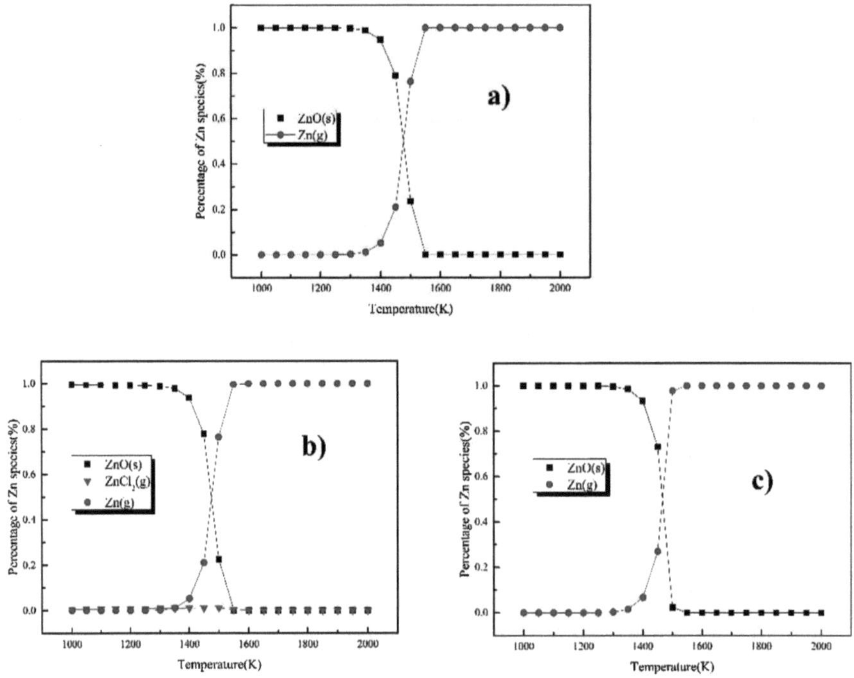

Fig. 4 The influence of different parameters on speciation of zinc **a** Basic system C, H, N, S, O; **b** +Cl, a = 1.2; **c** basic system + Cl + mineral contents

The trace element content in the SS is shown in the Table 2. It can be seen that the trace element in SS is notably high, especially zinc and manganese. Thus, the elements of zinc and manganese are selected for the thermodynamic equilibrium study and the results are shown in Figs. 4 and 5. It can be seen that, the equilibrium composition for zinc mainly depends on the temperature. ZnO (s) in only stable species in the equilibrium at the temperature between 1000 and 1400 K. With temperature further increase, the proportion of ZnO (s) sharply decreases with the increase of Zn (g). All of the zinc exists as vapor phase when the temperature above 1800 K. The presence of chlorine can lead to the formation of $ZnCl_2$ (g) but in a small proportion. Furthermore, it can be seen that zinc shows no interaction behavior with mineral phase. The average temperature in cement kiln is about 1723 K. Therefore, zine is predicted as a volatile element in cement kiln according

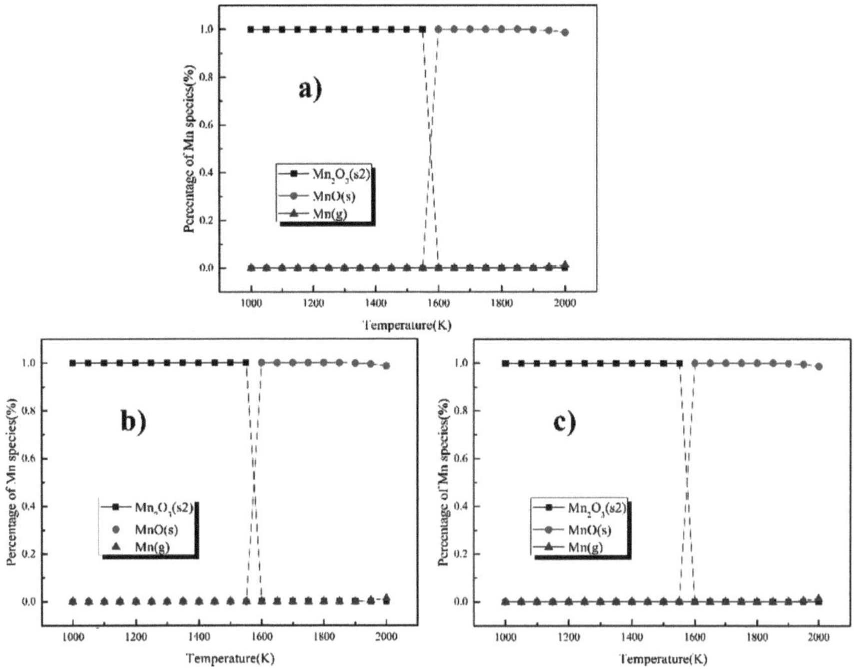

Fig. 5 The influence of different parameters on speciation of manganese **a** basic system C, H, N, S, O; **b** +Cl, a = 1.2; **c** basic system + Cl + mineral contents

to the simulated result. As for manganese, it can be seen that manganese shows great condensed potential during cement clinker manufacturing process. When the temperature is under 1500 K, almost of manganese is in the form of Mn_2O_3 (s2). With the temperature further increase, Mn (s) becomes the most stable species. The introductions of chlorine and mineral phase have no obvious influence on the species of manganese.

Conclusion

From our present study, we can draw the conclusion that the appropriate amount of SS additions can favor for the C_3S formation, while the excessive amount can cause the impediment effect. Besides, the addition of SS can effectively increase the liquid phase content during the cement clinker calcination process. SS contains high contents of trace elements, especially for zinc and manganese. According to the thermodynamic equilibrium calculation, zinc tends to exist as vapor phase, while manganese shows the great condensed potential during the cement clinker calcinations process.

Acknowledgements Zuotai Zhang and Zhenzhou Yang designed the experiments; Zhenzhou Yang conducted the experiments, analyzed the data, and wrote the manuscript; Zuotai Zhang advised the experiments and helped editing the manuscript.

References

1. L. Jiang et al., Co-pelletization of sewage sludge and biomass: the density and hardness of pellet. Bioresour. Technol. **166**, 435–443 (2014)
2. R.P. Singh, M. Agrawal, Potential benefits and risks of land application of sewage sludge. Waste Manag. **28**(2), 347–358 (2008)
3. J.M. Hong et al., Life cycle assessment of sewage sludge co-incineration in a coal-based power station. Waste Manag. **33**(9), 1843–1852 (2013)
4. M.B. Folgueras, M. Alonso, R.M. Diaz, Influence of sewage sludge treatment on pyrolysis and combustion of dry sludge. Energy **55**, 426–435 (2013)
5. T. Murakami et al., Combustion characteristics of sewage sludge in an incineration plant for energy recovery. Fuel Process. Technol. **90**(6), 778–783 (2009)
6. H.B. Zhou et al., Application of a recyclable plastic bulking agent for sewage sludge composting. Bioresour. Technol. **152**, 329–336 (2014)
7. J.A. Conesa et al., Pollutant formation and emissions from cement kiln stack using a solid recovered fuel from municipal solid waste. Environ. Sci. Technol. **45**(13), 5878–5884 (2011)
8. A. Zabaniotou, C. Theofilou, Green energy at cement kiln in cyprus—use of sewage sludge as a conventional fuel substitute. Renew. Sustain. Energy Rev. **12**(2), 531–541 (2008)
9. Y.Y. Zhang et al., Trace element partitioning behavior of coal gangue-fired CFB plant: experimental and equilibrium calculation. Environ. Sci. Pollut. Res. **22**(20), 15469–15478 (2015)
10. W. Xu et al., The utilization of lime-dried sludge as resource for producing cement. J. Clean. Prod. **83**, 286–293 (2014)
11. H.X. Li et al., Preparation of Portland cement with sugar filter mud as lime-based raw material. J. Clean. Prod. **66**, 107–112 (2014)
12. Y.M. Lin et al., Utilization of municipal sewage sludge as additives for the production of eco-cement. J. Hazard. Mater. **213**, 457–465 (2012)
13. B.B. Miller, R. Kandiyoti, D.R. Dugwell, Trace element behavior during co-combustion of sewage sludge with polish coal. Energy Fuels **18**(4), 1093–1103 (2004)

Valuable Metals and Energy Recovery from Electronic Waste Streams

Fiseha Tesfaye, Daniel Lindberg and Joseph Hamuyuni

Abstract E-waste management through traditional methods such as disposing in landfills, burning in incinerators or exporting abroad for disposal are no longer options due to the strict environmental regulations. Fortunately, the presence of valuable metals in the e-waste and increasing demand for the metals as well as complexities of the currently available primary raw materials make recycling an attractive and viable option both environmentally and economically. Moreover, it is efficient in terms of resource management by closing the loop of metals. Consequently, urban mining such as the recovery of precious metals from e-waste streams through sustainable recycling processes have emerged. The sustainable recycling practices address the scarcity of primary resources and reduce consumption of energy for metals production while managing environmental issues related to hazardous materials from the e-waste streams. In this paper, valuable metals recoveries from e-waste streams through pyrometallurgical and hydrometallurgical processes are critically reviewed. And, innovative ideas for different steps of the thermochemical processes in the valuable metals and energy recovery from the e-waste streams are discussed.

Keywords E-waste · Pyrometallurgy · Energy · Metals recovery · Environment management

F. Tesfaye (✉) · D. Lindberg
Johan Gadolin Process Chemistry Centre, Åbo Akademi University, Piispankatu 8, 20500 Turku, Finland
e-mail: fiseha.tesfaye@abo.fi

J. Hamuyuni
Department of Materials Science and Engineering, Aalto University School of Chemical Technology, Vuorimiehentie 2K, PO Box 16200, 00076 Aalto, Finland

L. Zhang et al. (eds.), *Energy Technology 2017*,
The Minerals, Metals & Materials Series, DOI 10.1007/978-3-319-52192-3_11

Introduction

The production of electrical and electronic equipment (EEE) is one of the fastest growing domains of the manufacturing industry in the world. In response to the fast advancement in technology, the demand for EEE has increased dramatically. The high diversity of EEE coupled with a rapid obsolescence of products, due to advancement in technology, and an increasing level of household equipment have led to a significant increase in electrical and electronic equipment waste (e-waste). For example, the e-waste flow is projected to grow from 3 to 5% per year in European Union (EU) [1]. Currently, from 20 to 25 million tons per year of e-waste are being generated globally with major share of EU, USA and Australasia. In EU, e-waste has been projected to increase by 45% between 1995 and 2020 [2]. Countries included in the Emerging Market Economies such as China and India are expected to become significant e-waste producers in the next decade [3].

The EU WEEE directive classifies e-waste as electrical or electronic equipment, which is discarded including all components which are part of the product at the time of discarding [1, 2]. The EU WEEE directive categorizes e-waste into ten different classes from large household appliances to automatic dispensers. E-waste has a complex composition of ferrous, non-ferrous, plastic and ceramic materials. E-waste is classified as hazardous material; therefore, it should be handled properly. However, the presence of valuable metals in e-waste such as Au, Ag, Pt, Ga, Pd, Ta, Te, Ge, and Se makes it attractive for recycling. Recovering materials from e-waste is more profitable than processing primary raw materials largely due to the savings in energy associated with e-waste recycling. According to Boliden Rönnskär (Skelleftehamn, Sweden), extracting metals from e-waste requires only from 10 to 15% of the total energy required in smelting and refining of metal ore concentrates. The e-waste recycling rate for the EU is significantly higher, 35%, than the e-waste recycling rate of the USA, 27% [4].

Currently, increasing demand for valuable metals, complexities of the available raw materials, and earth's intrinsic limitations pose a challenge to the entire valuable metals production system. The United Nations Environment Program (UNEP) is also calling for an urgent re-think of metals recycling practices as global demand for metals continues to soar. Consequently, urban mining such as the recovery of precious metals from electronic waste (e-waste) streams through sustainable recycling processes have been developing. The main purpose of this paper is to critically review energy efficient valuable metals recovery from e-waste streams by pyrometallurgical and hydrometallurgical processes. Main features of these processes including their advantages and disadvantages are described. And, innovative ideas for different steps of the thermochemical processes in the valuable metals and energy recovery from the e-waste streams are discussed.

Classification and Composition of E-waste

According to the Association of Plastics Manufacturers in EU (APME), average materials consumption in electric and electronic equipment are 38 wt% ferrous, 28 wt% non-ferrous, 19 wt% plastics, 4 wt% glass, 1 wt% wood, and 10 wt% others [5]. Thus, despite its common classification as a waste, e-waste constitutes considerable amounts of secondary resource.

Metals in e-waste can be grouped into precious metals—PMs (Au and Ag), platinum group metals—PGMs (Pd, Pt, Rh, Ir and Ru), base metals—BMs (Cu, Al, Ni, Sn, Zn and Fe), metals of concern—MCs (Hg, Be, In, Pb, Cd, As and Sb), and scarce elements—SEs (Te, Ga, Se, Ta and Ge) [6]. The initiation for the recovery of valuable metals is important for e-waste management, recycling, sustainability and resource conservation. The extraction of PMs, PGMs, and BMs from e-waste is a major economic drive due to their associated value. As shown in Table 1, printed circuit boards (PCBs) are the most *precious* part in e-waste streams. PCBs are found in electrical and electronics appliances such as televisions, computers, mobile phones, etc. For example, flat screens contain one or more PCBs equipped with electronic components and connectors. Considerable amounts of precious metals are contained both in the components and connectors as well as in the solders [7]. For example, PCB from an LCD television constitutes 575 mg of Ag, 138 mg of Au, and 44 mg of Pd [7]. In general, precious metals in PCB account for more than 80% of the total intrinsic value even though their composition is less than 1 wt% [8].

Generally, PCBs are composed of 40 wt% metals, 30 wt% plastics and 30% ceramics [9]. PCBs are coated with base metals (BMs) such as Sn, Ag or Cu to make them conductive. Polymers and industrial plastics are the other major constituents of PCBs that contain polyethylene, polypropylene, epoxies and polyesters. Large numbers and various kinds of small components are attached to PCBs. During the recycling process, generally, PCBs are crushed into smaller sizes (less than 1–2 mm) and various techniques including magnetic, electrostatic, electrowinning, and selective dissolution are implemented to separate the components

Table 1 Weight distribution of PMs, PGMs, and BMs in e-waste [6]

E-waste	Fe (wt%)	Al (wt%)	Cu (wt%)	Plastics (wt%)	Ag (ppm)	Au (ppm)	Pd (ppm)
TV-board	28	10	10	28	280	20	10
PCBs	7	5	20	23	1000	250	110
Mobile phone	5	1	13	56	1380	350	210
Portable audio	23	1	21	47	150	10	4
DVD-player	62	2	5	24	115	15	4
Calculator	4	5	3	61	260	50	5

[9–17]. The sustainable resource management demands the isolation of hazardous metals from e-waste and also optimization of the recovery of PMs. The loss of PMs during the recycling chain will adversely affect the process economy. The value distribution of PMs in PCBs and calculators is more than 80%. After PMs, Cu is the next highest valuable metal to be extracted from e-waste [6].

E-waste Recycling Technologies

There are three major reasons for e-waste processing: environmental concerns, energy savings and resource efficiency. E-waste recycling consists of three main steps: collection, preprocessing, and recovery of valuable materials and disposal of non-recyclable ones [18]. The recycling methodology broadly comprises of shredding, sorting, grading, compacting, baling, or processing segregated plastics and metal components, followed by separation, identification, and testing as relevant [19]. Each step is critical for the recovery of metals and the recycling economy. E-waste collection is facilitated by appropriate government policies, effective advertisement for public awareness, and by installing separate collection facilities at public places. The goal of the EU WEEE directive is to increase the collection rate of discarded electronic and electrical products from 65% by 2012 to 85% by 2016 [1, 2]. End of life electronic components are sorted at the collection facility where useable components are returned to the consumer supply chain.

Pre-processing of e-waste is one of the most important steps in the recycling chain. The discarded equipment are manually dismantled at collection facilities and individual components are isolated. During the early stage, housing, wiring boards and drives, and other components are disassembled. Mechanical processing is an integrated part of this stage where e-waste scrap is shredded into pieces using hammer mills [18]. Metals and non-metals are separated during this stage using techniques similar to that used in the mineral dressing, e.g., screening, magnetic, eddy current and density separation techniques. The recoveries of SEs, which exist as traces in the e-waste stream, require special recovery technology [19].

The final stage in the recycling chain of e-waste is the end processing, where the non-metal and metal fractions of e-waste are further processed. The non-metallic fractions of PCBs are mainly composed of thermoset resins and glass fibers. Thermoset resins cannot be re-melted due to their chain structure. However, chemical processes that include gasification, pyrolysis, supercritical fluid de-polymerization and hydrogenolytic degradation can be applied for producing chemical substances and fuels [20]. For example, the non-metallic fractions can be used in pyrometallurgical processes as fuels and reducing agents.

Recycling Through Pyrometallurgical and Hydrometallurgical Processes

Pyrometallurgical processing techniques, including conflagrating, smelting in a plasma arc furnace, drossing, sintering, melting, and varied reactions in a gas phase at high temperatures for recovering BMs and PMs from e-waste is the conventional method used in the past two decades. In the process, the crushed scraps are liquefied in a furnace or in a molten bath to remove plastics, in which the refractory oxides form a slag phase together with other metal oxides. Recently, Cui and Zhang [21] revised the process of metals recovery from e-waste streams, including hydrometallurgy and pyrometallurgical processes. They reported that hydrometallurgical processes are more predictable and controllable compared to pyrometallurgical processes. However, hydrometallurgical process routes have certain limitations in recovering PMs, at industrial scale. Optimal recovery of PMs can be achieved through pyrometallurgical process routes, this make them more economical and eco-efficient compared to the hydrometallurgical process routes [21].

For example, the Noranda process at Quebec, Canada, recycles about 1×10^5 tons of e-waste per year [22]. Materials entering the reactor are immersed in a molten metal bath at about 1250 °C, which is churned by a mixture of O_2-rich air (up to 39% oxygen), this effectively reduces energy consumption due to combustion of plastics and other inflammable materials in feeding. In the process, impurities including Fe, Pb, and Zn are converted to oxides, forming silica-based slag aided by the agitated oxidation zone, followed by cooling and milling of the slag for further recovery of metals prior to its disposal. PMs content of the Cu matte is removed before being transferred to the converters, which yield liquid blister copper. The PMs, including Au, Ag, Pt, and Pd together with SEs and Ni constitute about 0.9%, which are recovered through electrorefining process at the anodes.

Pyrometallurgical processing for the recovery of metals from e-waste is widely applied by Boliden Rönnskär (Skelleftehamn, Sweden) [23]. The scraps with high Cu content are processed in the Kaldo Furnace and around 1×10^5 tons of scraps including e waste was reportedly being processed in the Kaldo Furnace per year, as per an APME report during the year 2000. E-waste blended with lead concentrates is processed in a Kaldo rector with skip-hoist assisted feeding [24] and the required oxygen for combustion in oil-oxygen burner is provided through an oxygen lance in the system, while off-gases are subjected to additional combustion air at around 1200 °C. A standard gas handling system recovers thermal energy through the integrated steam network. The mixed Cu-alloy produced by the Kaldo furnace is processed in a copper converter for recovery of valuable metals (such as Cu, Ag, Au, Pd, Ni, Se, and Zn), while the dust content (containing Pb, Sb, In, and Cd) is subjected to other processing operations to recover the metals.

Hydrometallurgical Processes

Various investigators studied the extraction of PMs and BMs from e-waste using hydrometallurgical routes [15, 25–28]. These routes are based on traditional hydrometallurgical technology of metals extractions from their primary ores. Similar steps of acid or caustic leaching are employed for selective dissolution of PMs from e-waste. The pregnant solution is separated and purified for the enrichment of metal content thereby impurities are removed as gangue materials. The isolation of metal of interest is conducted through solvent extraction, adsorption and ion exchange enrichment processes. Finally, metals are recovered from solution through electrorefining (electrometallurgy) or chemical reduction processes [29–33]. Cui and Zhang [21] reviewed the processes for recovering metals from e-waste. Hydrometallurgy and pyrometallurgical processes were evaluated and discussed. It has been reported that hydrometallurgical processes have some benefits compared to pyrometallurgical processes because they are more exact, predictable and controllable [34].

Solvents especially halides, cyanides, thiourea and thiosulfates are used for the leaching of PMs from their primary ores. Process factors including pH, temperature and stirring control the dissolution of metals from their primary ores. The recovery of PMs from the leached solution is carried out by cementation, solvent extraction, adsorption on activated carbon, and ion exchange methods. Similar techniques could be employed for extracting metals from e-waste, however, its complex nature makes the process complicated compared to natural ores.

Park and Fray [8] proposed a hydrometallurgical method for recovering PMs from e-waste. Aqua regia was used as leachant and a fixed ratio of 1/20 between metals and leachant was exercised. Ag and Pd were extracted during the first stage with 98 and 93% recovery, respectively. For Au, a liquid-liquid extraction method was adopted with toluene and a recovery of 97% was reported [8]. It is noted that HNO_3, H_2SO_4, and HCl-based solutions are commonly employed for dissolving PMs from e-waste. From the leachants, PMs are recovered employing methods similar to those used in the mineral industry.

Limitations of Hydrometallurgy Route

Hydrometallurgical processes have been successfully used to recover PMs from e-waste. However, these processes have limitations that restrict their industrial scale application. These limitations include [21, 35, 36]:

- hydrometallurgical processes are slow and less profitable compared to the pyrometallurgical processes,
- mechanical processing of e-waste for efficient dissolution is time consuming, and causes 20% loss of PMs,
- there are risks of PM loss during dissolution and subsequent steps,

- leachant such as cyanide are hazardous and should therefore be used with high safety standards
- halide leaching needs special equipment made of stainless steel and rubbers due to strong corrosive acids and oxidizing conditions,
- the use of thiourea leachants is limited in gold extraction due to its high cost and consumption. Moreover, further developments are required to improve the current technology of thiourea-based gold leaching,
- the consumption of thiosulfate is comparatively higher and the overall process is slower, which limits its application for gold extraction from ores as well as from e-waste.

Pyrometallurgical Processes

State-of-the-art smelters and refineries can extract valuable metals and isolate hazardous substances efficiently. Such recycling facilities can close the loop of valuable metals and reduce environmental impact arising from large quantities of e-waste. Currently, e-waste recycling is dominated by pyrometallurgical routes [37], whereas the steel industry embraces the ferrous fractions for the recovery of Fe, and the secondary aluminum industry takes the aluminum fractions. Pyrometallurgical processes works with the steps of disassemble, separation/upgrading and purification that are fundamentally similar to those of mechanical or hydrometallurgical processes. Smelting in furnaces, incineration, combustion and pyrolysis are typical e-waste recycling processes by the pyrometallurgical process routes. The recovery of precious metals is not achieved by leaching, crushing or grinding, but by smelting in furnaces at high temperatures. In these processes, metals are sorted by exploiting their chemical and metallurgical properties, e.g., PMs are segregated into a solvent metal phase (Cu or Pb). Plastic components of e-waste cannot easily be recycled due to the mix of flame retardants, pigments and mixed types of plastics. However, smelting processes utilize the energy content of the plastics. Energy usage is reduced due to the combustion of plastics and other flammable materials in the e-waste feed, which partially substitute the role of coke as a reducing agent and energy source.

The metal fractions separated during the preprocessing of e-waste are composed of Fe, Al, Cu, Pb and PMs. After Fe and Al, Cu and Pb are the main constituents of a typical e-waste. Therefore, it is logical to send e-waste to smelters that accept Cu/Pb scrap. Currently, Cu and Pb smelters work as e-waste recyclers for the recovery of Pb, Cu and PMs. In these pyrometallurgical processes, e-waste/Cu/Pb scrap is feed into a furnace, whereby metals are collected in a molten bath and oxides form a slag phase layer. Details of the Cu smelting process are explained in Section “Copper Smelting Route”.

Copper Smelting Route

Primary and secondary copper smelters are adopted to recycle and extract PMs from e-waste. It is reported that due to their excellent gas cleaning units, copper smelting processes are more environmentally friendly compared to lead smelters that generate toxic fumes [38]. Copper smelting facilities near urban areas will minimize the cost of e-waste transportation, and therefore the recycling process costs can be lowered. In these processes, PMs are recovered via conventional electrorefining processes where they are segregated in slimes [38]. Usually, copper smelting routes including matte and black copper are used for e-waste recycling. In the primary copper smelting, copper matte (~40 to 60%) and blister copper (~98.5%) are produced. Furthermore, blister copper is refined to produce anode copper (~99%). In the black copper route (secondary copper smelting) crude copper is produced during a reduction process and is refined by oxidation in a converter. The black copper smelting process consists of reduction and oxidation cycles. Impurities are mostly segregated into the vapor phase and are discharged in the off gas [38].

Limitations of Pyrometallurgical Processes

Pyrometallurgical processes are generally more economical, eco-efficient and maximize the recovery of PMs, however, they have certain limitations [6, 21]:

- recovery of plastics is not possible because plastics replace coke as a source of energy,
- smelting cannot recover Al and Fe since they are oxidized and dissolved into the slag,
- smelting flame retardants and polyvinyl chloride (PVC) present in e-waste leads to the formation of dioxins, requiring special emission controls,
- a large investment is required for installing integrated e-waste recycling plants that maximize the recovery of valuable metals and also protect the environment by controlling hazardous gas emissions,
- instant burning of fine dust of organic materials can occur before reaching the metal bath. In such cases, agglomeration may be required to effectively harness the energy content and also to minimize the health risk posed by fine dust particles,
- ceramic components in feed material can increase the volume of slag generated in the blast furnaces, which increases the risk of losing PMs from BMs,
- partial recovery and purity of PMs are achieved by pyrometallurgical routes. Therefore, subsequent hydrometallurgical and electrochemical techniques are necessary to extract pure metals from BMs,
- process control and optimization of the smelting and refining processes are challenging due to complex feed materials,
- smelting cannot recover certain product components, such as chips or bare fiberglass boards.

Selected Industrial Processes for Metals Recovery from E-waste

Currently, industrial processes for recovering metals from e-waste are based on combined pyrometallurgical, hydrometallurgical and electrometallurgical processes. In pyrometallurgical processes, e-waste is blended with other materials and incorporated into primary/secondary smelting processes (e.g., into copper or lead smelters). Copper smelting is the main process route for e-waste recycling where PMs are collected in copper matte or black copper. In the final stage of Cu production, i.e., the electrorefining process, pure Cu metal is produced and the PMs are separated into slimes where they are recovered using hydrometallurgical processes. Currently, various industrial processes are used globally for extracting metals from e-waste, including the Umicore integrate smelting and refining facility, the Noranda process in Quebec, Canada, Boliden Rönnskär smelters in Sweden, Kosaka's recycling plant in Japan, the Kayser recycling system in Austria and the Metallo-Chimique N.V plants operated in Belgium and Spain.

Metals Recovery from E-waste at Rönnskär Smelters and Noranda Process

Boliden Rönnskär (Skelleftehamn, Sweden) is world's largest recycler of Cu and PMs from e-waste. The majority of the e-waste sent to Rönnskär comes from EU and North America [8]. Figure 1 shows the process flow sheet diagram of the Boliden Rönnskär smelter. A variety of scrap from the non-ferrous and electronic industry is introduced into the process at different stages depending on the purity and requirement of the final product. For instance, high Cu containing scrap is fed into the converting process directly, but the low-grade e-waste is fed into the Kaldo furnace (see Fig. 1). At Boliden Rönnskär (Skelleftehamn, Sweden) smelter, the annual recycling capacity of e-scraps is 1.2×10^5 tons. The feed material of the Kaldo converter consists of blended lead concentrates and e-waste that are combusted with the supply of oxygen and oil. The main steps in the Boliden Rönnskär smelters are drying, roasting, smelting, converting and refining, as schematically shown in Fig. 1.

The Kaldo furnace at Boliden Rönnskär, to which the e-waste is fed, produces a mixed Cu-alloy that is treated in a copper converter for the recovery of metals including Cu, Ag, Au, Pt, Pd, Ni, Se, and Zn. The volatile metals such as Pb, Sb, In, and Cd are segregated into the vapor phase that is recovered by a separate process. The off gases emission is treated for producing sulfuric acid and SO_2 gas.

The Noranda process is another commercial pyrometallurgical process for the recovery of metals from e-waste. The feed material for this process is composed of e-waste and copper ore concentrates. At this recycling facility, about 1×10^5 tons of e-waste is processed for metal recovery. A blend of e-waste and copper

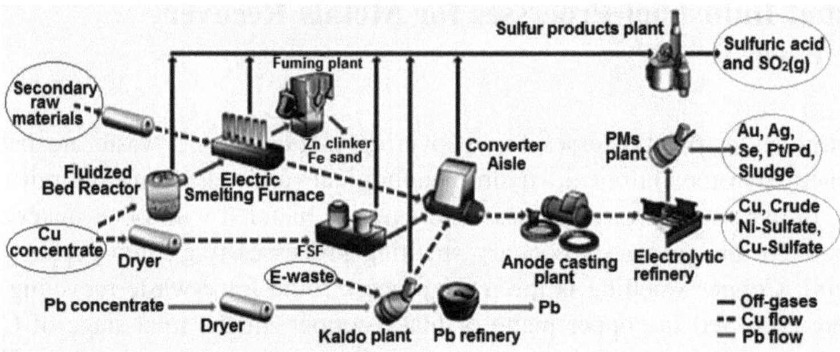

Fig. 1 Process flow sheet diagram of the Boliden Rönnskär (Skelleftehamn, Sweden) smelter, modified from [39]. The feed materials and useful end products are *encircled* with *red (left)* and *green (right) lines*, respectively

concentrate is fed into the molten bath at 1250 °C and the process temperature is maintained by injecting supercharged oxygen. Energy usage in the Noranda reactor is reduced due to the combustion of plastics in the e-waste feedstock. During the oxidation process, impurities including Fe, Pb and Zn are converted into oxides, and segregated into a silica-based slag. The slag is cooled and processed for the recovery of metals before disposal. PMs are segregated in liquid copper that is processed into a copper converter for higher purity. The blister copper is refined in the anode furnace and casted into anodes with 99.1% purity. The remaining residue (0.9%) contains PMs including Au, Ag, Pt, and Pd, and some other recoverable metals such as Ni, Se and Te. Finally, PMs are recovered through electrorefining process.

In general, PMs are recovered with a combination of pyrometallurgical and hydrometallurgical processes. It is worth noting that during mechanical separation of Fe, Al and plastics from e-waste, there is a risk of inevitably PMs losses. The PMs are closely tied up with non-ferrous metals and plastics in the PCBs. The acceptance of Fe, Al and plastics in the Cu smelting can enhance the overall recovery of PMs. Huisman and Stevels [40] reported that the direct smelting route for mobile phones is a more eco-efficient solution compared to the indirect smelting of disintegrated mobile phone components. However, batteries should be separated before smelting in the furnace [41]. Most of the integrated e-waste recycling plants directly use PCBs in the smelting furnaces, which have many advantages such as:

- maximization the overall segregation of PMs in the Cu fraction, causing the final recovery to be higher,
- partial replacement of coke with plastics as a source of energy during the smelting process.

Pyrometallurgical routes are beneficial to segregate and upgrade valuable metals in the BMs from e-waste that are further treated by hydrometallurgical routes to recover metals. Industrially, IsaSmelt, Kaldo, rotary and plasma arc furnaces are

Table 2 Summary of selected pyrometallurgical processes for recovering valuable metals from e-waste. TSL: top submerged lanced [42]

Industrial processes	Metals recovered	Main process features
Umicore's process	Au, Ag, Pd, Pt, Se, Ir, Ru, Rh, Cu, Ni, Pb, In, Bi, Sn, As, Sb	Isasmelt smelting, copper leaching & electrowinning and PMs refinery
Outotec's Ausmelt TSL and Kaldo Furnace	Zn, Cu, Au, Ag, In, Pb, Cd, Ge	Copper scrap and e-waste recycling
Boliden Rönnskär smelters	Cu, Ag, Au, Pd, Ni, Se, Zn, Pb	Smelting in Kaldo reactor, upgrading in Cu and high PMs recovery by refining
Noranda process	Cu, Au, Ag, Pt, Pd, Se, Te, Ni	Smelting of e-waste and Cu concentrate. Electrorefining for metal recovery
Boliden Rönnskär smelters tests	Cu and PMs	PC scrap feeding to Zn fuming process, plastics are used as reducing agents, PMs are segregated in Cu and are recovered at later stage
Umicore's trials	Au, Ag, Pd, Pt, Se, Ir, Ru, Rh, Cu, Ni, Pb, In, Bi, Sn, As, Sb	Plastics from e-waste is tested at energy and reducing agent during smelting
Dowa mining Kosaka Japan	Cu, Au, Ag	E-waste TSL smelting in secondary copper process
LS-Nikko's recycling facility, Korea	Au, Ag & PGMs metals	Recycling in TSL smelting followed by electrolytic refining
Day's patent	PMs, Pt and Pd	Smelting in plasma arc furnace at 1400 °C. PMs collected in BM. Ag and Cu used to collect metals
Aleksandrovich patent	PGMs and gold	Scrap combustion in a BM with carbon reduction
Aurubis recycling Germany	Cu, Pb, Zn, Sn and PMs	Smelting of Cu and e-waste in TSL, black Cu processing and electrorefining

used for collecting valuable metals in the BMs. A summary of selected industrial process is given in Table 2.

Summary and Conclusions

Traditional methods of managing e-waste which include disposing in landfills, burning in incinerators or exporting abroad for disposal are no longer options due to the strict environmental regulations. Fortunately, the presence of valuable metals in e-waste and increasing demand for the metals as well as complexities of the currently available primary raw materials make recycling an attractive and viable option both in terms of environment and economics. Moreover, it is efficient in terms of resource management by closing the loop of metals. Consequently, urban mining such as the recovery of PMs from e-waste streams through sustainable recycling processes have emerged. The sustainable recycling practices address the scarcity and complexity of primary resources and reduce consumption of energy while managing environmental issues related to hazardous materials from the e-waste streams. Therefore, recycling of e-waste is important for both resource and waste management. However, e-waste recycling is limited due to challenges such as insufficient collection facilities, higher transportation cost or lack of integrated and automatic smelting and refining facilities near urban areas.

Currently, pyrometallurgical, hydrometallurgical or a combination of both routes are used for recovering valuable metals from e-waste streams. Fundamentally, hydrometallurgical routes are similar to those used in the minerals processing industry, which include leaching and metal extractions from leachates. Pyrometallurgical process routes are more economical and eco-efficient particularly for the recovery of PMs, which result in considerable loss when applying hydrometallurgical processes. In the pyrometallurgical process routes the non-metallic fractions can be used as fuels and reducing agents. The thermal energy can be effectively recovered through integrated steam networks. Limitations in applying the pyrometallurgical processes include difficulties to recover Fe and Al, as they oxidize in the process, and the required strict control of emissions to avoid pollution. Both process routes have advantages and disadvantages, which therefore should be chosen for specific feed materials and desired products.

Acknowledgements The authors are grateful to the Academy of Finland for financial support. This work was made under the project "Chemistry of biomass impurities at reducing conditions in future thermal conversion concepts" as part of the activities of the Johan Gadolin Process Chemistry Center at Åbo Akademi University.

References

1. BIO Intelligence Service, in *Review of the scope of the directive 2012/19/EU on Waste Electrical and Electronic Equipment (WEEE)*, Final report prepared for European Commission–DG Environment (2013), p. 116
2. European Parliament, Directive 2002/96/EC of the European Parliament and of the Council of 27 January 2003 on waste electrical and electronic equipment (WEEE). Official J. Eur. Union **L37**, 24–38 (2003)
3. B.H. Robinson, E-waste: an assessment of global production and environmental impacts. Sci. Total Environ. **408**, 183–191 (2009)
4. Copper Studies, *E-Scrap: Urban Riches*, vol. 38 (Commodities Research Unit, London, UK, 2011)
5. APME (Association of Plastics Manufacturers in Europe), in *Plastics–A Material of Choice for the Electrical and Electronic Industry*, Plastics Consumption and Recovery in Western Europe, APME Report Code. 98-2004, (Brussels, Belgium, 1995)
6. C. Hagelüken, Improving Metal Returns and Eco-Efficiency in Electronics Recycling–A Holistic Approach for Interface Optimisation between Pre-Processing and Integrated Metals

Smelting and Refining. In: *Proceedings of the IEEE International Symposium on Electronics and the Environment*, Scottsdale, AZ, USA (2006)
7. M. Buchert, A. Manhart, D. Bleher, D. Pingel, Recycling critical raw materials from waste electronic equipment, Commissioned by the North Rhine Westphalia State Agency for Nature, Environment and Consumer Protection, Öko-Institut eV, (Freiburg, Germany, 2012), p. 80
8. Y.J. Park, D.J. Fray, Recovery of high purity precious metals from printed circuit boards. J. Hazard. Mater. **164**, 1152–1158 (2009)
9. I.O. Ogunniyi, M.K.G. Vermaak, D.R. Groot, Chemical composition and liberation characterization of printed circuit board comminution fines for beneficiation investigations. Waste Manag. **29**, 2140–2146 (2009)
10. J. Puckett, L. Byster, S. Westervelt, R. Gutierrez, S. Davis, A. Hussain, M. Dutta, *Exporting Harm—The High-Tech Trashing of Asia* (The Basel Action Network (BAN) Silicon Valley Toxics Coalition (SVTC), Seattle, WA, USA, 2002)
11. M.P. Luda, Recycling of printed circuit boards. Integr. Waste Manag. **2**, 285–299 (2010)
12. J. Moltó, R. Font, A. Gálvez, J.A. Conesa, Pyrolysis and combustion of electronic wastes. J. Anal. Appl. Pyrolysis **84**, 68–78 (2009)
13. J. Wienold, S. Recknagel, H. Scharf, M. Hoppe, M. Michaelis, Elemental analysis of printed circuit boards considering the ROHS regulations. Waste Manag. **31**, 530–535 (2011)
14. G. Jie,; L. Ying-Shun,; L. Mai-Xi, Product characterization of waste printed circuit board by pyrolysis, J. Anal. Appl. Pyrolysis **83**, 185–189 (2008)
15. H.M. Veit, A.M. Bernardes, J.Z. Ferreira, J.A. Tenório, C. de Fraga Malfatti, Recovery of copper from printed circuit boards scraps by mechanical processing and electrometallurgy. J. Hazard. Mater. **137**, 1704–1709 (2006)
16. J. Li, Z. Xu, Y. Zhou, Application of corona discharge and electrostatic force to separate metals and nonmetals from crushed particles of waste printed circuit boards. J. Electrost. **65**, 233–238 (2007)
17. H. Lu, J. Li, J. Guo, Z. Xu, Movement behavior in electrostatic separation: recycling of metal materials from waste printed circuit board. J. Mater. Process. Technol. **197**, 101–108 (2008)
18. C.E.M. Meskers, C. Hagelüken, S. Salhofer, M. Spitzbart, Impact of Pre-Processing Routes on Precious Metal Recovery from PCs, in *Proceedings of the European Metallurgical Conference (EMC)*, (Innsbruck, Austria, 2009)
19. V. Ari, A review of technology of metal recovery from electronic waste, in *E-Waste in Transition—From Pollution to Resource*, ed. by F.-C. Mihai, InTech, doi:10.5772/61569
20. J. Guo, Z. Xu, Recycling of non-metallic fractions from waste printed circuit boards: a review. J. Hazard. Mater. **168**, 567–590 (2009)
21. J. Cui, L. Zhang, Metallurgical recovery of metals from electronic waste: a review. J. Hazard. Mater. **158**, 228–256 (2008)
22. H. Veldbuizen, B. Sippel, Mining discarded electronics. Indus. Environ. **17**, 7–11 (1994)
23. L. Theo, Integrated recycling of non-ferrous metals at Boliden Ltd. Rönnskär smelter, in *Proceedings of IEEE International Symposium on Electronics and the Environment*, (1999), pp. 42–47
24. J. Leirnes, M. Lundstrom, in *Method for Working-Up-Metal-Containing Waste Products*, (US Patent. US4415360 (C22B 1/00), 1983)
25. Y. Chehade, A. Siddique, H. Alayan, N. Sadasivam, S. Nusri, T. Ibrahim, Recovery of gold, silver, palladium, and copper from waste printed circuit boards, in *Proceedings of the International Conference on Chemical, Civil and Environment Engineering (ICCEE)*, (Dubai, United Arab Emirates, 2012)
26. N. Dhawan, M. Kumar, V. Kumar, M. Wadhwa, Recovery of metals from electronic scrap by hydrometallurgical route, in *Proceedings of the Global Symposium on Recycling, Waste Treatment and Clean Technology (REWAS)*, (Cancun, Mexico, 2008), pp. 693–698
27. N. Dhawan, V. Kumar, M. Kumar, Recovery of metals from electronic scrap by hydrometallurgical route, in *Extraction and Processing Division (EPD) Congress; The Minerals, Metals and Materials Society*, (Warrendale, PA, USA, 2009), pp. 1107–1109

28. M. Delfini, M. Ferrini, A. Manni, P. Massacci, L. Piga, Antonio Scoppettuolo Optimization of precious metal recovery from waste electrical and electronic equipment boards. J. Environ. Prot. **2**, 675–682 (2011)
29. M. Sadegh Safarzadeh, M.S. Bafghi, D. Moradkhani, M. Ojaghi Ilkhchi, A review on hydrometallurgical extraction and recovery of cadmium from various resources. Miner. Eng. **20**, 211–220 (2007)
30. G.M. Ritcey, Solvent extraction in hydrometallurgy: present and future. Tsinghua Sci. Technol. **11**, 137–152 (2006)
31. B. Yang, Ion exchange in organic extractant system. Ion Exch. Adsorp. **10**, 168–179 (1994)
32. M. Shamsuddin, Metal recovery from scrap and waste. J. Metals **38**, 24–31 (1986)
33. L.L. Tavlarides, J.H. Bae, C.K. Lee, Solvent extraction, membranes, and ion exchange in hydrometallurgical dilute metals separation. Sep. Sci. Technol. **22**, 581–617 (1985)
34. V.M. Paretsky, N.I. Antipov, A.V. Tarasov, Hydrometallurgical method for treating special alloys, jewelry, electronic and electrotechnical scrap, in *Proceedings of the Minerals, Metals & Materials Society (TMS) Annual Meeting*, (Charlotte, NC, USA, 2004), pp. 713–721
35. G. Hilson, A.J. Monhemius, Alternatives to cyanide in the gold mining industry: what prospects for the future? J. Clean. Prod. **14**, 1158–1167 (2006)
36. S.R. La Brooy, H.G. Linge, G.S. Walker, Review of gold extraction from ores. Miner. Eng. **7**, 1213–1241 (1994)
37. H. Antrekowitsch, M. Potesser, W. Spruzina, F. Prior, Metallurgical recycling of electronic scrap, in *Proceedings of the EPD Congress*, (San Antonio, TX, USA, 2006); pp. 12–16
38. A. Anindya, D.R. Swinbourne, M.A. Reuter, R.W. Matusewicz, Distribution of elements between copper and FeO_x-CaO-SiO_2 slags during pyrometallurgical processing of WEEE. Miner. Process. Extr. Metall. **122**, 165–173 (2013)
39. B. Lindquist, 35 years of progress in environmental strategy and technology in non-ferrous metals mining and smelting industry, in *Presentation at the International Conference on Clean Technologies for the Mining Industries*, (Santiago, Chile, 2008), p. 10
40. J. Huisman, L.N. Stevels, Eco-efficiency of take-back and recycling, a comprehensive approach. IEEE Trans. Electron. Packag. Manuf. **29**, 83–90 (2006)
41. C. Hagelüken, Recycling of electronic scrap at Umicore's integrated metals smelter and refinery. Proc. EMC **59**, 152–161 (2005)
42. A. Khaliq, M.A. Rhamdhani, G. Brooks, S. Masood, Metal extraction processes for electronic waste and existing industrial routes: a review and Australian perspective. Resources **3**, 152–179 (2014)

Dry Granulation of Hot Metal and Heat Recovery from Off-Gas

Wenchao He, Xueqin Li, Xuewei Lv, Jie Qiu and Jie Dang

Abstract Iron powders are mainly produced by Hoganas process and water atomization process, while both of which have the disadvantages of energy consumption. To obtain iron powders with higher particle sphericity, more uniform particle size distribution and higher proportion of total iron (TFe), a novel method considering of energy conservation was developed in this paper. The hot metal was transformed into granules with air cooling in a rotary disc atomizer, iron powder particles and hot off-gas were collected respectively. It was found that the property of the obtained iron powders was mainly controlled by the flow rate of the hot metal, the diameter of the rotary disc and the rotary speed and the heat of off-gas has a high level of value as well.

Keywords Iron powder · Heat recovery · Rotary disc atomizer

Introduction

After ammonia and phosphate, titanium dioxide is the third-largest inorganic chemical products. Because of its superior physical and chemical properties, the titanium dioxide is considered to be the best white pigment and is widely used in chemical coating, paper, cosmetics and catalyst industry. As we all know, titanium dioxide is mainly produced by sulfuric acid process and chloride process. In comparison with sulfuric acid process, the chloride process has obvious advantages in technology, energy consumption and environmental protection. While in China, the production of titanium dioxide is dominated by sulfuric acid process due to issues such as technology and raw materials.

According to preliminary data from National Chemical Industry Productivity Center Titanium Dioxide Branch [1], the annual output of the titanium dioxide

W. He · X. Li · X. Lv (✉) · J. Qiu · J. Dang
College of Materials Science and Engineering, Chongqing University,
Chongqing 400044, China
e-mail: lvxuewei@163.com

L. Zhang et al. (eds.), *Energy Technology 2017*,
The Minerals, Metals & Materials Series, DOI 10.1007/978-3-319-52192-3_12

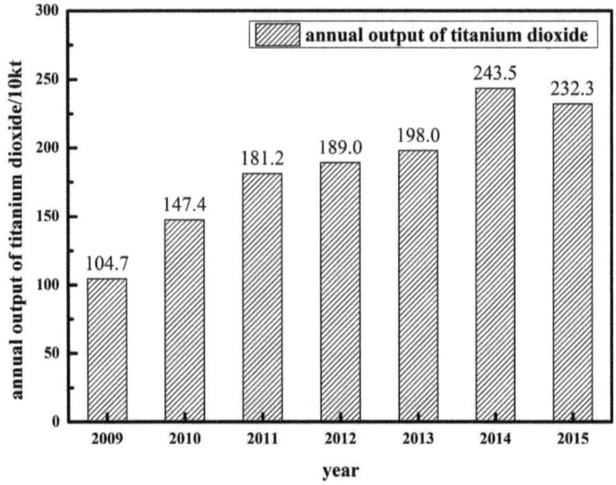

Fig. 1 The annual output of titanium dioxide from 2009 to 2015

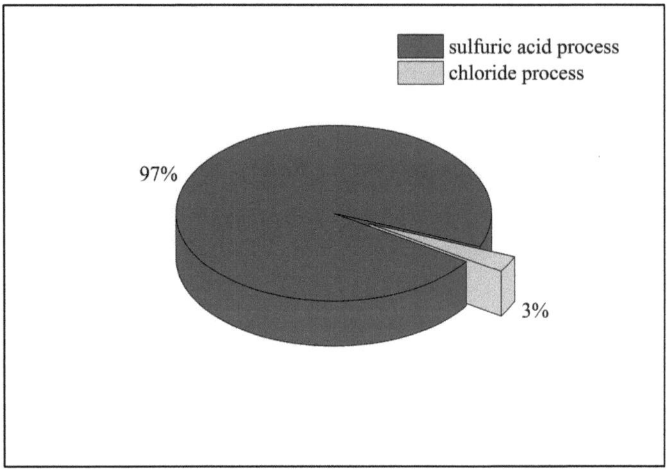

Fig. 2 The proportion of titanium dioxide produced by sulfuric acid process and chloride process in 2015

keeps increasing year by year, as shown in Fig. 1, and China instead of America becomes the biggest producer of titanium dioxide with the amounts of 243.5 million tons in 2014. In 2015, the annual output of the titanium dioxide was 232.3 million tons, only 3% of which was produced by chloride process, which is illustrated in Fig. 2.

Titanium is the one of the most abundant elements on earth, accounting for about 6.2% of the quantity of the Earth's crust. The titanium reserved in China accounts for one-third of the world total and mainly exists as titanomagnetite located in

Panzhihua [2, 3]. The raw materials used in sulfuric acid process are ilmenite with approximately 50% of TiO_2 and titanium slag with more than 70% of TiO_2. The sulfuric acid process can be divided into four steps: acidolysis, precipitation, hydrolysis and calcination. In order to get high-quality titanium dioxide, the Fe^{3+} in the titanium sulfate solution must be reduced to Fe^{2+} by iron powders. It is estimated that nearly 0.1–0.2 tons of iron powders are needed for 1 ton of titanium dioxide with ilmenite as raw materials, which means that hundreds of thousands of tons iron powders are used for reductant.

The Traditional Iron Powders Production Process

In the present, iron powders are mainly produced by Hoganas process and water atomization process, while both of which have the disadvantages of high energy consumption. Iron powders produced by Hoganas process with mill scale or iron concentrate powders as raw materials include a series of steps, carbon reduction in the tunnel kiln, crush, hydrogen reduction in the belt furnace, crush, screening and blending in order. In particular, the carbon reduction in the tunnel kiln lasts for 40 h at least at the temperature of 1473 K, which is energy consuming obviously. Iron powders produced by water atomization process with hot metal can be divided into a number of steps, water atomization by high-pressure water, dying, reduction, crush, screening and blending in order. The property of the iron powders is mainly controlled by the pressure of the water, such as the nozzle angle of high-pressure water, the flow rate of high-pressure water and the temperature of hot metal.

In comparison with Hoganas process, the water atomization process has a large number of advantages in a variety of respects. First of all, the water atomization process is more suitable for mass production due to its shorter process flow, higher efficiency and lower energy consumption. Secondly, the water atomization process has the advantage of wide raw material sources. And then, the property of the iron powders, particle shape, particle size and distribution etc., can be controlled by parameters of water atomization. Finally, the total iron (TF) content in the iron powder produced by water atomization accounts for more than 99.5%.

However, the hot metal containing large amounts of sensible heat is transformed into granules with high-pressure water impingement directly in the water atomization process, without any heat recovery. In addition, the water atomization process caused environmental problems such as excess water consumption and pollution of air and water. Additionally, extra energy is necessary to dry the iron powders. Therefore, an alternative and effective method is required to obtain iron powders with high properties and recover the sensible heat of hot metal.

Dry Granulation of Hot Metal and Heat Recovery from Off-Gas

In order to obtain iron powders with high properties and recover the sensible heat of hot metal, a novel method of employing rotary disc atomization with air cooling is developed in this paper. Previously, paraffin, glycerin and blast furnace slag have been used as raw materials in the rotary disc atomization experiments by many researchers to investigate the relationship between properties of particles and operating parameters [4–8]. In particular, Xie et al. [9–15] from Australia have been investigating the rotary disc atomization process since 2002 and intending to implement the technology in industry. Since the raw material is paraffin, glycerin and blast furnace slag, the previous research mentioned above paid little attention to oxidation degree of particles, which is one of the critical factors to evaluate whether the iron powder can be used as reductant or not.

The experimental apparatus of dry granulation of hot metal is schematically shown in Fig. 3. In this new process, first, the hot metal was granulated by a rotary disc atomizer. In the rotary disc atomization process, the hot metal was impinged onto the surface of the rotating disc at its center and then spread out nearly uniformly under the action of centrifugal force. The hot metal ejected from the edge of the rotating disc under the action of centrifugal forces, forming droplets. Droplets were rapidly cooled and solidified during flight traversing under air cooling and the

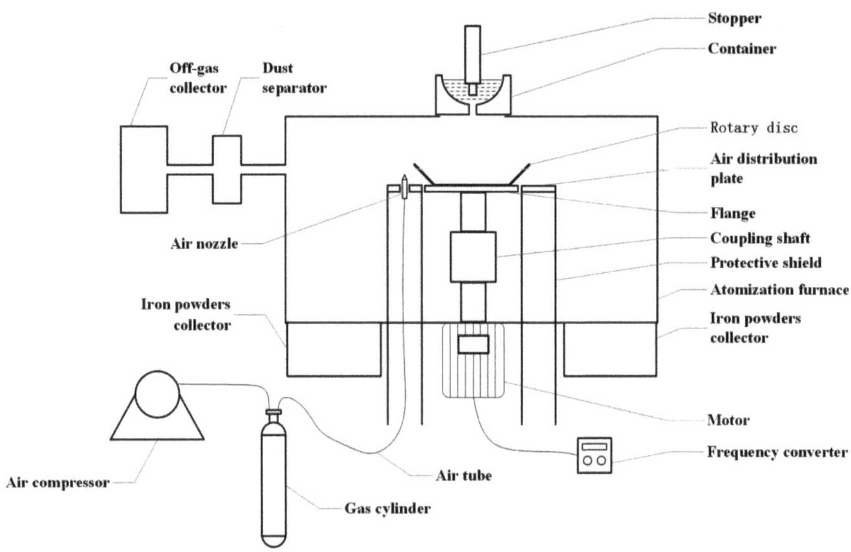

Fig. 3 The experimental apparatus of dry granulation of hot metal

temperature of iron powders declined to the bonding temperature or blow. Second, the iron powder particles sprayed against the furnace wall and then dropped onto the bottom of the pressure-tight atomizing furnace. Then the iron powder particles and hot off-gas were collected respectively. It was found that [16] the property of the obtained iron powders was mainly controlled by the flow rate of the hot metal, the diameter of the rotary disc and the rotary speed. Generally, the flow rate of the hot metal increased the size of the iron powders, while the diameter of the rotary disc and the rotary speed decreased the size of the iron powders.

While the temperature of obtained iron powder particles was nearly 1173 K, it contained a large amount of sensible heat. And during the atomization process, the surface of the iron powder particles were oxidized. According to Shin et al. [17], the reduction rate of iron oxides to metallic iron by hydrogen was found to be very fast, which was less than 10 min when using a pure hydrogen gas at 973 K. Thus, when the iron powder particles were collected and immediately introduced into a fluid bed filled with hydrogen, the reduction of iron oxides occurred on the surface of the particles, which will increase the purity of the powders obviously. The hot off-gas was collected through gas vent after filtration and dust cleaning. It is estimated that the temperature of the off-gas is at least 873 K, which has a high level of value.

In comparison with water atomization process, the rotary disc atomization process has a number of advantages: (1) the particle sphericity and particle size distribution can be controlled accurately by the flow rate of the hot metal, the diameter of the rotary disc and the rotary speed, (2) in the rotary disc atomization process, the iron powder particles were cooled with air throughout instead of water, which is energy conservation and environment protection, (3) there is no need to dry the iron powders with extra energy (4) the sensible heat of hot metal can be used to reduce iron oxides.

Conclusions

A novel method of dry granulation used to obtain iron powders and recover the sensible heat of hot metal was proposed in the current study. A series of experiments considering the flow rate of the hot metal, the diameter of the rotary disc and the rotary speed respectively will be designed to investigate the feasibility of the new process. The obtained iron powder particles are analyzed by XRD and TG to confirm the extent of surface oxidation, which is an important index to decide whether the iron powders are suitable for reductant or not.

Acknowledgements This study was supported by The National Natural Science Foundation of China (No. 0211002362009).

References

1. Z. Tao, The production of titanium dioxide decreased in China for the first time in 17 years. Iron Steel Vanadium Titanium 2016, **2** (2015)
2. S.L. Yang, J.F. Sheng, Technology of ilmenite melted into titanium slag and pig iron, (Metallurgy Industry Press, Beijing, 2006). p. 25
3. L.S. Li, Z.T. Sui, Physical chemistry behavior of enrichment selectivity of TiO_2 in Perovskite. Acta Phys. Chem. Sin. **17**, 845 (2001)
4. J. Liu, Y. Qingbo, P. Li, D. Wenya, Cold experiments on ligament formation for blast furnace slag granulation. Appl. Therm. Eng. **40**, 351–357 (2012)
5. J. Liu, Y. Qingbo, Q. Guo, Experimental investigation of liquid disintegration by rotary cups. Chem. Eng. Sci. **73**, 44–50 (2012)
6. D. Xie, S. Jahanshahi, T. Norgate, Dry granulation to provide a sustainable option for slag treatment, in *Sustainable Mining Conference Kalgoorlie*, (WA, Australia, 2010), pp. 22–28
7. D. Xie, B. Washington, T. Norgate, S. Jahanshahi, Dry granulation of slags–turning waste into valuable cement binde. CAMP-ISIJ **18**, 1088–1091 (2005)
8. D. Xie, Y. Pan, R. Flann, B. Washington, S. Sanetsis, J. Donnelley et al., Heat recovery from slag through dry granulation, in *1st CSRP Annual Conference* ed. by S. Jahanshahi, T. Rickards (Melbourne, Australia, CSIRO Minerals, 2007), pp. 29–30
9. D. Xie, S. Jahanshahi, T. Norgate, Dry granulation to provide a sustainable option for slag treatment, in *Sustainable Mining Conference*, (Kalgoorlie, WA, 2010), pp. 22–28
10. Dongsheng Xie, Sharif Jahanshahi, Waste heat recovery from molten slags. ICS **2008**, 674–677 (2008)
11. Y.H. Pan, P.J. Witt, D.S. Xie, CFD simulation of free surface flow and heat transfer of liquid slag on a spinning disc for a novel dry slag granulation process. Prog. Comput. Fluid Dyn. **10**(5–6), 292–299 (2010)
12. Yuhua Pan, P.J. Witt, D. Xie, CFD simulation of free surface flow and heat transfer of liquid slag on a spinning disc for a novel dry slag granulation process. Prog. Comput. Fluid Dyn. Int. J. **10**(5), 292–299 (2010)
13. B. Kuan, CFD modeling of liquid jet and cascade breakup in crossflows, in *7th International Conference on CFD in the Minerals and Process Industries*, (2009), pp. 1–6
14. Y. Pan, P. Witt, B. Kuan, D. Xie, Effect of flow and operating parameters on the spreading of a viscous liquid on a spinning disc. in *9th International Conference on CFD in the Minerals and Process Industries*, (2012), pp. 1–12
15. J. Sharif, Y. Pan, D. Xie, Some fundamental aspects of the dry slag granulation process. in *9th International Conference on Molten Slags, Fluxes and Salts*, (2012)
16. Y. PAN, P.J. Witt, B. Kuan, D. Xie, Effect of flow and operating parameters on the spreading of a viscous liquid on a spinning disc
17. S.M. Shin, D.W. Lee, J.Y. Yun et al., A study on the fabrication of iron powder from forging scale using hydrogen. Arch. Metall. Mater. **60**(2), 1547–1549 (2015)

The Energy Recovery of Livestock Waste in Taiwan

Esher Hsu and Chen-Ming Kuo

Abstract Energy recovery of livestock waste can not only reduce waste pollution but also generate renewable energy and further create economic benefits. Currently, most of livestock waste in Taiwan is treated as waste without efficiently recycled. This paper aims to explore and evaluate the energy recovery of livestock waste in Taiwan and further provide policy suggestions. Study results show that small farming scale of livestock industry in Taiwan leads to a lower efficiency of biogas energy recovery. Besides, the environment with relative low price of electricity and high invest cost for power generation plant is hard to provide incentive for implementation of energy recovery which has further caused difficulty on policy promotion for renewable energy recovery of livestock waste in Taiwan. Policy reform with increasing FITs and differentiated purchasing rate for biogas power is suggested to stimulate energy recycling of livestock waste in Taiwan.

Keywords Waste recycling · Energy recovery · Livestock waste · Fits

Introduction

Recycling methods of livestock waste for renewable energy recycling includes technology of biodiesel, biogas and solid biomass [1]. Energy recovery of livestock waste can effectively solve the problem of waste pollution from animal husbandry, improve self-sufficiency rate of energy supply to create more economic benefits, and further reduce greenhouse gas emissions towards sustainable agriculture [2, 3]. Energy recovery of livestock waste in Taiwan is implemented with *Renewable*

E. Hsu (✉)
Department of Statistics, National Taipei University, 67 Section 3,
Min-Sheng East Rd, Taipei 104, Taiwan
e-mail: hsu@mail.ntpu.edu.tw

C.-M. Kuo
Department of Mechanical Engineering, I-Shou University, 1 Section 1,
Syuecheng Rd, Dashu, Kaohsiung 84008, Taiwan

L. Zhang et al. (eds.), *Energy Technology 2017*,
The Minerals, Metals & Materials Series, DOI 10.1007/978-3-319-52192-3_13

Energy Development Act in 2009 to promote a sustainable development of livestock industry, and to reduce the costs of energy supply. *Operating Guidelines for Biogas power generation with Demonstration Project Grants* was initialized *by Bureau of Energy Department in 2008* to promote the energy recycling of livestock waste in Taiwan. The power companies are required to buy the electricity at the FITs (feed-in tariffs) from renewable sources to comply with the *Renewable Energy Development Act* and to provides an incentive for farmers and company as well. The wholesale energy power purchase rates for pig farms with pigs more than 5000 heads is NT$2.0615/kWh in 2010. The purchase rates were increased in each amending of FITs. The *three-year subsidy policy for the large scale pig farms in Pingtung and Changhua counties* in 2013 is NT$2.8/kWh. Because purchasing rate for biogas power is much less than the average market price and the small farming scale of livestock and technology problems, the renewable energy recycling from livestock waste is inefficiency. Currently, most of livestock waste in Taiwan is treated as waste without efficiently recycled. This paper aims to explore and evaluate the energy recovery of livestock waste in Taiwan and further provide policy suggestions.

Livestock Waste Recycling in Taiwan

Based upon the official statistics in 2013, farming scales of livestock industry are shown as in Table 1. Pig and Chicken industries are the most important animal husbandry in Taiwan. The farming scale in average is 679 heads per farm for pig industry and 16,132 heads per farm for chicken industry. The detailed distributions of livestock industry in Tables 1, 2, and 3 show that small scale of livestock farming pattern in Taiwan. Only 1.44% of pig farms with pigs more than 5000 heads (shown as in Table 2). This means that only 1.44% of pig farms qualify for

Table 1 Livestock statistics in Taiwan (2013)

Livestock	Farms	Heads	Heads/Farm
Cattle (including Cow)	1740	140,431	81
Sheep	2054	1,59,807	78
Deer	693	21,261	31
Horse	100	1046	10
Rabbit	30	10,939	365
Pig	8557	5,806,237	679
Chicken	5551	89,550,746	16,132
Duck	2794	9,290,527	3325
Geese	928	1,750,611	1886
Turkey	85	112,569	1324
Ostrich	57	2040	36

Source Ref. [7]

Table 2 Distribution of farming scale for pig farms

Farming Scale (Heads)	Farms	Heads	Average heads per farm	Farms	Heads
	Farms	%	Heads	%	(heads/farm)
Total	8557	100.00	5,806,237	100.00	679
~99	3137	36.66	89,721	1.55	29
100–199	1019	11.91	150,410	2.59	148
200–299	502	5.87	121,798	2.10	243
300–499	771	9.01	305,309	5.26	396
500–999	1549	18.10	1,164,952	20.06	752
1000–1999	1087	12.70	1,522,028	26.21	1400
2000–2999	212	2.48	508,521	8.76	2399
3000–4999	157	1.83	595,876	10.26	3795
5000–9999	75	0.88	537,868	9.26	7172
10,000–19,999	38	0.44	553,383	9.53	14,563
20,000~	10	0.12	256,371	4.42	25,637

Source Ref. [7]

Table 3 Distribution of farming scale for chicken farms

Farming Scale (Heads)	Farms	Heads	Average heads per farm	Farms	Heads
Total	5551	100.00	89,550,746	100.00	16,132
~2999	1469	26.46	420,462	0.47	286
3000–9999	810	14.59	5250,890	5.86	6483
10,000–19,999	1481	26.68	20,667,538	23.08	13,955
20,000–29,999	905	16.30	21,002,044	23.45	23,207
30,000–39,999	385	6.94	12,711,691	14.19	33,017
40,000–49,999	211	3.80	9,093,818	10.15	43,099
50,000~	290	5.22	20,404,303	22.79	70,360

Source Ref. [7]

the subsidy of energy recycling from Fits in 2010. Total waste from animal husbandry has been reduced yearly since 2016 and most of the livestock waste is treated for compost used as a plant fertilizer without efficiently recycled shown as in Figs. 1 and 2.

Energy Recovery of Livestock Waste in Taiwan

Currently, new government intends to promote renewable energy. The policy goal is increasing from 4705 kWh in 2015 to 12,513 kWh in 2025, and 17,250 kWh in 2030. The promotion is in favor of solar (photovoltaic) energy. Policy goal for the percentage of photovoltaic energy to total renewable energy is from 24% in 2015

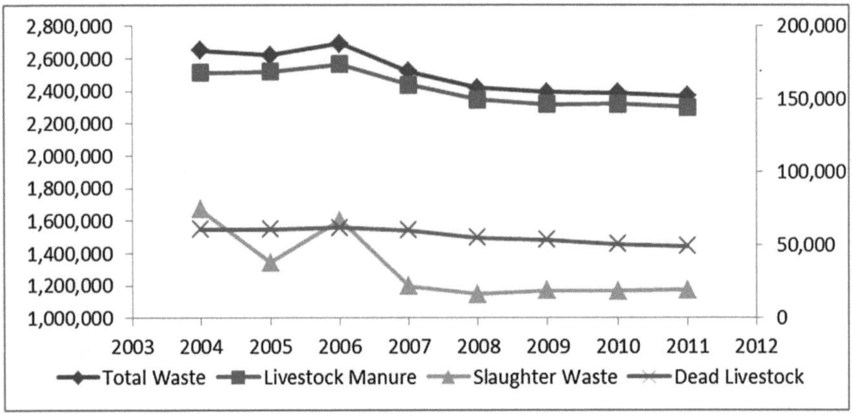

Fig. 1 Livestock waste components. *Source* Ref. [7]

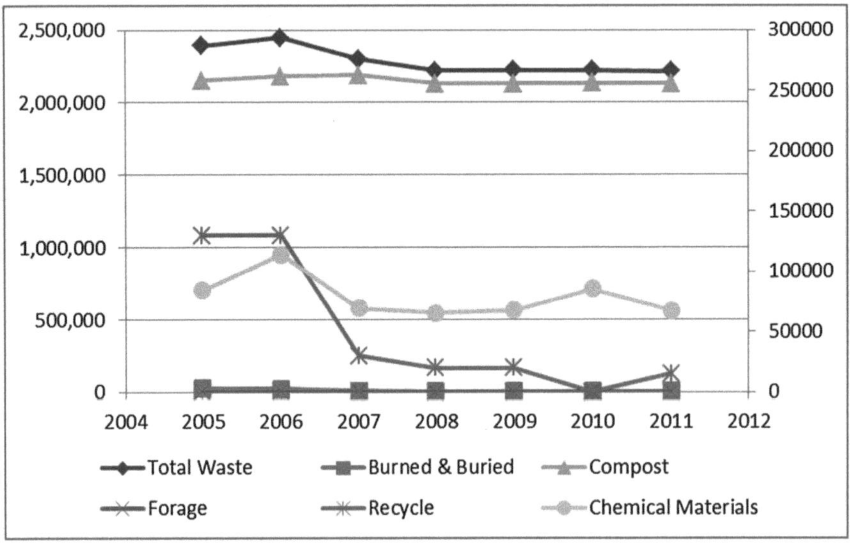

Fig. 2 Treatment methods of livestock waste. *Source* Ref. [7]

increased to 50% in 2030; Whereas, the percentage of energy from biomass (including biogas) to total renewable energy is reduced from 16% in 2015 to 6% in 2030.

The *Renewable Energy Development Act* (REDA, issued in 2009) provides for a system of fixed-in tariffs for electricity generated from renewable energy sources. Table 4 shows the policy targets of cumulative installed capacity of renewable energy. The policy goal of cumulative installed capacity was increased in each amending of REDA to reach the goal of increasing the share of renewable energy

Table 4 Policy targets of cumulative installed capacity of renewable energy

Year	Solar energy (Photovoltaic)	Biogas	Solid Waste
2010	22	25	790
2015	420	29	848
2020	1020	29	925
2025	2500	31	1369
2030	3100	31	1369

Unit 1000 KW

sources in electricity supply. For biogas, the policy goal is (1) 25,000 KW by 2010, (2) 29,000 KW by 2020, and (3) 31,000 KW by 2030.

Table 5 provides an overview of the Taiwan fixed-in-tariffs (FITs) for the facilities commissioned in 2013. Feed-in-tariffs [4] are subject to changes, either by regular amending or by legislative changes. Please note the information in Table 5 may not have been updated to reflect the latest changes. For biogas, subsidy supported based upon the *Operating Guidelines for Biogas power generation* with *Demonstration Project Grants* from *Bureau of Energy Department* in *2008*. The subsidy is increased in each amending of FITs. In 2010, the wholesale energy power purchase rate for biogas plan operated by pig farms with pigs more than 5000 heads is NT$2.0615/kWh; whereas, in 2013, based on the *Three-year subsidy policy for the large scale pig farms in Pingtung and Changhua counties*, the purchase rates are NT$2.6338/kWh for biogas plant without in anaerobic digestion and NT$3.3803/kWh for biogas plant in anaerobic digestion. In average, the purchase rate is NT$2.8/kWh [5].

The renewable energy recovery of livestock waste is hard to promote in Taiwan due to small farming scale and no incentive for implementation. Besides, low electricity price and high invest cost for power generation plant also lead to a lower efficiency of biogas energy recovery in Taiwan.

Table 5 Purchasing rate of renewable energy in Taiwan (2015)

Item	Type	Scale	Rate (NT$/kWh)
Solar energy (Photovoltaic)	On roof	1–10 kW	6.8633
		10–100 kW	5.7378
		100–500 kW	5.3627
		500 kW ~	5.1935
	On floor		4.8845
Biogas	Not in anaerobic digestion		2.6338
	In anaerobic digestion		3.3803
Solid waste			2.8240

Source Ref. [5]

For example, the biogas plant which using material from chicken farm with chicken 720,000 units and generated manure 80 tons per day, could generate electricity power 16,800 kWh per day (net electricity power 5,000,830 kWh per year). The biogas plant generates electricity power from chicken manure [6] at the farm through anaerobic digestion technology followed by a desulfurization purification process to effectively reduce CO2 emissions from chickening farming by 75% (information provided by Shih An farm in Taiwan). The power generating cost is around NT$7.7/kWh, but it can only receive FITs for NT$2.6338/kWh. Although the power company (Tai power) is required to buy the electricity at the FITs (feed-in tariffs) from renewable sources, the cost is much higher than the return for biogas plant. Therefore, current FITs did not provide incentive for livestock farm or company to invest for biogas plant. Comparing with other renewable energy, the FITs for biogas is much lower than others (see Table 5 for the details). In order to increase the demands of livestock waste for input of power plan, the FITs for biogas has to be increased and differentiated based upon the components of raw material.

Conclusion

Energy recovery of livestock waste can not only reduce waste pollution but also generate renewable energy and further create economic benefits. Currently, most of livestock waste is treated as waste without efficiently recycled. Study results show that small farming scale of livestock industry in Taiwan leads to a lower efficiency of biogas energy recovery. Besides, the environment with relative low price of electricity and high invest cost for power generation plant is lack of incentive for implementation of energy recovery and has further caused difficulty on policy promotion for renewable energy recovery of livestock waste in Taiwan. In order to make the recycling more efficient and more environment-friendly, the study results provide two suggestions for amending livestock waste recycling policy in Taiwan

(i) Developing a differentiated purchasing rate for biogas power based upon the portion of manure and slurry used and the scale of the biogas plant to promote energy recycling
(ii) FITs last for at least 10 years. The 10 years profit guarantee could provide a great incentive for farmers and company.

Policy reform with increasing FITs, 10 years FITs guarantee, and differentiated purchasing rate for biogas power is suggested to stimulate energy recycling of livestock waste in Taiwan which is suggested for future reform on amending recycling policy of livestock waste.

References

1. M. Ragwitz, A. Held, G. Resch, T. Faber, C. Huber, R. Hass, *Monitoring and evaluation of policy instruments to support renewable electricity in EU member states (summary)* (Umweltbundesamt, German, 2006)
2. S. Dagnall, H.L. Jon, P. David, Resource mapping and analysis of farm livestock manures—assessing the opportunities for biomass-to-energy schemes. Bioresour. Technol. **71**, 225–234 (2002)
3. European Commission, 2030 framework for climate and energy policies (2014). Retrieved 29 Nov 2014, from http://ec.europa.eu/clima/policies/2030/index_en.htm
4. V. Lauber, L. Mez, Three decades of renewable electricity policies in Germany. Energy Environ. **15**(4), 599–623 (2004)
5. Bureau of Energy Department, (2015). http://www.foreverpower.com.tw/news20141227.cfm
6. S. Luostarinen, A. Normak, M. Edström, *Overview of Biogas Technology* (2011)
7. COA, (2014). http://agrstat.coa.gov.tw/sdweb/public/inquiry/InquireAdvance.aspx

Life Cycle Assessments of Incineration Treatment for Sharp Medical Waste

Maryam Ghodrat, Maria Rashidi and Bijan Samali

Abstract Treatment of sharp medical waste (waste disposable syringes) produced in hospitals or health care facilities in an environmentally sound ways have raised concerns relating to public health and occupational safety. Life cycle assessment (LCA) is a decision-supporting tool in waste management practice; but relatively little research has been done on the evaluation of sharp medical waste treatment from a life cycle perspective. Our study assesses the environmental performances of medical waste incineration as a type of dominant technology for specific medical waste of average composition. Inventory models were used for waste incineration and residues landfill. Background data were derived from modelling performed in HSC Chemistry thermochemical package linked with GaBi environmental assessment tool. Two scenarios have been considered and compared: Waste disposable syringes partially replacing coke in Electric Arc Furnace (EAF) steelmaking (WSI) and conventional incineration with pure metallurgical coke without adding waste syringes (CI). The results of this study could support the medical waste hierarchy and indicate that from a life cycle perspective, replacing part of metallurgical coke with waste syringes in electric arc furnace steelmaking leads to fairly significant reduce in most of environmental impact categories.

Keywords LCA · Waste syringes · Incineration · EAF

Introduction

The management of wastes arising from health care establishmcnts is a matter of continuing concern to workers in the waste industry and the general public. These concerns are based upon the potential for spread of infection, the risk of injury, chemical toxicity and environmental damage. Sharps waste is a form of medical

M. Ghodrat (✉) · M. Rashidi · B. Samali
Centre for Infrastructure Engineering, School of Computing, Engineering and Mathematics, Western Sydney University, Sydney, Australia
e-mail: m.ghodrat@westernsydney.edu.au; mariam.godrat@gmail.com

L. Zhang et al. (eds.), *Energy Technology 2017*,
The Minerals, Metals & Materials Series, DOI 10.1007/978-3-319-52192-3_14

waste composed of used sharps, which includes any device or object used to puncture or lacerate the skin (syringes, needles, scalpels, razor blades, broken glass etc.). Sharps waste is classified as biohazardous waste and raised serious concerns in Australia because of some improper disposal practices accorded to them. There are two major treatment methods for medical waste one is incineration and the other is non-incineration technique [1]. Incineration technique which is the main focus of this study refers to high-temperature treatment technologies. Non-incineration technique on the other hand refers to the low temperature treatment, including chemical, radiation or biological treatment technologies [2]. Traditionally, incineration has been utilized as a key treatment method for Medical waste, especially sharp waste. That is due to the fact that incineration has many benefits including purification and sterilization of the wastes along with decreasing the volume of the waste. Recently, at least in New South Wales and Victoria, there have been significant increases in landfill levies, raising the price of non-recycled medical waste. Hence incineration has become the main method for disposal of sharp medical waste in Australia. Many hospitals are employing waste collection and recycling companies who have on-site incineration for treatment of the collected hospital wastes but because of the current strict pollution control regulations, some of the incinerators had to be shut down, especially in New South Wales. In this study the possibility of using existing electric arc furnace steelmaking for treatment of waste disposable syringes has been investigated. Electric Arc Furnace (EAF) steelmaking uses different carbon based materials as foaming agents [3]. Depending on cost and availability, coal or metallurgical cokes are two of the conventional foaming agents that are injected into the furnace. There are some other alternative carbon sources which could partially replace coke such as rubber, high density polyethylene (HDPE) and waste plastics. These waste materials may possibly react with gas and slag phases that lead to devolatilization and iron oxide reduction reactions [4]. Research on the use of waste plastics in a commercial electric arc furnace has been reported by Asanuma [5] and Sahajwalla [6–8]. Some other studies have reported the use of polymeric materials in EAF steelmaking [4, 9] and showed improvements in incineration performances and higher carbon/slag interactions with the addition of plastics [10]. The usage of waste disposable syringes in EAF steelmaking might bring advantages, on one side from the energy supply point of view and on the other side from their carbonaceous material applications for slag foaming. According to the series of research done by Sahajwalla as polyolefins, such as PE and polypropylene (PP) are produced from petroleum, they have calorific value of about 47 MJ/kg, which is higher than most of conventional coals and cokes [10]. Since carbon and hydrogen are the major components of waste disposable syringes, these have the potential to be used as an alternative to partly substitute coke in electric arc furnace steelmaking. On the other hand, it is clear that coke making is one of the major sources of greenhouse gas releases in the steelmaking process; therefore it is important that the coke consumption reduced further to cut the emission of greenhouse gases. A considerable reduction in coke consumption could be achieved by injecting supplementary carbonaceous materials such as waste syringes. In this study waste disposable syringes are considered to partially replace coke in electric arc furnace. A thermodynamic modelling approach has been implemented. Mass and energy balances were obtained from HSC Chemistry thermochemical package. Two scenarios have been modelled: Incineration of waste syringes partially replaced coke in EAF (WSI) and conventional incineration in EAF with pure metallurgical coke (CI). The particular focus is to compare the two scenarios in terms of environmental impact. Until now, a large number of analytical tools have been employed to compare waste treatment strategies, including cost–benefit analysis and life cycle assessment (LCA) [11–13]. In fact LCA is a quantitative means that could offer decision makers information about potential environmental burdens caused by different alternatives. Limited number of research has been conducted on life cycle assessment of medical waste [14–16] but there is no research that has specifically analysed the effect of using waste syringes in EAF as part of reductant. The present study hence assessed this effect from a life cycle perspective.

Methodology

The assessments in this study were based on results from modelling performed in HSC Chemistry thermochemical package. HSC Chemistry Sim 8 linked with GaBi environmental assessment tool [17] is used to quantify waste syringes treatment through incineration. GaBi is designed for assessment of waste management systems, and can be used for life cycle assessment and economical assessments of the system. It aggregates the quantified resource consumptions, emissions to air, soil, surface water and groundwater into several impact categories. The impact category used in this assessment is the updated ReCiPe category global warming potential based on the ReCiPe method [18].

Modelling Approach

As mentioned in Sect. “Methodology”, the approach of the modelling is based on the mass and energy balance obtained from HSC Chemistry thermodynamic package. The modelling has been performed from a technology data perspective, meaning that the technology that have been modelled covers modelling of the usage of waste syringes in an electric arc furnace as a test of the idea of partial replacement of metallurgical coke. The modelling was performed from a gate to gate perspective, excluding collection and transport of the waste as well as the treatment and material and energy substitution. The mass balance of the modelled incineration processes obtained from HSC Chemistry is shown in Figs. 1 and 2, respectively.

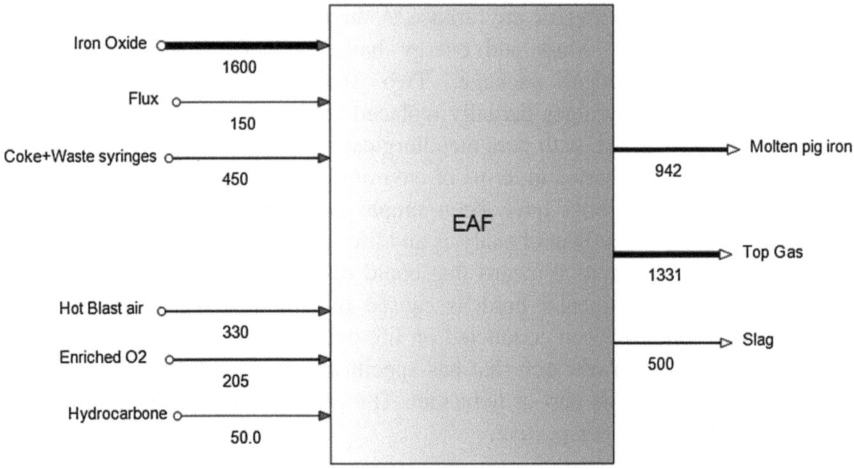

Fig. 1 Mass balance of the incineration process in EAF, WSI scenario (the values are in kg)

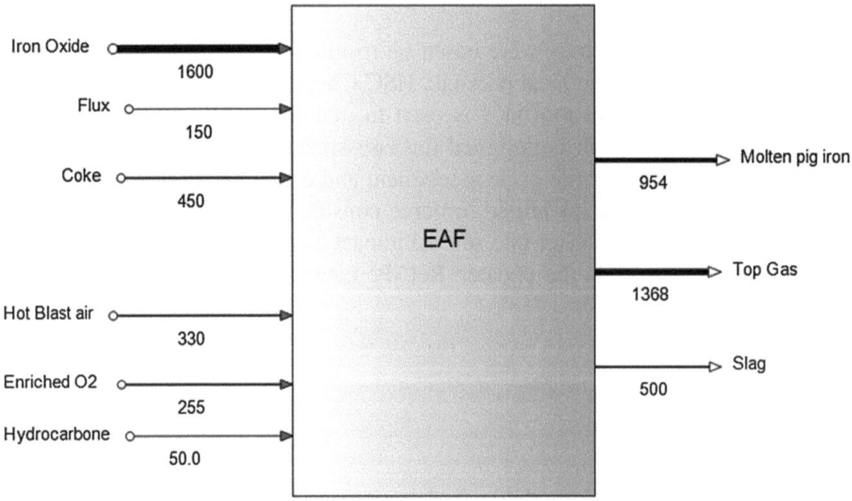

Fig. 2 Mass balance of the incineration process in EAF, CI scenario (the values are in kg)

Life Cycle Assessment (LCA)

A life cycle assessment study is a tool that enables the assessment of environmental burdens related to a product, process, or service by classifying energy and materials utilized in them and the emissions that go to the environment; furthermore it leads to finding opportunities for environmental improvements plans [19–21]. LCA methodology is considered as one of the most effective tools for detecting and evaluating the environmental impacts related to waste management options [22–24]. Some LCA models with the aim of determining how the system variations affect the environmental impacts through scenario analysis, do also exist [25]. The methodology used in this paper is based on thermodynamic modelling. First mass and energy balances are obtained from HSC Chemistry thermochemical package linked with GaBi environmental software [17]. GaBi is applied in order to examine system behaviour under different circumstances, such as material and energy requirements, environmental impacts and ecological footprint. Such strategy is utilized because the LCA of a product or plan should be an indication of a fairly precise evaluation of the product with reference to its impacts on the environment and on human health, and should lead to an overall environmental and ecological assessment. Therefore, a range of specific and selected environmental influences is assessed, but some other aspects such as economic and social parameters, are not taking into account in this study.

Material and Method

In this study, waste syringes treatment scenario involving incineration in a small Electric Arc Furnace (EAF) is evaluated based on the pyrometallurgical technique. In this scenario, waste disposable syringes have been used to partially replace coke as the carbon source in EAF steelmaking. As carbon and hydrogen are the major components of waste syringes, these clearly have the potential to be used as an alternative to coke in electric arc furnace. This scenario is then compared with the conventional incineration process in the same EAF using pure metallurgical coke as the reduction agent of iron oxide. The boundary of both scenarios is shown in Fig. 4. The environmental impacts of the two scenarios were evaluated through LCA. The purpose of the LCA is to ensure the environmental sustainability of the used technique and the feasibility of the application of waste syringes as a replacement for coke in EAF.

Goal and Scope Definition

The goal of this study is to compare incineration of a blend of coke and waste syringes in an electric arc furnace with that, using pure metallurgical coke without adding waste syringes to the feed.

On the other hand the key aim of this study is to assess the environmental impact of the use of plastics and polymer in waste syringes to partially replace coke as the carbon source in Electric Arc Furnace steelmaking. The disposable syringes subjected to the incineration are manufactured by many companies. The elementary composition of these syringes (i.e. the fraction of C, H, N, S and O) assumed for the assessment is derived from Ref [26] and shown in Table 1. The oxygen is 16.13%

Table 1 Elemental analysis of waste syringes

Sample	C (wt%)	H (wt%)	N (wt%)	S (wt%)	O (wt%)	Cl (wt%)	GCV (MJ/kg)
Waste syringes	72.2	11.46	0.18	0.03	16.13	–	42.24

in the elemental analysis of waste syringe. The nitrogen and sulphur in the waste syringe samples possibly exist because of some impurities. Base on [26] the disposable syringes are made from mixed plastics, in which the main element is polypropylene and the rest are polyethylene and poly vinyl chloride.

Functional Unit

The functional unit for WSI scenario is defined based on disposal of 269 kg of waste syringes with the specified composition in Table 1 blended with 1600 kg of ferrous scrap, 150 kg flux and 181 kg coke. A volume of 270 m^3 preheated air is blasted through tuyeres near the bottom of the furnace. 50 kg hydrocarbon additives and oxygen are also introduced through the tuyeres. A representative material balance is presented in Fig. 3.

System Boundary

System boundaries were set by application of a gate-to-gate approach. Two scenarios are considered and compared in this study, namely, waste syringes partially replacing metallurgical coke in process of EAF steelmaking (Waste SyringesIncineration—WSI) and conventional incineration in EAF steelmaking process with pure metallurgical coke (Conventional Incineration—CI).

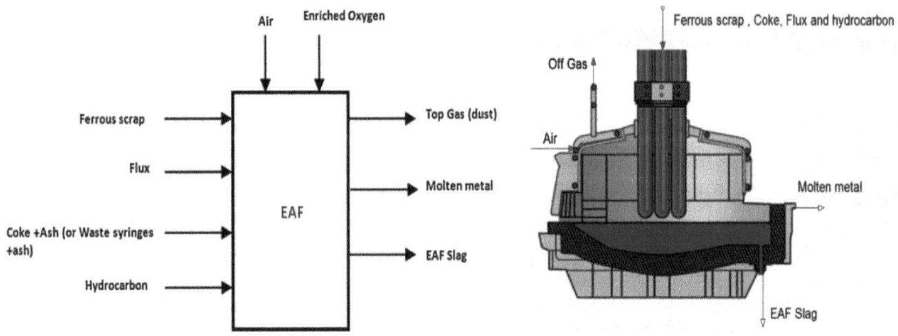

Fig. 3 Representative material balance for electric arc furnace

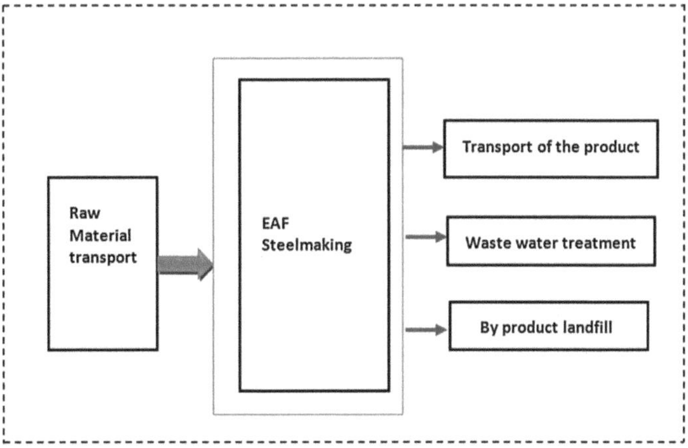

Fig. 4 System boundary for both considered scenarios (WSI and CI)

Both scenarios involve raw materials and energy consumption; road transportation of raw materials to the site; direct air, water, and soil emissions during the treatment processes and waste disposal (i.e., on-site wastewater, landfill disposal of the by-product i.e., slag). To simplify the LCA analysis of the selected scenario, the process of waste collection to the treatment site is excluded. The infrastructure (i.e., construction and equipment) process is also excluded because of the lack of information from considered waste treatment sites. The geographical scope of the study is regional and the study focused on conditions and technologies for 2016. Figure 4 shows system boundaries for both scenarios.

Data Source

A Mass and energy input-output model of two scenarios has been conducted by HSC Chemistry Sim 8. The pyrometallurgical route comprises processing of 2.5 tonnes of raw materials including ferrous scrap, flux, coke, ash, hydrocarbon, oxygen enriched and blast preheated air for conventional incineration process (CI) and for the second scenario (WSI); 75% coke has been replaced by waste syringes. These technology models provide the inventory of inputs and outputs to the process needed for the subsequent environmental assessment based on LCA methodology. In this paper, we will discuss how a thermodynamic analysis tool can be used to assess selected incineration process and then use the result to compare and analyse the environmental impact via GaBi database.

Life Cycle Impact Assessment Methodology

In this study, Life cycle impact assessment (LCIA) results were calculated at midpoint level by using the ReCiPe method [18] because the fate exposure of this model is consistent with multimedia modelling. ReCiPe is a developing LCIA method attempting to combined midpoint and endpoint impact assessment (as with IMPACT 2002+). ReCiPe 2009 comprises two sets of impact categories with linked sets of characterisation factors. Eighteen impact categories are addressed at the midpoint level and at the endpoint level. Most of these midpoint impact categories are further converted and combined into three midpoint categories. The complete characterization factors and detailed methodology for ReCiPe are obtainable on the website of Institute of Environmental Science in Leiden University of Nederland (http://www.cml.leiden.edu/research/industrialecology/researchprojects/finished/recipe.html).

Results and Discussion

Life Cycle Inventory (LCI) Analysis

After determining the goals and boundaries of the selected system, the LCI is the second phase of the LCA. This phase involved compiling energy, water and materials used in different processes for the systems being studied. According to the systems' boundaries and functional units, the inventories of energy and materials needed for the assessment are normalized and compared to the alternative process. Figure 5 shows the flow chart diagram for both scenarios. Table 2 summarizes the main inputs and outputs and the mass balance per process. As mentioned earlier, data for energy and emissions for the considered processes were obtained from thermodynamic modelling results linked with environmental analysis from GaBi software.

The LCIA of each scenario is provided in Table 3. These results were obtained from environmental impact modelling in GaBi. Most of the impact categories show higher values for the second scenario (CI). Note that in LCIA analysis negative values reveal the benefits obtained for the environment, while positive values represent the environmental impacts (burdens or costs to the environment).

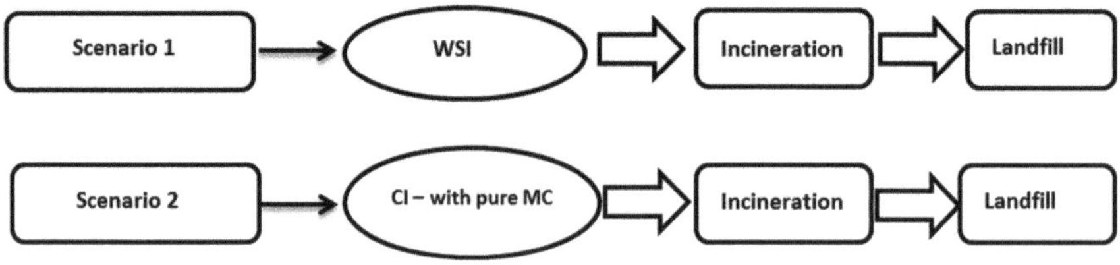

Fig. 5 Flowchart diagram for both scenarios, (*MC* : Metallurgical coke)

Table 2 Summary of inventories per scenario (values refer to the functional unit of 2.5 tonne total input materials)

Scenario	Stage	Flow	Unit	Value
WSI	Input	Coke + Waste syringes	kg	450
		Ferrous scrap	kg	1600
		Air	m^3	269.5
		Electricity	kWh	75.86
		Flux	kg	150
		Oxygen enriched	m^3	154
		Hydrocarbon	kg	50
	Output	Molten pig iron	kg	941.6
		Slag	kg	499.87
		Top gas	kg	1330.74
CI	Input	Coke	kg	450
		Ferrous scrap	kg	1600
		Air	m^3	269.5
		Electricity	kWh	785.5
		Flux	kg	150
		Oxygen	m^3	192
		Hydrocarbon	kg	50
	Output	Molten pig iron	kg	953.78
		Slag	kg	500.3
		Top gas	kg	1368

WSI Waste syringes incineration, *CI* Conventional incineration

Table 3 Environmental impact of WSI and CI scenarios

Impact category	Unit	WSI	CI
Climate change	kg CO_2 eq.	2.84	3.71
Terrestrial acidification	kg SO_2 eq.	6.04e−4	3.46−3
Freshwater eutrophication	kg P eq.	8.26e−5	8.3e−5
Ozone depletion	kg CFC-11 eq.	2.69e−12	2.82e−12
Fossil depletion	kg oil eq.	0.0495	0.26
Freshwater ecotoxicity	kg 1,4-DB eq.	4.6e−3	4.57e−3
Human toxicity	kg 1,4-DB eq.	0.582	0.632
Ionising radiation	kg U235 eq.	2.13e−3	2.24e−3
Marine ecotoxicity	kg 1,4-DB eq.	1.41e−3	1.44e−3
Marin eutrophication	kg N-Equiv.	9.99e−4	1.76e−3
Metal depletion	kg Fe eq.	6.6	6.51
Particulate matter formation	kg PM10 eq.	5.81e−4	1.61e−3
Petrochemical oxidant formation	kg NMVOC	0.0127	0.0185
Terrestrial ecotoxicity	kg 1,4-DB eq.	2.26e−3	6.27e−3
Water depletion	m^3	0.0408	0.155

Processes which involved waste utilization (such as the one we considered here: WSI) entail environmental burdens related to the fuel and electricity consumptions. The results obtained for the incineration process of waste syringes that partially replace metallurgical coke in EAF steelmaking is more satisfactory than the conventional incineration in EAF with pure metallurgical coke. As can be seen from data presented in Table 3 and Fig. 6 all impact categories (especially "climate change") show lower burdens for the environment in WSI scenario in comparison to that in CI scenario. This is due to the fact that utilization of waste plastic and polymer as a supplementary raw material to coke helps to improve iron oxides reduction, particularly with regard to the role of off-gas components on the reactions [3]. The lower environmental burden of WSI in comparison to CI, along with preventing these wastes to be dumped into the landfill, depicts clear benefits for all environmental impacts described in the selected LCIA methodology (ReCipe).

Figure 6 gives an overview of characterization results, relating to WSI and CI scenarios. As can be seen from Fig. 6, WSI scenario yields better environmental scores particularly for climate change, terrestrial acidification, marine eutrophication, particulate matter formation and Terrestrial ecotoxicity. As for eutrophication, CI scenario shows 4.4 times higher impact score in comparison to WSI scenario. For Terrestrial acidification on the other hand the impact score of CI scenario is even higher (5.72 times) and for the Particulate matter formation again CI scenario shows more than a threefold higher impact value compared to WSI. Energy consumption in WSI scenario is 10 times lower than that of CI which indicates the potential benefit of using waste syringes as part of fuel in electric arc furnace steelmaking.

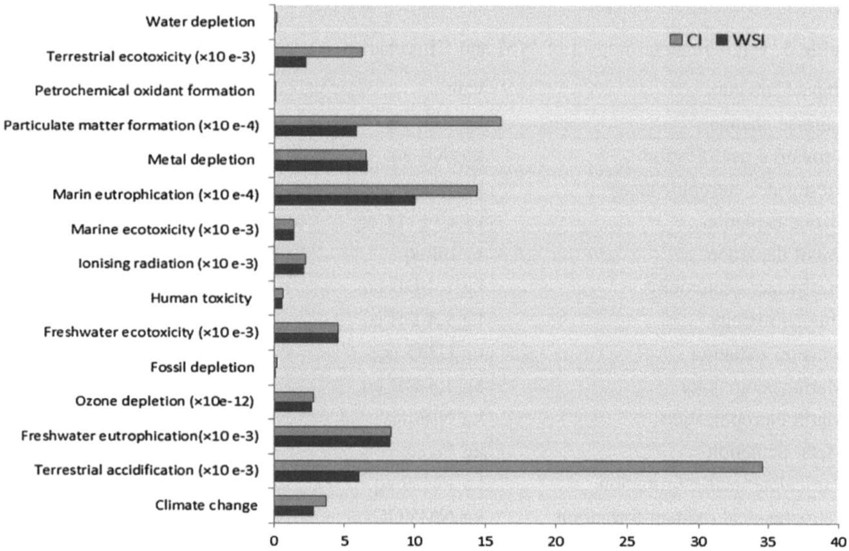

Fig. 6 LCIA results for WSI and CI scenarios

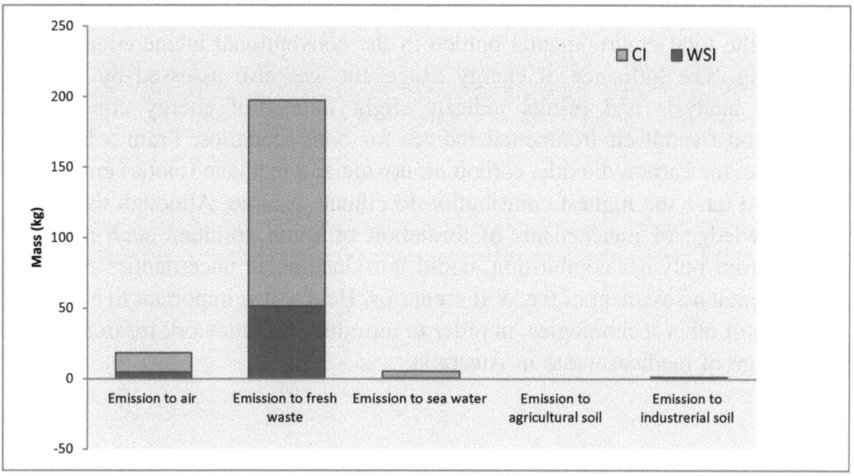

Fig. 7 ReCiPe endpoint analyses from GaBi for the model depicted in Figs. 1 and 2

An endpoint evaluation can also be made as shown in Fig. 7, permitting a detailed analysis of different options. On the other word, Fig. 7 shows the ReCiPe endpoint analysis from GaBi for the model depicted in Figs. 1 and 2. It is worthwhile to note that the combination which has been fed to the furnace in WSI scenario is basically a compound mixture of materials that yields these results in a specific way; therefore generalisation of the results is not advisable for the similar processes.

Conclusions

In the present study, life cycle assessment was used to evaluate incineration process in EAF steelmaking with waste syringes partially relacing metallurgical coke and compared with the conventional incineration process in EAF with pure metallurgical coke only. A thermodynamic based analysis was carried out for both scenarios with particular focus on calculating environmental impact. The results indicate that all environmental impacts are significantly reduced by replacing part of coke with waste syringes in electric arc furnaces. The emissions from the interaction of slag with metallurgical coke showed a higher rate of growth when compared to those from the waste syringes-coke blends. This could be attributed to the hydrogen available in the polymeric materials in waste syringes. The CO_2 and CO content in the off-gas decreased by 71 and 76% respectively with replacing 75% of coke with waste syringes. Generally, the usage of end-of-life waste syringe materials for carbon/slag and carbon/metal reactions in the EAF steelmaking process is better than conventional incineration of coke in EAF. Under long-term time frame, emission to air including inorganic emission, particle and heavy metals to air dominate the total environmental burden in the conventional incineration in EAF steelmaking. The influence of energy usage cut was also assessed by applying sensitivity analysis and results indicate slight impact of energy consumption reduction on overall environmental indices for both scenarios. From a life cycle perspective, the carbon dioxide, carbon monoxide and methane (biotic) emission in CI scenario have the highest contribution to climate change. Although the insufficient knowledge of mechanisms of formation of some pollutant such as dioxin emerges from polymer combustion, could introduce some uncertainties in overall environmental assessment of the WSI scenarios. Hence, it is important to extend the evaluation of other technologies, in order to introduce a framework for treatment of whole range of medical waste in Australia.

References

1. C. Jiang et al., Application of best available technologies on medical wastes disposal/treatment in China (with case study). Procedia Environ. Sci. **16**, 257–265 (2012)
2. Health care without harm treatment technologies: a resource for hospital administrators, facility managers, health care professionals, environmental advocates, and community members. Washington, DC (2001)
3. J.R. Dankwah et al., Reduction of FeO in EAF steelmaking slag by metallurgical coke and waste plastics blends. ISIJ Int. **51**(3), 498–507 (2011)
4. V. Sahajwalla et al., Recycling plastics as a resource for electric arc furnace (EAF) steelmaking: combustion and structural transformations of metallurgical coke and plastic blends. Energy Fuels **24**(1), 379–391 (2009)
5. M. Asanuma et al., Development of waste plastics injection process in blast furnace. ISIJ Int. **40**(3), 244–251 (2000)
6. V. Sahajwalla et al., Recycling waste plastics in EAF steelmaking: carbon/slag interactions of HDPE—coke blends. Steel Res. Int. **80**(8), 535–543 (2009)
7. S. Gupta, V. Sahajwalla, J. Wood, Simultaneous combustion of waste plastics with coal for pulverized coal injection application. Energy Fuels **20**(6), 2557–2563 (2006)
8. V. Sahajwalla et al., Environmentally sustainable EAF steelmaking through introduction of recycled plastics and tires: laboratory and plant studies. Iron Steel Technol. **6**(4), 43–50 (2009)
9. V. Sahajwalla et al., Recycling end-of-life polymers in an electric arc furnace steelmaking process: fundamentals of polymer reactions with slag and metal. Energy Fuels **26**(1), 58–66 (2011)
10. M. Zaharia et al., Recycling of rubber tires in electric arc furnace steelmaking: simultaneous combustion of metallurgical coke and rubber tyres blends. Energy Fuels **23**(5), 2467–2474 (2009)
11. O. Eriksson et al., Municipal solid waste management from a systems perspective. J. Cleaner Prod. **13**(3), 241–252 (2005)
12. W.P. Clarke, Cost-benefit analysis of introducing technology to rapidly degrade municipal solid waste. Waste Manage. Res. **18**(6), 510–524 (2000)
13. W. Zhao et al., Comparative life cycle assessments of incineration and non-incineration treatments for medical waste. Int. J. Life Cycle Assess. **14**(2), 114–121 (2009)
14. M. Eckelman et al., Comparative life cycle assessment of disposable and reusable laryngeal mask airways. Anesth. Analg. **114**(5), 1067–1072 (2012)

15. M. Overcash, A comparison of reusable and disposable perioperative textiles: sustainability state-of-the-art 2012. Anesth. Analg. **114**(5), 1055–1066 (2012)
16. M.S. Hossain et al., Clinical solid waste management practices and its impact on human health and environment–a review. Waste Manage. **31**(4), 754–766 (2011)
17. PE, GaBi software and database. Available at: http://www.gabi-software.com (2011)
18. M. Goedkoop et al., ReCiPe 2008. A life cycle impact assessment method which comprises harmonised category indicators at the midpoint and the endpoint level 1 (2009)
19. Society of Environmental Toxicology and Chemistry, *Guidelines for Life-Cycle Assessment: A Code of Practice*, ed. by F. Consoli. From the SETAC workshop held at Sesimbra, Portugal, 31 Mar–3 Apr 1993 (Setac, 1993)
20. K. Christiansen et al., *Nordic Guidelines on Life-Cycle Assessment* (Nordic Council of Ministers, 1995)
21. A. Astrup Jensen, Life cycle assessment (LCA): a guide to approaches, experiences and information sources. No. 6. European Communities (1998)
22. G. Finnveden, Methodological aspects of life cycle assessment of integrated solid waste management systems. Resour. Conserv. Recycl. **26**(3), 173–187 (1999)
23. R. Clift, A. Doig, G. Finnveden, The application of life cycle assessment to integrated solid waste management: part 1—methodology. Process Saf. Environ. Prot. **78**(4), 279–287 (2000)
24. F. Cherubini, S. Bargigli, S. Ulgiati, Life cycle assessment of urban waste management: energy performances and environmental impacts. the case of Rome, Italy. Waste Manage. **28**(12), 2552–2564 (2008)
25. J. Winkler, B. Bilitewski, Comparative evaluation of life cycle assessment models for solid waste management. Waste Manage. **27**(8), 1021–1031 (2007)
26. A. Dash, S. Kumar, R.K. Singh, Thermolysis of medical waste (Waste Syringe) to liquid fuel using semi batch reactor. Waste Biomass Valorization **6**(4), 507–514 (2015)

Part V
Energy Technologies: Poster Session

Effect of Granularity on Pretreatment of Coke With Microwave Irradiation

Qing-hai Pang, Zhi-jun He, Jun-hong Zhang, Wen-long Zhan, Teng-fei Song and Zhe Ning

Abstract In order to investigate the influence of granularity on the gasification property of coke after microwave irradiation, comparison of ordering of carbon structure and internal microstructure of granular coke before and after microwave pretreatment was carried out. It is shown in X-ray diffraction results that the diffraction intensity of the coke sample significantly increased after 40 s irradiation under 264 W microwave power, which indicated the decrease in reactivity of coke.

Keywords Granularity · Microwave radiation · Gasification reactivity

Introduction

The sharp decrease in price of steel in recent years has led to a serious degradation in quality of ironmaking raw materials in the consideration of cost. The metallurgical properties of raw material were seriously reduced due to the inferior grade of ore, which caused the unstable running state of blast furnace (BF). Therefore, improving the metallurgical performance of raw materials based on practical raw material condition has become the primary task for ironmaking researchers [1, 2].

The reactions in BF mainly take place between the reducing gas and fix bed of iron-bearing materials. Thus, the permeability of gas and liquid in the burden layer is essential to BF melting production. It is a critical kinetic condition in indirect reduction process and dominates the utilization ratio of fossil fuels. Metallurgical coke serve as a source of reducing agent and heat as well as the supporting spine of

Q. Pang (✉) · Z. He · J. Zhang · W. Zhan · Z. Ning
School of Materials and Metallurgy, University of Science and Technology Liaoning, Anshan 114051, China
e-mail: edikitty@126.com

T. Song
School of Ecological and Metallurgy, University of Science and Technology Beijing, Anshan 100083, China

L. Zhang et al. (eds.), *Energy Technology 2017*,
The Minerals, Metals & Materials Series, DOI 10.1007/978-3-319-52192-3_15

burden in BF. Consequently, the physical and chemical properties of coke are of great importance to the smooth operation of iron production [3, 4].

According to our previous investigations [5–7], microwave irradiation may increase the ordering of coke carbon structure and the chemical stability of coke. Hence, the activity of carbon in coke in the gasification with CO_2 can be reduced after microwave treatment. As a result, the severe physical degradation in coke structure can be eased, and the strength of burden column can be increased.

Experimental

Samples

A metallurgical coke which was utilized in practical BF ironmaking process was selected as the sample of this investigation, the composition of which was shown in Table 1.

Microwave Irradiation

A coke sample was crushed and sieved into different granularity zones. The granularity zones chosen for this study is shown in Table 2. Pulverized coke under 0.074 mm was utilized to determine the proper condition for microwave treatment, such as microwave power and irradiation time.

The operating frequency of microwave oven is 2450 ± 25 MHz. Microwave powerd and irradiation time can be adjusted from the front panel. The structure of industrial microwave oven was shown in Fig. 1.

X-ray Diffraction

X-ray diffraction was widely utilized in phase structure determination of chemical compound in materials, by which the combination pattern of chemical bond can be revealed Meanwhile, with the help of this technology, the important information on

Table 1 Proximate analysis of metallurgical coke

Proximate analysis, wt%			Ultimate analysis, wt%				
FC_d	A_d	V_d	C_d	H_d	N_d	O_d	S_d
84.88	13.45	1.67	85.26	0.38	0.41	0.33	0.62

FC fixed carbon; *V* volatile; *A* ash; *M* moisture; *d* dried; *C* carbon; *H* hydrogen; *N* nitrogen; *O* oxygen; *S* sulfur

Table 2 Proximate analysis of metallurgical coke

1	2	3	4	5	6
~0.074 mm	5–6.5 mm	8–10 mm	12–15 mm	18–20 mm	25 mm

Fig. 1 Schematic diagram of industrial microwave oven

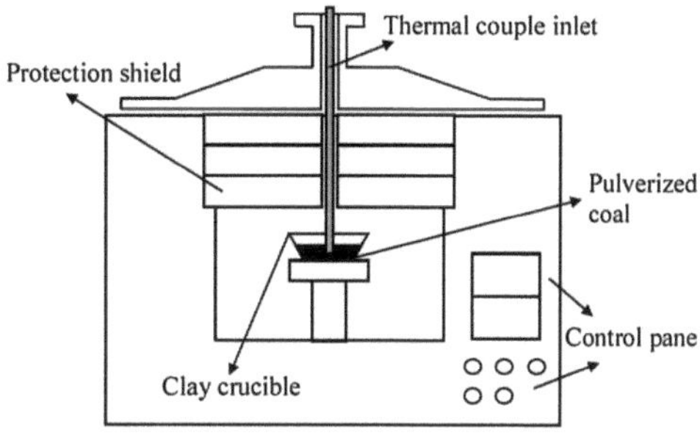

physical and chemical properties can also be obtained. In order to explore the mechanism of variation in combustion properties after microwave treatment, Raman spectroscopy was introduced to investigate the structure of carbon matrix in PC before and after microwave irradiation.

Results and Discussion

Determination of Microwave Pretreatment Condition

CP under 0.074 mm was chosen as the sample to be irradiated under different microwave powers and treating time, and subsequently the carbon structure of irradiated CP samples were analyzed with X-ray diffraction. Consequently, the optimal condition for the carbon structure ordering of CP sample can be determined. X-ray diffraction spectra of different treated CP samples are shown in Fig. 2, and detailed parameters of peaks are exhibited in Table 3.

Fig. 2 X-ray diffraction spectra of coke samples: **a** Raw coke; **b** 132 W, 40 s; **c** 132 W, 80 s; **d** 264 W, 40 s; **e** 264 W, 40 s

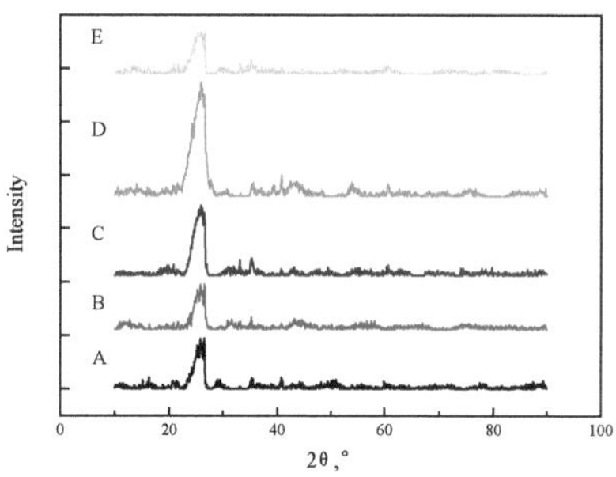

Table 3 Parameters of diffraction peaks of CP samples

Sample	Raw coke	132 W, 40 s	132 W, 80 s	264 W, 40 s	264 W, 80 s
Intensity, counts	93	84	132	212	79
Width, °	3.93	3.84	4.38	10.47	4.57

Spectra in Fig. 2 show that microwave irradiation has a great influence on the diffraction intensity of CP samples. The diffraction intensity decreased slightly with the 40 s irradiation under 132 W microwave power. When irradiation time extends to 80 s under the same microwave power, there is a significant increase in intensity of diffraction peak. Besides, no significant variation in peak width of samples was noticed with the treatment under 132 W microwave power. Under 264 W microwave conditions, 40 s irradiation caused a sharp increase in the intensity of diffraction peak, and simultaneously the same phenomenon also occurred in the peak width of C, which indicates the dramatic ordering of carbon structure after 40 treatment under 264 W microwave power. However, both the diffraction intensity and peak width decreased to the original level of raw CP sample with the extension of irradiation time to 80 s, which indicates that long time microwave irradiation is not useful to reduce reactivity of coke.

Influence of CP Granularity on Microwave Treatment

According to the variation in carbon structure of 0.074 mm CP sample, the influence of CP granularity on microwave treatment was tested with 40 s under 264 W microwave power. The X-ray diffraction spectra of CP samples with different granularity arc shown in Fig. 3.

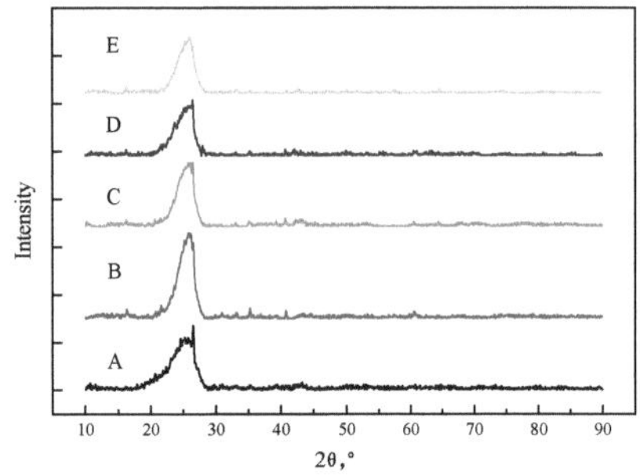

Fig. 3 Diffraction spectra of CP samples: **a** Raw coke; **b** 5 mm; **c** 10 mm; **d** 15 mm; **e** 20 mm

Calculation of Microstructure Parameters

PeakFit, a curve processing software, was utilized to analyze the variation in CP microstructure parameters. Consequently, the diffraction peak can be separated into two symmetrical peaks, by which the microstructure parameters can be obtained through calculation. The fitting and separated peaks are shown in Fig. 4.

The lateral size (L_a) and stacking height (L_c) of the crystallite were calculated by the application of Scherrer equations, whose expression can be deduced as follow:

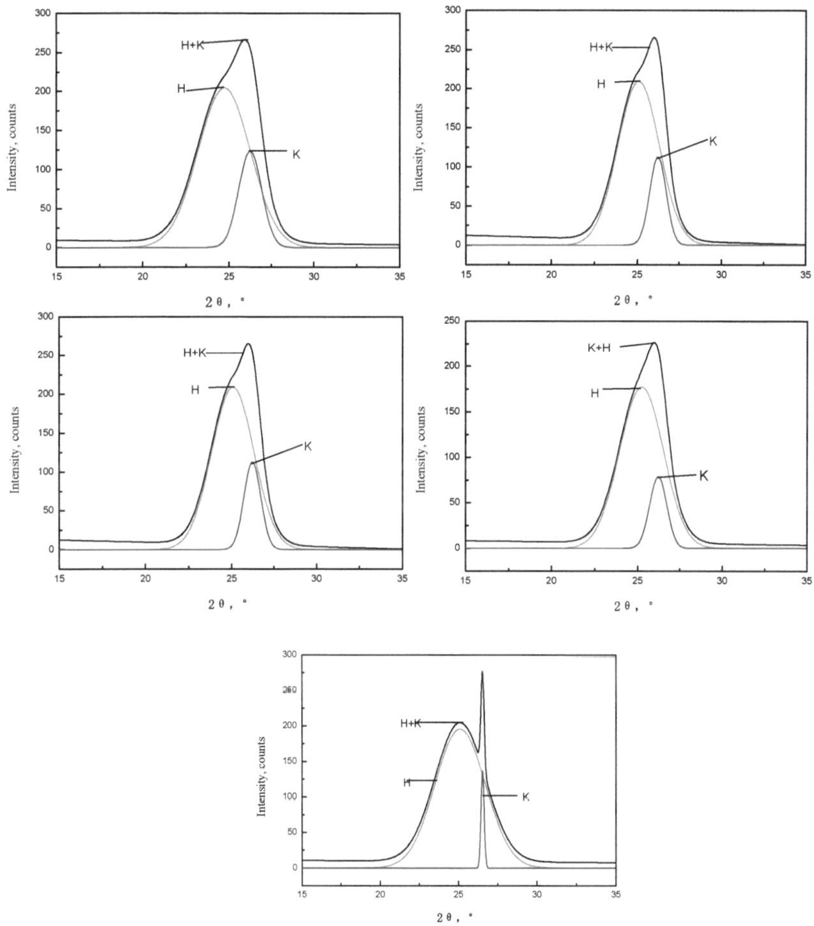

Fig. 4 Fitting curves of diffraction peaks

(1) Calculation of F and G band structural parameters

$$d_{002,F} = \frac{\lambda}{2\sin\theta_{002,F}} \qquad d_{002,D} = \frac{\lambda}{2\sin\theta_{002,D}} \tag{1}$$

$$L_{c,F} = \frac{k \cdot \lambda}{\beta_{002,F} \cdot \cos\theta_{002,F}} \qquad L_{c,D} = \frac{k \cdot \lambda}{\beta_{002,D} \cdot \cos\theta_{002,D}} \tag{2}$$

where λ is the wave length of X ray, its value is 1.5418 Å; k is the correction factor, whose value is 0.94; $\theta_{002,D}$ and $\beta_{002,F}$ respectively represent the diffraction angle and peak width at half height of band; $\theta_{002,D}$ and $\beta_{002,D}$ respectively represent the diffraction angle and peak width at half height of D peak; $d_{002,F}$ and $L_{c,F}$ respectively represent the crystalline interplanar spacing and Stacking height of F peak; d_D and $L_{c,D}$ respectively represent the crystalline interplanar spacing and Stacking height of D peak;

(2) Calculation of F and G band areas

$$X_F = \frac{S_F}{S_F + S_D} \qquad X_D = \frac{S_D}{S_D + S_F} \tag{3}$$

where S_F and S_D respectively represent the band areas of F and D band; X_F and X_D respectively represent the area proportion of F and D band.

(3) Structural parameter calculation of CP samples

$$d_{002,a} = X_F \cdot d_F + X_D \cdot d_D \tag{4}$$

$$L_{c,a} = X_F \cdot L_{c,F} + X_D \cdot L_{c,D} \tag{5}$$

where d_{002} and L_c respectively represent the average crystalline interplanar spacing and stacking height of CP samples. Microstructure parameters of different granularity CP samples were obtained through above calculation and shown in Table 4.

Table 4 Microstructure parameters of CP samples

Granularity (mm)	$d_{002,H}$, Å	$L_{c,H}$, Å	$d_{002,K}$, Å	$L_{c,K}$, Å	X_H, %	X_K, %	$d_{002,a}$, Å	$L_{c,a}$, Å
0.074	3.599	0.412	3.393	0.938	79.0	21.0	3.524	0.522
5.000	3.550	0.579	3.397	1.280	82.3	17.7	3.523	0.703
10.000	3.585	0.491	3.403	1.225	77.7	22.3	3.544	0.655
15.000	3.528	0.490	3.398	1.244	85.0	15.0	3.509	0.603
20.000	3.545	0.397	3.358	6.274	95.8	4.2	3.537	0.644

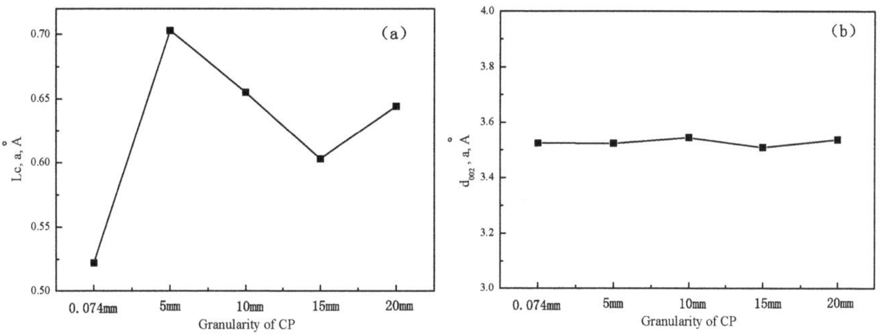

Fig. 5 Variation of $L_{c,a}$ (**a**) and $d_{002,a}$ (**b**) with CP granularity

It can be seen from Fig. 5 that granularity has led to different variations in microstructure parameter of CP sample after microwave irradiation. The value of $L_{c,a}$ dramatically increased when granularity of CP increased from 0.074 mm to 5 mm. With the further increase in CP granularity, a significant decrease in value of $L_{c,a}$ was noticed. However, no obvious variation in $d_{002,a}$ value was observed after microwave irradiation, namely microwave treatment has no effect on the crystalline interplanar spacing of coke. Consequently, the value of $L_{c,a}/d_{002,a}$ was dominated by the variation of $L_{c,a}$, which means the extent carbon structure ordering is more significant with the CP granularity of 5 mm.

Conclusion

Based on the results of this investigation, the following conclusions can be obtained:

Variation in carbon structure of CP samples indicates the ordering behavior of CP under microwave fields. Calculation results of structural parameters of CP samples show that the layer stack height of CP is dramatically increased after microwave irradiation, which implies the decrease in reactivity of CP in gasification reactions.

Different granularity of CP samples may lead to various degrees in carbon structure ordering of coke in microwave fields. The optimum condition for microwave treatment of coke is 5 mm granularity and 40 s irradiation under 264 W microwave power.

Acknowledgements This work is financially supported by item sponsored by National Science Foundation of China (NO. 51474124, NO. 51504131). Simultaneously. Thank you for the assistance of Dr. Song Teng-fei from the University of Science and Technology, Beijing.

References

1. N. Masaaki, N. Seiji, K. Kenji, Development of production and utilization technology of coke with high strength and high reactivity. ISIJ Int. **96**, 17 (2010)
2. S. Nomura, T.G. Callcott, Maximum rates of pulverized coal injection in ironmaking blast furnaces. ISIJ Int. **51**, 1033 (2011)
3. H.M. Wang, C.L. Wang, X.H. Zhou, Y. Wang, The relation between metallurgical properties of coke and regular B.F. performance, Laigang. Sci. Technol. **139**, 64 (2009)
4. S. Pusz, M. Krzesińska, L. Smędowski, J. Majewska, Changes in a coke structure due to reaction with carbon dioxide. Int. J. Coal Geol. **81**, 287 (2009)
5. Q. Pang, J. Zhang, R. Mao et al., Mechanism of effect of microwave modification on pulverized coal combustion properties. J. Iron Steel Res. Int. **21**, 312–320 (2014)
6. Q. Pang, J. Zhang, R. Mao et al., Kinetic analysis on combustion characteristic of microwave irradiated pulverized coal. J. Iron Steel Res. **25**, 11–18 (2013)
7. Q. Pang, J. Zhang, Y. Jin et al., Effect of microwave modification on pulverized coal combustion properties. Metalurgia Int. **17**, 5–10 (2012)

Effect of Microwave and Ultrasonic Coupling Treatment on Granularity and Microstructure of Pulverized Coal

Zhi-jun He, Ji-hui Liu, Qing-hai Pang, Jun-hong Zhang, Wen-long Zhan and Zhe Ning

Abstract The influence of microwave irradiation and ultrasonic vibration on granularity of 0.074 mm pulverized coal was investigated in this research. It is indicated in the results that the granularity of coal decreased after microwave treatment, which led to the increased in specific area of pulverized coal in combustion reaction and consequently optimized the kinetic condition of combustion reaction. Higher ultrasonic power may lead to better performance of PC in combustion. The optimum condition for ultrasonic treatment is 10 min under 350 W microwave power.

Keywords Microwave radiation · Ultrasonic · Granularity

Introduction

The worldwide economic crisis has caused a severe influence on the demand of steel products, which has led to a sharp decrease in the profit of iron and steel enterprises. The production of crude steel is mainly based in the blast furnace (BF) and converter process in China, which is also called long process. Besides, blast furnace is the biggest energy consumer in entire steel production process. How to control the cost of the BF melting ironmaking process according to current steel market has become the primary task for all ironmaking researchers [1].

Pulverized coal injection (PCI), an effective way to replace metallurgical coke with pulverized coal to perform as reduction agent and heat source as well as the carbonizing agent, may significantly reduce the consumption of coke and namely the cost of the BF ironmaking process. However, the combustion of pulverized coal (PC) in the raceway of BF has become the limitation of PCI. Incomplete combustion of PC may even deteriorate the permeability of liquid and gas created by

Z. He · J. Liu · Q. Pang (✉) · J. Zhang · W. Zhan · Z. Ning
School of Materials and Metallurgy, University of Science and Technology Liaoning, Anshan 114051, China
e-mail: edikitty@126.com

L. Zhang et al. (eds.), *Energy Technology 2017*,
The Minerals, Metals & Materials Series, DOI 10.1007/978-3-319-52192-3_16

metallurgical coke, which may influence the smooth operation of BF. Therefore, the efficient combustion of PC in BF is of great importance [2].

Our previous investigations show that microwave irradiation has a positive effect on the combustion of PC, which is attributed to the increase of high activity radicals produced by microwave irradiation [3–5]. However, the obscure variations in granularity of PC have proven to be dynamically ineffective to the combustion of PC though cracks were observed in microstructure of PC after irradiation. Hence, ultrasonic was introduced to enhance the pulverization of microwave irradiated PC, by which the specific surface area can be increased and kinetically optimized for the PC combustion.

Experimental

Samples

Bituminite for practical PCI in certain iron and steel enterprise were selected as the sample of this investigation, the composition of which is shown in Table 1.

Microwave and Ultrasonic Coupling Treatment

Coals were pulverized and sieved under 0.074 mm as specimen. PC Sample of 100 g was loaded in a clay crucible and placed into the chamber of the microwave oven, the operating frequency of which was 2450 ± 25 MHz. Microwave power and irradiation time could be adjusted from the front panel. The structure of the industrial microwave oven is shown in Fig. 1.

Ultrasonic Treatment

In order to study the effect of ultrasonic on physical properties of PC, PC samples were loaded into a beaker and placed into an ultrasonic water bath environment for 10 min treatment under 350 W ultrasonic powers.

Table 1 Proximate analysis of pulverized coals

Proximate analysis, wt%			Ultimate analysis, wt%				
FC_d	A_d	V_d	C_d	H_d	N_d	O_d	S_d
57.80	38.17	4.03	61.19	3.82	0.64	0.27	19.53

FC fixed carbon; *V* volatile; *A* ash; *M* moisture; *d* dried; *C* carbon; *H* hydrogen; *N* nitrogen; *O* oxygen; *S* sulfur

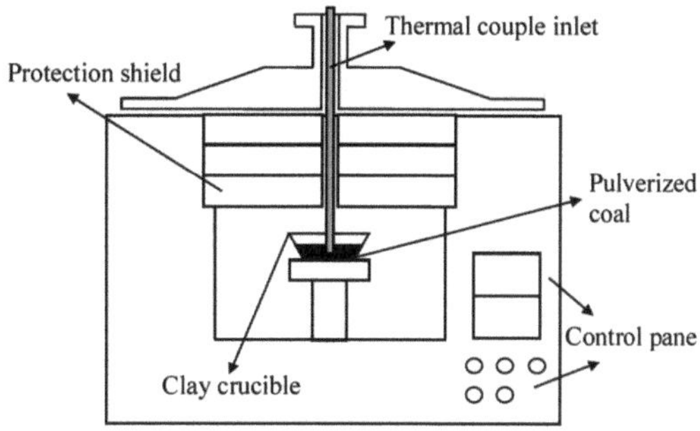

Fig. 1 Schematic diagram of industrial microwave oven

Results and Discussion

Combustion Test of Microwave Treated PC

Our previous studies have proven the positive effect of microwave irradiation on combustion properties of PC. Therefore, the effect of microwave treatment on PC combustion will not be discussed this research. Microwave-irradiated PC will be regarded as the raw material for ultrasonic treatment. The combustion of PC was tested, and burnout was calculated by ash balance method. A diagram of the device for PC combustion test is shown in Fig. 2, and the burnout rates of different PC are displayed in Fig. 3.

It can be seen that the burnout rate of PC increased after ultrasonic treatment while higher ultrasonic power may lead to superior burnout rate. Simultaneously, the burnout rate increased sharply with the extension time of ultrasonic treatment initially and became stable after 10 min. Therefore, the optimum condition for ultrasonic treatment of PC is 10 min under 350 W.

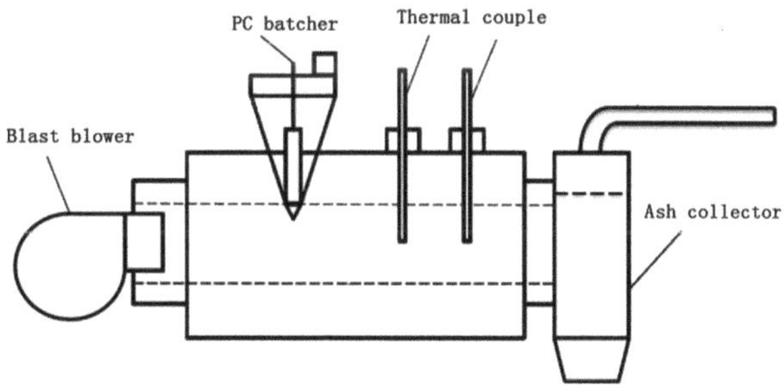

Fig. 2 Diagram of PC combustion device

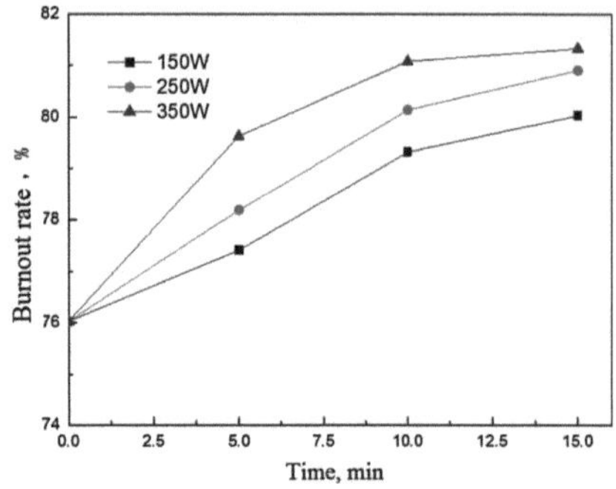

Fig. 3 Burnout rate of microwave and ultrasonic coupling treated PC samples

Granularity Analysis

Variations in PC particle size distribution before and after microwave and ultrasonic treatment were analyzed in order to investigate the effect of treatment on the physical structure of PC. The distribution of PC particles is shown in Fig. 4.

It is indicated in Fig. 4 that microwave irradiation led to the slight pulverization of PC, by which the particles around 10 μm decreased and the minimum particle size of PC was decreased. However, the content of PC pulverization was too slight to ameliorate the kinetic condition of PC combustion. With further treatment of

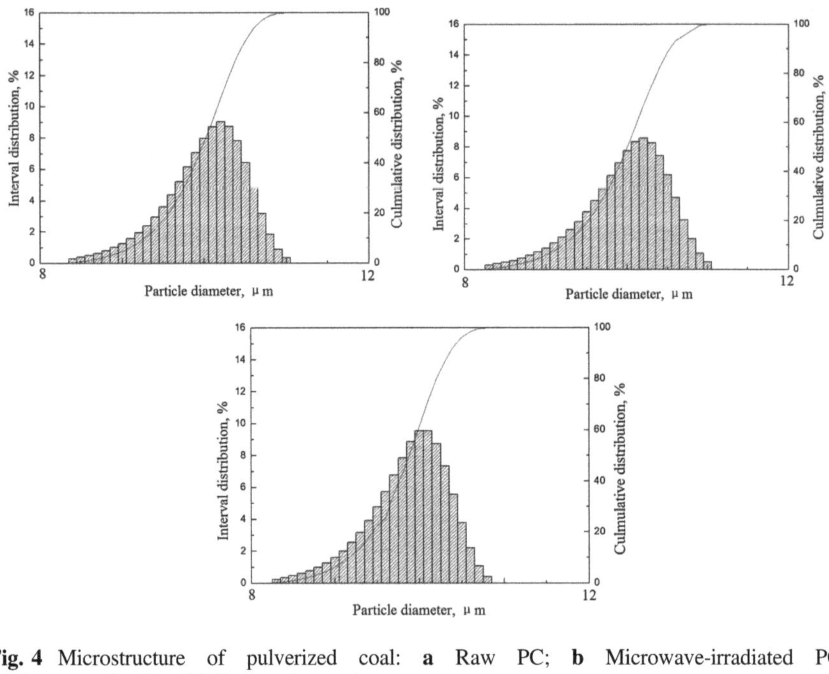

Fig. 4 Microstructure of pulverized coal: **a** Raw PC; **b** Microwave-irradiated PC; **c** Microwave-irradiated PC after ultrasonic treatment

ultrasonic on microwave-treated PC, a significant pulverization effect was noticed. Though no variation occured in the minimum particle size of PC, the maximum particle size of PC was decrease while the particle proportion around 10 μm was substantially increased. Therefore, microwave and ultrasonic coupling treatment may effectively pulverize the PC particles and optimize the combustion kinetics of PC.

Microstructure Observation

The microstructure of PC samples was observed in order to analyze the variation of microwave and ultrasonic coupling treatment on particle shape and distribution of PC. The SEM photos of samples are shown in Fig. 5.

Microstructure comparison among different treated PC samples showed that no difference between raw and microwave irradiated PC can be observed. The size and the distribution of particles were very similar after microwave irradiation. However, the proportion of smaller particles seemed to be increased after ultrasonic treatment.

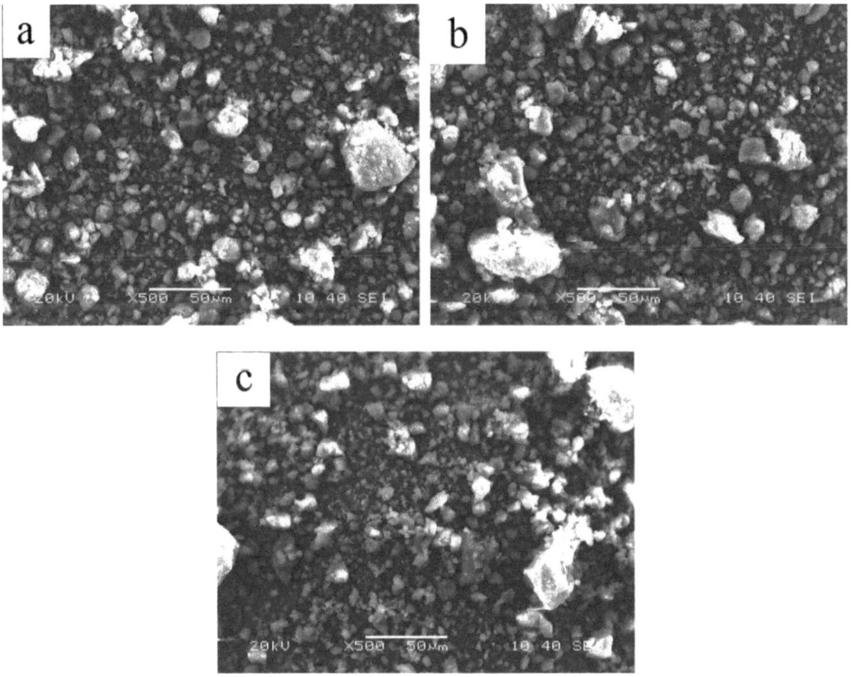

Fig. 5 Microstructure of pulverized coal: **a** Raw PC; **b** Microwave-irradiated PC; **c** Microwave-irradiated PC after ultrasonic treatment

Conclusion

Based on the results of this investigation, the following conclusions can be obtained:

Higher ultrasonic power may lead to better performance of PC in combustion. The combustion property of microwave-treated PC increased after ultrasonic treatment and the increment rise with treating time in the initial 10 min. The burnout rate becomes stable with the further treatment of ultrasonic, and the optimum condition for ultrasonic treatment is 10 min under 350 W power.

Ultrasonic treatment is effective for the pulverization of microwave-irradiated PC and consequently decreases the average granularity of PC, by which the kinetic condition of PC combustion can be optimized.

Acknowledgements This work is financially supported by item sponsored by National Science Foundation of China (NO. 51474124, NO. 51504131).

References

1. Q. Pang, J. Zhang, C. Qi et al., K_2CO_3 catalysis on the reactivity of top charged coke and stamp charged coke. Int. J. Miner. Metall. Mater. **20**, 17–27 (2013)
2. J. Zhang, G. Wang, J. Shao, Y. Chen, T. Yang, Pulverized coal combustion of nitrogen free blast furnace. J. Iron. Steel Res. Int. **20**, 1–8 (2013)
3. Q. Pang, J. Zhang, R. Mao et al., Mechanism of effect of microwave modification on pulverized coal combustion properties. J. Iron. Steel Res. Int. **21**, 312–320 (2014)
4. Q. Pang, J. Zhang, R. Mao et al., Kinetic analysis on combustion characteristic of microwave irradiated pulverized coal. J. Iron Steel Res. **25**, 11–18 (2013)
5. Q. Pang, J. Zhang, Y. Jin et al., Effect of microwave modification on pulverized coal combustion properties. Metal. Int. **17**, 5–10 (2012)

Influence of Sodium on Coke Microstructure in Different Reaction Atmosphere

Zhijun He, Wenlong Zhan, Junhong Zhang, Qinghai Pang, Sen Zhang and Chen Tian

Abstract Alkali metal is one of the key critical factors that determine the coke degradation. Previous studies on the influence of alkali metal on coke structure were mostly focus on the catalysis of alkali carbonates for coke gasification reaction. Besides, the difference of alkali metal and coke gasification effects on coke degradation was still not clear. Highly reactive coke developed by Nippon Steel was proved to increase the reaction efficiency and to decrease CO_2 emission effectively. Therefore, conventional coke and highly reactive coke adsorbed 5% sodium were studied in this paper. Coke degradation tests were undertaken in atmosphere of N_2 and CO_2. SEM, EDS and XRD were employed to analyze samples. Results showed that the degradation of conventional coke strength by sodium was stronger than highly reactive coke in the atmosphere of N_2. The effect of CaO catalyst on the coke reactivity has been interpreted as a change in highly reactive coke microstructure and strength.

Keywords Alkali metal · Coke · Degradation · Microstructure · CO_2

Introduction

Coke is the main raw material of blast furnace (BF), performing many roles, e.g., a thermal role to provide heat energy, a chemical role to act as a reducing agent, a carburization role to provide soluble carbon in hot metal, and a mechanical role to maintain the permeability for upward flowing gases [1, 2]. Increasing environmental concerns associated with coke making are influencing the economy of the BF route. Pulverized coal injection (PCI) technology has been developed to replace coke. At lower coke rates, mechanical properties of coke such as high-temperature strength become increasingly important.

Z. He · W. Zhan (✉) · J. Zhang · Q. Pang · S. Zhang · C. Tian
University of Science and Technology Liaoning, Anshan 114051, China
e-mail: zhanwenlong288@163.com

L. Zhang et al. (eds.), *Energy Technology 2017*,
The Minerals, Metals & Materials Series, DOI 10.1007/978-3-319-52192-3_17

Under the driven of cost orientation in the current blast furnace process, coke is believed to display an accelerated degree of degradation. Many factors influence coke degradation, e.g., the mechanical stress, the solution loss reaction, the alkaline attack, the thermal gradient, and the shock impact by high-speed blast [3]. Extensive researches have been carried out on the coke quality exposed to the alkalis. It was confirmed that coke reactivity could be improved by the catalytic effect of alkalis [4–8]. Zhang et al. [9] studied the changes of coke microstructure and strength with alkalis adsorption. Compared with the original coke, the main reason causing the cracks and multidimensional expansion of microcrystals was the kalsilite formed by alkalis and graphite crystal. Previous studies were mainly focused on the influence of alkalis on coke properties in N_2 atmosphere. As the solution loss reaction by CO_2 is another factor influencing coke degradation, the effect of alkalis on the coke characteristics under CO_2 atmosphere is a subject of key interest and further research. Few results have been reported regarding the interaction of alkalis with coke under CO_2 atmosphere. While coke with high reactivity-high strength developed by Nomura et al. is promising as a means of improving the reaction efficiency [10–12], the influence of alkalis on high reactivity-high strength coke is worth further studying.

In this article, two cokes, conventional coke of BF and high reactivity coke, are investigated for comparison. The detailed mechanisms about the influence of sodium on the coke properties under different reaction atmosphere are established.

Experimental

Two types of cokes with different reactivity were used. The high reactivity coke was prepared by the carbonization of caking coals/5 mass% CaO compound mixtures. Proximate analysis, CRI and CSR of the two cokes are shown in Table 1.

Previous research usually soaked cokes in solutions with different concentration of alkalis carbonates for a range of times, and then the soaked cokes were dried and tested to analyze the variation of properties. This method can only investigate the influence of alkalis as carbonates on coke properties. As alkalis are primarily present as a gaseous phase in the high temperature zone of a BF, the influence of alkali vapors on the coke properties instead of alkali carbonates could be more reasonable. A series of adsorption-sodium experiments were carried out. A 100 g

Table 1 Proximate analysis, CRI and CSR of coke samples

Items	Proximate analysis				Hot strength	
	Volatile matter	Ash	Moisture	Fixed carbon	CRI	CSR
Conventional coke	0.84	11.93	2.83	84.40	24.90	61.90
High reactivity coke	0.60	10.75	3.21	85.44	33.60	58.81

sample of coke with a particle size of 19–21 mm was placed atop mixtures of activated carbon and Na_2CO_3 in a corundum crucible. A porous corundum spacer separated coke samples and Na_2CO_3/carbon powder mixtures. The covered crucible was heated in a muffle furnace to 1100 °C at a heating rate of 10 °C/min. N_2 and CO_2 were pumped into the furnace for 2 h, respectively. And then the furnace was cooled down slowly to room temperature.

Results and Discussion

The Influence of Sodium on Coke Properties

Figures 1, 2, 3 and 4 shows the hot performance results of coke in the atmosphere of N_2 and CO_2, respectively.

The coke reactivity increased in different degree after adsorption-sodium in the atmosphere of N_2 as shown in Fig. 1. Besides, the catalytic effect on conventional coke was better than the high reactivity coke in the case of same adsorption rate. As mentioned before, CaO was added in caking coal to produce highly reactivity coke.

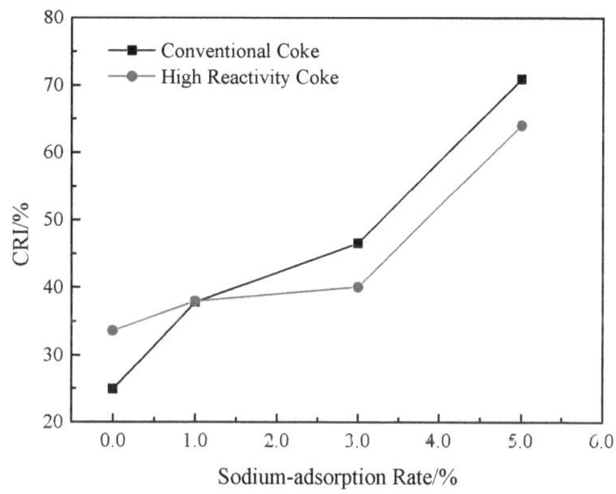

Fig. 1 Relationship between CRI and sodium-adsorption rate in the atmosphere of N_2

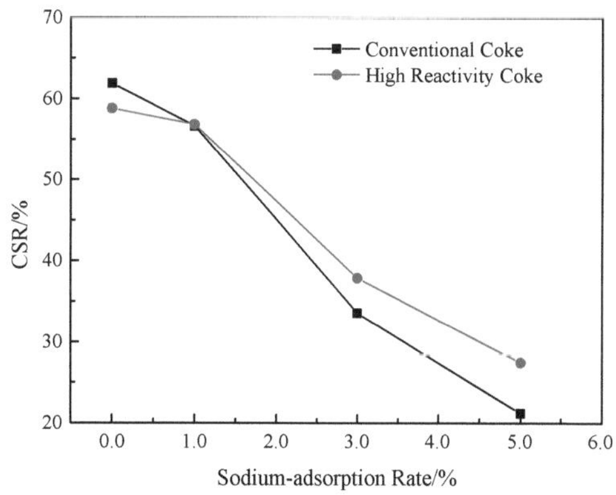

Fig. 2 Relationship between CSR and sodium-adsorption rate in the atmosphere of N_2

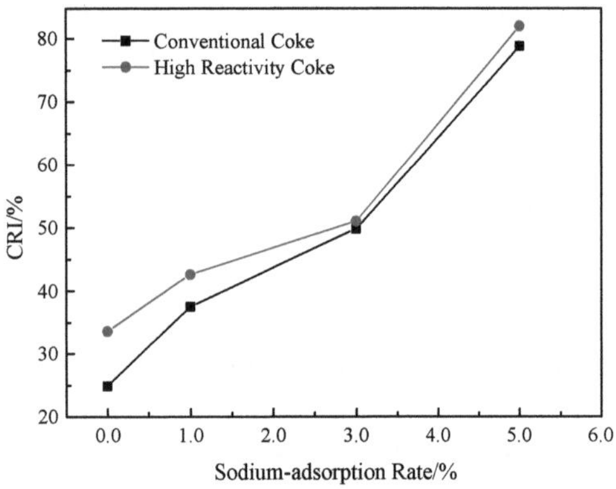

Fig. 3 Relationship between CRI and sodium-adsorption rate in the atmosphere of CO_2

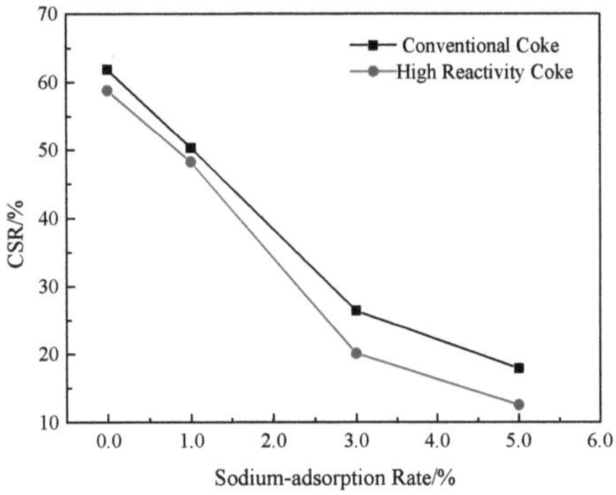

Fig. 4 Relationship between CSR and sodium-adsorption rate in the atmosphere of CO_2

CaO penetrated in the coke matrix through pores and was absorbed on the pore walls, which prevented the alkalis vapors absorption. Figure 2 shows that the CSR of high reactivity coke was higher than conventional coke. It was consistent with the former result.

Figures 3 and 4 shows that the CRI of two cokes was higher in the atmosphere of CO_2 than that of N_2, and the CSR was corresponding lower.

Influence of Sodium on Coke Microstructure

SEM of Two Cokes After Sodium-Adsorption in the Atmosphere of N_2

Figure 5 shows the SEM and EDS results of conventional coke after sodium-adsorption.

The mass weight of sodium reached to 27.53% in coke texture, which was quite higher than the other mineral matters. It was indicated that sodium vapor were absorbed on the coke surface and would not react with coke minerals in the

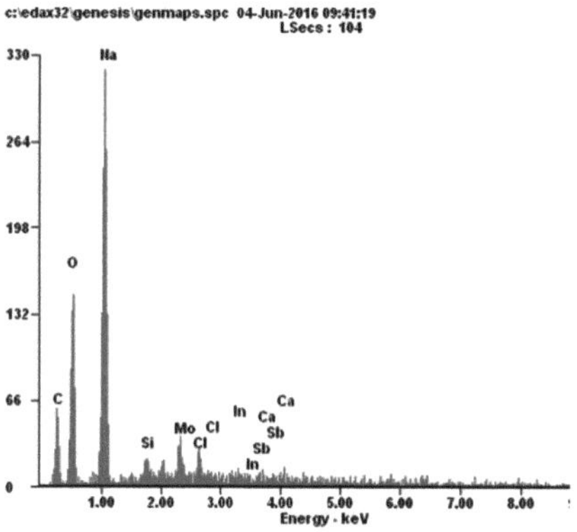

Element	*Wt%*	*At%*
CK	25.45	40.65
OK	24.81	29.75
NaK	27.53	22.97
SiK	02.11	01.44
MoL	07.40	01.48
ClK	02.91	01.57
InL	05.42	00.91
SbL	02.68	00.42
CaK	01.68	00.80
Matrix	Correction	ZAF

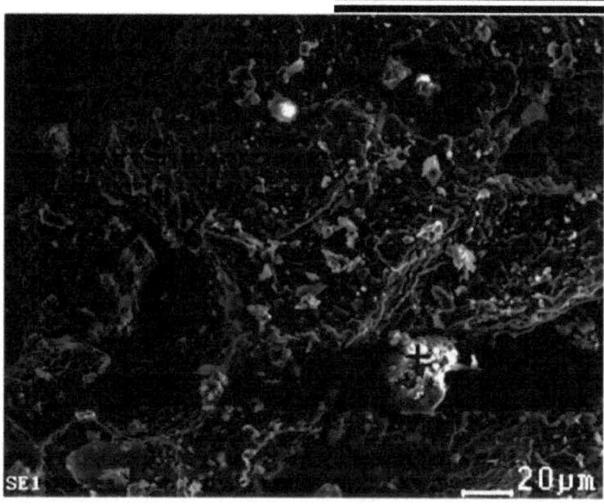

图 5 N_2气氛下普通焦炭吸附 Na 后的扫描电镜图像及能谱分析

Fig. 5 SEM and EDS of conventional coke after sodium-adsorption in the atmosphere of N_2

atmosphere. However, it would be existed as oxides or carbonates finally in the sampling process.

Figure 6 shows the SEM and EDS results of high reactivity coke after sodium-adsorption.

The mass weight of sodium was only 4.67% in coke texture, while the component of Ca was much higher. CaO was used as the additives to produce the high reactivity coke. Previous researches on reactions of coke carbons with alkalis suggest that coke degradation is essentially the result of intercalation of alkalis into the carbon structure. For the high reactivity coke, however, the molecular force between Ca^{2+} and carbon layers was much stronger than that of Na^{+}, indicating that the existed intercalation of CaO between carbon layers could prevent alkalis vapors penetration and further intercalation.

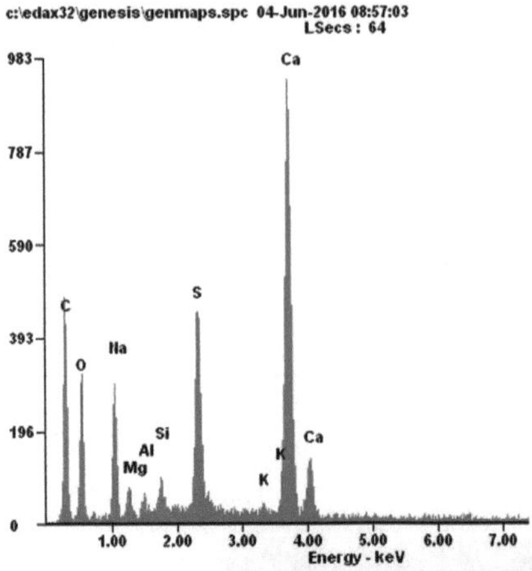

Element	*Wt%*	*At%*
CK	33.78	53.36
OK	16.73	19.84
NaK	04.67	03.86
MgK	01.06	00.83
AlK	00.68	00.48
SiK	01.17	00.79
SK	08.51	05.03
KK	00.76	00.37
CaK	32.64	15.45
Matrix	Correction	ZAF

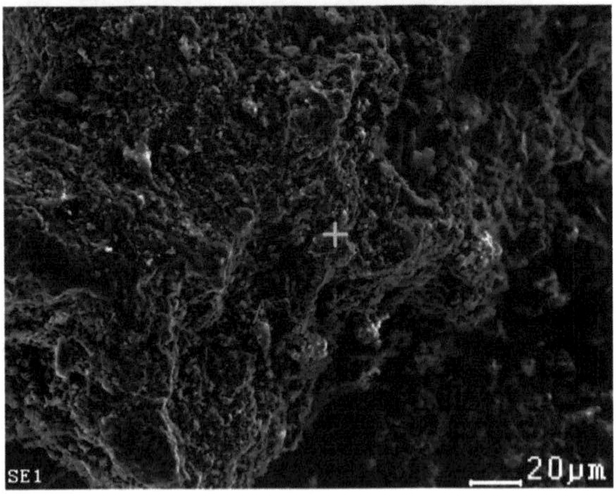

Fig. 6 SEM and EDS of high reactivity coke after sodium-adsorption in the atmosphere of N_2

SEM of Two Cokes After Sodium-Adsorption in the Atmosphere of CO_2

Figures 7 and 8 shows the SEM results of the two cokes after sodium-adsorption in the atmosphere.

It can be seen that the pore wall of conventional coke became thicker with volume expansion after the penetration of sodium vapor. On the other hand, severe faults and erosion grooves have even appeared on the surface of high reactivity coke texture after solution loss reaction. For the conventional coke, the sodium vapor penetrates in the coke matrix through pores and is absorbed on the pore walls. As the solution loss reaction proceeding with the rising temperature, micro cracks are formed that allow sodium vapor to further penetrate into the coke carbon matrix. For the high reactivity coke, the formed cracks and accelerated reaction rate could provide new paths for CO_2 diffusion. The interaction of sodium vapor with micro

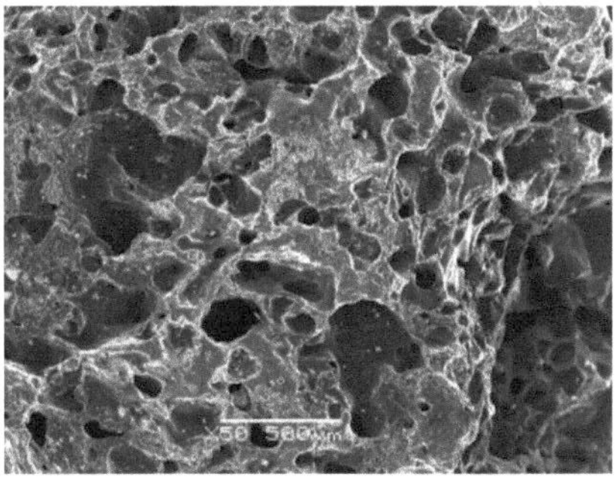

Fig. 7 SEM of conventional coke after sodium-adsorption in the atmosphere of CO_2

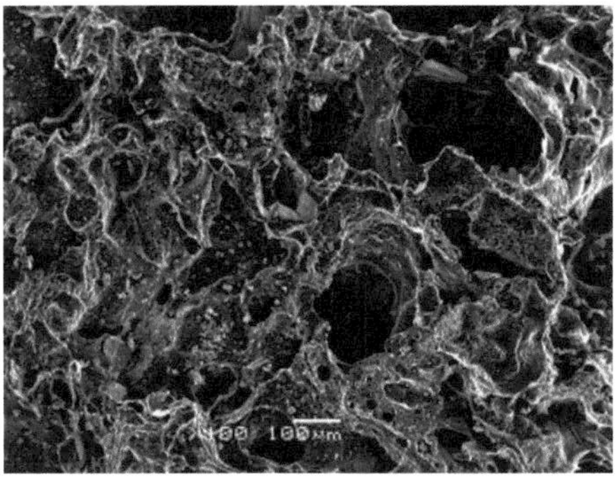

Fig. 8 SEM of high reactivity coke after sodium-adsorption in the atmosphere of CO_2

crystals of coke leads to a further loss of the coke structure integrity, making the much severer coke degradation.

XRD Patterns of the Two Cokes After Sodium-Adsorption

Figures 9 and 10 shows the SEM results of the two cokes after sodium-adsorption in the atmosphere of N_2 and CO_2, respectively.

It can be seen that the (002) peak height and FWHM of high reactivity coke were both higher than those of conventional coke in the atmosphere of N_2. Compared with the high reactivity coke, the graphite peak of conventional coke became much broader and less prominent after the penetration of sodium vapors, indicating disorder and a possible intercalation of sodium atoms between carbon layers. Meanwhile, the results showed a reverse trend in the atmosphere of CO_2, indicating a much severer degradation of high reactivity coke.

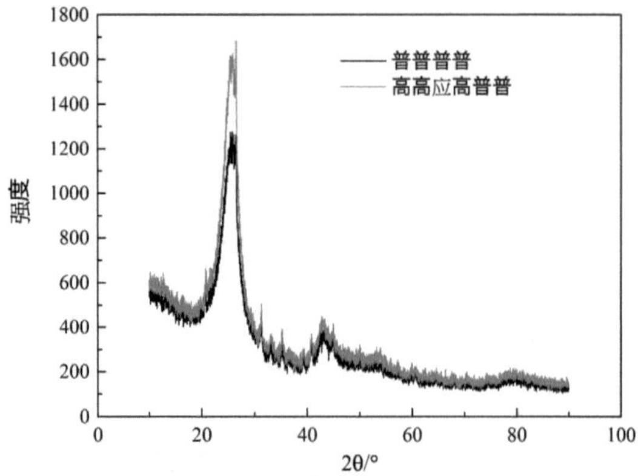

图9 N_2气氛下两种焦炭吸附碱金属 Na 后的 XRD 图谱

Fig. 9 XRD patterns of two cokes after sodium-adsorption in the atmosphere of N_2

Fig. 10 XRD patterns of two cokes after sodium-adsorption in the atmosphere of CO_2

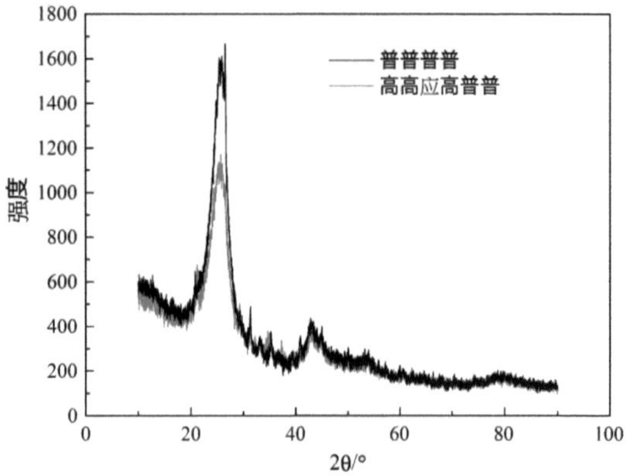

Conclusions

Results showed that the degradation of conventional coke strength by sodium was stronger than highly reactive coke. The less severe degradation of highly reactive coke was attributed to CaO distributed on the coke surface, inhabiting alkali vapors diffusion. However, the phenomenon in CO_2 was contrary to the former. Changes of coke strength were related to the solution loss reaction of highly reactive coke starts to react at a lower temperature. Therefore, sodium atom could easily enter into carbon matrix, causing coke microstructure expansion and fracture. The effect of CaO catalyst on the coke reactivity has been interpreted as a change in highly reactive coke microstructure and strength.

References

1. K. Li et al., The evolution of structural order, microstructure and mineral matter of metallurgical coke in a blast furnace: a review. Fuel **133**, 194–215 (2014)
2. M. Lundgren et al., The evolution of structural order as a measure of thermal history of coke in the blast furnace. Metal. Mater. Trans. B **45**(2), 603–616 (2014)
3. S. Gupta et al., Effect of CO_2 gasification on the transformations of coke minerals at high temperatures. Energy Fuels **21**(2), 1052–1061 (2007)
4. T. Hilding et al., Degradation behaviour of a high CSR coke in an experimental blast furnace: effect of carbon structure and alkali reactions. ISIJ Int. **45**(7), 1041–1050 (2005)
5. H. Zhao, S. Cheng, New cognition on coke degradation by potassium and sodium in alkali enriched regions and quantificational control model for BF. J. Univ. Sci. Technol. Beijing **34** (3), 333–341 (2012)
6. B.K. Chan, K.M. Thomas, H. Marsh, The interactions of carbons with potassium. Carbon **31** (7), 1071–1082 (1993)
7. S. Gupta et al., Coke graphitization and degradation across the tuyere regions in a blast furnace. Fuel **113**, 77–85 (2013)
8. H. Zhao et al., Degradation effect of potassium and sodium on coke. J. Univ. Sci. Technol. Beijing **35**(4), 438–447 (2013)
9. K. Li et al., Influence of alkaline (Na, K) vapors on carbon and mineral behavior in blast furnace coke. Fuel **145**, 202–213 (2015)
10. S. Nomura et al., Improvement in blast furnace reaction efficiency through the use of highly reactive calcium rich coke. ISIJ Int. **45**(3), 316–324 (2005)
11. M. Sato et al., Prediction of next-generation ironmaking process based on oxygen blast furnace suitable for CO_2 mitigation and energy flexibility. ISIJ Int. **55**(10), 2105–2114 (2015)
12. M. Natio et al., Development of production and utilization technology of z with high strength and high reactivity. Tetsu to Hagane **96**(5), 201–208 (2010)

Part VI
Deriving Value from Challenging Waste Materials: Recycling and Sustainability Joint Session: Deriving Value from Challenging Waste I

Maximizing the Values of Steelmaking Slags

Naiyang Ma

Abstract Along with steel production, various steelmaking slags are constantly generated at considerably high rates. In general, steelmaking slags are not categorized into hazardous materials and do not cause any environmental issues. Therefore, recycling of steelmaking slags is often aimed at creating values by using the slags to replace high-cost raw materials. Chemical compositions and particle size distributions of slag products from steelmaking slags will determine what raw materials the slag products can replace. Consequently, maximizing the values of steelmaking slags will require to produce slag products with the highest values and to process the slags with the lowest costs. In this paper, several slag processing processes are assessed based on maximization of the values of the steelmaking slags.

Keywords Steelmaking slags · Recycling · Slag processing

Introduction

Due to its superior price/performance ratio, steel remains one of the most influential man-made materials in the world. However, along with the steel production, various steelmaking slags are constantly generated at considerably high rates. In general, these steelmaking slags are not categorized into hazardous materials and do not cause any significant environmental issues [1, 2]. Therefore, they are allowed to be stored on the ground as long as the slags are identified as usable materials.

Recycling of steelmaking slags is to use the slags as alternative materials of virgin natural resources or other high-cost raw materials. Since steelmaking slags are rich in iron, calcium and magnesium, they can be often internally recycled to replace steel/iron scrap for iron units in the ironmaking and steelmaking process,

N. Ma (✉)
ArcelorMittal Global R&D—East Chicago Laboratories, 3001 E Columbus Dr., East Chicago, IN 46312, USA
e-mail: naiyang.ma@arcelormittal.com

L. Zhang et al. (eds.), *Energy Technology 2017*,
The Minerals, Metals & Materials Series, DOI 10.1007/978-3-319-52192-3_18

and to replace iron ores for iron units and limestones and dolomite for flux units in the sintering and ironmaking process [3–5]. They can also be externally recycled to substitute limestone/lime in remediation of acidic soil and in treatment of acid mine drainage [6, 7]. Due to their cementitious property, the steelmaking slags can be utilized in cement production [8]. The steelmaking slags are widely accepted as alternative materials of crushed stones and sands in construction [9, 10]. Generally speaking, the steelmaking slags can be utilized as three major classes of alternative materials for their iron units, calcium and magnesium units and stone units.

Here then comes a question about what slag products should be made when processing the steelmaking slags. Since the steelmaking slags generally do not cause significant environmental concerns, recycling of the steelmaking slags is often aimed at creating values. Slag products as alternative materials of different raw materials generally have quite different values. For instance, steel scrap has a much higher value than stones. Therefore, it is crucial to process the steelmaking slags into high-value slag products with low processing costs in order to maximize the total values of the steelmaking slags.

Slag Products and Their Values

As described in the previous section, steelmaking slags can be utilized as alternative materials for their iron units, calcium and magnesium units and stone units. However, in slag processing industry, though there exist various processes, some quite complicated and others rather simple, the steelmaking slags are usually separated into only two categories of slag products: "metallic" and "non-metallic." Each category of the slag products is then separated into various size fractions. The "metallic" slag products are presumed to be iron-rich and they are alternative materials of steel/iron scrap and iron ores. The "non-metallic" slag products are supposed to be iron-lean and they are alternative materials of limestones and dolomite as fluxes, and alternative materials of crushed stones and sands as construction materials.

Prices of some virgin natural commodities and other high-cost raw materials to be replaced by the steelmaking slags are presented in Table 1.

From Table 1, one can see that iron-containing commodities have rather higher values than fluxes and construction materials; steel scrap is much more expensive

Table 1 USGS estimated commodity prices in 2015 [11]

Commodity	Unit price, \$/tonne
Steel scrap	228
Fluxed pellets	84
Iron ore fines	56.43
Limestone and dolomite (fluxes)	10.46
Crushed stone (construction)	10.46
Sand and gravel (construction)	7.72

Table 2 Iron contents and size ranges of various steelmaking slag products

Slag product	Total iron (%)	Size range (mm)
"A" scrap	>85	>75
"B" scrap	>60	12.5–75
"C" scrap	>40	<12.5
Large size "Non-metallic" slag	<20	>25
Small size "Non-metallic" slag	<20	<25

than iron ores; prices of fluxes and construction materials are quite close. It is clear that production of more iron-rich slag products from the steelmaking slags will make the slags more valuable.

Whether a slag product can be utilized as an alternative material of commodities listed in Table 1 is determined by chemical compositions and particle sizes of the slag product. In Table 2, iron contents and particle sizes of common slag products are presented [12–15]. Correlation of slag products with iron content and particle size in the table is also graphically plotted in Fig. 1. It can be seen that for iron-bearing slag products, "metallic" slag products with larger particle sizes correspond to higher iron concentrations in the slag products. However, for "non-metallic" slag products, size differences do not result in great differences in iron concentrations in the slag products.

The steelmaking slag products contain iron, fluxes (CaO and MgO) and impurities (S, P, SiO_2, Al_2O_3, etc.). When the steelmaking slag products are utilized in the ironmaking and steelmaking process, iron and fluxes are beneficial, but the impurities are harmful. Beneficial components will have positive effects on values of the slag products while harmful components will have negative impacts on the values. Status of the beneficial components in the slag products can also have very strong influences on the values. For instance, when being used in the sintering,

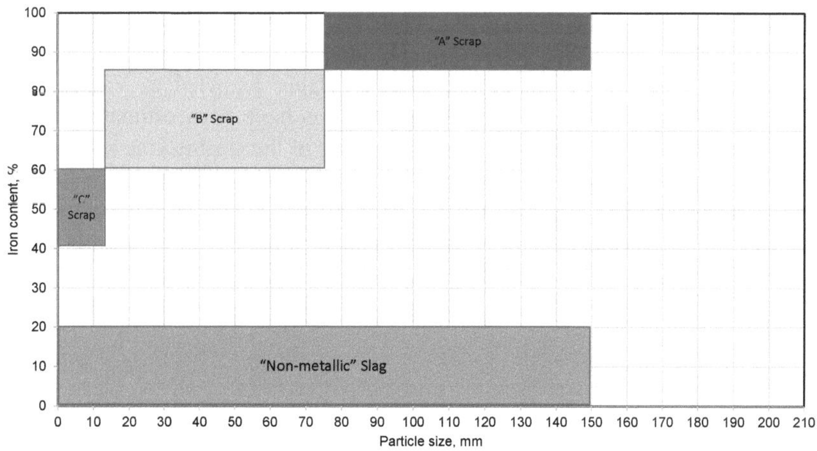

Fig. 1 Correlations among steelmaking slag products, iron contents and size ranges

Table 3 Applications and values of steelmaking slag products

Slag product	Commodity to be replaced	Where to use	Total iron (%)	CaO (%)	Value ($/tonne)
"A" scrap	Steel scrap	BOF/EAF/Melters	85	NA	208.39
"B" scrap	Pellets + Fluxes	BF	60	14	84.16
"C" scrap	Iron ore fines + Fluxes	Sintering	40	24	41.33
Large Size "Non-metallic" slag	Crushed stone/pellets + Fluxes	Construction/BF	20	40	10.46/35.30
Small size "Non-metallic" slag	Sands and gravels/fluxes + Iron ore fines	Construction/ Sintering	20	35	7.72/25.38

ironmaking and steelmaking, metallic iron in the steelmaking slags will have a higher value than iron oxides, and free CaO and MgO will be more valuable than calcium silicates and calcium carbonate. Opposite to using the steelmaking slag products in the ironmaking and steelmaking process, if using them in construction, metallic iron and free CaO and MgO in the slag products become troublesome components since metallic iron may damage crushers and cut tires, and free CaO and MgO may cause damages of road base due to expansion. However, S, P, SiO_2 and Al_2O_3 in slag products are generally not harmful components in construction applications. As a result, the values of the steelmaking slag products will be quite different when they are utilized in different processes.

Since value calculation in this paper is for demonstration purpose, in the following discussion, only iron and CaO values are considered based on iron equivalence and CaO equivalence, but effects of component status and impurities are neglected.

Assume that concentrations of total iron in steel scrap, pellets and iron ore fines are 93, 62 and 62%, respectively. CaO in limestone is 51%. Take the boundary iron content in Table 2 for each steelmaking slag product, and CaO contents in the slag products are assumed based on the most available knowledge. Values of the steelmaking slag products in different applications have been estimated and presented in Table 3. It is clear that internal recycling of the steelmaking slag products in the sintering, ironmaking and steelmaking process consistently create higher values than external use of the slag products as stones and sands; values of the steelmaking slag products are proportional to concentrations of total iron in the slag products; and slag products with larger sizes have higher values.

Total Values of Steelmaking Slags

As discussed before, iron, especially metallic iron has the highest value in steelmaking slags. In BOF and EAF steelmaking, at the end of each heat, molten slag samples are taken for analysis after liquid steel is tapped. Obviously, analysis of the molten slag samples provides actual compositions of the raw steelmaking slags to be processed.

At ArcelorMittal USA, sampled molten steelmaking slag is crushed into powder, pressed into pellets and analyzed with XRF analyzers. Statistic parameters of iron concentrations in slag samples from several steelmaking plants are presented in Table 4. The data in Table 4 show that on average, iron concentrations in raw steelmaking slag samples are around 23%. However, the iron content in steelmaking slags varies greatly, from under 10% to over 35%.

In the following discussion, assume raw slag contains 23% iron. Assume that after processed, all iron goes to only "A" scrap or only "B" scrap or only "C" scrap, and "non-metallic" slag products do not contain any iron. Assume the "non-metallic" products are either used as stones or landfilled with a cost of $40 per tonne. Slag processing cost is not counted. The total slag values are then calculated and presented in Table 5.

Table 4 Descriptive statistics of iron concentrations in raw slag samples

Statistic parameters	Plant 1	Plant 2	Plant 3	Plant 4
Mean	23.31	22.24	24.25	23.67
Median	23.10	22.08	23.94	22.38
Mode	27.33	21.90	23.71	21.31
Standard deviation	3.01	3.50	4.17	3.64
Range	23.41	28.16	20.60	15.41
Minimum	13.90	8.60	14.69	18.52
Maximum	37.30	36.77	35.29	33.93
Count	5939	1018	713	29

Table 5 Total values of steelmaking slags

Metallics	"A" scrap only	"B" scrap only	"C" scrap only
Fe in metallics	85	60	40
Yield of metallics (%)	27.06	38.33	56.43
Yield of non-metallics (%)	72.94	61.67	43.57
Total slag value if non-metallics is recycled as stones ($/tonne)	64.02	38.71	28.21
Total slag value if non-metallics is landfilled ($/tonne)	27.21	7.6	6.76

The calculation shown in Table 5 tells that under various conditions, production of scrap with high iron content and large particle size results in high total slag values. Maximization of steelmaking slag values should be in the direction of maximizing production of "A" scrap, then "B" scrap, and then "C" scrap, maximizing extraction of iron, minimizing the production of "non-metallic" slag products, and minimizing the processing costs.

Assessment of Processes of Slag Processing

There are various processes for processing steelmaking slags. The three representative categories are: conventional air cold process followed by iron recovery, modification of slag at hot stage, and dry granulation of molten slag.

The conventional process of treating steelmaking slag generally includes the following procedures: haul the slag pots to a slag yard after the steelmaking slag (molten or solidified) is fed into the slag pots, pour the hot slag on the ground, water and stir the slag, move the colder slag to a slag pile for further cooling, and process the cold slag by magnetic separation and screening with or without crushing. There are many variations of the conventional process of treating the steelmaking slags. Some have many repetitive steps of crushing-magnetic separation-screening while others possess rather simple processing stages [10]. The slag is separated into "metallic" and "non-metallic" products with different size fractions. The iron in the slag is extracted into various scrap products.

Due to disadvantages of high levels of free lime and free magnesium in "non-metallic" steelmaking slag products produced in conventional treatment process, various new technologies have been developed and tested. One is modification of slag by injecting oxygen and sand into molten slag at high temperature [16–18]. Another is dry granulation of molten slag [19–21]. These two new technologies are both only treating molten slag with good fluidity. Slag modification process cannot recover iron at all and the product can only be used as an alternative material of stones with low values. Products of dry granulation include iron concentrate and sand which have lower values due to their small particle sizes (<5 mm). Therefore, compared to conventional process, slag modification and dry granulation are demoting values of steelmaking slags.

Looking over processes treating steelmaking slags, conventional process remains in the position where maximum slag value can be created. However, there certainly exist many potentials to optimize the conventional process of treating steelmaking slags. One direction is to improve recovery of iron from the slags and to reduce iron loss in the "non-metallic" slag. Possible optimization methods may include hot-stage separation of iron from slag components, hot-stage disintegration of slags into appropriate particle sizes, hot-stage removal of impurities like S and P, and high-efficiency approach of separating iron and iron oxides from slag components.

Concluding Remarks

Steelmaking slags can be recycled for their iron units, flux units and stone units. Their values are determined by what virgin natural resources and other high-cost raw materials they can replace. Using as alternative materials of iron-bearing resources can have higher values than as alternative materials of limestones, dolomite, crushed stones and sands. For "metallic" slag products, large particles correspond to high levels of iron contents and hence high values. For "non-metallic" slag products, larger particles often have higher values than small particles. Extraction of more iron from steelmaking slags into "metallic" products will make whole slags more valuable. The conventional process of treating steelmaking slags remains in the position to create the highest values of the slags. Directions of optimizing values of steelmaking slags are hot-stage processing for separating iron from the slags, fragmenting slags into appropriate particle sizes and removing impurities from "metallic" products.

References

1. AISI, *Steel Technology Roadmap—Environmental Leadership* (US Department of Energy, 2013) http://energy.gov/sites/prod/files/2013/11/f4/roadmap_chap4.pdf, p. 94
2. J.L. Wintenborn, J.J. Green, *Steelmaking Slag: A Safe and Valuable Product* (National Slag Association, 1998) http://www.acobrasil.org.br/siderurgiaemfoco/CCABrasil/NSA%20Risk%20Assessment%20Summary.pdf, p. 2
3. AISI, *Steel Technology Roadmap—Iron Unit Recycling* (US Department of Energy, 2013) http://energy.gov/sites/prod/files/2013/11/f4/roadmap_chap3.pdf, pp. 61–62, 65
4. R. Remus et. al., *Best Available Techniques (BAT) Reference Document for Iron and Steel Production* (Publications Office of the European Union, Luxembourg, 2013) http://eippcb.jrc.ec.europa.eu/reference/BREF/IS_Adopted_03_2012.pdf, p. 40
5. International Iron and Steel Institute, *The Management of Steel Industry By-products and Waste* (International Iron and Steel Institute, Brussels, Belgium, 1994), pp. 4.4–4.8, 5.3–5.4
6. National Slag Association, A guide for the use of steel slag in agriculture and for reclamation of acidic lands. http://www.nationalslag.org/sites/nationalslag/files/ag_guide909.pdf
7. P. Ziemkiewicz, J. Skousen, The use of steel slag in acid mine drainage treatment and control. *Green Lands* (1998), pp. 46–56
8. P.E. Tsakiridis et al, Utilization of steel slag for Portland cement clinker production. J. Hazard. Mater. **152**(2), 805–811 (2008)
9. National Slag Association, Common use of steel slag. http://www.nationalslag.org/common-uses-slag
10. K. Horll et al, *Processing and reusing technologies for steelmaking slag* (Report No. 104). Nippon Steel (2013)
11. USGS, *Mineral Commodity Summaries 2016* (USGS, Reston Virginia, 2016) http://minerals.usgs.gov/minerals/pubs/mcs/2016/mcs2016.pdf), p. 86, 90, 91, 142, 156
12. H. Alanyali et al., Application of magnetic separation to steelmaking slags for reclamation. Waste Manage. **26**, 1133–1139 (2006)
13. N.-Y. Ma, J.B. Houser, Recycling of steelmaking slag fines by weak magnetic separation coupled with particle size screening. J. Clean. Prod. **82**, 221–231 (2014)

14. Y. Topkaya, N. Sevinc, A. Gunaydın, Slag treatment at kardemir integrated iron and steel works. Int. J. Miner. Process. **74**, 31–39 (2004)
15. H. Yi et al., An overview of utilization of steel slag. Procedia Environ. Sci. **16**, 791–801 (2012)
16. D. Durinck et al., Hot stage processing of metallurgical slags. Resour. Conserv. Recycl. **52**, 1121–1131 (2008)
17. R.M. Santos et al., Stabilization of basic oxygen furnace slag by hot-stage carbonation treatment. Chem. Eng. J. **203**, 239–250 (2013)
18. M. Kuehn, H. Mots, P. Drissen, Chances for improving value added from by-products and residues of steelmaking. La Revue De Metallurgie—CIT **105**(6), 327–332 (2008)
19. S. One et al, *Blast granulation system of BOF slag and its products* (Report No. Overseas 38). Nippon Kokan (1983)
20. S. Cha, Economic aspects of slag processing with slag atomization technology. Paper presented at the 3rd global slag conference and exhibition, Istanbul, Turkey, 19–20 Nov 2007, pp. 5–18
21. H. Purwanto et al., Characteristics of glass beads from molten slag produced by rotary cup atomizer. Mater. Trans. **45**(12), 3286–3290 (2004)

Recycling of Zinc from the Steelmaking Dust in the Sintering Process

Piotr Palimaka, Stanislaw Pietrzyk and Michal Stepien

Abstract During the steel scrap melting process in an electric arc furnaces, the generated dust may contain a significant amount of zinc. The content of zinc in case of melting galvanized steel scrap is variable and ranges from 15 to 40 wt%. This dust cannot be re-used for the steelmaking process. On the other hand the zinc content is too low to treat the steel dust as a feedstock in pyrometallurgy of zinc. When they are disposed in landfills they have adversely effect the environment. This work presents selected results of research and development of new technology, which was developed in order to utilization steelmaking dust in the sintering process. One of the products (zinc-rich) can be used as a high-quality raw material for production of zinc by pyro- or hydrometallurgical processes. The second one product from the shaft furnace can be used as a feedstock for the steelmaking process.

Keywords Dust recycling · Steelmaking dust utilization · Zinc recovery

Introduction

Protection of the steel surface with zinc coating is a commonly used anti-corrosive technique. About 50% of the world production of zinc is used for the galvanizing treatment of steel parts, which prolongs the service life 3–4 times. Hence, a large amount of the galvanized steel scrap is formed, which is melted in electric arc furnaces (EAF). In this process, zinc is vaporized and oxidized. Due to the high melting point of ZnO (1957 °C), it is transferred to the particulate matter in the form of fine dust. On the average, in the EAF process of steel production, per one tonne of the product falls 15–25 kg of dust formed in this process; the final amount of dust depends on the feedstock type and the technology used [1]. This dust

P. Palimaka (✉) · S. Pietrzyk · M. Stepien
Faculty of Non-Ferrous Metals, AGH—University of Science and Technology, PL 30-059 Krakow, Poland
e-mail: palimaka@agh.edu.pl

L. Zhang et al. (eds.), *Energy Technology 2017*,
The Minerals, Metals & Materials Series, DOI 10.1007/978-3-319-52192-3_19

contains 15–40% Zn and approximately 20% Fe as well as other elements such as cadmium, nickel, chromium, manganese, coal, tin, antimony and copper. In the natural environment, all these metals are transferred to the surface and ground water, degrading the environment [2]. Moreover, accumulating in the bodies of plants and animals end up in human bodies. Because they do not decompose, once absorbed have the ability to bioaccumulate [3]. Therefore their storage creates a serious threat to the natural environment.

Waste generated during the metallurgical process should be returned to the same process or to other processes. Owing to this, the waste-free technology can become a reality. During re-melting of galvanized steel scrap, the dust generated can not be recycled and returned to the steelmaking process because of the previously mentioned high zinc content. On the other hand, this content is not sufficiently high to treat the dust as a raw material applicable in the processes of zinc metallurgy (it can be used only as an additive).

Current processing technologies for metallurgical dust are focused on the pyro- and hydrometallurgical methods. Pyrometallurgical processes rely mainly on high-temperature reduction of zinc contained in the dust and its re-oxidation in gaseous phase. One of the most known and used processes is currently the Waelz kiln process [4, 5]. There are more than 14 plants in the world using this process [6]. Other, very similar process, both in commercial and pilot plant operation process are: ALLMET, AUSMELT, METWOOL, PRIMUS, ZINCOX, OXYCUP, CONTOP, ENVIROPLAST, TECNORED, OXICUP, REDSMELT [7–12]. Hydrometallurgical processes rely on leaching of zinc waste and separation of zinc in the process of electrolysis, by extraction through ion exchange, precipitating the insoluble zinc compounds or forcing their crystallization [13]. Currently, developed industrial technologies include: ZINCEX, EZINEX, RECUPAC [14, 15]. There are still many studies in the laboratory stage [16–21].

Despite the availability of so many processes for the recovery of zinc from metallurgical dust, none of them can solve the problem comprehensively, i.e. combine in a single process the recovery of zinc from waste and the manufacture of an iron-carrying feedstock for the blast furnaces. The development and implementation of such a method would bring many ecological and economical advantages. It would also mean a step forward in the global trends of environmental protection by pointing out the possibility of using a waste-free technology. The challenge related with this issue has been taken by the staff of researchers from the AGH—University of Science and Technology within the framework of a national project implemented in cooperation with industrial partners.

Experiments and Results

This paper presents the selected laboratory tests for high-temperature reduction of ZnO during sintering of briquettes containing steelmaking dust. The studies constitute one of the stages in the development of a new, waste-free technology for the

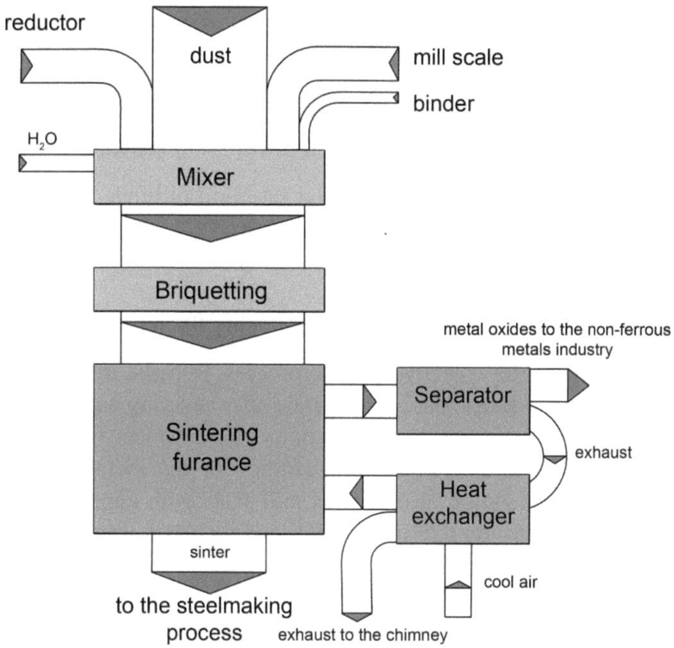

Fig. 1 Flow of materials and agents in the devices used for the proposed technological process

management of metallurgical dust with high zinc content, originating from electric arc furnaces. Figure 1 shows the flow of materials and agents in the devices used for the proposed technological process.

In this technology, a mixture comprising adequate quantities of dust, mill scale, coke breeze, and a binder is moistened in the mixing process and then subjected to briquetting. Briquettes are next undergoing the reduction in a shaft furnace, where in the temperature range of 1100–1200 °C the zinc oxide is reduced with carbon contained in the briquettes and CO which is formed by the Boudouard reaction. Zinc reduced to the form of gas is discharged together with other process gases and undergoes re-oxidation outside the sintering furnace. Then it is removed from the furnace in the form of dust. The dust can be used as a high-quality commercial product and a raw material for the zinc production. In the processes conducted in a shaft furnace, the reaction of reduction also includes the lead oxide, but because of its low content in the processed steelmaking dust, it should assume the form of fine droplets in the briquette rather than occur as a separate phase. A small portion of lead can move to the gaseous phase. The process of reduction should also cover iron oxides, reduced to FeO and Fe_3O_4. The exhaust gases are directed to the heat exchanger and then to the furnace stack. To carry out the process it is necessary to supply the air and exhaust gases with proper oxygen content, such that will ensure in the sintering process the required amount of heat resulting from the combustion of carbon contained in the briquettes and a reducing atmosphere in the reducing zone. The residue from the process in the form of sintered briquettes will be discharged by means of a screw conveyor and used in the iron and steel metallurgy.

The selection of guidelines for the project and preliminary studies were carried out on the metallurgical dust from one of the national steel mills. The mixture for the briquetting process was selected in such a way as to provide:

- the required integration of briquettes (the selection of binder),
- the required strength before and after sintering,
- the required amount of reducing agent to obtain a high rate of the ZnO reduction,
- the required amount of fuel capable of producing the required temperature.

In accordance with the objectives of the process, to make the material suitable for re-use in the steel industry, the content of Fe after sintering and reduction should be minimum 50% and that of zinc maximum 2%. To meet the requirement of adequate iron content after sintering, in addition to the metallurgical dust, the mixture for briquetting should also contain mill scale with approx. 70% Fe. As a binder, a two-component mixture of $Ca(OH)_2$ and molasses was used. Its task was to ensure full integration of grains during pressing and high mechanical strength of briquettes. Additionally, the selected ingredients were not acting as a ballast in later metallurgical processes. Two types of reducing agents were tested in the studies, i.e. coke and coal dust. Carbon in the sintering process acts as a reductant and a source of heat that is needed to ensure proper temperature of the process. The amount of carbon was chosen in such a way as to provide adequate progress in the reaction of the reduction of metal oxides, to make up for the negative energy balance of chemical reactions and provide heating of the substrate to an adequate sintering temperature. The compositions of mixtures prepared for briquetting are shown in Table 1. The steelmaking dust contained 33.2 wt% Zn, 22.6 wt% Fe, 9 wt% Cl, 3 wt% Pb, 2.5 wt% Ca, 2 wt% Si, 1.5 wt% Mn and 1 wt% C.

Mixtures were pressed into the form of briquettes without parting plane and marked in further part of the study as briquettes B1, B2 and B3 formed from mixtures M1, M2 and M3, respectively. Their compressive strength, understood as a force causing their destruction, was 300, 340 and 470 for briquettes B1, B2 and

Table 1 The compositions of mixtures prepared for briquetting

Component	Mixture		
	M1	M2	M3
	Content, wt%		
Dust	45.4	45.1	45.8
Mill scale	38.25	38.34	38.9
Reductor	5.4 (coke breeze)	6.32 (coal dust)	5.49 (coke breeze)
Molasses	4.43	4.49	4.51
$Ca(OH)_2$	2.66	2.69	2.71
H_2O	4.8	3.4	2.9

B3, respectively. Figure 2 shows a model of the roll forming briquettes without parting plane and the shape of a single briquette.

In order to determine the removal ratio of zinc, the briquettes were subjected to the process of sintering in a laboratory chamber furnace. For this purpose, they were placed in a graphite crucible and covered with a cover also made of graphite. Thus a reducing atmosphere was created in the crucible, necessary for the reduction of zinc oxides. Zinc reduced to the gaseous form, after leaving the space of the crucible was oxidized with oxygen present in the furnace chamber and deposited on the ceramic lining of the furnace as a yellow solid matter. Tests were conducted at 1000, 1100 and 1200 °C and at each of those temperatures, three different times of 15, 30 and 45 min were additionally applied. Figure 3 shows the appearance of briquettes before and after the sintering at 1200 °C for 30 min.

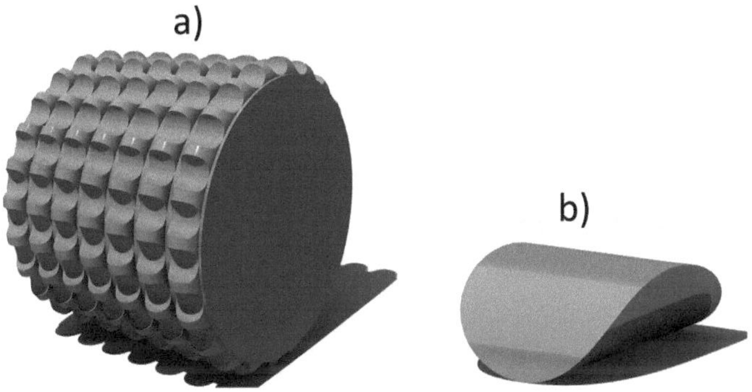

Fig. 2 Model of the roll forming briquettes without parting plane (**a**) and the shape of a single briquette (**b**)

Fig. 3 The appearance of briquettes before and after the sintering at 1200 °C for 30 min

Table 2 Values of the relative weight loss and zinc removal ratio

Temp. °C		Briquettes B1		Briquettes B2		Briquettes B3	
	Time, min	Relative weight loss,%	Zn removal ratio,%	Relative weight loss,%	Zn removal ratio,%	Relative weight loss,%	Zn removal ratio,%
1000	15	24.72	48.82	13.80	20.79	21.17	54.97
	30	27.89	70.27	17.51	29.74	29.47	69.91
	45	31.78	72.73	22.00	33.55	32.82	83.25
1100	15	30.92	80.16	19.58	37.80	30.90	56.17
	30	38.55	91.23	24.87	55.34	43.76	73.41
	45	41.76	96.39	27.71	55.82	43.36	86.28
1200	15	39.74	90.92	26.26	56.90	38.82	93.77
	30	44.92	95.31	26.80	60.88	43.52	97.97
	45	45.93	96.44	28.73	60.91	43.76	98.28

After the experiment, the briquettes were cooled, weighed, and then ground to a form of particulate and analyzed for the zinc content on an ARL QUANT'X X-ray spectrometer. The results of the analysis allowed for the determination of zinc removal ratio (η) from the briquettes (Eq. 1), to establish next optimum conditions of the sintering process.

$$\eta, \% = \frac{m_{(0)} - m_{(t)}}{m_{(0)}} * 100 \tag{1}$$

where:

$m_{(0)}$ initial content of zinc in the sample (before reduction), g
$m_{(t)}$ final content of the zinc in the sample (after reduction), g

The measured values of mass changes expressed as a relative weight loss in samples and the calculated values of the zinc removal ratio are summarized in Table 2.

Analyzing the events possibly taking place during sintering, it was found that the observed changes in the weight of briquettes before and after the sintering might be due to several, simultaneously occurring processes, such as:

- the reduction of oxides of zinc, lead and iron,
- the evaporation of gaseous zinc and partly of Pb,
- the evaporation of H_2O,
- partial decomposition and evaporation of binder,
- the combustion of fuel (coke or coal dust depending on the type of mixture).

Detailed analysis of the obtained results listed above leads to the conclusion that the weight loss in briquettes is adequate to the zinc removal ratio. However, considering the occurrence of other above mentioned processes, running in parallel with the reduction of ZnO, it is impossible to estimate process effectiveness based on the weight loss in briquettes only (hence results also the need to repeat the

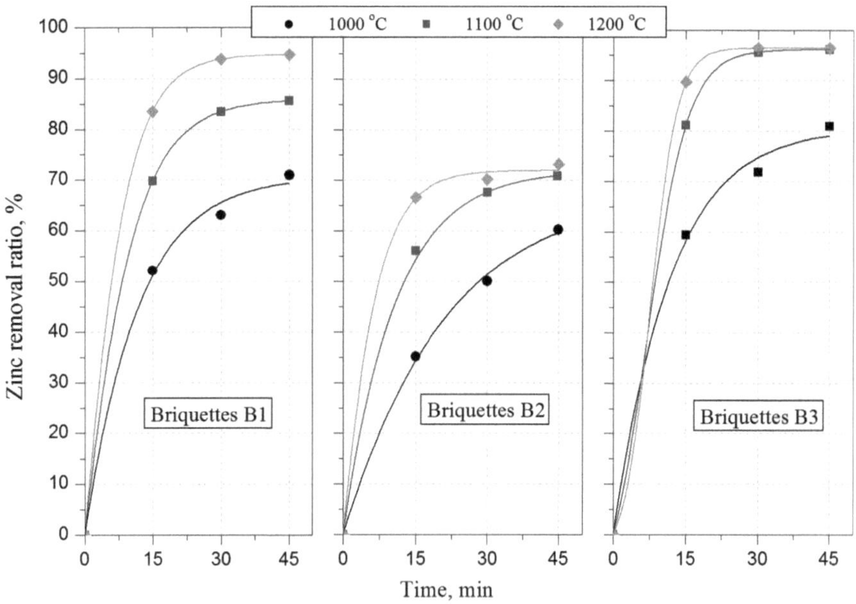

Fig. 4 Zinc removal ratio as a function of time

analysis after each experiment). The calculated values of the zinc removal ratio are illustrated in Fig. 4 for each of the conducted experiments.

It was found that briquettes sintered at 1100 and 1200 °C were compact and sintered within their whole volume; they retained their shape and had adequate mechanical strength. Although they were not tested for strength properties, their disintegration in a steel mortar during the preparation for analysis after sintering has proved to be very difficult. Examining the obtained results, it can be concluded that the zinc removal ratio during sintering depends on both temperature and time of the sintering process. Ensuring adequate residence time of briquettes (at least 30 min) in the zone of a temperature close to 1200 °C should provide, in the case of using coke breeze as a reducing agent, the zinc removal ratio equal to 95%. This means that the required content of Zn < 2% and Fe > 50% will be satisfied, and the sinter can be used as a feedstock for blast furnaces.

Conclusions

Summing up the results of studies of the high-temperature sintering process, it can be concluded that the zinc removal ratio depends on the following factors:

- **Type of reducing agent**. It has been assumed that the reactions of the reduction of metal oxides inside the briquettes occur by both direct reduction with carbon

(2) and indirect reduction with gaseous CO (3) formed in the Boudouard reaction (4).

$$MeO + C = Me + CO \quad (2)$$

$$MeO + CO = Me + CO_2 \quad (3)$$

$$CO_2 + C = 2CO \quad (4)$$

The results have clearly indicated the stronger effect of coke breeze as a reducing agent. Coke breeze was used in the briquettes B1 and B3. This effect is associated with a higher content of carbon in this agent, higher reactivity and lower level of impurities (tar, dust and sulphur).

- **Temperature**. The strongest effect of temperature was visible after 15 min of the experiment duration for each of the tested mixtures. After this time, the temperature increase from 1000 to 1200 °C has increased the zinc removal ratio by approx. 30%. However, after 30 and 45 min, the effect has weakened considerably and, depending on the mixture composition, has yielded the result of only 15–25%.
- **Time of reduction**. At lower temperatures, obtaining high zinc removal ratio takes longer time. Even after 45 min this level did not exceed 80% in the case of briquettes comprising coke breeze as a reducing agent. At a temperature of 1200 °C, practically after 15 min, the reduction of zinc oxides has assumed a high yield of up to 85–90%. This applied to briquettes containing in their composition coke breeze as a reducing agent.
- **Contact of reactants**. Proper and effective contact of the reactants is important in the case when the reactions of reduction also occur in the solid phase (direct reduction). From this point of view, the applied briquetting process seems to be the best solution.
- **The content of H_2O**. Higher moisture content in the mixture can have a beneficial effect on the sintering process, due to the presence of pores generated by moisture evaporation, which facilitate transport of reactants to the reaction zone (CO gas) and the discharge of products, i.e. of the gaseous zinc, from the reaction zone. In this case, moisture content in briquettes was observed to have no major effect on the zinc removal ratio. The main purpose of the additional moistening of the mixture was to obtain briquettes of an adequate strength, and this goal was achieved.

References

1. A. Guézenneca, J. Huber, F. Patisson, P. Sessiecq, J. Birat, D. Ablitzer, Dust formation in electric arc furnace: birth of the particles. Powder Technol. **157**, 2–11 (2005). doi:10.1016/j.powtec.2005.05.006

2. P. Ostrowska, K. Mierzwa, Recovery of zinc from selected metallurgical waste. Hutnik-WH **64**, 369–373 (2007)
3. Z. Wozniacki, T. Telejko, R. Kenig, Sintering as the method of utilization of steelmaking dusts with a high content of zinc oxides. Hutnik-WH **81**, 166–171 (2014)
4. K. Mager, U. Meurer, B. Garcia-Egocheaga, N. Goicoechea, Recovery of zinc oxide from secondary raw materials: New developments of Waelz process, in *Recycling of Metals and Engineered Materials*. (The Minerals, Metals and Materials Society, 2000), pp. 329–344
5. Steel Dust Recycling, Waelz Kiln Technology. Available: http://www.globalsteeldust.com/waelz_kiln_technology [06/10 2016]
6. N.G. Gandiaga, B.G.E. Vergara, *Proceedings of the "Rewas" 99: Global symposium on recycling, waste treatment and clean technology, TMS, INASMET*, San Sebastian, (1999) pp. 1511–152
7. Zunkel A D Recovering zinc and lead from electric arc furnace dust: a technology status Report, in *Recycling of Metals and Engineered Materials*, ed. by D.L. Stewart, J.C. Daley, R. L. Stephens (Wiley, Hoboken, NJ, USA, 2000) doi: 10.1002/9781118788073.ch21
8. J.L. Roth, R. Frieden, T. Hansmann, J. Monai, M. Solvi, PRIMUS, a new process for recycling by-products and producing virgin iron. Rev. Met. Paris **98**, 987–996 (2001)
9. M. Holtzer, A. Kmita, A. Roczniak, The recycling of materials containing iron and zinc in the oxycup process. Arch. Foundry Eng. **15**, 126–130 (2015)
10. N.A. Barcza, D.G.C Robertson, A.F.S Schoukens, Enviroplas technology for the recovery of lead and zinc from lead blast furnace slags, in *Proceedings of International lead and zinc study group 6th International Conference "Recycling lead and zinc into the 21th century*, Madrid, Spain (1995)
11. M.A.V. Abdel-Latif, Fundamentals of zinc recovery from metallurgical wastes in the enviroplast process. Miner. Eng. **15**, 945–952 (2002)
12. W. Lu, D. Huang, The evolution of ironmaking process based on coal containing iron ore agglomerates. ISIJ Int. **41**, 807–812 (2001)
13. M. Jha, V. Kumar, R. Singh, Review of hydrometallurgical recovery of zinc from industrial waste. Resour. Conserv. Recycl. **33**, 1–22 (2001). doi:10.1016/S0921-(00)00095-1
14. G. Diaz, C. Martin, C. Frias, F. Sanchez, Emerging application of ZINCEX and PLACID technologies. JOM 85–98 (2001)
15. M. Olper, The EZINEX process—a new way and advanced way zinc from a for electrowinning zinc from a chloride solution, in *World Zinc, '93* ed. by I.G. Matthew (Australian Institute of Mining and Metallurgy, Victoria, Australia 1993), pp. 491–494
16. T. Nakamura, E. Shibata, U.T. Takasu, H. Itou, Basic consideration on EAF dust treatment using hydrometallurgical processes. Resour. Process. **55**, 144–148 (2008)
17. Z. Youcai, R. Stanforth, Integrated hydrometallurgical process for production zinc from electric arc furnace dust in alkaline medium. J. Hazard. Mater. **B80**, 223–240 (2000)
18. N. Leclerc, E. Meux, J.-M. Lecuire, Hydrometallurgical recovery of zinc and lead from electric arc furnace dust using monotrilotriacetate anion and hexahydrated ferric chloride. J. Hazard. Mater. **B91**, 257–270 (2002)
19. D.M. Lenz, F.B. Mmartins, Lead and zinc selective precipitation from leach electric arc furnace dust solutions. Revista Materia **12**, 503–509 (2007)
20. M. Cruells, A. Roca, C. Nunez, Electric arc furnace flue dusts: characterization and leaching with sulphuric acid. Hydrometallurgy **31**, 213–231 (1992)
21. P. OustadakisS, P.E. Takiridis, A. Katsiapi, S. Agatzinileonardou, Hydrometallurgical process for zinc recovery from electric arc furnace dust (EAFD), Part I: characterization and leaching by diluted sulphuric acid. J. Hazard. Mater. **179**, 1–7 (2010)

Direct Preparation of Metal Doping Ni–Zn Ferrite from Zn-Containing Electric Arc Furnace Dust by Calcination Method

Hui-gang Wang, Min Guo and Mei Zhang

Abstract Metal-doped Ni–Zn ferrite with spinel structure was directly synthesized from treated zinc-containing electric arc furnace dust (Zn-containing EAFD) by solid state reaction method, realizing the transformation of Zn-containing EAFD from solid waste to high value-added material. A certain mass of $NiCl_2 \cdot 6H_2O$ was added to the treated Zn-containing EAFD before calcination. Then the effects of the mass ratio of treated Zn-containing EAFD to $NiCl_2 \cdot 6H_2O$ ($R_{TZE/N}$, g g^{-1}) and calcination temperature on the synthesis and magnetic properties of as-synthesized samples were systematically investigated by X-ray Diffraction, Raman Spectroscopy and Physical Property Measurement System. It is indicated that $R_{TZE/N}$ and calcination temperature had important influences on the synthesis of single-phase spinel ferrite and their magnetic properties. Pure metal-doped Ni–Zn ferrite, which exhibited good magnetic properties of higher saturation magnetization (Ms, 60.5 emu g^{-1}) and lower coercivity (Hc, 49.8 Oe), was obtained when the calcination temperature was controlled at 1100 °C for 2 h with $R_{TZE/N}$ of 1:0.9.

Keywords Zn-containing EAFD · Solid state reaction · Metal-doped Ni–Zn ferrite · Magnetic property

Introduction

Zinc-containing electric arc furnace dust (Zn-containing EAFD), a solid waste produced during the process of steelmaking, contained many valuable metals such as iron (Fe), Zn, lead (Pb), chromium (Cr), manganese (Mn), etc. [1, 2]. Up to now, main efforts have been taken on only extracting Zn from Zn-containing EAFD by

H. Wang · M. Guo (✉) · M. Zhang
School of Metallurgical and Ecological Engineering,
University of Science and Technology Beijing, Beijing 100083, China
e-mail: guomin@ustb.edu.cn

H. Wang
e-mail: wanghuigang0822@126.com

L. Zhang et al. (eds.), *Energy Technology 2017*,
The Minerals, Metals & Materials Series, DOI 10.1007/978-3-319-52192-3_20

pyrometallurgical and hydrometallurgical processes. While, increasing metals recovery and making full use of the valuable metals are still big challenges for the comprehensive utilization of Zn-containing EAFD.

Spinel ferrites such as MFe_2O_4 (M: nickel (Ni), Zn, Mn, magnesium (Mg), etc.) has attracted considerable interests and efforts due to its stability, novel magnetic/electric property, and great applications [3, 4]. Among them, Ni–Zn spinel ferrite with high saturation magnetization (Ms) and low coercivity (Hc) has been widely used as soft magnetic materials, thus, obtaining high Ms and low Hc has always been the focus of the study. Solid state reaction method is considered as the conventional and economical method for the synthesis of spinel ferrites. Yadoji et al [5] and Rezlescu et al. [6] obtained Ni–Zn ferrite and pointed at that chemical composition and suitable doping of calcium (Ca), Mn, Mg, etc. into Ni–Zn ferrite could improve its Ms value to some extent.

Based on the above analysis, it is sure that the chemical composition plays a key role in determining the magnetic properties, and ferrites with multi-components may improve the magnetic properties. Considering the fact that Fe, Zn, Pb, Cr, Mn, etc. coexisted in the Zn-containing EAFD, so, by adjusting the components of dust, it is reasonable to assume that the mole ratio of Fe to M (Zn, Pb, Mn, Mg etc.) of around 2.0 can be obtained, which leads to direct synthesis of spinel ferrites from the dust. However, none of the effort has been taken on this issue till now.

In this paper, pure metal-doped Ni–Zn ferrite with spinel structure was directly synthesized from Zn-containing EAFD by using solid state reaction method. The influences of the added mass of $NiCl_2{\cdot}6H_2O$ and calcination temperature on the synthesis of metal-doped ferrite were study by XRD and Raman spectroscopy. The synthesis mechanism of Ni–Zn ferrite was also discussed. In addition, the magnetic properties of as-synthesized samples were tested. Under the optimum preparing conditions, the single-phase metal-doped Ni–Zn ferrite with high Ms and low Hc was obtained successfully. This paper may explore a novel and simple pathway for the comprehensive utilization of Zn-containing EAFD.

Experimental

Material

Analytical reagent grade $NiCl_2{\cdot}6H_2O$ and hydrochloric acid (HCl) (36–38 wt%) were purchased from Sinopharm Chemical Reagent Co., Ltd., China. Zn-containing EAFD was supplied by Tianjin Pipe (Group) Corporation. The treated Zn-containing EAFD was obtained by the leaching of HCl solution (0.5 mol L^{-1}). The chemical compositions and major mineralogical phases of Zn-containing EAFD and treated Zn-containing EAFD are presented in Table 1 and Fig. 1.

Figure 1a indicated that two major phases were identified as franklinite ($ZnFe_2O_4$) and magnetite (Fe_3O_4), and minor $CaCO_3$, SiO_2 and KCl. After

Table 1 Chemical compositions of Zn-containing EAFD and treated Zn-containing EAFD

Composition	Fe	Zn	Ca	Si	K	Cl	Mn	Pb	Mg	Al	Cr
Zn-containing EAFD (wt%)	37.29	7.79	5.33	2.21	3.11	2.63	1.18	1.16	0.72	0.44	0.16
Treated Zn-containing EAFD (wt%)	49.92	8.65	0.42	2.83	0.12	0.05	1.50	1.06	0.58	0.47	0.20

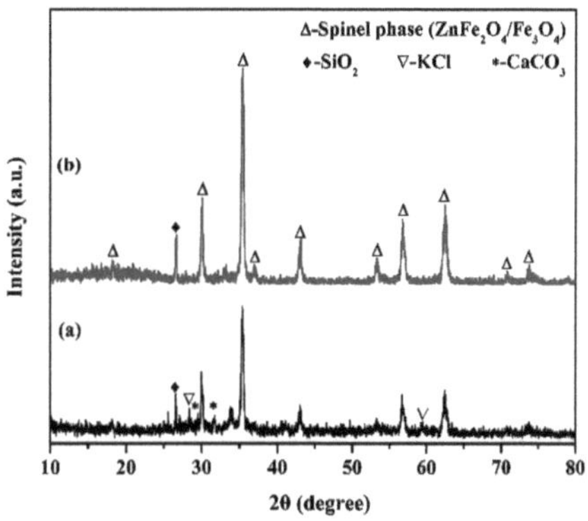

Fig. 1 XRD patterns of (*a*) Zn-containing EAFD and (*b*) treated Zn-containing EAFD

treatment, Ca, K and Cl etc. were mostly removed (Table 1) and the main phases were only $ZnFe_2O_4$ and Fe_3O_4 as presented in Fig. 1b. The mole ratio of Fe to M (Zn, Ca, Mn, Pb and etc.) in the treated Zn-containing EAFD was about 4.0 which was larger than 2.0, suggesting that the amount of M was insufficient to synthesize single phase ferrite.

Experimental Procedure

Treatment of Zn-Containing EAFD

Zn-containing EAFD (20 g) and 0.5 mol L^{-1} HCl solution (200 mL) were mixed in a 500 mL erlenmeyer flask, then electromagnetically stirred at 1000 rpm for 10 h at 25 °C. Next, solid-liquid separation was conducted with a speed of 3000 rpm for 5 min. Subsequently, the treated Zn-containing EAFD was dried at 105 °C for 24 h for further experiments.

Synthesis of Metal-Doped Ni–Zn Ferrite

1.0 g treated Zn-containing EAFD was mixed with a certain amount of $NiCl_2{\cdot}6H_2O$ (0.6, 0.7, 0.8 and 0.9 g). After mixed adequately in mortar by grinding, the

Fig. 2 General flow sheet of synthesizing of metal-doped Ni–Zn ferrites from Zn-containing EAFD

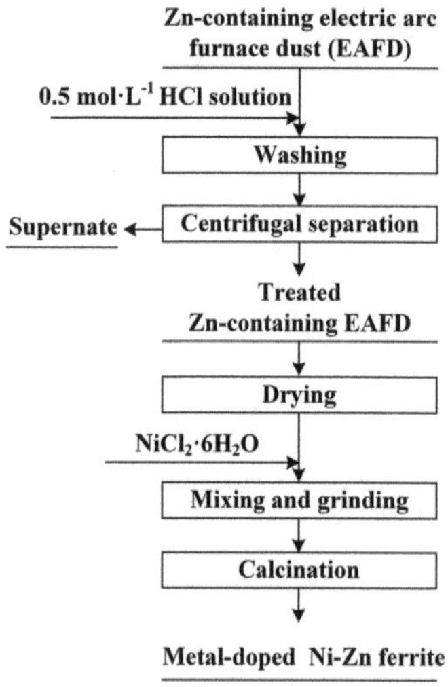

homogeneous mixture was transferred into an alumina crucible, and placed in a muffle furnace. Next, the temperature rising process began and the soaking time was 2.0 h at 800, 900, 1000 and 1100 °C. The flow diagram is illustrated in Fig. 2.

Analysis and Characterization

The chemical compositions and mineralogical phases of the Zn-containing EAFD, treated Zn-containing EAFD and as-prepared samples were carried out by X-ray Fluorescence (XRF-1800, Japan) and X-ray Diffraction (XRD, Japan, Rigaku), respectively. A laser confocal Raman spectrometer (JY-HR800, France, Jobin Yvon Company) was used to analyze the structural variation of as-prepared samples at room temperature. The magnetic properties of the formed samples were tested by Physical Property Measurement System [PPMS, America, 9T (EC-II)].

Results and Discussion

Synthesis Mechanism of Metal-Doped Ni–Zn Ferrite

During the calcination process, the added $NiCl_2·6H_2O$ melts and dehydrates simultaneously with temperature up to around 50 °C, then it hydrolyzes completely and generates nickel oxide (NiO) with temperature increasing to about 220 °C according to the result of Thermogravimetry-Differential Scanning Calorimetry (TG-DSC). The formed NiO would react with the main contents such as Fe_3O_4 and

$ZnFe_2O_4$ in treated Zn-containing EAFD in the air to produce spinel ferrite (Ni, Zn) Fe_2O_4. The main reactions can be described as follows:

$$NiCl_2{\cdot}6H_2O(s) = NiCl_2{\cdot}6H_2O(l) = NiCl_2 + 6H_2O \tag{1}$$

$$NiCl_2 + H_2O = NiO + 2HCl(g) \tag{2}$$

$$\begin{aligned} &x\mathrm{NiO} + y\mathrm{Fe_3O_4} + \frac{y}{4}\mathrm{O_2} + z\mathrm{ZnFe_2O_4} \\ &= (\mathrm{Ni}_x\mathrm{Zn}_z)\mathrm{Fe}_{2x+2z}\mathrm{O}_{4x+4z} + \frac{3y-2x}{2}\mathrm{Fe_2O_3} \\ &(3y > 2x, \mathrm{Fe/M} > 2.0) \end{aligned} \tag{3}$$

$$\begin{aligned} &x\mathrm{NiO} + y\mathrm{Fe_3O_4} + \frac{y}{4}\mathrm{O_2} + z\mathrm{ZnFe_2O_4} \\ &= (\mathrm{Ni}_x\mathrm{Zn}_z)\mathrm{Fe}_{2x+2z}\mathrm{O}_{4x+4z} \quad (3y = 2x, \mathrm{Fe/M} = 2.0) \end{aligned} \tag{4}$$

Moreover, in terms of spinel structural conditions, it facilitates the metal doping and ions diffusion because of the existence of clearance vacancies and lattice imperfections within the unit cell of the ferrite, which is easy to obtain a single phase spinel ferrite [7]. So, tiny Ca, Mn, Pb, Cr, etc. existed in the treated Zn-containing EAFD may be dissolved in the (Ni, Zn)Fe_2O_4 forming single-phase metal-doped Ni–Zn ferrite MFe_2O_4 (M: Ni, Zn, Ca, Mg, Mn, Pb; Fe: Fe, Al, Cr). Based on the above analysis of reactions (3 and 4), the mass ratio of treated Zn-containing EAFD to $NiCl_2{\cdot}6H_2O$ ($R_{TZE/N}$, g g^{-1}) and calcination temperature played important roles in determining the synthesis of single-phase metal-doped spinel ferrite.

Effect of $R_{TZE/N}$

Effect of $R_{TZE/N}$ on the Synthesis of Metal-Doped Ni–Zn Ferrite

Figure 3A gives the XRD patterns of as-synthesized samples with different $R_{TZE/N}$. It can be observed that (Zn, Ni)Fe_2O_4 and tiny α-Fe_2O_3 were both found with decreasing $R_{TZE/N}$ from 1:0.6 to 1:0.7. When $R_{TZE/N}$ was controlled at 1:0.8 or 1:0.9, all diffraction peaks were the standard diffraction peaks of spinel ferrite (JCPDS: 01-087-2337) with cubic structure, suggesting that single-phase Ni–Zn ferrite could be successfully synthesized, and all the other phases, i.e. Si-containing, Ca-containing phases, and etc., entered into the lattice of the ferrite.

In addition, it can be also seen from Fig. 3B that with $R_{TZE/N}$ varying from 1:0.6 to 1:0.9, there was a continuous shift in 2θ towards a higher angle side, further confirming the elements existed in the Zn-containing EAFD were doped into the obtained ferrites. The lattice parameters *a*, average grain sizes *d* and X-ray density

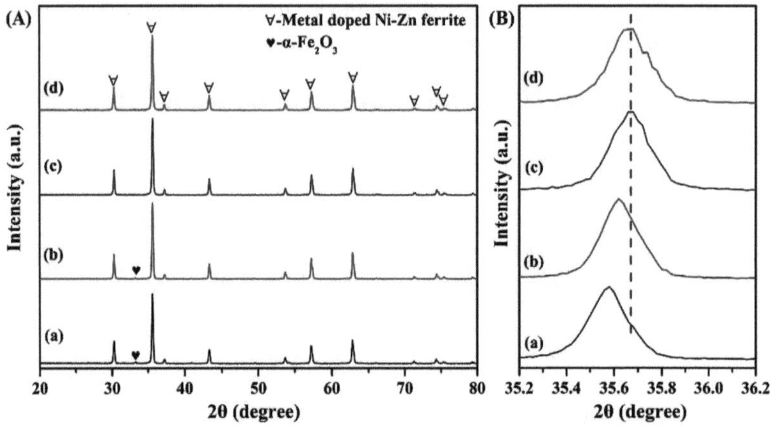

Fig. 3 **A** XRD patterns of as-synthesized samples and **B** shift in 2θ from 35.2° to 36.2° with varying $R_{TZE/N}$ (*a*) 1:0.6; (*b*) 1:0.7; (*c*) 1:0.8; (*d*) 1:0.9 calcined at 1000 °C for 2 h

d_x of as-synthesized ferrites were calculated according to the following Eqs. (5), (6) and (7) [8], and the results are summarized in Table 2.

$$a = \frac{\lambda}{2\sin\theta}\sqrt{h^2 + k^2 + l^2} \tag{5}$$

$$d = \frac{0.9\lambda}{\beta\cos\theta} \tag{6}$$

$$d_x = \frac{8M}{Na^3} \tag{7}$$

where a is lattice constant, nm; d is average grain size, nm; λ is wavelength of X-ray radiation (0.15406 nm); 2θ is the position of the highest diffraction peak (3 1 1) in the experimental XRD patterns; and β is the full width at half maximum (FWHM) of the diffraction peak (3 1 1) in the experimental XRD patterns; M is the molecular weight; N is Avogadro's number, d_x is the X-ray density, g cm^{-3}.

From Table 2, it can be seen that with increasing amount of Ni from 1:0.6 to 1:0.9, the lattice constant a and the average grain size d of the synthesized ferrites decreased from 0.8353 to 0.8343 nm, and from 46.64 to 43.87 nm, while the X-ray density (d_x) increased from 5.325 to 5.348 g cm^{-3}.

Table 2 Characteristic parameters of as-synthesized ferrites with different $R_{TZE/N}$

$R_{TZE/N}$	a (nm)	d (nm)	d_x (g cm^{-3})
1:0.6	0.8353	46.64	5.325
1:0.7	0.8348	46.36	5.337
1:0.8	0.8344	44.19	5.344
1:0.9	0.8343	43.87	5.348

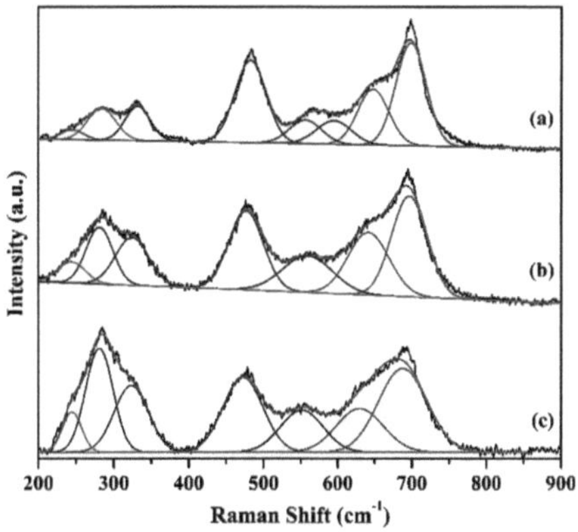

Fig. 4 Raman spectra of as-synthesized samples with different $R_{TZE/N}$ calcined at 1000 °C for 2 h. (*a*) 1:0.7, (*b*) 1:0.8, (*c*) 1:0.9

The room temperature Raman spectra was carried out to understand the structural transition of as-synthesized samples and the results are illustrated in Fig. 4 and Table 3.

As for the three samples, it can be seen from Fig. 5 and Table 3 that Raman peaks located at ~690 and ~640 cm^{-1} are assigned to $A_{1g}(1)$ and $A_{1g}(2)$ modes, reflecting the stretching vibration of Fe–O and M–O chemical bond in tetrahedral sites. Raman shifts at ~555, ~475, ~280 and ~245 cm^{-1} are assigned to T_{2g} modes and ~330 cm^{-1} are assigned to E_g modes reflecting the bending vibration of A–O chemical bond in tetrahedral sites in spinel structure. When $R_{TZE/N}$ varied from 1:0.8 to 1:0.9 (Fig. 5b, c), the pure spinel ferrites were successfully synthesized. However, when $R_{TZE/N}$ was controlled at 1:0.7 (Fig. 5a), Raman active peak of 595 cm^{-1} assigning to α-Fe_2O_3 [9] appeared. These results can also be confirmed by Fig. 4.

Table 3 Raman parameters of as-prepared samples with different $R_{TZE/N}$

Assignment	Raman modes (cm^{-1})			Vibrations	Ions involved
	1;0.7	1:0.8	1:0.9		
$A_{1g}(1)$	698	696	689	Symmetric stretching	A–O (Fe–O)
$A_{1g}(2)$	648	642	631	Symmetric stretching	A–O (M–O)
	595				
$T_{2g}(3)$	557	558	551	Asymmetric stretching	A–O
$T_{2g}(2)$	482	477	472	Asymmetric stretching	A–O
E_g	332	324	323	Bending	A–O
T_{2g}	284	281	281	Asymmetric stretching	A–O
T_{2g}	245	244	245	Asymmetric stretching	A–O

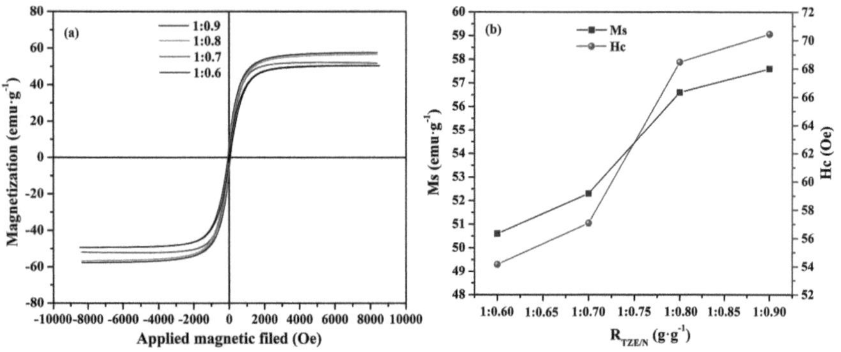

Fig. 5 **a** Room temperature hysteresis loops, **b** variation of the Ms and Hc values of synthesized samples with different $R_{TZE/N}$, calcined at 1000 °C for 2 h

Effect of $R_{TZE/N}$ on the Magnetic Properties of Metal-Doped Ni–Zn Ferrite

The hysteresis loops at room temperature are shown in Fig. 5. It can be seen that with increasing Ni content from 1:0.6 to 1:0.7, the Ms value increased slightly from 50.6 to 52.3 emu g^{-1} due to the decreasing content of nonmagnetic α-Fe_2O_3 in the synthesized samples (Fig. 3). Further increasing Ni content to 1:0.8 and 1:0.9, the Ms increased to 56.6 and 57.6 emu g^{-1}, respectively. This may be mainly ascribed to the formation of pure metal-doped Ni–Zn ferrites and doping of many ions. And the Ms variation trend o can be explained by Neel's two sublattice models [10].

As for the Hc value of the ferrite which is caused by displacement and rotation of domain wall influenced by anisotropy during the process of magnetization. With increasing Ni content from 1:0.6 to 1:0.9, Hc value increased from 54.2 to 70.4 Oe as shown in Fig. 5. This phenomenon may be attributed to the anisotropy increasing within the magnetic lattice and particle size decreasing [11].

Effect of Calcination Temperature

Effect of Calcination Temperature on the Synthesis of Metal-Doped Ni–Zn Ferrite

Figure 6 presents the XRD patterns under different calcination temperatures. Figure 6A indicated that all the diffraction peaks were the standard diffraction peaks of spinel ferrites, suggesting that the pure metal-doped Ni–Zn ferrites were obtained. Moreover, the diffraction peaks became sharper and narrower with increasing calcination temperature (Fig. 6B), further confirming the improved crystallinity of the synthesized ferrites.

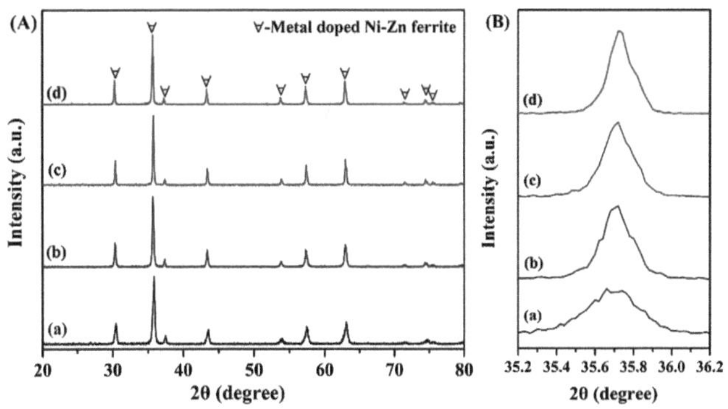

Fig. 6 **A** XRD patterns and **B** their partial enlargement drawing of 2θ from 35.2° to 36.2° with $R_{TZE/N}$ of 1:0.9 calcined for 2 h at (*a*) 800 °C, (*b*) 900 °C, (*c*) 1000 °C, (*d*) 1100 °C

Effect of Calcination Temperature on the Magnetic Properties of Metal-Doped Ni–Zn Ferrite

The magnetic property of the formed ferrites at room temperature is shown in Fig. 7. It can be seen that the Ms increased from 43.5 to 60.5 emu g^{-1}, while the Hc decreased from 93.0 to 49.8 Oe with the change of calcination temperature from 800 to 1100 °C. The enhanced Ms may be mainly ascribed to the increased grain size, improved crystallinity (Fig. 6B), and more magnetic Ni^{2+} ions at B-site. And the decreased Hc may depended on the larger grain size of the obtained ferrites with the increasing of calcination temperature. In addition, the formed single-phase metal-doped Ni–Zn ferrite synthesized from Zn-containing EAFD exhibited better magnetic properties (high Ms of 60.5 emu g^{-1} and low Hc of 49.8 Oe) compared with those synthesized from pure reagents [12, 13]. This might be due to the

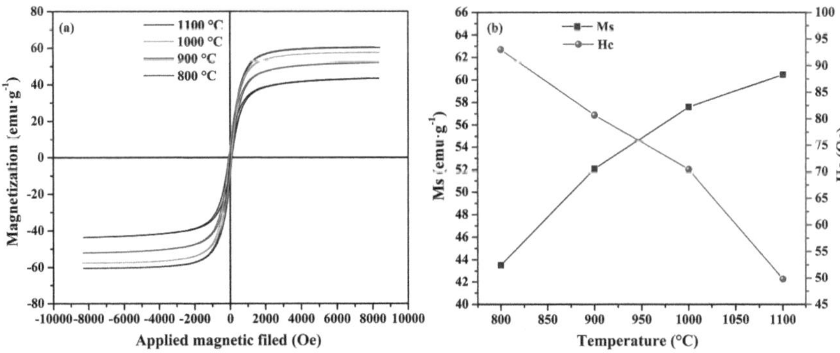

Fig. 7 **a** Room temperature hysteresis loops, **b** variation of the Ms and Hc values of synthesized ferrites under different calcination temperatures, $R_{TZE/N}$ of 1:0.9 for 2 h

multi-components doping of magnetic or nonmagnetic ions including Ni (2 μB), Mn (5 μB), Al (0 μB), Ca (0 μB), Mg (0 μB), Pb (4 μB) etc. into the obtained ferrites.

Conclusions

In this paper, single-phase metal-doped Ni–Zn ferrite with spinel structure was directly synthesized by only adding solid $NiCl_2{\cdot}6H_2O$ into Zn-containing EAFD using one-step solid state reaction method for the first time. $R_{TZE/N}$ and the calcination temperature had great influences on the synthesis and magnetic properties of single-phase metal-doped Ni–Zn ferrite. Under the optimum conditions (1100 °C, 2 h, $R_{TZE/N}$ = 1:0.9), pure metal-doped Ni–Zn ferrites, which showed higher Ms of 60.5 emu g^{-1} and lower Hc of 49.8 Oe, were obtained. This proposed process realized the transformation of the dust from solid waste to high value-added spinel ferrite.

Acknowledgements This work was financially supported by the National Basic Research Priorities Program of China (No. 2014CB643401), the National Natural Science Foundation of China (No. 51672025), and Shanxi Collaborative Innovation Center of High Value-added Utilization of Coal-related Wastes.

References

1. F. Kukurugya, T. Vindt, T. Havlík, Behavior of zinc, iron and calcium from electric arc furnace (EAF) dust in hydrometallurgical processing in sulfuric acid solutions: thermodynamic and kinetic aspects. Hydrometallurgy **154**, 20–32 (2015)
2. R.A. Shawabkeh, Hydrometallurgical extraction of zinc from Jordanian electric arc furnace dust. Hydrometallurgy **104**(1), 61–65 (2010)
3. K.-S. Lin et al., Synthesis and characterization of nickel ferrite nanocatalysts for CO_2 decomposition. Catal. Today **174**(1), 88–96 (2011)
4. G. Srinivasan et al., Microwave and MM-wave magnetoelectric interactions in ferrite-ferroelectric bilayers. Eur. Phys. J. B **71**(3), 371–375 (2009)
5. P. Yadoji et al., Microwave sintering of Ni–Zn ferrites: comparison with conventional sintering. Mater. Sci. Eng., B **98**(3), 269–278 (2003)
6. E. Rezlescu et al., Effect of substitution of divalent ions on the electrical and magnetic properties of Ni-Zn-Me ferrites. IEEE Trans. Magn. **36**(6), 3962–3967 (2000)
7. M. Liu et al., The influence of Ca-doping to the strength of sintered Ni-Cu-Zn ferrites. Mater. Rev. **23**(10), 52–54 (2009)
8. S. Amiri, H. Shokrollahi, Magnetic and structural properties of RE doped Co-ferrite (RE=Nd, Eu, and Gd) nano-particles synthesized by co-precipitation. J. Magn. Magn. Mater. **345**, 18–23 (2013)
9. I.R. Beattie, T.R. Gilson, The single-crystal Raman spectra of nearly opaque materials. Iron (III) oxide and chromium (III) oxide. J. Chem. Soc. A **l**, 980–986 (1970)
10. M.Y. Li, G. Li, *Physical chemistry of ferrites* (Science Press, Beijing, 1978)

11. C. Caizer, M. Stefanescu, Magnetic characterization of nanocrystalline Ni–Zn ferrite powder prepared by the glyoxylate precursor method. J. Phys. D **35**(23), 3035–3040 (2002)
12. M.A. Gabal, R.M. El-Shishtawy, Y.M. Al, Angari, Structural and magnetic properties of nano-crystalline Ni–Zn ferrites synthesized using egg-white precursor. J. Magn. Magn. Mater. **324**(14), 2258–2264 (2012)
13. D.H. Bobade, S.M. Rathod, M.L. Mane, Sol–gel auto-combustion synthesis, structural and enhanced magnetic properties of Ni^{2+} substituted nanocrystalline Mg–Zn spinel ferrite. Phys. B **407**(18), 3700–3704 (2012)

Part VII
Deriving Value from Challenging Waste Materials: Recycling and Sustainability Joint Session: Deriving Value from Challenging Waste II

Hydrometallurgical Processing of Copper Smelter Dust for Copper Recovery as Nano-particles: A Review

D.O. Okanigbe, A.P.I. Popoola and A.A. Adeleke

Abstract In view of the steady depletion of primary sources of copper and the increased global demand for refined copper, it becomes necessary to explore some secondary sources for possible extraction of copper. The waste copper smelter dust (CSD) is a rich secondary resource for copper as shown by the chemical composition of the South African Palabora coppers smelter plant CSD that assayed 18.02, 13.36, and 3.4 wt% copper, iron and sulphur; respectively. Studies on CSD have focused majorly on either dust characterization or treatment, while hydrometallurgical extraction without pretreatment and with pretreatment using techniques such as oxidative roasting are also considered quite attractive. The challenge of iron dissolution during the leaching stage in these processes necessitates adequate purification of the leach liquor before the extraction of the metal as nano-particles. Hence, this review examined the theories relating to the characterization and treatment of CSD for copper recovery as nanoparticles; with factors having a bearing on the treatment process such as kinetics considered with the aim of providing scientific basis for the research.

Keywords Extraction · CSD · Assay · Characterization · Hydrometallurgy · Nano-particles · Computational modeling

D.O. Okanigbe · A.P.I. Popoola (✉)
Department of Chemical, Metallurgical and Materials Engineering,
Tshwane University of Technology, Pretoria 0183, South Africa
e-mail: okanigbedo@tut.ac.za; odanielifeoma@yahoo.com

A.A. Adeleke (✉)
Department of Materials Science and Engineering,
Obafemi Awolowo University, Ile-Ife 220002, Nigeria
e-mail: abrahamadeleke@gmail.com

L. Zhang et al. (eds.), *Energy Technology 2017*,
The Minerals, Metals & Materials Series, DOI 10.1007/978-3-319-52192-3_21

Introduction

The United States of America (U.S.A) was the leading copper consumer and is credited with roughly 16% of the total consumption of world refined copper, which approximates to 2.4 million tons as at 1980 [1]. In 2002, the U.S.A was surpassed by China as the world's leading consumer of refined copper. The thriving economy in China further added to its increase in annual refined copper consumption during the 8 years from 1999 to 2007 as illustrated on Fig. 1 [2].

Consequently, it is predicted that by 2025, the global demand for refined copper will increase to 29.5 million metric tons from the 22.4 million metric tons predicted for 2015 [3]. This increasing global demand for refined copper has impacted negatively on the primary sources of copper, thus making secondary copper processing an ever-increasing importance within the global copper industry [4].

Recently, the copper industry has witnessed significant growth in the recovery of copper and other metal values from secondary materials [5–7] such as:

(a) Metallurgical wastes—copper smelter dust, low grade slags, residues, anode slimes, oxide residues etc.
(b) Industrial wastes—bars, pipes, wire, copper sheeting, ship screws, etc.
(c) Consumer wastes—bronze and brass applications.
(d) Electrical and electronic waste (e-waste)—domestic electrical, computer, audio-visual, and
(e) Telecommunication appliances

At present, the disposal of metallurgical wastes such as copper smelter dust (CSD) is becoming costly owing to increasingly stringent environmental regulations [8]. Furthermore, the physical and chemical nature of CSD (such as the content of heavy metals) causes it to be classified as hazardous waste [9]. While the South African National Standard (SANS) annual target level of dust is 300 mg/m^2/day and the industrial action level of dust is 1200 mg/m^2/day [10]. Table 1 presents a breakdown of the CSD deposition rates around the Palabora

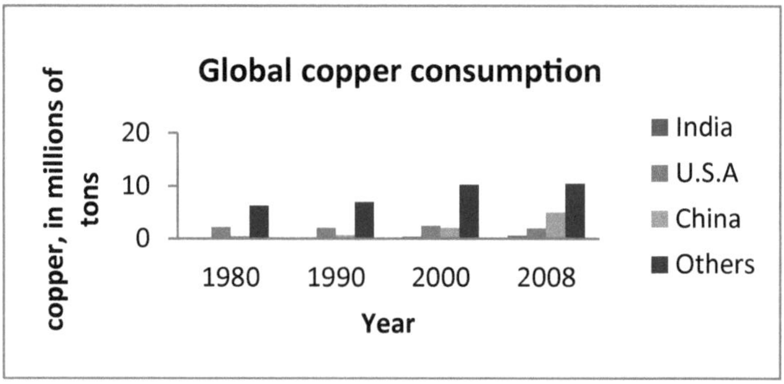

Fig. 1 Global consumption of copper from 1980 to 2008 (adapted from [2])

Table 1 Breakdown of monthly dust deposition rates 23 June 2009 to 23 June 2010 [11]

Parameter	Smelter	Vermiculite operation plant	Phalaborwa town
Monthly	12	12	9.00
Minimum (mg/m^2/day)	528.58	3175.99	101.37
Maximum (mg/m^2/day)	2643.58	2643.58	430.39
Average (mg/m^2/day)	1376.57	4890.71	224.01

mining company's (PMC) smelter and in its vermiculite operations plant also in Phalaborwa Town [11]. In view of the above, there has been an increasing interest in developing processes for the recovery of metals from these wastes [12].

Often, pyrometallurgical and hydrometallurgical processes are employed for treating such residues [12]. But the pyrometallurgical processes require investment in expensive equipment, results in high energy consumption and thus, leads to the production of worthless residues [13–16]. As a result, much focus has been on hydrometallurgical processes [12].

The leach-solvent extraction-electrowinning process option has been the traditional hydrometallurgical route of recovering copper from its source as copper cathodes [17, 18]. But recovering the copper value as nano-particles is more attractive, because of its vast applications in areas such as heat transfer and microelectronics and coupled with the fact that it is significantly less expensive than silver and gold; thus making it a worthy alternative to noble metals in some applications [19, 20]. Currently, a novel approach at laboratory scale was reported by Darezereshki and Bakhtiari [21]; it combines hydrometallurgical techniques with nano-technology to recover the oxide of copper as precipitate of nano-particles from leach solution of CSD. Yet; additional work is required to enhance the process to be able to produce metallic copper nano-particles with acceptable grade and recovery from this metallurgical waste.

Hence, in the light of the above mentioned, this review will be looking at the theoretical background, the characterization and treatment of CSD, the recovery of copper as nano-particles from leach solution of CSD.

Theoretical Background

Copper Smelter Dust (CSD)

Copper smelter dust is the product consisting of SO_2, N_2, O_2, water vapor, heavy metals and other impurities recovered from exhaust gas streams found in furnaces, flues and settling chambers as a result of roasting, smelting and converting operations from copper refining processes as illustrated on Fig. 2 [17, 22, 23]. The basic elements and their concentrations, are a function of the following: the characteristics of the sulphureted ores that are processed, the species and their concentration, the reactor type used in the different stages of the process together with the conditions

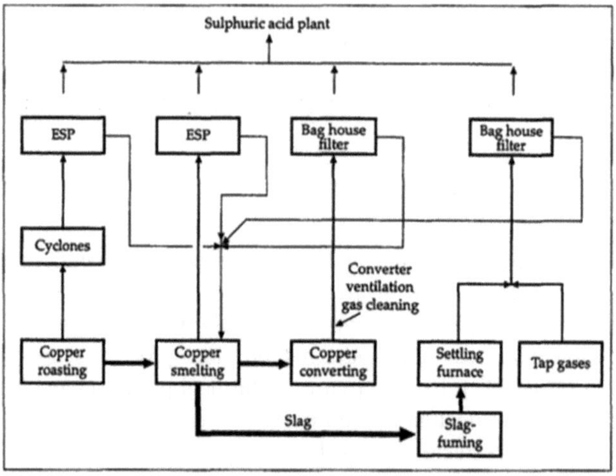

Fig. 2 Schematic drawing of the locations of the gas cleaning devices in a copper-plant (adapted from [22])

under which these stages operate [24]. All these factors add up to influence the chemical and mineralogical characteristics of the copper smelter dusts (Tables 2, 3, 7 and 8), thus giving a reason for their variability and as a result, the processes chosen for their treatment, are based on the recovery of valuable elements and the stabilization of undesirable elements present in the copper smelter dusts [25].

Mechanism of Dust Formation

There are two kinds of dust that forms in metallurgical smelting processes, namely mechanically formed dust which consists of small particles of the charging material that get carried away with the process gas and the chemically formed dust, which consists of vaporized components that condense into particles as the process off-gas temperature cools down. Mechanical dust forms as a result of the entrainment of small solid and/or liquid particles from charged material into process off-gas. Mechanical entrainment involves the elutriation by the process gas of small solid and liquid particles that travel at a relatively high velocity through the process [26, 27]. Process reactions can cause the charging material to collapse or fragment into smaller dust particle.

Chemical dust formation is caused by the vaporization of components from the process, followed by the condensation of the components from the process gas at lower temperatures in the off-gas system [27]. Materials with low melting points such as As, Sb, their oxides and sulphide, often have high vapour pressure and tend to vaporise during smelting, forming chemical dust. Chemical dust has a very fine particle size because of the nucleation and growth processes occurring as the

Table 2 Typical chemical compositions of CSD in literatures as determined using different analytical techniques

Reference	Chemical composition of CSD (wt%)										Analytical technique(s)
	Cu	Fe	Zn	S	Bi	As	Cd	Pb	Sb	Other	
Oriol et al. (2011)	4.0	–	18	>7.0	1.1	20.50	9.5	8.0	1.2	–	ICP-AES, ICP-MS
Vítková et al. [38]	41.64	29.61	0.33	12.98	2.30	0.43	0.03	0.33	–	12.35	ICP-AES, HPLC
Bakhtiari et al. [15, 16]	35.5	15.3	–	12.2	–	–	–	–	–	55.5	AAS
Alguacil et al. [17]	24.5	14.0	0.15	–	–	0.9	–	0.08	–	60.37	AAS, ICP-MS
Montenegro et al. [8]	10.8	0.8	15.6	10.4	3.5	19.4	–	7.8	0.1	31.6	ICP-AES
Morales et al. [39]	27.0	11.0	5.80	7.50	0.21	13.0	0.16	1.50	–	33.83	XRF
Ha et al. [43]	10.9	1.6	7.8	–	1.9	7.1	1.3	14.2	0.1	55.1	ICP-AES
Wu et al. [42]	2.69	0.44	38.63	–	–	–	–	–	–	58.24	XRF
Vakylabad et al. [40, 41]	33.67	21.19	–	9.02	–	0.80	–	6.64	–	28.68	AAS
Qiang et al. [12]	7.53	–	40.21	–	–	–	–	6.62	–	45.64	ICP-OES, IC

Table 3 Typical chemical compositions of CSD from Sarcheshmeh Copper Mine, Iran; Mufulira copper smelter, Zambia and Chilean copper smelting plant, Chile, as determined using different analytical techniques

Elements	CuO	Fe_2O_3	S	ZnO	PbO	Bi_2O_3	CaO	MgO	TiO_2	SiO_2	Al_2O_3	Others
wt%	18.02	13.36	3.44	0.27	0.12	0.02	3.52	2.86	1.11	33.06	22.19	1.95

process off-gas cools down. In copper smelting processes, easily volatile components include Ag, As, Pb, Sb, and Zn. Copper can also vaporize in the hot combustion zone of the furnaces. Oxygen enrichment increases the combustion temperature in the flash furnace and as a result, more copper is vaporized. Vaporized copper particles are very fine and easily drift out of the furnace with the off-gas.

Uses of Copper Smelter Dust (CSD)

Although there is no report in literature of application of copper nano-particles obtained from leach solution of CSD; except for its recovery from this waste [21]. Yet, copper nano-particles, due to their excellent physical and chemical properties and cheap production method, have been of great interest. Copper nano-particles have extensive applications as heat transfer systems [28], antimicrobial materials [29, 30], super strong materials [31, 32], sensors [33, 34] and catalysts [35, 36].

Studies on CSD can be divided into two groups; those focused on dust characterization [22, 37] and those focused on dust treatment [8, 15–17, 38–41], even though the two are often considered together.

Characterization of Copper Smelter Dust

Chemical Composition of Copper Smelter Dust

There are about six analytical techniques reported in literatures that have been used to determine the chemical composition (CC) of CSDs and these has been summarized into a table of analytical methods and relevant research findings (Table 2).

Therefore, the findings from the use of these analytical techniques showed that overall, this metallurgical waste often contains contaminants (Tables 2 and 3) such as antimony (Sb), bismuth (Bi), arsenic (As), lead (Pb), zinc (Zn), cadmium (Cd), selenium (Se), tellurium (Te) and compounds thereof [8, 12, 42, 43]. But it often contains significant amount of copper (Tables 2 and 3), titanium and alumina as in the case of PMC's CSD (Table 3).

Particle Size of Copper Smelter Dust

The major techniques reportedly adopted to determine the particle size distribution (PSD) of this metallurgical waste has been summarized on Table 4. As can be seen on Table 4, the marlven particle size analyzer (MPA) was used to determine the PSD of a CSD from a smelter in South Korea; nevertheless, results using this techniques tend to appear coarser than those of other methods. Austin and Shah [44] proposed a method for inter-conversion of laser diffraction (MPA) and sieve-size distributions (screening), and a simple conversion can be developed by regression for materials of constant characteristics. Additionally, the outcomes can depend on the relative refractive indices of the solid particles and liquid medium (usually,

Table 4 Particle size of copper smelter dust determined using different methods

S. No.	MPA	Wet sieving	SEM-SE/BSE	Cyclosizer	Reference
1	–	80% less than 38 μm	–	Coarse and fine fractions	Vakylabad et al. [41]
2	–	70% of particles were finer than 80 μm (d_{70} = 80)	–	–	Bakhtiari et al. ([15, 16]; 2011)
3	5–50 μm	–	–	–	Ha et al. [43]
4	–	–	Spherical particles and other irregular particles with size ranging from 1–10 μm	Coarse and fine fractions	Morales et al. [39]

Table 5 Showing the results of particle size distribution of the waste copper smelter dust of PMC

Size range (μm)	Mass (g) retained	Mass (%)	Cumulative mass (%) undersize	Cumulative mass (%) oversize
+300	51.00	12.39	87.61	12.39
−300 to +212	5.70	1.39	86.22	13.78
−212 to +150	4.50	1.09	85.13	14.87
−150 to +106	3.90	0.95	84.18	15.82
−106 to +75	6.90	1.68	82.50	17.50
−75 to +53	10.00	2.43	80.07	19.93
−53	329.50	80.07		

Table 6 Methods reportedly used in literature to determine PSD of CSD (adapted from Wills 2003)

Method	Wet or dry	Fractionated sample	Approximate size range (μm)
Test sieving	Both	Yes	5–100,000
Laser diffraction	Both	No	0.1–2000
Optical microscopy	Dry	No	0.2–50
Elutriation	Dry	Yes	5–45

Table 7 The mineralogical composition of the CSD from PMC determined using XRD

S. No.	Minerals	Chemical formulae of minerals	Weight %
1	Cuprospinel	$Cu_{6.88}Fe_{17.12}O_{32.00}$	24.34
2	Chalcopyrite	$Cu_{4.00}Fe_{4.00}S_{8.00}$	7.95
3	Mullite Syn	$Al(Al_{0.69}\ Si_{1.22}\ O_{4.85})$	42.97
4	Gypsum	$Ca(SO_4)(H_2O)_2$	11.69
5	Quartz low	SiO_2	11.45
6	Magnetite	$Mg_{0.04}Fe_{2.96}O_{4.00}$	1.60

though not necessarily, water), and even particle shape. Furthermore, on one hand the most commonly used method for PSD determination of CSDs is the wet screening method (Table 4), probably because of its simplicity. While on the other hand the cyclosizer is often used to classify the residues produced after the dissolution studies of CSDs, into coarse and fine fractions (Table 4).

Hence, the average size of this CSD as determined by these aforementioned methods fall within the size range of 5–50 μm (Tables 4 and 5) as have been reported in literature [39, 43]. Furthermore, the CSD is said to be an ultrafine material requiring the use of the wet aspect (Table 6) of these techniques for its PSD analysis. This is in order to minimize losses through clogging and prevent dusting issues that could lead to environmental pollution [45].

Mineralogical Composition of Copper Smelter Dust

Depending on where this dust was collected (Fig. 2), often the main minerals of the CSD are magnetite, cuprospinel, chalcopyrite, chalcanthite, delafosite,

Table 8 Mineralogical composition of copper smelter dust determined by XRD, SEM/EDS AND TEM/EDS

S. No.	Cu	Fe	Zn	Pb	As	Other	Technique	Reference
1	$CuSO_4$, CuS	–	ZnS, $ZnSO_4$	$PbSO_4$	As_2O_3	–	XRD	Oriol et al. (2011)
2	$CuSO_4.5H_2O$, $CuFeO_2$	Fe_3O_4	–	–	–	–	XRD	Vítková et al. [38]
3	Cu-Ca-SO_4, $CuFeO_2$, $CuSO_4$	–	–	–	–	–	TEM	Vítková et al. [38]
4	$CuSO_4.5H_2O$ $Cu_3(SO_4)(OH)_4$,$CuFe_2O_4$, $CuFeO_2$, Cu_2S, $CuFeS_2$, Cu_5FeS_4, Cu(Fe)S	Fe_3O_4 Fe-Cu-Co-(As)	–	–	–	SiO_2, $CaSO_4.2H_2O$, $CaCO_3$	XRD, SEM/EDS, TEM/EDS	Ettler et al. 47
5	$CuFe_2O_4$, $CuSO_4$, CuO.$CuSO_4$, CuO, Cu_3AsS_4, $CuFeO_2$	Fe_3O_4	$ZnSO_4$	As_4O_6 As_4O_6	–	–	XRD	Morales et al. [39]
6	Water soluble phase content 5.43%, Oxide phase content 1.54%, Sulphide phase content 4.21%	–	–	–	–	others content 0.15%	XRD	Xu [46]
7	$CuSO_4.5H_2O$, CuO.$3H_2O$, $CuFeO_4$	–	–	–	–	–	XRD	Alguacil et al. (2013)
8	Cu_2S $CuFeS_2$, CuS, Cu_5FeS_4	–	–	–	–	–	XRD, SEM/EDS	Vakylabad et al. ([41; 2011)
9	CuO	–	ZnO	PbO	–	$AlCl_3$	XRD	Qiang et al. [12]

Key: *XRD* x-ray diffractometer, *SEM/EDS* scanning electron microscopy/energy dispersive spectrometer, *TEM/EDS* transmission electron microscopy/energy dispersive spectrometer chalcocynite, zincosite and claudetite (Tables 7 and 8). The major technique used to determine the minerals in this dust is the XRD (Oriol et al. 2011; [39, 46]); although seldom used SEM/EDS and TEM/EDS for this purpose [40, 41, 47]. Below are the brief descriptions of the techniques used in the mineralogical analysis of CSD's, as reported in literature; also including the earlier discussed SEM, under particle size determination.

Hence, for an improvement on the recovery of copper from this metallurgical waste, a comprehensive physical-chemical-mineralogical characterization of the CSD is needed [37]. Since most characterization requirements are mainly based on current operating practices, there is the need to use different analytical techniques to appropriately characterize the CSD in order to generate more robust data for decision making.

Treatment of Copper Smelter Dust for Copper Recovery

Enrichment of Copper Smelter Dust

Gravity Separation Method

The gravity methods of separation are used to treat a great variety of materials ranging from heavy metals such as lead sulphide (specific gravity of 7.5) to coal (specific gravity of 1.3) at particle sizes in some cases below 50 μm (Wills 2003). So therefore, for an effective separation, there must be a marked density difference between the mineral(s) of interest and the gangue(s) minerals; thus obeying the concentration criterion, which is $\geq \pm 2.5$ (wills 2003). Also bearing in mind that the motion of a particle in a fluid is not just dependent on the specific gravity alone; but also on its particle size and shape [48].

Going through Table 9, one would clearly observe the marked difference in specific gravity (S.G.) of some of the minerals contained in the CSD—Chalcopyrite (S.G. = 4.30) and Bismuthinite (S.G. = 7.20), Tenorite (S.G. = 6.50) and Arsenolite (S.G. = 3.20) and Gunningite (S.G. = 3.87) for which separation is possible (i.e. $\geq \pm 2.5$) with this method; but with a particle sizes range from 10 to 38 μm (Fig. 3; Tables 4 and 5); only the modern gravity techniques (Knelson and Falcon concentrators) have proven to be efficient for beneficiation of materials in this particle size range of up to 50 μm(Wills 2003).

Magnetic Separation Method

Here, the main physical property of note is the magnetic susceptibility; separation using this method is possible, for the characteristic mineralogical composition of most CSDs −0.70–1.30 amps for Chalcopyrite, 0.50–1.00 dc amps for cuprospinel,

Table 9 Common minerals found in CSD and their specific gravity (84)

S. No.	Elements	Typical forms	Mineral	Specific gravity (SG)
1	Copper	Cu_2S	Chalcocite	5.50–5.80
		CuS	Covellite	4.60–4.80
		$CuFeS_2$	Chalcopyrite	4.10–4.30
		Cu_5FeS_4	Bornite	5.06–5.08
		$CuSO_4.5H_2O$	Chalcanthite	2.12–2.30
		$CuFeO_2$	Delafossite	5.41
		$CuFeO_4$	Cuprospinel	5.00–5.20
		CuO	Tenorite	6.50
		Cu_3AsS_4	Energite	4.40–4.50
2	Arsenic	As_2S_3	Arsenolite	3.87
3	Bismuth	Bi_2S_3	Bismuthinite	6.80–7.20
4	Cadmium	CdS	Greenockite	4.80–4.90
5	Iron	Fe_2O_3	Hematite	5.26
		FeS_2,	Pyrite	4.95–5.10
		Fe_3O_4	Magnetite	5.17–5.18
6	Lead	$PbSO_4$	Anglesite	6.30
7	Antimony	Sb_2S_3	Stibnite	4.63
8	Zinc	(Zn,Fe)S $ZnSO_4$	Sphalerite Gunningite	3.90–4.20 3.20

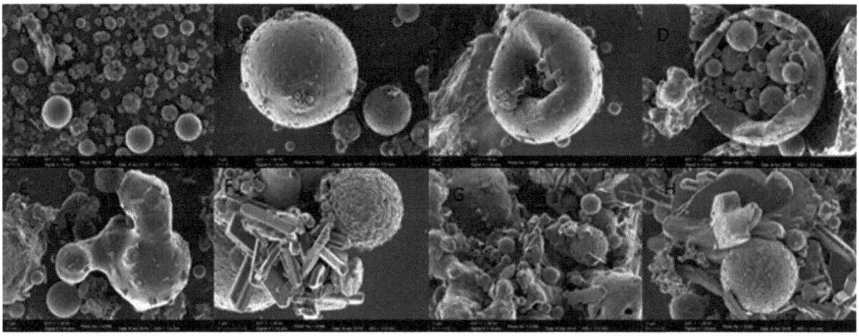

Fig. 3 **a**, **b**, **e–h** shows the variation in particle sizes in this dust sample, **c**, **d** shows how smaller particles with sizes of approximately 2 μm are embedded in larger particle of about 8–10 μm

0.01 amps Magnetite and 0.10–0.30 amps for Hematite (Table 10). Yet, the particle size remains an important factor to consider, if this method is to be used to separate the contained minerals from its gangue; considering the fine-ultrafine nature of the CSD, dusting will definitely be an issue to deal with should this method be used to upgrade it. Otherwise, the wet approach will be better suited for upgrading this metallurgical waste, like the wet high intensity magnetic separator (WHIMS).

Table 10 Magnetic property of minerals in CSD (85)

Minerals	Typical form	Magnetic susceptibility Total range (AMPS)	Magnetic susceptibility Best range (AMPS)	No.
Copper	Covellite	0.80 to >1.70	>1.70	2
	Chalcopyrite	0.30–1.70	0.70–1.30 dc	8
	Bornite	0.30–1.10	0.60–0.90	4
	Chalcanthite	0.60–1.20	0.80–1.00	3
	Delafossite	–	–	–
	Cuprospinel	0.10–1.10	0.50–1.00 dc	3
	Tenorite	–	–	–
	Energite	>1.70	>1.70	3
Arsenic	Arsenolite	1.00 to >1.70	>1.70	3
Bismuth	Bismuthinite	>1.70	>1.70	2
Cadmium	Greenockite	>1.70	>1.70	1
Iron	Hematite	0.25–0.50	0.10–0.30	11
	Pyrite	>1.70	>1.70	9
	Magnetite	<0.01 to 0.05	0.01	6
Lead	Anglesite	>1.70	>1.70	9
Antimony	Stibnite	>1.70	>1.70	6
Zinc	Sphalerite	1.30 to >1.70	>1.70	8
	Sphalerite (Fe)	0.20–1.20	0.60–1.00 dc	6
	Gunningite	0.90 to >1.70	1.10–1.20	1

Table 11 Thermal and Electrical conductivity of elements in CSD (49)

S. No.	Element	Typical forms	Thermal Conductivity (J/m s K)	Electrical Conductivity (1/mohm cm)
1	Copper	Covellite Chalcopyrite Bornite Chalcanthite Delafossite Cuprospinel Tenorite Energite	401.00	595.800
2	Arsenic	Arsenolite	50.20	30.030
3	Bismuth	Bismuthinite	7.92	9.363
4	Cadmium	Greenockite	96.90	146.413
5	Iron	Hematite Pyrite Magnetite	72.80	102.987
6	Lead	Anglesite	35.30	48.431
7	Antimony	Stibnite	24.40	25.641
8	Zinc	Sphalerite gunningite Zincosite	116.00	169.033

Electrostatic Separation Method

This is another minerals processing technique that depends on the difference in electrical conductivity (Table 11). Here, the difference between the electrical conductivity of copper and the gangue minerals in the CSD—595.800 Cu, 9.363 Bi, 102.987 Fe, 48.431 Pb and 25.641 Sb is significant [49]; and separation is possible on this basis; but what is of note is that the sample must be very dry for a good result, while bearing in mind that it is also particle size sensitive.

Froth Flotation Method

Froth flotation is a highly versatile method for physically separating particles based on differences in the ability of air bubbles to selectively adhere to specific mineral surfaces in mineral/water slurry. Froth flotation can be adapted to a broad range of mineral separations, as it is possible to use chemical treatments (Table 12) to selectively alter mineral surfaces so that they have the necessary properties for the separation. It is particularly useful for processing fine-grained ores that are not

Table 12 Chemical reagents to be used in metallurgical test work [86]

Manufacture	Chemical name	Active ingredient	Water soluble	Use
Cytec	7261A	Polyacrylamide	x	Depressant of metal sulfides
Cytec	7262	Polyacrylamide	x	Depressant of metal sulfides
Cytec	3894	Dialkyl thionocarbamate	–	Selective collector for Cu
Cytec	404	Mercaptobenzothiazole	x	Cu collector for tarnished Cu minerals
Nalco	TX-15281	Thio-ester	–	Selective collector for Cu
Nalco	TX-15155	Thio-ester	–	Selective collector for Cu
fisher	DG-13	Activated carbon 13	–	Flotation reagent collector
SNF	Sodium isopropyl xanthate	Xanthate	x	Sulphide collector
cytec	Aero froth 65	Polyglycol	x	Flotation frother
cytec	Methyl isobutyl carbinol	Alcohol	x	Flotation frother
Fisher	Calcium hydroxide	$Ca(OH)_2$	x	Base, flotation pH modifier
	Sulphur dioxide	$SO_2(g)$	–	Acid, flotation pH modifier

Table 13 The mineralogy of the tailings from froth flotation of PMC's CSD as determined using XRD

S. No.	Minerals	Chemical composition	wt%
1	Titanomagnetite	$Fe^{2+}(Fe^{3+},Ti)_2O_4$	7.1
2	Mullite	$Al(Al_{0.69}\ Si_{1.22}\ O_{4.85})$	45.5
3	Quartz	$SiO_{2.00}$	20.2
4	Bernardite	$Ti_{4.00}\ As_{20.00}\ S_{32.00}$	14.1
5	Christite	$Ti_{4.00}Hg_{4.00}As_{4.00}S_{12.00}$	4.0
6	$S_8O_4N_4C_{32}H_{36}$	$S_{8.00}O_{4.00}N_{4.00}C_{32.00}H_{36.00}$	9.1

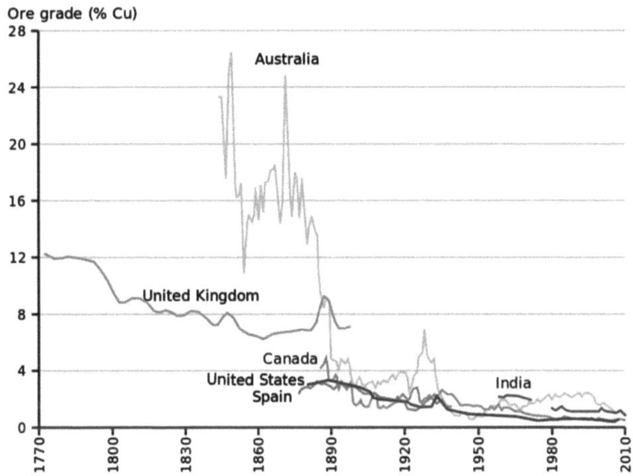

Fig. 4 Graph of the decreasing rate of copper in ores (Wikipedia, the free encyclopedia)

amenable to conventional gravity concentration [50]. The mineralogical composition of the tailings obtained from the froth flotation of PMC's CSD is presented on Table 13; it reveals significant amounts of titanium compounds (25.20 wt%) on one hand, silicon and aluminum oxide as mullite (45.50 wt%) on the other; thus, suggesting a further beneficiation, for an efficient processing of the CSD.

Roasting Pretreatment

The decreasing rate of copper in ores now requires pre-treatment of ores (Fig. 4). Thus suggesting that direct leaching and the associated challenge of iron dissolution in the process presents the need for thorough purification and controlled methodology prior to recovery of copper in the subsequent stage [51].

Hence, the hydrometallurgical extraction of copper from its sulphide sources, often involves the use of the roast-leach process option. This process option is carried out by first subjecting the copper source to either low temperatures (partial roasting) or high temperatures (dead roasting) with the hope of producing dissolvable calcine for copper recovery [52].

Kinetics and Mechanism of Roasting CSD

The essence of chemical kinetics is to study the rates of chemical reactions. The kinetics of oxidative and sulphation roasting is a function of the transport of reactant gas to the unreacted solid and the rate of reaction between the solid and gas. This in turn is a function of the following parameters:

(i) Temperature of the system,
(ii) The particle size,
(iii) The porosity of the condensed product phase,
(iv) The shape of the concentrate particles, and the
(v) The chemical composition and mineralogy.

The CSD of PMC is composed of the Cu–Fe–S–O system (Table 7); it can be upgraded first by any of the appropriate mineral processing methods and/or pre-treated (either the concentrate (e.g. from froth flotation) or the CSD) via oxidative roasting to make it amenable to leaching for the recovery of its contained copper value. The thermal analysis of the CSD of PMC was carried out using both the thermogravimetric and differential thermogravimetric (DSC-TGA) to determine the thermal behavior of this CSD (Figs. 5 and 6), they reveal vaporization of water of crystallization (100–120 °C) and series of decomposition (500, 750 °C) marked by weight losses and gain; probably as a result of the formation of sulphates and oxides of copper, iron, aluminum, titanium etc.

Dissolution and Extraction of Copper from Copper Smelter Dust

Dissolution of Copper from Copper Smelter Dust

The treatment of CSD has been studied in detail by several authors ([23, 43, 17]; Wu et al. 2014). Based on the processes put forward at the pilot and/or industrial level, in line with the reasons specified, the processes can be classified as:

(A) Roasting of the dusts and leaching [53, 54],
(B) Smelting and leaching [55, 56],
(C) Roasting [57, 58] and Smelting [59, 60].

Hence, a good number of the studies concentrate on hydrometallurgical treatments of the dusts, using acid or alkaline reagents [61–68].

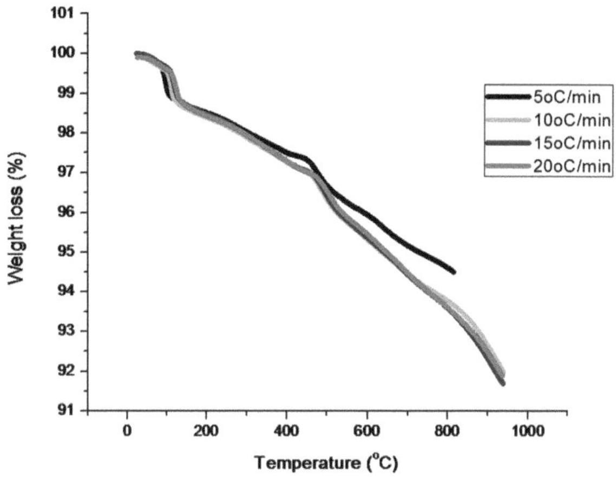

Fig. 5 Graph of weight loss (%) versus temperature (°C) at heating rates of 5, 10, 15, 20 °C/min determined by a Thermogravimetric Analyzer (TGA)

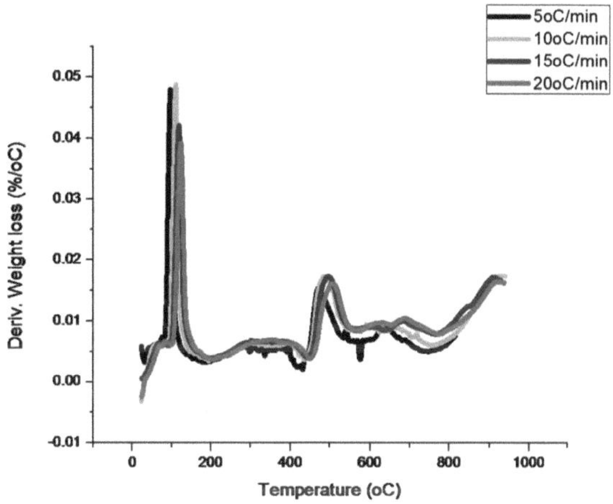

Fig. 6 Graph of derived weight loss (%) versus temperature (°C) at heating rates of 5, 10, 15, 20 °C/min determined by a Differential Scanning Calorimeter (DSC)

Solvent Extraction of Copper from Copper Smelter Dust

For metallurgical application esters and ketones of phosphoric acid with straight or branched chain alcohols are found particularly suitable. Also alkyl amines are used, in all cases dissolved in a suitable solvent such as kerosene. The extraction may occur either by neutral ionic pairs being absorbed by solvent and forming a loose chemical bond with the organic molecule, or by a cation or an anion being bonded to the molecule in exchange for a hydrogen cation or a chloride or nitrate anion. Solvent extraction has found particular application in the extraction of copper from copper smelter dust [21]; thus resulting in a 50% purity improvement of leach solution (Tables 14 and 15).where possible (mineral processing), this option can be avoided, but for the effect of the toxic chemicals to the environment (cliskov et al. 2013).

Table 14 Chemical composition of solution from acid leaching of CSD [21]

Element	Cu	Fe	Ag	Mg	Mn	Mo	Pb	Sb
Concentration (mg L^{-1})	20,925.4	6322.3	4.67	112.5	7.47	118.5	16.74	96.07

Table 15 Chemical composition of solvent extraction leaching solution of CSD (21)

Element	Cu	Fe	Ag	Mg	Mn	Mo	Pb	Sb
Concentration (mg L^{-1})	12,806.9	84.3	1.9	53.9	7.47	0.38	1.55	–

Electrowinning of Copper from CSD

The electrowinning of copper ions derived from dissolution, or solvent extraction is an important contributor to the global copper product supply. The process of electrolysis for copper was first advanced in the late 19th century and notwithstanding much advancement in technology, the principles and rudimentary equipment remain the same [69].

Kinetics of Dissolution of Copper from CSD

On the basis of thermodynamics, the final equilibrium state for a reaction may be predicted, but thermodynamics gives no information about the rate at which this equilibrium is approached [48]. Hence, the yield of such a reaction is a function of firstly, the position of the equilibrium between the reacting substances and secondly, the rate at which equilibrium is attained [70]. Bakhtiari et al. [71] studied the bio-leaching kinetics of copper from CSD. Based on the data obtained, it was deduced that at pH 1.8 and the pulp density less than 7%, the dissolution of copper followed shrinking core kinetic model and the process was limited by diffusion of lixiviant. However with a pulp density of 7%, the process was reported to be reaction limited.

Recovery of Copper as Nano-particles from Purified Leach Solution of CSD

Chemical Reduction Method

The chemical reduction of copper salts is the easiest, simplest and the most commonly used production method for copper nano-particles. During the production of metallic copper nanoparticles good control of shapes and sizes is achievable using chemical reduction of copper precursors [72, 73]. Conventionally, in the chemical reduction techniques, a copper salt is reduced by a reducing agent such as sodium borohydride [74], hydrazine [75], ascorbate [76], glucose [77], polyol [78] as well as isopropyl alcohol with cetyltrimethylammonium bromide [79].

The Solvothermal Method (Thermal Decomposition Method)

This method involves the selection of a proper precursor and localized heating. There are several reports of copper nano-particles being produced using this method. Zhang et al. [80] produced copper nano-particles with the average diameter of 10 nm from the thermal decomposition of copper oxalate precursor. CuO nano-particles were also synthesized from the thermal decomposition of the precursors: brochantite $Cu_4(SO_4)(OH)_6$, and posnjakite $Cu_4(SO_4)(OH)_6.H_2O$ [71, 81].

Hence, the production of metallic copper nano-particles with good control of shapes and sizes is achievable using chemical reduction method [72, 73]. The grade of the copper nano-particles is important, so much that a method that guarantees simplicity, very low grain size, the presence of a single phase and production of high purity nano-crystals, together with high crystallinity and ecofriendliness attributes, such as the thermal decomposition method will be better suited for this purpose [21, 82].

Thermochemistry

Computational Thermochemistry

Thermochemical data are a vital factor in the safe and successful upgrade of chemical processes in the chemical industry. Despite collection of experimental thermochemical data for many molecules, there are quite a number of species for which there are no data. Furthermore, the data in the collections are sometimes incorrect. Experimental measurements of thermochemical processes are often expensive and difficult, so it is highly desirable to have computational methods that can make reliable predictions (Curtiss and Pople 1999). Howbeit, the FactSage 7.0 computer package, which consists of a series of information, calculation and manipulation modules that enable one to access and manipulate compound and solution databases. With the various modules running under Microsoft Windows one can perform a wide variety of thermochemical calculations and generate tables, graphs and figures of interest to chemical and physical metallurgists [83].

Present and Future Challenges of CSD

The known resource recovery process options involves capital and operating costs greater than can be right by the revenue of the recovered metal. Thus, the copper smelting industry shows increasing interest in developing processes for the recovery of the copper value from this metallurgical waste. As a result, it would be beneficial to have a processing method by which impurities could be removed from the CSD, while sustaining an acceptable level of revenue metals in the resulting concentrate. It would also be gainful if this could be achieved in an economic manner in order to justify the treatment of stockpiled dust.

Conclusion

The deduction from discussions leading to this point, can be summarized as this; that the waste copper smelter dust plays host to substantial amount of copper value, majorly in the form of oxides and often in small but appreciable amount as sulphides as indicated by XRD results particularly with respect to that from PMC. The bulk of this metallurgical waste occurs in the size range below 53 μm as can be seen from the SEM and wet screening results. The recovery of the copper value contained in this metallurgical dusts can either be by leaching directly or using the roast-leach process options, in which case, the waste material will be pre-treated by roasting to transform the copper sulphides to sulphates or oxides after which leaching with either water or dilute acid to take the contained copper into solution will be carried out. Solvent extraction treatment of the leached solution will also be considered to purify the solution before recovery of the copper value as nanoparticles; bearing in mind the desirability of copper nanoparticles as good substitutes to noble metals; so that recovering the CSD's contained copper value as nano-particles is believed a better approach of converting this metallurgical waste to product. Furthermore, all of these can be achieved in an economical manner by reducing the time, cost and manpower required, through effective pre-experimental searching for optimal conditions, composition etc. by thermodynamic calculations.

Recommendation

In order to ensure an efficient processing of the waste CSD, it is strongly recommended that an intermediate mineral processing stage be introduced to the current hydrometallurgical processing of CSD, where possible; so as to reduce/remove the contained impurities and consequently enhance the chances of obtaining substantial amount of copper as nanoparticles with desired grade.

References

1. A.A. Baba, K.I. Ayinla, F.A. Adekola, M.K. Ghosh, O.S. Ayanda, R.B. Bale, A.R. Sheik, S. R. Pradhan, A review on novel techniques for chalcopyrite ore processing. Int. J. Min. Eng. Mineral Process. **1**(1), 1–16 (2012)
2. USGS, Facts about Copper: copper uses, resources, supply, demand and production information 2009, http://eology.comusgsuses-of-copper/. Accessed 15/12/2010
3. R.T. Jones, P.J. Mackey, An overview of copper smelting in Southern Africa (2015)
4. J. Wood, S. Creedy, R. Matusewicz, M. Reuter, Secondary copper processing using Outotec Ausmelt TSL technology, in *Proceedings of MetPlant* (2011), pp. 460–467
5. M. Reuter, A. van Schaik, Thermodynamic metrics for measuring the "sustainability" of design for recycling. JOM **60**(8), 39–46 (2008)
6. European Commission, European Dioxin Inventory—Secondary Copper Production [online] (2009). Available from http://ec.europa.eu/environment/dioxin/pdf/stage1/seccopper.pdf. Accessed 23 Feb 2011. A. Umer et al., Selection of a suitable method for the synthesis of copper nanoparticles. Nano **7**(05), 1230005 (2012)
7. J.B. Wang, C.H. Hung, C.H. Hung, G.P. Chang-Chien, Polychlorinated dibenzo-p-dioxin and dibenzofuran emissions from an industrial park clustered with metallurgical industries. J. Hazard. Mater. **161**(2), 800–807 (2009)
8. V. Montenegro, H. Sano, T. Fujisawa, Recirculation of Chilean copper smelting dust with high arsenic content to the smelting process. Mater. Trans. **49**(9), 2112–2118 (2008)
9. Regulation EC, No 1907/2006 of the European Parliament and of the Council of 18 December 2006, concerning the Registration. Evaluation, Authorisation and Restriction of Chemicals (REACH), establishing a European Chemicals Agency, amending Directive, vol. 45 (1999), pp. 1–849
10. SANS, *South African National Standard: Ambient Air Quality—Limits for Common Pollutants* (2005)
11. U. Neveling, Palabora Mining Company Annual report on ambient air quality monitoring (2011)
12. L. Qiang, I.S. Pinto, Z. Youcai, Sequential stepwise recovery of selected metals from flue dusts of secondary copper smelting. J. Clean. Prod. **84**, 663–670 (2014)
13. I.S. Pinto, H.M. Soares, Selective leaching of molybdenum from spent hydrodesulphurisation catalysts using ultrasound and microwave methods. Hydrometallurgy **129**, 19–25 (2012)
14. I.S. Pinto, H.M. Soares, Recovery of molybdates from an alkaline leachate of spent hydrodesulphurisation catalyst–proposal of a nearly-closed process. J. Clean. Prod. **52**, 481–487 (2013)
15. F. Bakhtiari, H. Atashi, M. Zivdar, S.S. Bagheri, Continuous copper recovery from a smelter's dust in stirred tank reactors. Int. J. Miner. Process. **86**(1), 50–57 (2008)
16. F. Bakhtiari, M. Zivdar, H. Atashi, S.S. Bagheri, Bioleaching of copper from smelter dust in a series of airlift bioreactors. Hydrometallurgy **90**(1), 40–45 (2008)
17. F.J. Alguacil, I. Garcia-Diaz, F. Lopez, O. Rodriguez, Recycling of copper flue dust via leaching-solvent extraction processing. Desalin. Water Treat. **56**(5), 1202–1207 (2015)
18. G.A. Kordosky, Copper recovery using leach/solvent extraction/electrowinning technology: forty years of innovation, 2.2 million tonnes of copper annually. J. S. Afr. Inst. Min. Metall. **102**(8), 445–450 (2002)
19. N.N. Hoover, B.J. Auten, B.D. Chandler, Tuning supported catalyst reactivity with dendrimer-templated Pt-Cu nanoparticles. J. Phys. Chem. B **110**(17), 8606–8612 (2006)
20. Y. Niu, R.M. Crooks, Preparation of dendrimer-encapsulated metal nanoparticles using organic solvents. Chem. Mater. **15**(18), 3463–3467 (2003)
21. E. Darezereshki, F. Bakhtiari, Synthesis and characterization of tenorite (CuO) nanoparticles from smelting furnace dust (SFD). J. Min. Metall. B: Metall. **49**(1), 21–26 (2013)
22. C. Samuelsson, G. Carlsson, Characterization of copper smelter dusts. CIM Bull. **94**(1051), 111–115 (2001)

23. Y. Chen, T. Liao, G. Li, B. Chen, X. Shi, Recovery of bismuth and arsenic from copper smelter flue dusts after copper and zinc extraction. Miner. Eng. **39**, 23–28 (2012)
24. D.K. Steele, K.S. Gritton, S.B. Odedirk, *Treatment of Copper Smelting and Refining Wastes* (US Department of the Interior, Bureau of Mines, 1994)
25. A.K. Biswas, W.G. Davenport, *Extractive Metallurgy of Copper: International Series on Materials Science and Technology*, vol. 20. Elsevier (2013)
26. D.R. Swinbourne, E. Simak, A. Yazawa, V. Melbourne, Accretion and dust formation in copper smelting-thermodynamic considerations, in *Sulfide Smelting* (2002), pp. 247–259
27. E. Miettinen, Thermal conductivity and characteristics of copper flash smelting flue dust accretions. Teknillinen korkeakoulu (2008)
28. J.A. Eastman, S.U.S. Choi, S. Li, W. Yu, L.J. Thompson, Anomalously increased effective thermal conductivities of ethylene glycol-based nanofluids containing copper nanoparticles. Appl. Phys. Lett. **78**(6), 718–720 (2001)
29. R.K. Guduru, K.L. Murty, K.M. Youssef, R.O. Scattergood, C.C. Koch, Mechanical behavior of nanocrystalline copper. Mater. Sci. Eng., A **463**(1), 14–21 (2007)
30. Y. Wang, M. Chen, F. Zhou, E. Ma, High tensile ductility in a nanostructured metal. Nature **419**(6910), 912–915 (2002)
31. X. Kang, Z. Mai, X. Zou, P. Cai, J. Mo, A sensitive nonenzymatic glucose sensor in alkaline media with a copper nanocluster/multiwall carbon nanotube-modified glassy carbon electrode. Anal. Biochem. **363**(1), 143–150 (2007)
32. K.B. Male, S. Hrapovic, Y. Liu, D. Wang, J.H. Luong, Electrochemical detection of carbohydrates using copper nanoparticles and carbon nanotubes. Anal. Chim. Acta **516**(1), 35–41 (2004)
33. Y. Guo, W. Meyer-Zaika, M. Muhler, S. Vukojević, M. Epple, Cu/Zn/Al xerogels and aerogels prepared by a sol–gel reaction as catalysts for methanol synthesis. Eur. J. Inorg. Chem. **2006**(23), 4774–4781 (2006)
34. M.L. Kantam, V.S. Jaya, M.J. Lakshmi, B.R. Reddy, B.M. Choudary, S.K. Bhargava, Alumina supported copper nanoparticles for aziridination and cyclopropanation reactions. Catal. Commun. **8**(12), 1963–1968 (2007)
35. J.A. Rodriguez, P. Liu, J. Hrbek, J. Evans, M. Perez, Water gas shift reaction on Cu and Au nanoparticles supported on CeO_2 (111) and ZnO (000$ bar 1$): intrinsic activity and importance of support interactions. Angew. Chem. **119**(8), 1351–1354 (2007)
36. C. Pecharromán, A. Esteban-Cubillo, I. Montero, J.S. Moya, E. Aguilar, J. Santarén, A. Alvarez, Monodisperse and corrosion-resistant metallic nanoparticles embedded into sepiolite particles for optical and magnetic applications. J. Am. Ceram. Soc. **89**(10), 3043–3049 (2006)
37. E. Balladares, U. Kelm, S. Helle, R. Parra, E. Araneda, Chemical-mineralogical characterization of copper smelting flue dust. Dyna **81**(186), 11–18 (2014)
38. M. Vítková, V. Ettler, J. Hyks, T. Astrup, B. Kříbek, Leaching of metals from copper smelter flue dust (Mufulira, Zambian Copperbelt). Appl. Geochem. **26**, S263–S266 (2011)
39. A. Morales, M. Cruells, A. Roca, R. Bergó, Treatment of copper flash smelter flue dusts for copper and zinc extraction and arsenic stabilization. Hydrometallurgy **105**(1), 148–154 (2010)
40. A.B. Vakylabad, A comparison of bioleaching ability of mesophilic and moderately thermophilic culture on copper bioleaching from flotation concentrate and smelter dust. Int. J. Miner. Process. **101**(1), 94–99 (2011)
41. A.B. Vakylabad, M. Schaffie, M. Ranjbar, Z. Manafi, E. Darezereshki, Bio-processing of copper from combined smelter dust and flotation concentrate: A comparative study on the stirred tank and airlift reactors. J. Hazard. Mater. **241**, 197–206 (2012)
42. J.Y. Wu, F.C. Chang, H.P. Wang, M.J. Tsai, C.H. Ko, C.C. Chen, Selective leaching process for the recovery of copper and zinc oxide from copper-containing dust. Environ. Technol. **36** (23), 2952–2958 (2015)
43. T.K. Ha, B.H. Kwon, K.S. Park, D. Mohapatra, Selective leaching and recovery of bismuth as Bi_2O_3 from copper smelter converter dust. Sep. Purif. Technol. **142**, 116–122 (2015)
44. L.G. Austin, I. Shah, A method for inter-conversion of Microtrac and sieve size distributions. Powder Technol. **35**(2), 271–278 (1983)

45. B.A. Wills, T. Napier-Munn, Wills' mineral processing technology: an introduction to the practical aspects of ore treatment and mineral recovery. Butterworth-Heinemann (2015)
46. Z.F. Xu, L.I. Qiang, H.P. Nie, Pressure leaching technique of smelter dust with high-copper and high-arsenic. Trans. Nonferrous Met. Soc. China **20**, s176–s181 (2010)
47. V. Ettler, M. Vítková, M. Mihaljevič, O. Šebek, M. Klementová, F. Veselovský, P. Vybíral, B. Kříbek, Dust from Zambian smelters: mineralogy and contaminant bioaccessibility. Environ. Geochem. Health **36**(5), 919–933 (2014)
48. T. Rosenqvist, Principles of Extractive Metallurgy. Tapir Academic Press (2004)
49. The encyclopedia of earth, 2015: https://en.wikipedia.org/wiki/Encyclopedia_of_Earth
50. S.K. Kawatra, T.C. Eisele, Depression of Pyrite Flotation by Yeast and Bacteria. *Mineral Biotechnology: Microbial Aspects of Mineral Beneficiation, Metal Extraction, and Environmental Control* (2001), p. 3
51. S. Prasad, B.D. Pandey, Alternative processes for treatment of chalcopyrite—a review. Miner. Eng. **11**(8), 763–781 (1998)
52. F. Magagula, High temperature roasting of sulphide concentrate and its effect on the type of precipitate formed. Doctoral dissertation, 2012
53. Z.B. Yin, E. Caba, L. Barron, D. Belin, W. Morris, M. Vosika, R. Bartlett, Copper extraction from smelter flue dust by lime-roast/ammoniacal heap leaching. *Residues and Effluents: Processing and Environmental Considerations* (1992), pp. 255–267
54. B. Gorai, R.K. Jana, Z.H. Khan, Electrorefining electrolyte from copper plant dust. Mater. Trans. **43**(3), 532–536 (2002)
55. M.P. Smirnov, V.T. Khvan, G.A. Bibenina, R.P. Kefilyan, N.I. Il'yasov, Complex treatment of lead and rhenium containing sulfate dusts from copper plants. Tsvetn. Met. **6**, 3–6 (1984)
56. Z.W. Zhang, W. Lu, F. Zheng, Separation and recovery of copper and zinc from flue dust. Huanjing Kexue **11**(6), 1012–1016 (1992)
57. M. Carter, E.R. Vance, L.P. Aldridge, M. Zaw, G. Khoe, Immobilization of arsenic trioxide in cementitious materials, in *Australasian Institute of Mining and Metallurgy* (1994), pp. 275–280
58. Y. Fu, L. Jiang, D. Wang, Removal of arsenic from copper smelter flue dust by calcinations. Yelian Bufen **6**, 14–16 (2000)
59. Z. Yu, Process for bismuth recovery from the flue dust of copper smelting. Huaxue Shijie **28** (10), 465–468 (1987)
60. H. Mochida, O. Iida, *Copper smelter flue dust treatment* (Kokai Tokkyo Koho, Jap, 1988), p. 3
61. H.H. Law, M.P. Bohrer, P. O'Hara, Recovery of metals from copper smelter furnace flue dust. *Residues and Effluents: Processing and Environmental Considerations* (1992), pp. 295–310
62. R.S. Kunter, W.E. Bedal, Chloride-process treatment of smelter flue dusts. JOM **44**(12), 35–38 (1992)
63. G.J. Roman-Moguel, G. Plascencia, J. Pérez, A. García, Copper recycling from waste pickling solutions. JOM **47**(10), 18–19 (1995)
64. P.J. Gabb, J.P. Evans, Kennecott Utah Copper Corporation, Hydrometallurgical processing of impurity streams generated during the pyrometallurgy of copper. U.S. Patent 5,616,168, 1997
65. A. Robles, A.E. Serna, M. Sandez, Alkaline arsenic leaching from smelter flue dust and leaching solution regeneration, in *Copper99-Cobre99, International Conference on TMS-AIME*, Warrendale, PA (1999), pp. 261–272
66. E. Vircikova, M. Havlik, Removing as from converter dust by a hydrometallurgical method. JOM **51**(9), 20–23 (1999)
67. J.J. Ke, R.Y. Qin, Arsenic removal and bismuth recovery from copper smelter flue dust (2000)
68. M.I. Martín, A. López-Delgado, F.A. López, A.G. Coedo, M.T. Dorado, F.J. Alguacil, Treatment of copper converter flue dust for the separation of metallic/non-metallic copper by hydrometallurgical processing. J. Chem. Eng. Jpn. **36**(12), 1498–1502 (2003)
69. N.T. Beukes, J. Badenhorst, Copper electrowinning: theoretical and practical design. J. South Afr. Inst. Min. Metall. **109**(6), 343–356 (2009)
70. O. Kubaschewski, C.B. Alcock, P.J. Spencer, Materials Thermodynamics (1993)

71. F. Bakhtiari, E. Darezereshki, One-step synthesis of tenorite (CuO) nano-particles from $Cu_4(SO_4)(OH)_6$ by direct thermal-decomposition method. Mater. Lett. **65**(2), 171–174 (2011)
72. X. Song, S. Sun, W. Zhang, Z. Yin, A method for the synthesis of spherical copper nanoparticles in the organic phase. J. Colloid Interface Sci. **273**(2), 463–469 (2004)
73. S. Kapoor, T. Mukherjee, Photochemical formation of copper nanoparticles in poly (N-vinylpyrrolidone). Chem. Phys. Lett. **370**(1), 83–87 (2003)
74. M. Aslam, G. Gopakumar, T.L. Shoba, I.S. Mulla, K. Vijayamohanan, S.K. Kulkarni, J. Urban, W. Vogel, Formation of Cu and Cu_2O nanoparticles by variation of the surface ligand: preparation, structure, and insulating-to-metallic transition. J. Colloid Interface Sci. **255**(1), 79–90 (2002)
75. H. Zhu, C. Zhang, Y. Yin, Novel synthesis of copper nanoparticles: influence of the synthesis conditions on the particle size. Nanotechnology **16**(12), 3079 (2005)
76. Y. Wang, P. Chen, M. Liu, Synthesis of well-defined copper nanocubes by a one-pot solution process. Nanotechnology **17**(24), 6000 (2006)
77. S. Panigrahi, S. Kundu, S.K. Ghosh, S. Nath, S. Praharaj, S. Basu, T. Pal, Selective one-pot synthesis of copper nanorods under surfactantless condition. Polyhedron **25**(5), 1263–1269 (2006)
78. B.K. Park, S. Jeong, D. Kim, J. Moon, S. Lim, J.S. Kim, Synthesis and size control of monodisperse copper nanoparticles by polyol method. J. Colloid Interface Sci. **311**(2), 417–424 (2007)
79. A.A. Athawale, P.P. Katre, M. Kumar, M.B. Majumdar, Synthesis of CTAB–IPA reduced copper nanoparticles. Mater. Chem. Phys. **91**(2), 507–512 (2005)
80. X. Zhang, D. Zhang, X. Ni, J. Song, H. Zheng, Synthesis and electrochemical properties of different sizes of the CuO particles. J. Nanopart. Res. **10**(5), 839–844 (2008)
81. E. Darezereshki, F. Bakhtiari, A novel technique to synthesis of tenorite (CuO) nanoparticles from low concentration $CuSO_4$ solution. J. Min. Metall. B **47**(1), 73–78 (2011)
82. K. Byrappa, Hydrothermal processing, in *Kirk-Othmer Encyclopedia of Chemical Technology* (2005)
83. C.W. Bale, P. Chartrand, S.A. Degterov, G. Eriksson, K. Hack, R.B. Mahfoud, J. Melançon, A.D. Pelton, S. Petersen, FactSage thermochemical software and databases. Calphad **26**(2), 189–228 (2002)
84. Wikipedia 2014 https://en.wikipedia.org/wiki
85. S. Rosenblum, I.K. Brownfield, *Magnetic Susceptibilities of Minerals*. US Department of the Interior, US Geological Survey (2000)
86. Z.A. Zanetell, Penalty element separation from copper concentrates utilizing froth flotation (2007)
87. O. Font, N. Moreno, G. Aixa, X. Querol, R. Navia, *Copper Smelting Flue Dust: A Potential Source of Germanium*

Part VIII
Deriving Value from Challenging Waste Materials: Recycling and Sustainability Joint Session: Deriving Value from Challenging Waste III

Chromium Removal from Iron-Rich Waste Generated During Processing Lateritic Nickel Ores

Hong Vu, Tomas Frydl, Petr Dvorak, Jana Selucka and Petra Starkova

Abstract Waste, generated during high pressure leaching of lateritic nickel ores, contains about 50% of iron in form of magnetite but high chromium content (~3%) does not allow its extensive utilization as the raw material for pig iron production in blast furnace. The chromium removal from the waste was carried out by two methods: (1) oxidative alkali roasting with NaOH and/or Na_2CO_3, followed by water leaching and (2) oxidative leaching in KOH sub-molten salts solutions. The results show that about 90.1% of Cr was removed when the waste was roasted at 900 °C for 3 h with 33.1 wt% of NaOH. The removal of 88.2% of Cr was achieved if the waste was roasted at 900 °C for 3 h with 40 wt% of Na_2CO_3. The sodium content in the water-leached residues was reduced from 5% to about 0.5% by reaction with CaO in aqueous solutions at 80 °C for 2 h. About 70% of Cr was removed if the waste was leached at 190 °C in 73 wt% KOH for 8 h.

Keywords Chromium removal · Oxidative alkali roasting · Oxidative leaching · Sub-molten salts

---

H. Vu (✉) · P. Dvorak · J. Selucka · P. Starkova
University of Chemistry and Technology Prague, Prague, Czech Republic
e-mail: vun@vscht.cz

P. Dvorak
e-mail: dvorakp@vscht.cz

J. Selucka
e-mail: seluckaj@vscht.cz

P. Starkova
e-mail: starkova.petra@vscht.cz

T. Frydl
Aero Vodochody Aerospace a.s., Odolena Voda, Czech Republic
e-mail: frydl.t@seznam.cz

L. Zhang et al. (eds.), *Energy Technology 2017*,
The Minerals, Metals & Materials Series, DOI 10.1007/978-3-319-52192-3_22

Introduction

Hydrometallurgical processing of nickel-cobalt lateritic ores at Sered in Slovakia generated approximately 9 millions tones of leaching residues that contain mainly iron around 45–50% beside a number of minor metals such as chromium, nickel, cobalt and aluminum [1]. Due to their fine particle-size and the heavy metals content, the leaching residues have caused significant environmental problems in the surrounding area. Utilization of this leaching residues as raw material for iron making has not realized owning to the presence of impurities, especially chromium. Macasek et al. [1] proposed their utilization as a semi-natural magnetic sorbent for treatment of radioactive contaminated soils and sediments. But the feasible desorption process and reutilization of the sorbent remains unsolved. Similar lateritic residue from the ammonium carbonate technological (CARON) process carried out at the Comandante Ernesto CheGuevara plant (Moa, Holguín, Cuba) was processed by bioleaching methods using Acidithiobacillus thiooxidans DMS 11478 to recover the heavy metals [2]. Under the optimal conditions 86% Co, 83% Mg, 72% Ni, 62% Fe and 23% Cr were extracted. Ema and Harada [3] proved experimentally methods that allowed obtaining low-grade Cr-containing iron and dechromiumed iron from the nickel leached residue from a lateritic processing plant. Recovered Cr-containing iron can be used to make abrasive materials and dechromiumed iron with 0.06% of Cr can be used to make common steel.

Traditionally chromium in form of soluble hexavalent compounds is obtained from chromite ores by pyrometallurgical way that consists of oxidation roasting, water leaching and multi-stage evaporation crystallization [4, 5]. At first chromite ore is converted into water-soluble chromate by oxidative roasting with sodium carbonate (Na_2CO_3) to produce sodium chromate (Na_2CrO_4). The sodium chromate (Na_2CrO_4) produced is dissolved in water and then converted into marketable sodium dichromate ($Na_2Cr_2O_7$). Various investigations on soda-ash roasting have been also reported [6–9]. Recent new liquid-phase oxidization technology, in which chromite ore is oxidized with highly concentrated KOH sub-molten salts media under ambient pressure and separated from leach liquors as $K_2Cr_2O_7$, has been reported in the works of Zhang and her colleagues [10–14]. In this new technology, the chromium extraction surpassed 99% after 8 h reaction when chromite ore was dissolved in 80 wt% KOH sub-molten salt solution at 320 °C. The extraction of chromium from becomes more efficient because KOH sub-molten salt media has high mean ion activity coefficient, high boiling point and excellent performance in flow and transfer [15]

In this paper both traditional oxidative alkali roasting and new oxidative leaching in sub-molten salts media were used to remove chromium from nickel lateritic leaching residue. The experiments were carried out to determine the influence of roasting temperature, the amount of NaOH or Na_2CO_3 in the roasting mixtures and roasting time on the Cr removal. The oxidative leaching of the leaching residues was performed to obtain the dependence of extraction of chromium on leaching temperature, concentration of KOH, ratio of KOH to amount of leaching residue, and leaching time.

Materials and Experiments

Materials

Approximately 200 kg of nickel lateritic leaching residue sampled from the waste dump at Sered, Slovakia was dried in an oven at 95 °C for 2 days and pulverized. The prepared sample was subjected to chemical, mineralogical and morphological analysis. The chemical analysis found that the leaching residue contains mainly Fe (~48%), followed by Si (~7%) and Al (~4.5%), as shown in Table 1. The Cr content in the leaching residue is ~3% and a significant amount of Ni is also detected in the sample. The leaching residue consisted of particles with size range from 0.2 to 61.1 μm with the mean diameter of 16.7 μm. According to the results of XRD analysis, the sample consisted of four main phases: dominant magnetite,

Table 1 Chemical composition of the leaching residuet

Ni (wt%)	Cr (wt%)	Fe (wt%)	Al (wt%)	Si (wt%)	Ca (wt%)	Mg (wt%)
0.44	3.09	47.85	4.52	7.18	4.10	2.77

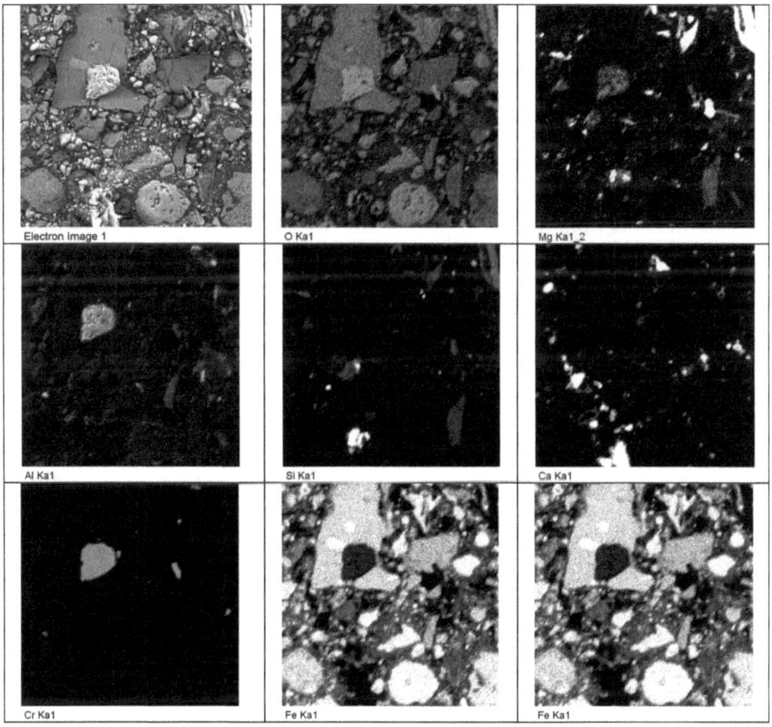

Fig. 1 SEM/EDX micrographs of nickel lateritic leaching residue

followed with quartz and calcite, and minor gypsum. Mineral phases containing Cr were not detected, probably due to its low content. SEM/EDX analysis showed that Cr is present together with O, Mg and Al in the same particle, indicating the presence of Cr-containing picotite (Fig. 1).

Equipment and Experimental Procedures

The prepared leaching residue was mixed with NaOH or $NaCO_3$ of technical grade at different weight ratio. The mixture was sintered in a muffle furnace at different temperatures for various time intervals. During sintering air was blown into the furnace at flow rate of 18 l/min. The sinter was milled in a ball mill and subjected to water leaching. The water leaching experiments were performed in a 100 mL glass reactor fitted with a condenser and a thermometer, heated with a mantle controlled by a regulator and agitated with a magnetic stirrer. The leaching temperature was maintained at 80 °C; liquid to solid ratio (l:s) was 10:1; and the solutions were mixed at speed of 840 rpm for 120 min. After leaching tests finished, the leach residues were separated by filtration, water-washed, dried at 95 °C and subjected to analysis.

The oxidative leaching experiments were carried out in a 250 mL glass reactor fitted with a condenser, a thermometer and a gas introduction pipe, heated with a mantle controlled by a regulator and agitated with a magnetic stirrer. At first KOH of technical grade was slowly mixed with distilled water in the glass reactor and the mixture was heated to a required temperature. Then a required amount of leaching residue was added to the sub-molten solution and oxygen was introduced to the solution at the flow rate of 0.2 l/min. The solutions were mixed at speed of 840 rpm. After leaching was completed, the solutions were diluted 3 times with distilled water to prevent the crystallization of KOH. After that the leach residues were separated by filtration, water-washed, dried at 95 °C and subjected to analysis.

Analysis

Solution and solid samples were analyzed using a GBC atomic absorption spectrophotometer (model GBC 932 plus) to determine metals composition. For elemental analysis of solid samples, approximately 0.25 g of samples dissolved in 20 ml of aqua regia. The solutions were then diluted to 100 mL. Mineralogical analysis was undertaken with X-ray diffraction (PANalytical's X'Pert PRO) using Cu radiation. SEM-EDX analysis was carried out using Vega 3 Tescan Scanning Electron Microscope The particle size distribution of the leaching residue was measured by a laser sizer (Fritsch, Analysette 22 NanoTec) using Fraunhofer approximation software.

Results and Discussions

Oxidative Alkali Roasting

The results from oxidative roasting of the nickel lateritic leaching residue with KOH showed that with increasing roasting temperature up to 900 °C the Cr content in the final product obtained after water leaching sinters decreased from 0.7 to 0.3%. As roasting temperature passed 900 °C the Cr content increased with increasing temperature, as shown in Fig. 2. XRD pattern analysis indicates that with increasing temperature more hematite appeared thanks to oxidation of magnetite, as illustrated in Table 2. If roasting temperature was higher than 850 °C reactions between sodium, iron and silicon oxides occured, that led to the formation of sodium-iron silicate—$Na_{0.8}Fe_{0.8}Si_{0.2}O_2$. It is probable that at temperature above 900 °C the roasting mixture partly melted and the liquid melt layer prevented oxygen diffusion into the reacted particles and that led to the decrease of formation of soluble hexavalent chromium compounds.

Figure 3 shows the influence of NaOH content in roasting mixtures on the degree of Cr removal at 900 °C. It is evident that with increasing NaOH content from about 16 to 33% the Cr content decreased from 0.35 to only 0.16%,

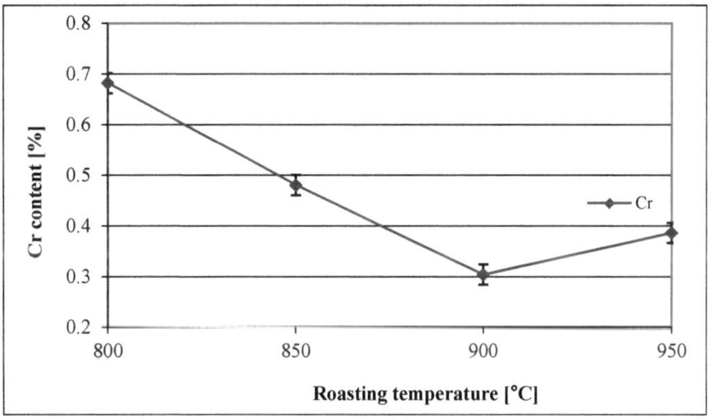

Fig. 2 Influence of roasting temperature on Cr removal: 23.1 wt% NaOH, 3 h

Table 2 Influence of roasting temperature on mineralogical composition of leached residues: 23.1 wt% NaOH, 3 h

Temperature(°C)	Magnetite	Hematite	$Na_{0.8}Fe_{0.8}Si_{0.2}O_2$	Calcite
	Semiquantity (%)			
800	24	38	32	6
850	24	47	29	–
900	35	25	40	–
950	27	13	61	–

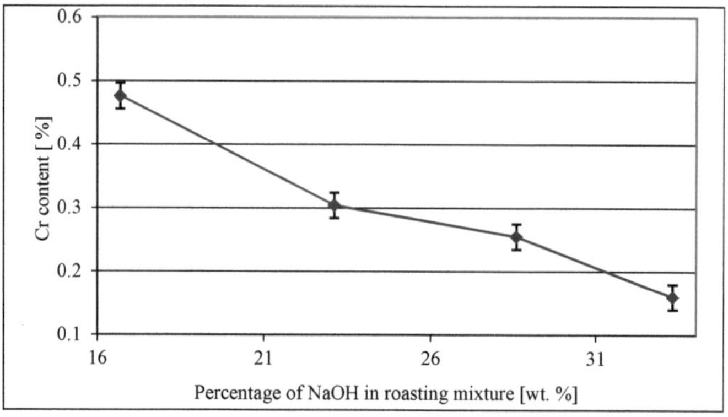

Fig. 3 Influence of ratio of NaOH/residue on Cr removal: 900 °C, 3 h

Table 3 Influence of NaOH amount in roasting mixture on mineralogical composition of leached residues: 900 °C, 3 h

Amount of NaOH (%)	Magnetite	Hematite	$Na_{0.8}Fe_{0.8}Si_{0.2}O_2$	Calcite	Albite
	Semiquantity (%)				
16.7	15	41	26	5	12
23.1	35	25	40	–	–
28.6	30	13	46	11	–
33.3	Amorphous phase				

corresponding to the removal degree at 90.1%. XRD pattern analysis showed that at NaOH content of 33% the final product was amorphous, indicating total decomposition of the original leaching residue (Table 3).

As illustrated in Table 4, increasing roasting time did not significantly change the mineralogical composition of the final product but affected the degree of chromium removal. As roasting time increased up to 3 h, the removal degree also increased as chromium content in the final product decreased about 50% (Fig. 4). With roasting time increased to 4 h, the removal degree decreased slightly. It is probable that with prolonging roasting time insoluble chromium compounds were formed.

Table 4 Influence of roasting time on mineralogical composition of leached residues: 900 °C, 23.1 wt% NaOH

Roasting time (h)	Magnetite	Hematite	$Na_{0.8}Fe_{0.8}Si_{0.2}O_2$	Calcite
	Semiquantity (%)			
1	22	35	43	–
2	34	20	43	3
3	35	25	40	–
4	30	24	45	1

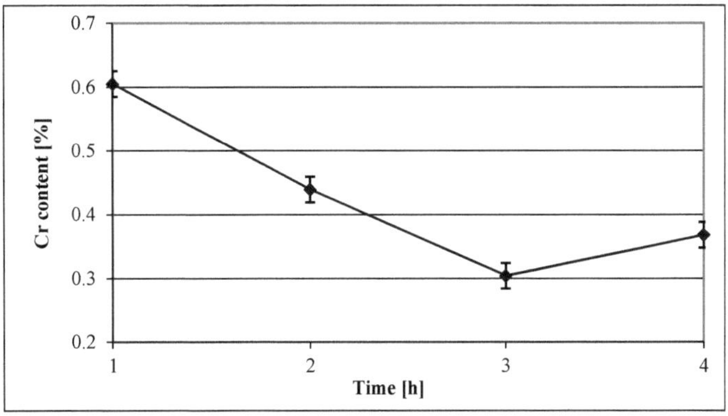

Fig. 4 Influence of roasting time on Cr removal: 900 °C, 23.1 wt% NaOH

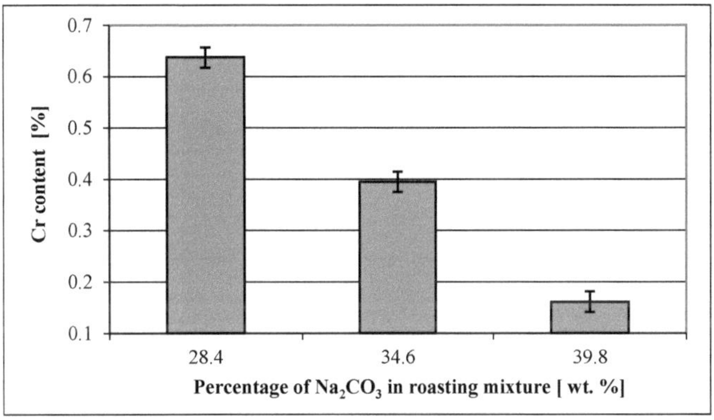

Fig. 5 Influence of ratio of Na_2CO_3/residue on Cr removal: 900 °C, 3 h

The same chromium removal degree was achieved when NaOH was substituted by Na_2CO_3 for roasting tests carried out at 900 °C for 3 h. The amount of Na_2CO_3 added was calculated according to the molar amount of sodium in NaOH. With percentage of Na_2CO_3 in the roasting mixture of 40% only 0.17% of Cr remained in the final product (Fig. 5).

The final products from water leaching of sinters contained high sodium content of about 5%. The sodium removal was performed by reaction of the final products with CaO in aqueous solutions at 80 °C for 2 h. With increasing weight ratio between CaO and final products ranging up to 1.6:10 the sodium content decreased to only 0.5%. The refined products contained on average 55% Fe, 1.5% Al, 10% Ca and 1.7% Mg.

Oxidative Leaching in Sub-molten KOH

The influence of leaching temperature on residual chromium content in the final leaching residue from sub-molten KOH leaching is shown in Fig. 6. It was observed that with increasing temperature up to 160 °C and from 180 °C the residual chromium content decreased. Temperatures ranging from 160 to 180 °C did not influence the degree of chromium removal. It is probable that at temperature up to 180 °C only chromium bound in less stable phases was extracted. Temperature higher than 180 °C is needed to break down more stable spinel phases.

The influence of concentration of KOH on residual chromium content in the final leaching residue was shown in Fig. 7. The residual chromium content increased slightly from 2.7 to 2.83% when the KOH concentration increased from 56 to 60%.

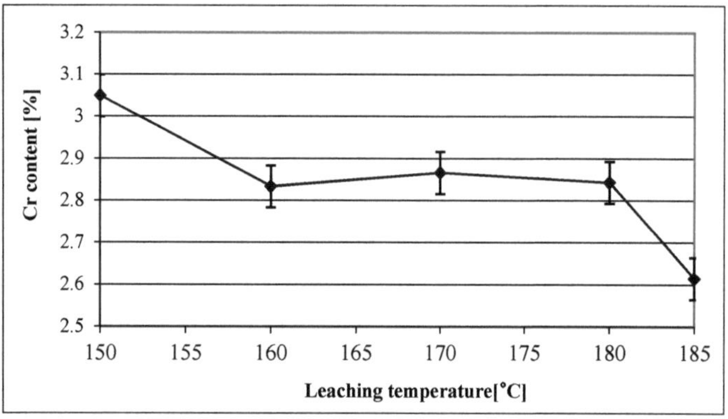

Fig. 6 Influence of ratio of leaching temperature on Cr removal: 5 h, KOH/residue ratio 3.45:1, 60 wt% KOH

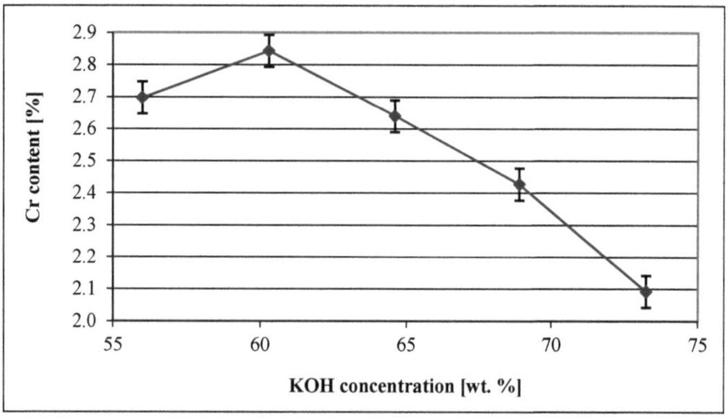

Fig. 7 Influence of KOH concentration on Cr removal: 5 h, KOH/residue ratio 3.45:1, 180 °C

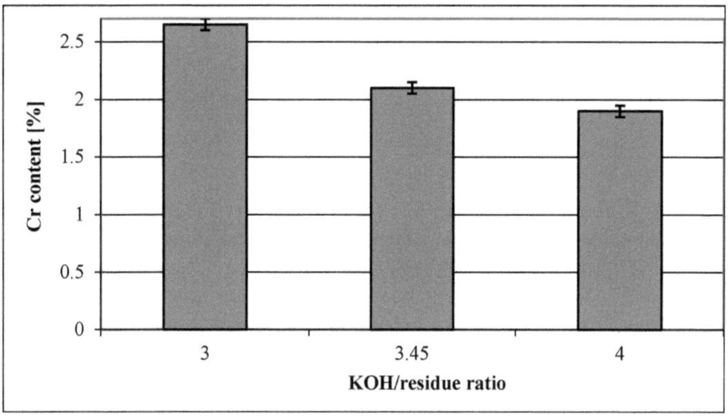

Fig. 8 Influence of KOH/residue ratio on Cr removal: 5 h, 73 wt% KOH, 180 °C

Increasing KOH concentration from 60 to 73% led to a gradual decrease of residual chromium content to 2.1%. It is probable that at KOH concentration of 60%, the solution changes to sub-molten and from that concentration increasing KOH concentration increases activity of O^{-2} anions, which participate directly in oxidation of Cr (III).

The influence of KOH/residue ration on residual chromium content was illustrated in Fig. 8. Increasing KOH/residue ratio decreased the residual chromium content. Excess of KOH results in increasing fluidity of leaching mixtures and mass transfer of oxygen, which leads to increased chromium oxidation and its dissolution. The final leaching residues from sub-molten KOH leaching consist of mainly magnetite, hematite and amorphous phases.

Based on the previous results new leaching conditions were chosen as followed: leaching temperature 190 °C, KOH concentration of 75%, KOH/residue ration 4:1, leaching time 8 h. Under these conditions the final leaching residue was obtained with the following elemental composition: 0.9% Cr, 2.05% Al and 8.3% K.

Conclusions

Oxidative alkali roasting process and oxidative leaching in sub-molten KOH were used to remove chromium from nickel lateritic leaching residue. The results of experiments showed that the roasting process is effective for chromium removal. About 90.1% of Cr was removed when the leaching residue was roasted at 900 °C for 3 h with 33.1 wt% of NaOH. The removal of 88.2% of Cr was achieved if the waste was roasted at 900 °C for 3 h with 40 wt% of Na_2CO_3. In this study the maximal chromium removal of 70% was achieved when the leaching residue was leached at 190 °C and KOH/residue ratio 4:1 in 73 wt% KOH for 8 h. Higher leaching temperature and longer leaching time could increase the chromium removal.

Acknowledgements The work was supported by the Research intention MSM No. 223100002.

References

1. F. Macasek, J. Kufcakova, P. Rajec, S. Hredzák, The sered leached nickel ores residues—a magnetic sorbent for fission radionuclides and lead contaminants of solids. Chem. Pap., Slovak Acad. Sci. **58**(3), 163–169 (2004)
2. G. Cabreraa, J.M. Gómez, I. Hernández, O. Coto, D. Canter, Different strategies for recovering metals from CARON process residue. J. Hazard. Mater. **189**, 836–842 (2011)
3. S. Ema, T. Harada, Recovery and utilization of iron resources contained in laterite nickel leached residue. Int. J. Miner. Process. **19**(1–4), 127–143 (1987)
4. M.P. Antony, V.D. Tathavadkar, C.C. Calvert, A. Jha, The soda-ash roasting of chromite ore processing residue for the reclamation of chromium. Metall. Mater. Trans. B **32**(6), 987–995 (2001)
5. R. Weber, B. Rosenow, H.-D. Block, N. Lonhoff, Process for the preparation of sodium dichromate, U.S. Patents, 1993
6. U. Arndta, M. Batza, R. Bellinghausena, H.D. Blocka, H. Helkera, N. Lonhoffa, H.H. Morettoa, H.G. Nieder-Vahrenholza, H. Rinkesa, B. Spreckelmeyera, R. Webera, Method for manufacturing alkaline chromates from chromium ore. Miner. Eng. **9**(11), 1183 (1996)
7. C. Arslan, G. Orhan, Investigation of chrome(VI) oxide production from chromite concentrate by alkaline fusion. Int. J. Miner. Process. **50**(1–2), 87–96 (1997)
8. Z. Kowalski, C. Mazanek, Sodium chromate-material flow analysis and technology assessment. J. Clean. Prod. **6**, 135–142 (1998)
9. V.D. Tathavadkar, M.P. Antony, A. Jha, The soda-ash roasting of chromite minerals: kinetics considerations. Metall. Mater. Trans. **32B**(4), 593–602 (2001)
10. Z. Sun, Y. Zhang, S.L. Zheng, Y. Zhang, A new method of potassium chromate production from chromite and KOH-KNO3-H2O binary sub-molten salt system. AIChE J. **55**, 2646–2656 (2009)
11. H.B. Xu, S.L. Zheng, Y. Zhang, Z.H. Li, Z.K. Wang, Oxidative leaching of a Vietnamese chromite ore in highly concentrated potassium hydroxide aqueous solution at 300 °C and atmospheric pressure. Miner. Eng. **18**, 527–535 (2005)
12. Y. Zhang, Z.H. Li, T. Qi, S.L. Zheng, H.Q. Li, H.B. Xu, Green manufacturing process of chromium compounds. Environ. Prog. **24**, 44–50 (2005)
13. S. Zheng, Y. Zhang, Z. Li, T. Qi, H. Li, H. Xu, Green metallurgical processing of chromite. Hydrometallurgy **82**, 157–163 (2006)
14. G. Chen, X. Wang, H. Du, Y. Zhang, J. Wang, Sh Zheng, Y. Zhang, A clean and efficient leaching process for chromite ore. Miner. Eng. **60**, 60–68 (2014)
15. Y. Zhang, Z. Li, H. Xu, S. Zheng, T. Qi, H. Li, Sub-molten salt environmentally benign technology, in *Proceedings of the 100th AICHE Annual Meeting*, Philadelphia (2008)

Removal of Magnesium from Liquor Produced by Nickel Mining by Crystallization

Kristine Bruce Wanderley, Denise Crocce Romano Espinosa and Jorge A. Soares Tenório

Abstract Mining of nickel ores generates large volumes of waste that must be removed in order to eliminate water contamination and reduce environmental impacts created by waste barriers. In this context, the need for innovations concerning the recovery of magnesium present in the liquor produced by the leaching of nickel laterite with sulfuric acid arises. Therefore, this work aims to develop a study regarding the crystallization of magnesium from a synthetic liquor solution using a batch autoclave system at high temperature. Agitation speed was kept constant at 1000 rpm throughout the 3 h of batch reaction for temperatures of 230 and 200 °C. Aliquots taken at each hour were analyzed by ion chromatography to measure magnesium concentration. The highest magnesium removal was observed for temperature 230 °C. X-ray diffraction (XRD) and scanning electron microscopy (EDS-SEM) analysis proved that the product formed was magnesium sulphate monohydrate crystals. Thermogravimetric analysis coupled with a mass spectrometer was used for the evaluation of the thermal stability of magnesium sulphate monohydrate. It was found that MgO and SO_2 were formed by the decomposition of the salt at 1100 °C.

Keywords Monohydrate · High temperature · Thermal decomposition · Laterite ore

Introduction

Nickel mining involves leaching of the ore with sulphuric acid, resulting in the formation of an aqueous solution containing dissolved metals such as magnesium, manganese, chromium, aluminum, cobalt, nickel, copper and zinc as well as nickel [1]. These metals need to be removed in order for the process to be environmentally

K.B. Wanderley · D.C.R. Espinosa · J.A. Soares Tenório (✉)
Chemical Engineering Department, Polytechnic School, University of São Paulo,
Rua do Lago, 250, 2nd Floor, São Paulo, SP 05508-080, Brazil
e-mail: jtenorio@usp.br

L. Zhang et al. (eds.), *Energy Technology 2017*,
The Minerals, Metals & Materials Series, DOI 10.1007/978-3-319-52192-3_23

friendly and economic. The aqueous solution suffers a neutralization step to precipitate metals and thus resulting in a final solution composed of mainly $MgSO_4$, which also needs to be removed from the mother solution.

Studies regarding the removal of magnesium from mother liquor require the addition of a reagent such as a hydroxide [2]. However, addition of a reagent is not only costly but it also results in the formation of products with market no value and with no use in the process.

Therefore, a new innovate method is the removal of magnesium from solution by crystallization. The driving force of crystallization is supersaturation and there are three techniques to achieve supersaturation: cooling, evaporative and addition of an anti-solvent [3, 4].

The solubility of most salts increases with temperature however, magnesium sulphate presents an inverse solubility. Thus, to achieve supersaturation, evaporative crystallization was used. Figure 1 presents the solubility curve for the $MgSO_4$-Water system and it can be seen that the solubility increases until approximately 78 °C and then decreases abruptly. Notice that the inflection points indicate the formation of a less hydrated salt. It can be concluded from Fig. 1 that the monohydrate form is stable at temperatures higher than 78 °C and presents the lowest solubility at approximately 230 °C. Therefore, high temperature crystallization method was chosen in order to minimize the solubility of magnesium sulphate in the solution, forcing crystals with one molecule of hydration to precipitate.

The formation of a monohydrate crystal is desired as it contains low crystal water and therefore less energy is needed to decompose it. The products of the thermal decomposition of the monohydrate crystal are MgO and SO_2 [5, 6]. These components may be reutilized internally in the process: MgO to neutralize the

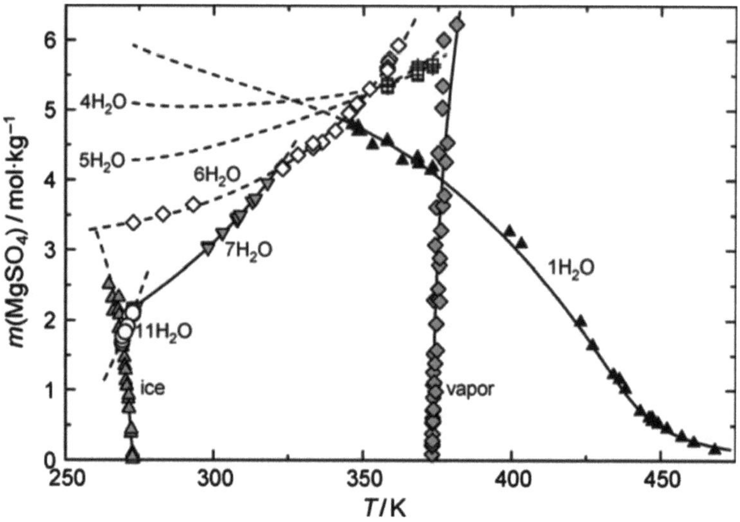

Fig. 1 Solubility curve for the $MgSO_4$-H_2O system [5]

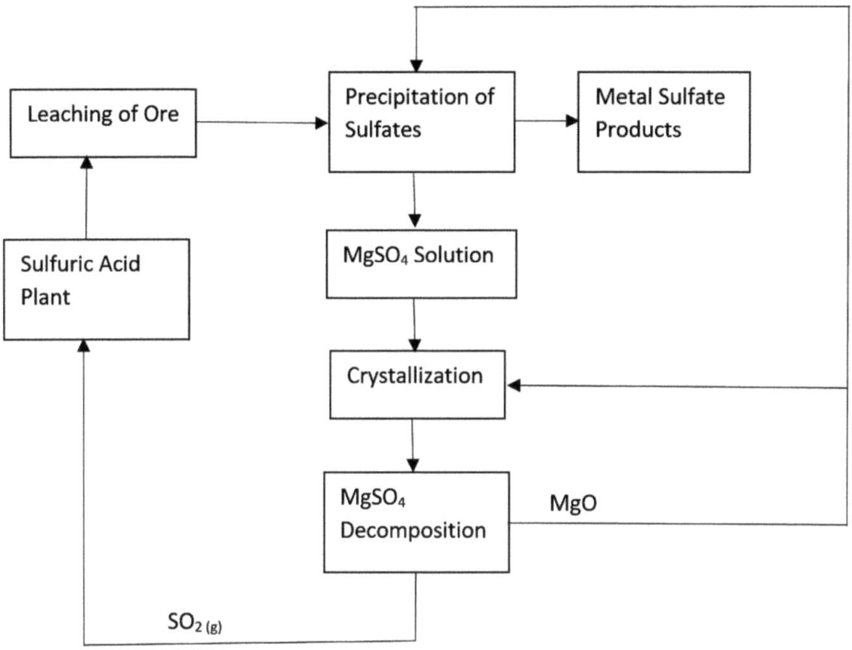

Fig. 2 Schematic flow diagram of the process

sulfate solution in the precipitation step and SO_2 in the production of sulphuric acid [7]. This allows generation of the feed materials (MgO and H_2SO_4) and lowers the total cost of the process in addition to reducing environmental issues. A schematic flow diagram of this process is shown in Fig. 2.

The thermal decomposition of magnesium sulphate monohydrate consists firstly on the dehydration of the salt around 300 °C, forming anhydrate $MgSO_4$ [8]. At around 1100 °C decomposition of $MgSO_4$ occurs and SO_3 is formed [9]. However, at this temperature SO_3 is not stable and spontaneously forms SO_2 and O_2, as can be seen from the reactions represented by Eqs. 1 and 2.

$$MgSO_4 \xrightarrow{1100\,^{\circ}C} MgO_{(s)} + SO_{3\,(g)} \tag{1}$$

$$SO_{3(g)} \xrightarrow{1100\,^{\circ}C} SO_{2\,(s)} + {}^{1}\!/_{2}O_{2\,(g)} \tag{2}$$

This way, the $MgSO_4$ solution produced by the leaching of the ore may be crystallized to produce monohydrate crystal which in turn may suffer thermal decomposition to produce MgO and SO_2. SO_2 could be forwarded to a sulfuric acid plant to produce sulfuric acid. This allows the waste $MgSO_4$ solution to be recycled internally in the process as MgO and H_2SO_4 instead of being discarded in the environment.

Materials and Methods

A solution with a concentration of 8 g/L of magnesium was prepared using magnesium sulphate heptahydrate and deionized water. This concentration corresponds to the concentration found in the leaching liquor of nickel limonite ore. Batch experiment were carried out using a Parr reactor. 200 mL of the solution was placed inside the reactor. Agitation speed was maintained at 1000 rpm throughout the 3 h of operation at a temperature of 230 and 200 °C. Every hour the valve was opened and an aliquot of 10 mL was taken for ion chromatography analyses for magnesium concentration.

Thermogravimetric–mass spectrometry study was also carried out to evaluate the thermal stability of the crystal. An 8 mg sample of $MgSO_4.H_2O$ was heated from room temperature to 1200 °C at 10 °C/min on the TG-MS system under atmospheric nitrogen flow.

Results and Discussion

Removal of Magnesium from Mother Liquor and Product Characterization

The aliquots taken were analyzed by ion chromatography for magnesium concentration and indicate that the temperature of 230 °C favored the removal of magnesium from solution and thus was the optimum temperature for crystallization amongst the investigated. Also, as can be seen from Fig. 3, at temperature 200 °C,

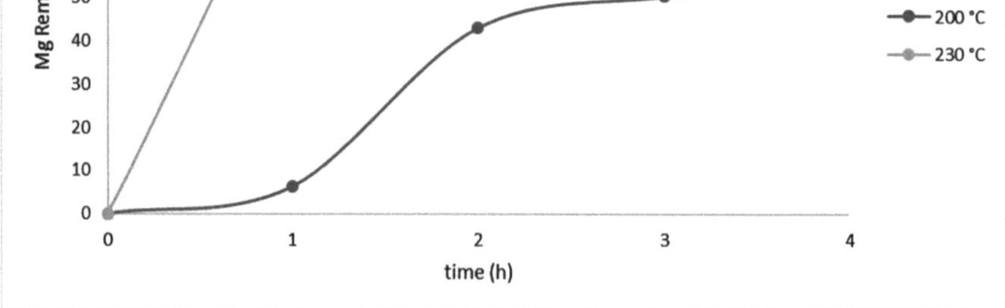

Fig. 3 Magnesium % removal for temperatures 230 and 200 °C

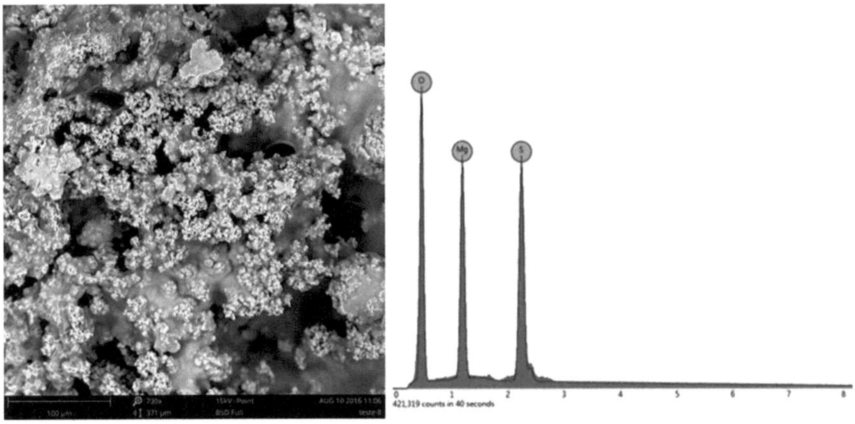

Fig. 4 SEM-EDS of crystals

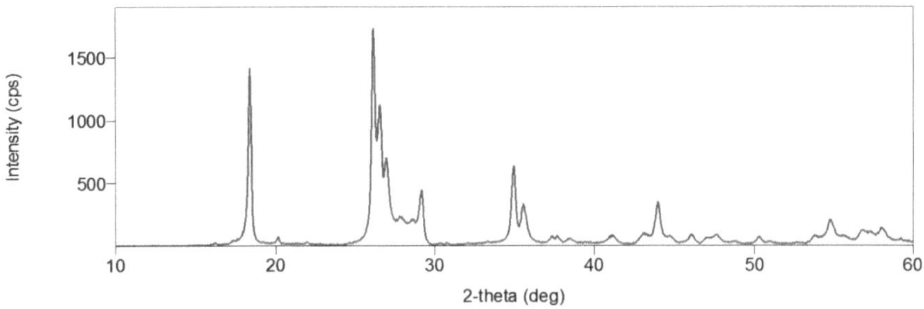

Fig. 5 X-ray diffraction (XRD) data for crystals formed at 230 °C

magnesium removal increases as the resident time increases. However, for 230 °C magnesium removal after 2 h was very similar to 3 and 4 h with approximately 90% of removal from solution.

In order to know the composition of the product a sample was taken and analyzed by SEM-EDS. As the crystals formed at 230 °C presented higher magnesium removal from solution, samples of this product were taken for further analyses. Figure 4 shows the SEM-EDS of the crystal. Three peaks corresponding to O, Mg and S may be seen and the SEM image shows irregular shaped crystals clustered together. The XRD peaks shown in Fig. 5 confirm the formation of a product composed of only magnesium sulphate monohydrate crystals.

Decomposition of Magnesium Sulphate Monohydrate

A study on the decomposition of magnesium sulphate monohydrate was also done. As can be seen in Fig. 6, there is an initial mass decrease at around 300 °C, possibly indicating water removal from the hydrated salt resulting in the formation of the anhydrate form. Between 300 °C and approximately 1000 °C there is no change in mass and then at 1000 °C the mass decreases sharply due to total decomposition of the salt. From Fig. 6 it is clear that decomposition is complete at 1100 °C and as the temperature continues to increase until 1200 °C the mass remains constant again, indicating that the decomposition was completed at 1110 °C. The total decomposition time was 2 h.

Therefore, the two decrease in mass are due to dehydration and decomposition and each occur in only one single step at a certain temperature. Notice that the mass decrease for decomposition is much steeper than dehydration, as the initial salt contained only one water of hydration. These results are similar to those found by [10].

The thermogravimetric analyzer was coupled with a mass spectrometer to investigate the relationship of mass loss of magnesium sulphate monohydrate and the formation of typical volatiles. As can be seen from Fig. 7, the curve representing SO_2 shows the evolution of this gas with the increase of temperature. The curve has its highest peak at around 1100 °C corresponding to the release of SO_2, the same temperature of decomposition of $MgSO_4$ as can be seen by the TG curve. As the temperature increases, SO_2 ion current decreases immediately indicating that all SO_2 was had been released as decomposition of $MgSO_4$ had been completed.

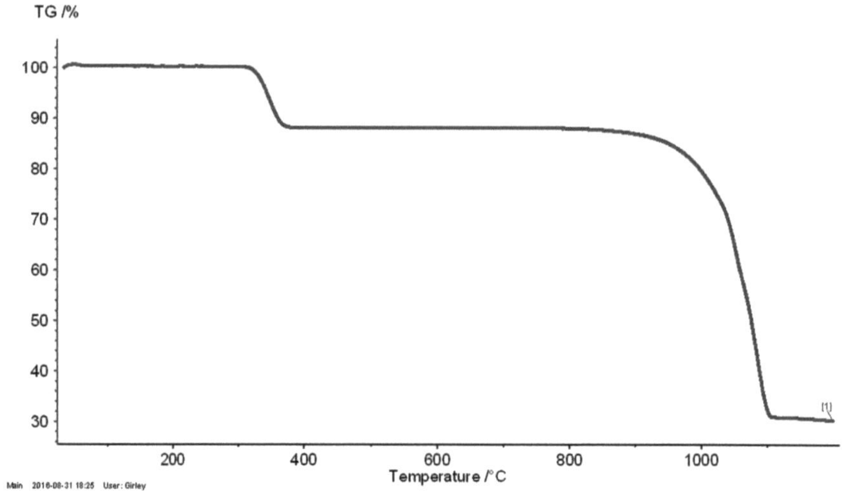

Fig. 6 TG measurements of the decomposition of $MgSO_4 \cdot H_2O$ in flowing N_2

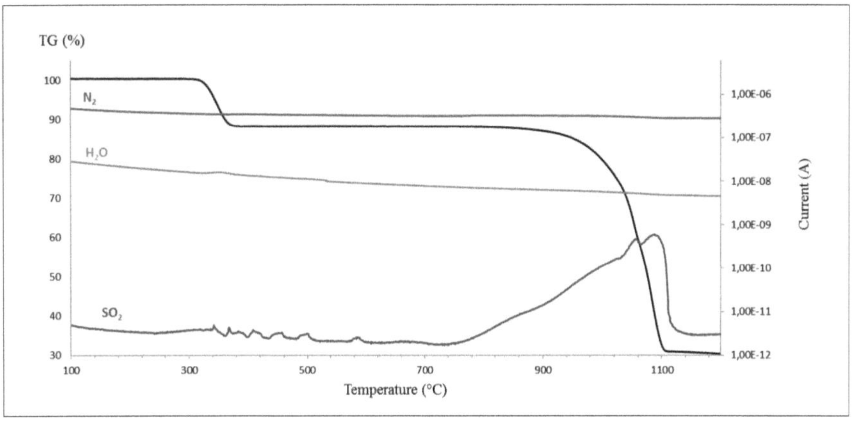

Fig. 7 TG and results of mass spectroscopy showing the release volatiles during the decomposition of $MgSO_4$

For the curve representing water, only a slight decrease can be observed. This may be due to the fact that the product contained only one molecule of hydration therefore no major signal peaks may be observed.

Conclusion

The methodology used was able to produce magnesium sulphate monohydrate crystals form solution. It may be concluded that 230 °C was the best temperature magnesium removal from solution as monohydrate crystals. Up to 90% of magnesium was removed at 230 °C at a resident time of 3 h. The TG-MS coupled analysis proves that the decomposition temperature of $MgSO_4$ to be 1110 °C. The mass spectrometer signals indicate that the mass loss steps are due to the release SO_2 and is in perfect correlation with the TG step. Therefore, the products of decomposition are SO_2 and MgO.

Thus, the process allows generation of feed materials such as MgO and SO_2 which can be sent to a sulfuric acid plant for acid production which in turn may be used internally in the system. This lowers the total cost of the process in addition to reducing environmental issues.

References

1. T. Karidakis, S. Agatzini-Leonardou, P. Neou-Syngouna, Removal of magnesium from nickel laterite leach liquors by chemical precipitation using calcium hydroxide and the potential use of the precipitate as a filler material. Hydrometallurgy **76**(1–2), 105–114 (2005)

2. S. Liu, G. Nancollas, The crystallization of magnesium hydroxide. Desalination **12**, 75–85 (1973)
3. A. Lewis et al., in *Industrial Crystallization: Fundamentals and Applications* (Cambridge University Press, Cambridge, 2015), p. 14
4. Toshiharu Irisawa, *Crystal Growth Technology* (Applied Science Pucblishers, William Andrew, 2003), pp. 25–54
5. M. Steiger, K. Linnow, D. Ehrhardt, M. Rohde, Decomposition reactions of magnesium sulfate hydrates and phase equilibria in the $MgSO_4$–H2O and Na^+–$Mg_2{}^+$–Cl^-–$SO_4{}^{2-}$–H_2O systems with implications for Mars. Geochim. Cosmochim. Acta **75**, 3600–3626 (2011)
6. M.N. Scheidema, T. Pekka, Decomposition thermodynamics of magnesium sulfate. Ind. Eng. Chem. Res. **50**(16), 9550–9556 (2011)
7. H. Kay, Treatment of nickelferous oxidic materials for the recovery of nickel values. US Patent, 3466144, 1969
8. J. Paulik, F. Paulik, M. Arnold, Dehydration of magnesium sulphate heptahydrate investigated by quasi isothermal-quasi isobaric TG. Thermochimico Acta **50**, 105–110 (1981)
9. K.H. Lau, D.L. Hildenbrand, D. Cubicciotti, Thermal Decomposition of some Metal Sulfates. Div. Fuel Chem. **21**, 48–54 (1976)
10. M.N. Scheidema, T. Pekka, Decomposition thermodynamics of magnesium sulfate. Ind. Eng. Chem. Res. **50**(16), 9550–9556 (2011)

Synthesis of Magnesium Oxide from Ferronickel Smelting Slag Through Hydrochloric Acid Leaching-Precipitation and Calcination

M.Z. Mubarok and A. Yudiarto

Abstract At the present paper, synthesis of magnesia from ferronickel smelting slag is discussed. The first stage was extracting magnesium from the slag by leaching in HCl solution. Magnesium extraction of 97% was obtained from the slag leaching with HCl 8 molar, solid:liquid ratio 1:40 (g/ml) and temperature 80 °C after 24 h. Solution purification was performed by adjusting solution's pH at 95 °C with addition of MgO powder. At pH range range of 3–4, co-dissolved iron was removed as akagenite (β-FeOOH) precipitate. Dissolved nickel precipitation was detected at pH range of 4–6, while dissolved Mg and Al started to precipitate in the forms of pyroaurite ($Mg_6Fe_2CO_3(OH)_{16}{\cdot}4H_2O$) and hydrotalcite ($Mg_6Al_2CO_3(OH)_{16}{\cdot}4H_2O$) at pH range of 8–9. High purity of brucite ($Mg(OH)_2$) was obtained by Mg precipitation at pH range of 11–12, under stirring speed of 300 rpm and stirring time of 4 h. Calcination of brucite at temperature of 1000 °C completely converted brucite to periclase (MgO) powder after 240 min.

Keywords Ferronickel slag · Magnesium · Leaching · Precipitation · Calcination

Introduction

Ferronickel (FeNi) alloy is one of primary products of nickel laterite ore processing, especially from saprolite ore type. Commercial productions of ferronickel from nickel laterite ore in industries are mostly done by using ELKEM Technology that principally involves drying in a rotary dryer, calcination in a rotary kiln and reduction smelting of the calcine in electric smelting furnace followed by refining

M.Z. Mubarok (✉)
Department of Metallurgical Engineering, Faculty of Mining and Petroleum Engineering, Institute Technology of Bandung, Jl. Ganesha, 10, Bandung 40132, Indonesia
e-mail: zaki@mining.itb.ac.id

A. Yudiarto
PT. Antam, Tbk, Jl. Letjen TB Simatupang No. 1 Lingkar Selatan, Tanjung Barat, Jakarta 12530, Indonesia

L. Zhang et al. (eds.), *Energy Technology 2017*,
The Minerals, Metals & Materials Series, DOI 10.1007/978-3-319-52192-3_24

stage. Saprolite ore used as the feed of the FeNi plant typically contains 1.8–2% nickel, 10–25% iron, 15–35% MgO and 30–50% SiO_2 [1]. During smelting, nickel and iron oxides are reduced to their metal states and form ferronickel alloy. Nickel and iron grade in the crude FeNi is dependent on the feed composition, reduction parameters and the smelter objective to meet market requirements. On the other hand, stable oxides in the ore feed such as MgO, SiO_2, Al_2O_3, P_2O_5 forms slag phase which has mass rate of approximately 60–70% from the calcine feed.

Indonesia is one of the countries with large nickel deposit in the form of laterite ore. Recently, the country has several ferronickel smelters that treat local nickel laterite ore. New FeNi smelters have been built and starting the operation to meet the market demand and to comply the government policy for local processing of mineral resources within the country. PT. Antam's ferronickel plant in Pomalaa, South East Sulawesi Province of Indonesia has been producing FeNi with nickel content of approximately 20% for more than 30 years. The plant uses ELKEM Technology that treats the saprolite ore in rotary dryer followed by calcination in rotary kiln, reduction smelting in electric furnace and refining in LD converter. Analysis of Pomalaa FeNi smelting slag shows that silicon and magnesium are the major component of the slag which mostly present in the forms of silicate compounds [2]. MgO contents of the FeNi smelting slag from Pomalaa are typically in the range of 25–30% [2]. In terms of MgO grade, FeNi smelting slag is comparable with the grade of MgO in dolomite mineral, which has been used for producing various magnesium compounds.

The extraction of magnesium from FeNi smelting slag has not been extensively studied yet. Commercial productions of magnesium metal and magnesium-bearing compounds are still commonly using primary sources such as magnesite ($MgCO_3$), dolomite ($CaMg(CO_3)_2$), brucite ($Mg(OH)_2$) and several salts such as carnallite ($KMgCl_3{\cdot}6H_2O$), bischofite ($MgCl_2{\cdot}6H_2O$), kainite ($MgSO_4$ $KCl{\cdot}3(H_2O)$), langbeinite ($K_2Mg_2(SO_4)_3$), as well as from seawater. The other raw materials for magnesium production which have not been widely treated are silicate minerals such as serpentine $(Mg,Fe)_3Si_2O_5(OH)_4$, olivine ($(Mg{\cdot}Fe)_2SiO_4$), and forsterite (Mg_2SiO_4) [3]. Several magnesium compounds that can be synthesized from the aforementioned raw materials are kieserite ($MgSO_4.xH_2O$), $MgCl_2$ salt, brucite ($Mg(OH)_2$) and MgO powder. Kieserite powder has been widely used for agricultural fertilizer ingredients. Magnesium chloride salt, $MgCl_2$, is a raw material for production of magnesium metal through fused salt electrolysis. Brucite is used for fire-retardant as an alternative for $Al(OH)_3$. Meanwhile, pure magnesium oxide has been used for manufacturing of magnesia brick as well as for specific purposes of neutralizing agent [4–6].

The leach-amenability of silicate minerals in hydrochloric acid solution has been investigated by several researchers [7–10]. It has been well recognized that silicate minerals are relatively difficult to be dissolved in HCl in comparison to oxides and hydroxides. Harris [10] suggested that magnesium silicate minerals can be dissolved in HCl with a concentration of 20–25% at temperature range of 50–105 °C. The best condition recommended was at temperature range of 80–90 °C under well controlled slurry pH of 0.7–0.8 to avoid gelatinous silica formation. Chen et al. [9]

reported the investigation results of serpentine ore leaching in HCl solution using a rotating disc and showed that magnesium dissolves quickly in 7 M HCl at a temperature of 95 °C. It was also concluded that the grain size of the ore does not affect the speed of the reaction, as indicated by the slight increase of Mg dissolution by the decrease of the ore particle size. Bakker [8] revealed that magnesium within silicate ore can be dissolved with concentrated HCl at a temperature of 90 °C and atmospheric pressure according the following reaction :

$$\left(Mg_{2-x-y}, Fe_x, Ni_y\right)SiO_{4(s)} + 4HCl_{(aq)} \rightarrow (2 - x - y)MgCl_{2(aq)} + xFeCl_{2(aq)} + yNiCl_{2(aq)} + 2H_2O_{(l)} + SiO_{2(s)} \quad (1)$$

At the present paper, leach-amenability of magnesium from ferronickel smelting slag in hydrochloric acid solution and recovery of magnesium from pregnant leach solution is discussed.

Materials and Methods

Sample Preparation and Characterization

Approximately 30 kg of FeNi smelting slag sample was received from PT. Antam's ferronickel smelting plant in Pomalaa, South East Sulawesi Province of Indonesia. The "as received sample" was dried at temperature of 200 °C for 24 h until the sample is completely dry. The sample was then milled using a disc mill and classified based on its particle size distribution by sieve series with opening sizes of 100 mesh, 150 mesh and 200 mesh. Each fraction was divided by using a riffle splitter to obtain 200 g of sample with particle size distributions of +100 mesh, −100 + 150 mesh, −150 + 200 mesh and −200 mesh. Chemical composition analysis of the slag sample was performed by X-Ray Fluoroscence (XRF) for each size fraction and the results are presented in Table 1. The result of XRF analysis shows that the FeNi smelting slag sample has MgO content of about 31% for all size fraction used in the investigation. Analysis with Scanning Electron Microscope

Table 1 Chemical composition of FeNi smelting slag sample with various particle size distributions

No	Particle size distribution (mesh)	Composition (%)							
		Ni	FeO	SiO_2	CaO	MgO	Cr_2O_3	Al_2O_3	Basicity
1	+100	0.09	9.40	52.83	1.95	30.64	0.71	2.20	0.62
2	−100 + 150	0.09	9.39	52.86	1.96	30.68	0.71	2.20	0.62
3	−150 + 200	0.10	9.50	53.02	1.97	30.70	0.72	2.21	0.62
4	−200	0.10	9.47	52.95	1.96	30.65	0.71	2.20	0.62

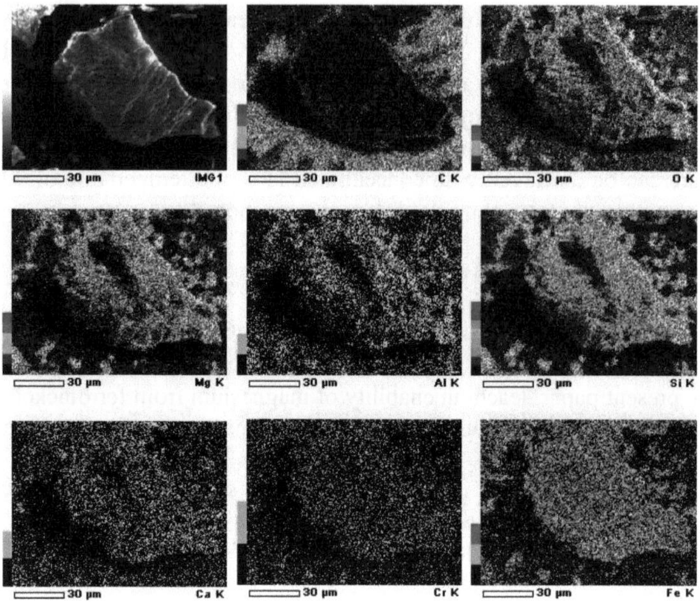

Fig. 1 Results of SEM-XRay mapping of FeNi smelting slag sample

(SEM) and Energy Dispersive X-Ray mapping was conducted to determine the distributions of magnesium and other metals in the FeNi slag. The result of SEM-X-Ray mapping is presented in Fig. 1. The SEM-X-ray mapping analysis result indicates that magnesium is uniformly distributed on the entire surface of the slag particle.

Slag composition at various particle size distributions is relatively uniform, in which silica is present as the major component of the slag. In addition to MgO, components in the slag that are supposed to consume acid are FeO, Al_2O_3, CaO, Cr_2O_3 and small amount of NiO. XRD analysis result of the slag sample is illustrated in Fig. 2. The XRD analysis revealed that the FeNi slag majorly consists of clinoenstatite ($MgSiO_3$) and forsterite (Mg_2SiO_4). This finding confirms the presumption that magnesium in ferronickel slag is mostly present in the forms of magnesium-silicates. The slag has basicity value of 0.62, which was calculated by the following equation:

$$basicity = \frac{(\%CaO) + (\%MgO)}{\%SiO_2} \tag{2}$$

where %CaO, %MgO and $\%SiO_2$ are the contents of those mentioned oxides (wt%) in the slag.

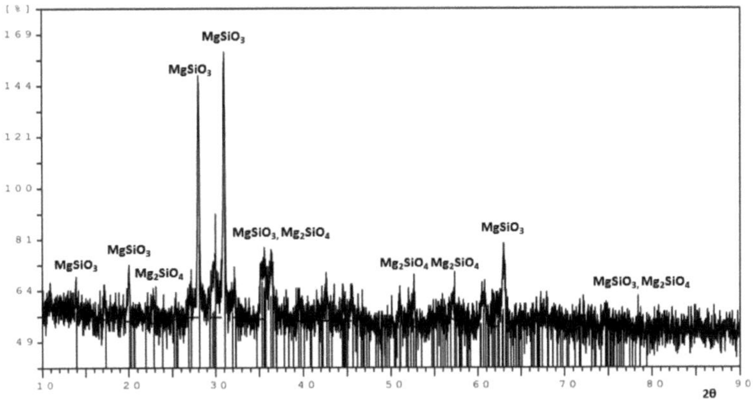

Fig. 2 XRD spectra of FeNi slag sample

Leaching Test

Series of batch agitation leaching tests at atmospheric pressure were carried out in a sealed five-neck reactor glass having a volume of 1 litre. The reactor was equipped with distillatory flash to condense vaporizing water and maintaining a constant volume of the solution. The agitation was performed by a magnetic stirrer which was integrated with a hot plate and temperature controller. The effects of hydrochloric acid concentration, slag particle size distribution, solid-liquid ratio (slurry density) and solution temperature on the magnesium extraction percentage were studied. Hydrochloric acid concentration was varied at 0.5, 1, 2, 4, 8 and 10 M under a constant temperature of 80 °C, solid:liquid ratio of 1:20 (g/ml) and slag particle size distribution of −150 + 200 mesh. Leaching temperature was varied at 65, 80 and 95 °C under a constant acid concentration of 8 M, solid:liquid ratio of 1:20 and particle size distribution of −150 + 200 mesh. The effect of slag particle size distribution on the magnesium extraction was evaluated by varying slag particle size distributions at +100 mesh, −100 + 150 mesh, −150 + 200 mesh and −200 mesh. Furthermore, the influence of solid:liquid ratio on magnesium extraction was investigated by varying solid:liquid ratios at 1:5, 1:10, 1:20 and 1:40 (g/ml). Leaching test was carried out for 24 h under a constant stirring speed of 450 rpm. The weight of the slag sample used was 50 g. Solution samples of 1 ml were periodically taken every 1, 2, 4, 8, 24 and 48 h after the starts of agitation for dissolved metals analysis using Atomic Absorption Spectrophotometer (AAS). The leaching residues were washed thoroughly and dried. Phases remaining in the leach residue were identified by X-ray Diffraction (XRD) analysis and the result was compared with the XRD spectra of the initial slag sample.

Solution Purification and Magnesium Recovery

Recovery of magnesium from pregnant leach solution (PLS) was carried out through 3 main stages: (a) impurities removal (b) precipitation of $Mg(OH)_2$ and (c) calcination of $Mg(OH)_2$ to produce MgO powder. Impurities removal was performed to precipitate co-dissolved iron, aluminium and chromium from PLS. The PLS having a volume of 500 ml generated by the leaching of FeNi smelting slag in 8 molar of HCl solution was heated until 95 °C and stirred with a magnetic stirrer bar. A pure analytical grade of MgO powder was added into the solution to increase the solution pH to ±1 for 1, 2, and 4 h. The precipitate obtained was separated by using filtration paper, dried and weighed for calculations of iron precipitation percentages. The solution filtrate was re-heated to 95 °C and pure MgO powder was added to increase the solution pH to 3–4 followed by precipitate filtration, drying and weighing of the precipitate. The same procedure was repeated by neutralizing the solution pH to 5–6. X-Ray Diffraction (XRD) analysis was conducted for identification of predominant compounds present in the precipitate. The second stage of Mg recovery from PLS was brucite precipitation ($Mg(OH)_2$) from purified $MgCl_2$ solution through three-stages of neutralization using NH_4OH solution. At the first stage, the solution pH was increased to 8–9 and $Mg(OH)_2$ precipitate was filtered, dried and weighed. Similar procedure was repeated by neutralizing the solution pH to 9–10 and 11–12. The precipitates obtained from each stage were characterized by XRD. At the last stage, MgO powder was synthesized by calcination of brucite at temperature of 1000 °C for 4 h.

Results and Discussion

Magnesium Leaching from FeNi Smelting Slag in Hydrochloric Acid Solution

Based on XRD analysis of the slag sample and prediction of predominant phases in the FeNi smelting slag by ternary diagram of FeO–MgO–SiO_2 (for MgO/SiO_2 ratio of 0.6 and FeO content 9.5%), reactions that may occur during the leaching of FeNi smelting slag in hydrochloric acid solution are as follows:

$$Mg_2SiO_{4(s)} + 4HCl_{(aq)} = 2MgCl_{2(aq)} + SiO_{2(s)} + 2H_2O_{(l)} \tag{3}$$

$$\begin{aligned} &FeMg(Si, Al)_2O_{6(s)} + 10HCl_{(aq)} + 3/2O_{2(g)} \\ &\quad = MgCl_{2(aq)} + FeCl_{2(aq)} + 2SiO_{2(s)} + 2AlCl_{3(aq)} + 5H_2O_{(l)} \end{aligned} \tag{4}$$

$$\begin{aligned} &(Ca, Mg, Fe)_2Si_2O_{6(s)} + 12HCl_{(aq)} + 2O_{2(g)} \\ &\quad = 2MgCl_{2(aq)} + 2FeCl_{2(aq)} + 2CaCl_{2(aq)} + 2SiO_{2(s)} + 6H_2O_{(l)} \end{aligned} \tag{5}$$

$$(\mathrm{Fe,Mg})\mathrm{SiO}_{3(s)} + 4\mathrm{HCl}_{(aq)} + 1/2\mathrm{O}_{2(g)} = \mathrm{FeCl}_{2(aq)} + \mathrm{MgCl}_{2(aq)} + \mathrm{SiO}_{2(s)} + 2\mathrm{H}_2\mathrm{O}_{(l)} \quad (6)$$

Referring to chemical composition of the FeNi slag sample and the above predicted chemical reactions, hydrochloric acid concentration was varied from 0.5 to 10 molar. Profiles of magnesium extraction as a function of time and hydrochloric acid concentrations at temperature of 80 °C, solid:liquid ratio of 1:20 (g/ml) and particle size distribution of −150 + 200 mesh are presented in Fig. 3. The highest magnesium extraction after 24 h was 64%, obtained at HCl concentration of 8 M. The investigations were then carried out under variations of particle size distribution, solid to liquid ratio and temperature. Profiles of magnesium extraction percentage as a function of leaching duration and particle size distribution at temperature of 80 °C, HCl concentration 8 M, solid:liquid ratio 1:20 are depicted in Fig. 4. As expected, at finer particle size of the slag, higher extraction percentage of magnesium was obtained. Significant increase of magnesium extraction percentage was exhibited by the decrease of particle size distribution from −150 + 200 mesh to -200 mesh, while particle size reduction from +100 to −100 + 150 and −150 + 200 mesh did not give significant impact on the extraction percentage of magnesium. At particle size distribution of coarser than −200 mesh, Mg extraction still tends to increase by further agitation beyond 48 h, while at particle size distribution of −200 mesh, Mg extraction reaches the maximum level at 81% after 24 h and no further enhancement of Mg extraction occurred by the extent of leaching duration to 48 h.

Profiles of Mg extraction percentage as a function of leaching duration and solid: liquid ratio (g/ml) at slag particle size distribution of −200 mesh, HCl concentration 8 M and temperature 80 °C are illustrated in Fig. 5. The experimental results show that the lower the ratio of solid mass to liquid volume, the higher Mg extraction was obtained. At lower solid to liquid ratio, the solution viscosity is lower and the mass transfer of reacting species to the interface of solid-liquid and the reaction products from the interface to the bulk of solution would be faster. Moreover, at lower solution viscosity, the likelihood of gelatinous silica formation is minimized.

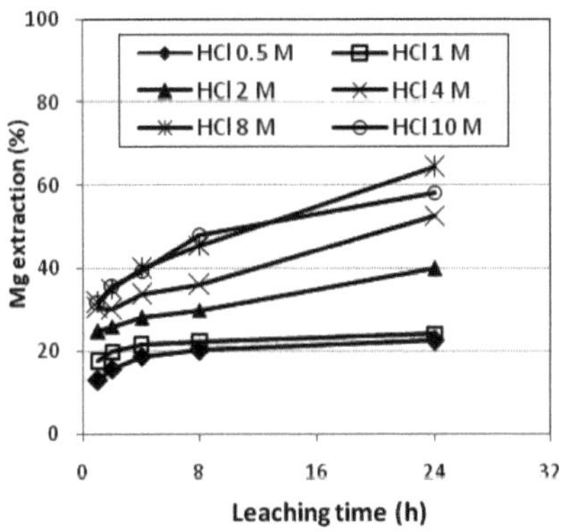

Fig. 3 Profiles of Mg-extraction as a function of time and HCl concentrations at 80 °C, solid/liquid ratio 1: 20 (g/ml) and particle size distribution −150 + 200 mesh

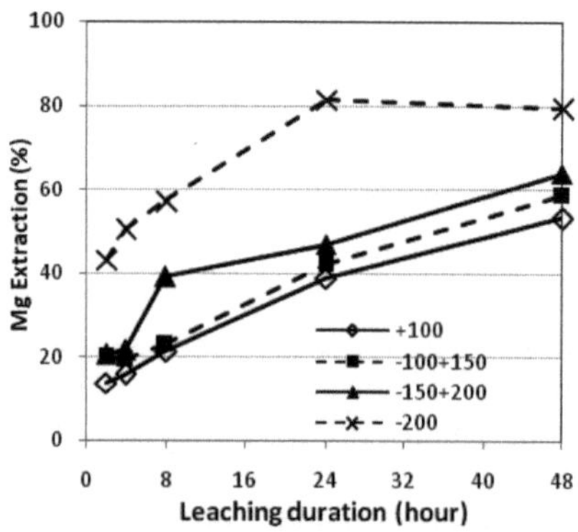

Fig. 4 Profiles of Mg extraction percentage as a function of leaching duration and particle size distribution at 80 °C, HCl 8 M, solid: liquid ratio 1:20 (g/ml)

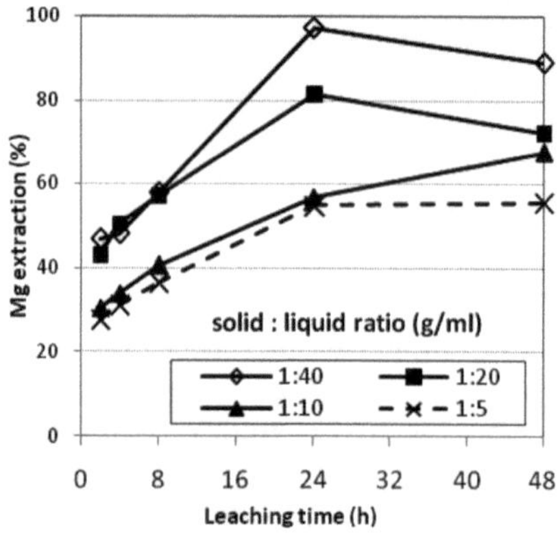

Fig. 5 Profiles of Mg extraction as a function of leaching duration and solid/liquid ratio at 80 °C, particle size distribution −200 mesh, HCl 8 M

The highest Mg extraction of 97% was obtained at solid:liquid ratio of 1:40 after 24 h. The further extension of leaching duration to 48 h resulted in a decrease of Mg extraction due to the precipitation of gelatinous silica that simultaneously entraps magnesium. The effect of temperature on Mg extraction was evaluated by varying solution temperature at 65, 80 and 95 °C under constants acid concentration of 8 M, solid:liquid ratio 1:20 and slag particle size distribution −200 mesh. Profiles of Mg extraction as a function of leaching duration and temperature of solution at acid concentration of 8 M, solid:liquid ratio 1:20 and particle size distribution −200 mesh are illustrated in Fig. 6. The increased of temperature from 65 to 95 °C improved Mg extraction significantly after 24 h. This is consistent with the general rule regarding the correlation between reaction speed and temperature, in which at higher temperatures the collisions of reactants take place more intensively and results in a faster reaction rate between magnesium-bearing compounds in the slag and the leaching agent. In case that the leaching kinetics is controlled by mass transfer, the enhancement of temperatures would accelerate the mass transfer rate of reacting species to the reaction interface.

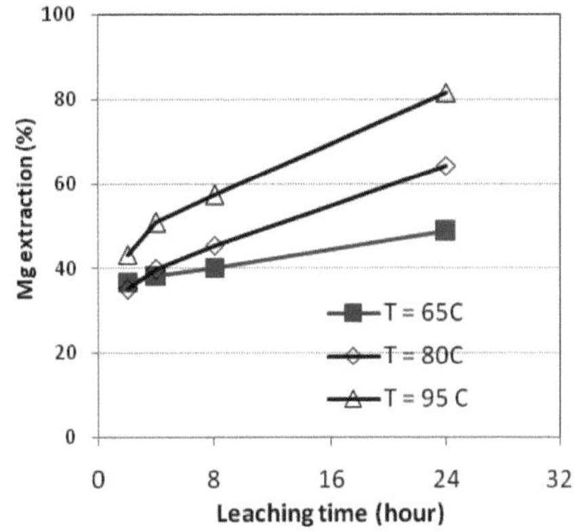

Fig. 6 Profiles of Mg extraction as a function of leaching duration and temperature at HCl 8 M, solid/liquid ratio 1:20 (g/ml) and particle size −200 mesh

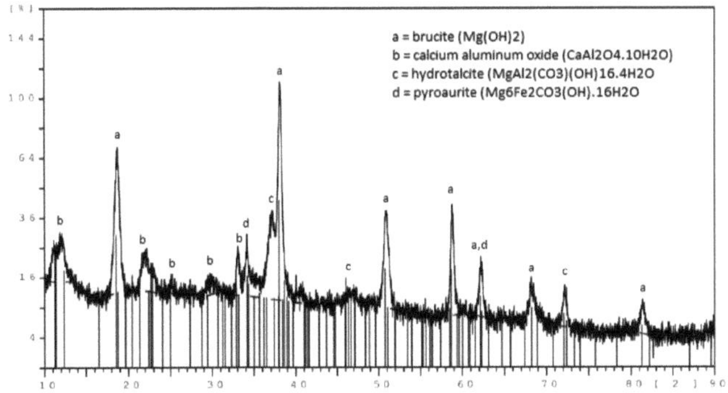

Fig. 7 XRD spectra of $Mg(OH)_2$ precipitate obtained by precipitation at pH 9–10

Precipitation of Impurities and Magnesium Recovery

XRD analysis of precipitates obtained from PLS neutralization at pH range of 1–4 identified iron precipitate in the form of akaganeite (β-goethite). This is in accordance with the previous studies by Ismael and Carvalho [11] which showed that at temperatures lower than 100 °C, dissolved iron will be precipitated mostly as akaganeite. Different to the precipitate obtained at pH 1, the akaganeite precipitate obtained at pH 3–4 has coarser grain and insoluble in water. After separation of iron-rich precipitate, $Mg(OH)_2$ precipitation was conducted at pH range of 9–11, 10–11 and 11–12. Brucite ($Mg(OH)_2$) precipitate started to form at pH 9–10. At this pH range, brucite precipitate still contains calcium-aluminum oxide, hydrotalcite ($Mg_6Al_2CO_3(OH)_{16}{\cdot}4H_2O$) and pyroaurite ($Mg_6Fe_2CO_3(OH)_{16}{\cdot}4H_2O$) as impurities. XRD analysis result of the precipitate obtained at pH 9–10 is shown in Fig. 7. High purity of $Mg(OH)_2$ precipitate was obtained at pH range of 10–11. No impurities were detected by XRD analysis of $Mg(OH)_2$ precipitate obtained at this pH range as can be seen in Fig. 8. Visually, the brucite precipitate obtained at

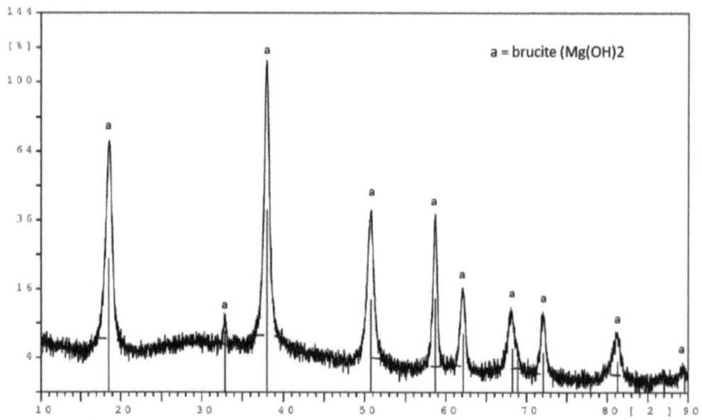

Fig. 8 XRD spectra of $Mg(OH)_2$ precipitate obtained by precipitation at pH 10–11

Fig. 9 Physical appearance of MgO powder obtained by calcination of $Mg(OH)_2$ precipitate at 1000 °C for 4 h

pH range of 10–11 has still a dull white colour, while bright white brucite powder was obtained by precipitation at pH range of 11–12. In addition to pH, characteristic of brucite precipitate is affected by temperature of precipitation. Moreover, the precipitation of brucite is also dependent on the ratio of $[OH^-]/[Cl^-]$ in solution. At higher $[OH^-]/[Cl^-]$ ratio, lower concentration of Mg^{2+} is required for $Mg(OH)_2$ precipitation, and vice versa [12].

$Mg(OH)_2$ calcination was carried out at temperature of 1000 °C for 4 h in a muffle furnace. Complete removal of crystalline water from brucite precipitate took place after 4 h. In general, calcination higher temperatures results in faster removal of crystalline water from brucite precipitate. Physical appearance of MgO powder and XRD analysis of the powder obtained by the calcination of $Mg(OH)_2$ at temperature of 1000 °C for 4 h are presented in Figs. 9 and 10, respectively.

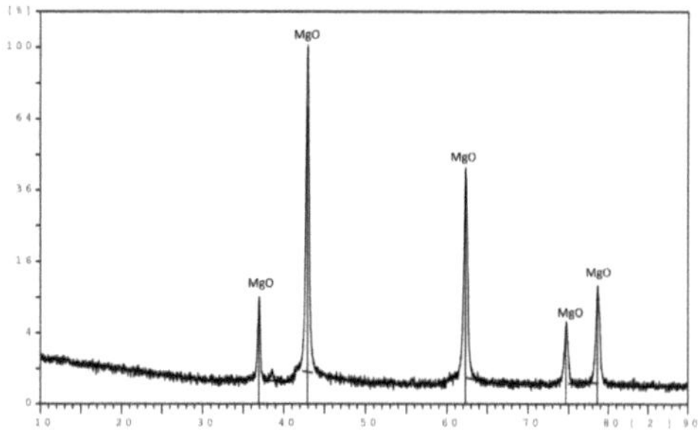

Fig. 10 XRD spectra of MgO powder obtained by calcination of $Mg(OH)_2$ precipitate at 1000 °C for 4 h

Conclusion

High purity MgO powder can be synthesized from ferronickel smelting slag through leaching in hydrochloric acid followed by impurities removal by neutralization, precipitation of $Mg(OH)_2$ and calcination of $Mg(OH)_2$. The presence of magnesium in silicate compounds in the slag (i.e. $MgSiO_3$ and Mg_2SiO_4) leads to low-amenability of Mg leaching in hydrochloric acid solution. Slag leaching with HCl concentration of 8 molar at solid:liquid ratio 1:40 (g/ml), slag particle size distribution −200 mesh and temperature 80 °C resulted in 97% Mg extraction after 24 h. Co-dissolved iron was completely removed at pH range of 3–4 as akagenite or β-goethite (β-FeO-OH) precipitate. High purity of brucite ($Mg(OH)_2$) was obtained by Mg precipitation from purified PLS using NH_4OH solution at pH range of 11–12, under stirring speed of 300 rpm and stirring time of 4 h. Calcination of brucite at temperature of 1000 °C completely converted brucite to periclase (MgO) powder after 4 h.

Acknowledgements The authors thank PT ANTAM for providing the sample of FeNi smelting slag used in the investigation and Program Penelitian Unggulan Perguruan Tinggi, DIKTI for financial support.

References

1. A.D. Dalvi, W.G. Bacon, R.C. Osborne, The past and the future of nickel laterites, in *PDAC 2004 International Convention*, Toronto (2004)
2. M.Z. Mubarok et.al., Extraction of magnesium and nickel from ferronickel smelting slag through leaching and solvent extraction (in Indonesian Language), in *Proceedings of the 9th Materials and Metallurgy Seminar*, Cilegon, Indonesia (Oct 2016)
3. F. Habashi, *Handbook of Extractive Metallurgy, Volume II: Primary Metals, Secondary Metals, Light Metals* (Wiley, 1997) pp. 981–1037
4. M.A. Shand, *The Chemistry and Technology of Magnesia* (Wiley, 2006)
5. M.D. Walter, Overview of flame retardants including magnesium hydroxide. Martin Marietta Magnesia Specialties (2015)
6. P. Ficara et al., Magnola: a novel commercial process for the primary production of magnesium. CIM Bull. **91**, 75–80 (1998)

7. Y.S. Chu, Y.R. Lim, H.B. Park, H. Song, J.K. Lee, S.H. Lee, Extraction of Mg ion and fabrication of Mg compound from ferro-nickel slag. J. Korean Ceram. Soc. **47**, 613–617 (2010)
8. J.S. de Bakker, The recovery of magnesium oxide and hydrogen chloride from magnesium chloride brines and molten salt hydrates. Dissertation of the Degree of Doctor of Philosophy, Queen's University, Kingston, Ontario, Canada, 2011
9. T.T. Chen, J.E. Dutrizac, C.W. White, Serpentine ore microtextures occurring in the Magnola magnesium process. JOM 20–23 (2000)
10. G.B. Harris, Process for the recovery of value metals from material containing base metal oxide. US Patent, No. US 7, 329, 396, B2 (2008)
11. M.R.C. Ismael, J.M.R. Carvalho, Iron recovery from sulphate leach liquors in zinc hydrometallurgy. Miner. Eng. **16**(1), 31–39 (2003)
12. C. Yamagata, J.O. Paschoal, Systematic precipitation of magnesium hydroxide using NH_4OH to preparing MgO-PSZ precursor powder, in *Proceedings of the 8th International Latin American Conference on Powder Technology* (2011), pp. 1874–1879

Thermodynamic Analysis of the Recycling of Aircraft AL Alloys

Senlin Cui and In-Ho Jung

Abstract Significant amount of commercial aircrafts will be retired in 2020, and the recycling of Al components in such aircrafts becomes an important issue in aerospace and Al industry. Recycling of Al from obsoleted Al products can effectively reduce the energy consumption and CO_2 emission which is more economical and ecofriendly. In the present study, thermodynamic analysis for the various recycling technologies of Al alloys were performed based on the comprehensive and accurate thermodynamic database. The recycling technologies reviewed in this work include the vaporization process to reduce Zn, precipitation process to reduce Fe, Mn, Si, Ti, Zr, flux treatment for Al_2O_3 and MgO, etc. Several key issues and limitation of the recycling of the aircraft Al were discussed. The recycling technologies discussed can also be utilized to recycle other Al alloys for common usage. All the thermodynamic calculations were performed using the FactSage FTlite database.

Keywords Aluminum alloy · Recycling · Aircraft · Thermodynamic calculation

Introduction

Thousands of end-life-use aircrafts have been setting in "graveyard". It is known that more than 70 wt% of a commercial airplane is made of Al [1]. The obsoleted aircraft itself is an important source of secondary Al production. Discarded aircraft can provide valuable source of high grade Al alloys. 2xxx and 7xxx series alloys are the predominant aircraft alloys, while 6xxx and 8xxx series have only limited usage. To a large extent, aircraft alloys can be classified into 2xxx and 7xxx series alloys which have very specified alloy composition. The most promising recycling process is that the aircraft Al alloys can be directly recycled into aerospace grade

S. Cui (✉) · I.-H. Jung
Department of Mining and Materials Engineering, McGill University, 3610 University Street, H3A 0C5 Montreal, QC, Canada
e-mail: cuisenlin@gmail.com

L. Zhang et al. (eds.), *Energy Technology 2017*,
The Minerals, Metals & Materials Series, DOI 10.1007/978-3-319-52192-3_25

alloys. Since 2024 and 7075 are the mostly used Al alloys in aircraft [2], the current work aims to analyze the possibility of recycling aircraft alloys to 2024 and 7075 alloys by thermodynamic calculations.

Recycling Process

Ideally, the aircrafts can be systematically disassembled. All the Al components are collected and classified. The wasted Al alloys were pre-sorted into 2xxx and 7xxx series alloys. Then the pre-sorted metals will be well cleaned, shredded, and re-melted to provide metal in a most useful way. The 2xxx series alloys can go back to 2024 and 7xxx series alloys back to 7075 alloy. The key determinants in this process are sorting and composition refinement. Sorting of alloys into 2xxx and 7xxx series is technically achievable, since 2xxx series alloy has high Cu content and 7xxx series alloy is rich in Zn. The controlling of the amount of different alloy elements and removal of certain unfavorable alloy elements are the key part.

Aircraft al Recycling Simulation

In real recycling process, 2xxx, 6xxx, and 7xxx series alloys in the best situation can be separated. However, it is hard to economically separate each series of alloy in details since the alloy composition is very close, for example 2024 and 2014. In the worst case, Al alloys are only separated from other alloys, for example, Ti alloy and steel, without further separation. Therefore, three possible situations need to be considered: a mixture of 2xxx series, a mixture of 7xxx series, and a mixture of all the aircraft aluminum alloys. For simplification, only few typical alloys are considered. The aircraft alloy sample includes 9 alloys. 7xxx series alloys include 7075, 7475, 7175, and 7050 with the most used 7075 alloy. 2xxx series alloys include 2024, 2014, 2026, and 2324 with the widely used 2024 alloy. While 6013 is taken as a representative of 6xxx series alloys.

The average composition of each alloy is used to prepare the alloy mixture. The amount of each alloy used to prepare the alloy mixture was shown in Table 2 (since the amount of each alloy used in aircraft is unknown and varies with aircraft model). The alloys named as "Mix 7xxx", "Mix 2xxx", and "Mix all" are the alloy mixture of 7xxx series alloy, 2xxx series alloy, and all series of aircraft alloys, respectively. The alloy mixture namely "Mix 7xxx", "Mix 2xxx", and "Mix all" are also shown in Table 1. It can be seen from the table that "Mix 7xxx" and "Mix 2xxx" alloy can be refined to 7075 and 2024 if the Zr can be effectively removed. All other elements are within or lower than the nominal composition of 7075 and 2024. The composition of all other alloying elements can be adjusted by adding certain amount of pure elements. For the "Mix all" alloy, only contents for Mg, Zn, and Zr exceed the nominal composition range of 2024. In the following section, the

Table 1 New alloy composition compared with 2024 alloy

Alloy	Cr	Cu	Fe	Mg	Mn	Si	Ti	Zn	Zr	Other	Al
7075	0.18–0.28	1.2–2	0–0.5	2.1–2.9	0–0.3	0–0.4	0–0.2	5.1–6.1	0	0–0.15	Bal.
2024	0–0.1	3.8–4.9	0–0.5	1.2–1.8	0.3–0.9	0–0.5	0–0.15	0–0.24	0	0–0.15	Bal.
Mix 7xxx	0.10917	1.9528	0.1	2.3139	0.06111	0.080556	0.041667	6.0056	0.063889		
Mix 2xxx	0.033333	4.2028	0.17444	1.1556	0.63889	0.30833	0.06	0.13361	0.05		
Mix all	0.077188	2.7604	0.14354	1.8365	0.30208	0.20333	0.052292	3.5574	0.057083		
De-Zr	0.077196	2.7607	0.14356	1.8367	0.30212	0.20336	0.052298	3.5578	0.051050	0	91.015
De-Zn	0.07999	2.8607	0.14875	1.7506	0.31305	0.21072	0.054191	0.21958	0.052898	0	94.310
De-TiZr	0.080133	2.8658	0.14902	1.6765	0.31361	0.21109	0.006048	0.21997	0.00015261	0	94.478
New alloy	0.0786	3.991	0.1467	1.6499	0.30864	0.48446	0.1443	0.21648	0.00015019	0	92.979
2024	0–0.1	3.8–4.9	0–0.5	1.2–1.8	0.3–0.9	0–0.5	0–0.15	0–0.24	0	0–0.15	Bal.

Table 2 Amount of each alloy used to prepare the alloy mixtures (in gram)

Alloy	7475	7150	7075	7050	2024	2014	2026	2324	6013
Mix 7xxx	50,000	40,000	30,000	60,000					
Mix 2xxx					40,000	50,000	60,000	30,000	
Mix all	50,000	60,000	70,000	100,000	40,000	50,000	60,000	3000	20,000

removal or refinement of the major alloying elements is first discussed. As a more complicated situation, the "Mix all" alloy is refined to 2024 alloy using thermodynamic calculations using FactSage software [3]. During the thermodynamic calculation, FTlite, FTsalt, FToxide, and FTPS thermodynamic databases are used, the details of these databases are described elsewhere [4]. It should be clarified that the reliability of the current calculation is critically depend on the quality of these databases.

Composition Control of al Melts

Removal of Mg (Ca)

Chlorination using Cl_2 is the most used way in industry to remove Ca and Mg in Al melts. To elucidate the Mg removal using Cl_2 gas, 100 g of Al-5 wt% Mg melts under the protection of 0, 5, 10, 20, 30, 40, 50, 60, and 100 g of equal moles of NaCl and KCl flux are simulated to remove Mg using Cl_2 gas. The simulation results are shown in Fig. 1. It can be seen from the figure that Mg can be effectively removed from Al melts to a level of 10^{-5} by Cl_2 gas at 700 °C. However, the addition of flux does not have apparent effect on Mg removal. In details, the remaining Mg level is higher than that without any flux when the flux amount is less than 40 g, and lower when the flux amount is more than 40 g. Cl_2 gas is very effective in removal of Mg in Al alloy. Flux is not necessary for the Mg removal process.

Removal of Li (Na and K)

Li can be removed from the Al melt by way of chlorination using Cl_2 or chlorine containing flux. During Li removal, the interaction between Mg and Li should be considered as the chlorination process will also inevitably affect Mg. The calculated composition profiles of Li and Mg in flux free Al-2 wt% Mg-0.1 wt% Li, Al-2 wt% Mg-0.2 wt% Li, and Al-2 wt% Mg-0.5 wt% Li alloys at 700 °C under varying Cl_2 amount are shown in Fig. 2. It can be seen from the diagram that Li concentration immediately drops as Cl_2 addition. When the Li concentration reaches about 0.01–0.001 wt%, Mg starts to react with Cl_2 gas. 7 g of Cl_2 can reduce Mg and Li

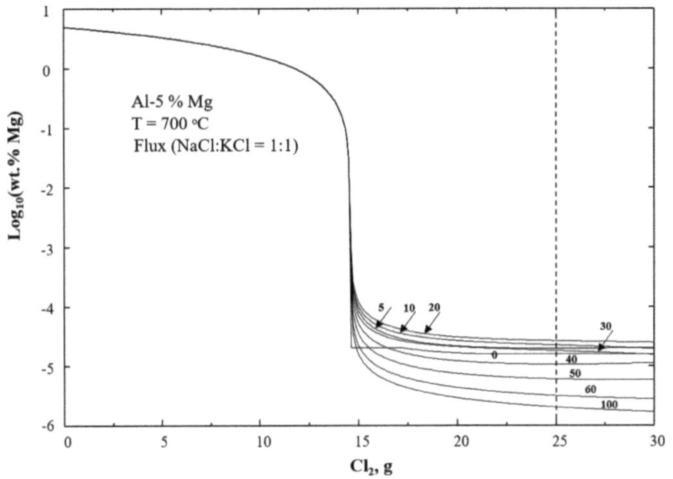

Fig. 1 Calculated Mg concentration profiles of an Al-5 wt% Mg alloy reacting with Cl_2 gas under the protection of different amounts of salt

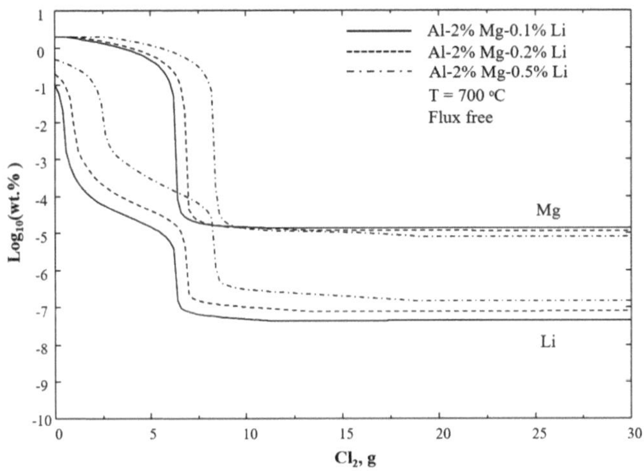

Fig. 2 The calculated concentration profiles of Li and Mg of Al-2 wt% Mg-0.1 wt% Li, Al-2 wt % Mg-0.2 wt% Li, and Al-2 wt% Mg-0.5 wt% Li as a function of Cl_2 amount

to a bottom value, further Cl_2 addition only slightly decrease both Li and Mg. Chlorination using Cl_2 gas is not a suitable way to remove Li from Al in presence of Mg. Our subsequent thermodynamic calculation indicated that flux of (KCl:NaCl: $AlCl_3$ = 4:4:2 in mole ratio) is effective in remove Li from Al melts without apparent affection on Mg. However, large amount of flux is required.

Removal of Zn

Even though less reactive than Al, Zn is considerably easy to remove by vaporization because of its high vapor pressure. The vapor pressure of Al is relatively stable and can be neglected. Our thermodynamic calculation indicate that Mg will not affect the Zn vaporization process under the optimum condition of low temperature and high pressure.

Removal of Fe, Si, Mn, and Cr

According to Ashtari et al. [5] and de Moraes [6], the Fe and Mn can be removed from Al melt by changing the Fe/Mn ratio to form α-$Al_{15}(Fe, Mn)_3Si_2$ (AlFeMnSi_alpha) phase followed by sedimentation and filtering process. According to them, more than 60% of Mn and Fe can be removed. Our thermodynamic calculation indicates that the α-$Al_{15}(Fe, Mn)_3Si_2$ phase is not forming when the Si content is lower than 2 wt%. Al alloy with Fe concentration within the α-$Al_{15}(Fe, Mn)_3Si_2$ primary phase region can be refined by adding Mn to form α-$Al_{15}(Fe, Mn)_3Si_2$ precipitates. This method is also applicable for the reduction of Mn in melt and works up to at least 20 wt% Si. It should be noted that this method can only be practically applicable for Al alloys with high percentage of Si.

Si can be removed by adding Ca in the melt to form $CaSi_2Al_2$ phase which is floated on the top layer of the melts [7]. Our thermodynamic calculation indicates that Ca addition can promote the formation of $CaAl_2Si_2$ when the Si content in the melt is higher than 6.43 wt%. The amount of Cr added into Al is lower than 3 wt% which is not an issue on alloy recycling. And no effective way to economically remove Cr from Al melt is found.

Removal of Ti and Zr

The solubility of Ti and Zr in Al is lower than 1 at.%. Ti is a common alloying element for both 7xxx and 2xxx series alloys, and Zr is added in certain specially alloy. Ti and Zr are very prone to B. The addition of B element in Al melt will promote the formation of TiB_2 and ZrB_2 type stable precipitates. The solubility of B in Al is less than 20 ppm. Similar to the removal of Fe, the TiB_2 and ZrB_2 precipitates can be filtered after formation.

Flux for Refinement

The flux used in Al recycling can either be in solid state or liquid state depend on the detailed refining process. Basically, liquid flux with a lower density than the melt ensures a better metal-salt settling. And solid flux with higher density than the melt can submerge in the melt much easily. The liquid flux used in Al refinement is normally equal moles of KCl and NaCl with the addition of certain amount of Na_3AlF_6, CaF_2, or $NaAlF_4$, as equal moles of KCl and NaCl gives a lowest melting point.

Removal of Inclusions

The Al_2O_3 phase may form in the Al melts during the refining process due to the presence of H_2O, O_2 from air or Argon gas during the de-zinc process. In industry, NaF and KF are added to the refining flux to remove Al_2O_3 inclusion. Our thermodynamic calculation indicates that the oxidation might only happen with Mg in the Al refining process and the KF and NaF addition to the flux is very useful in preventing oxidation.

Refining the Alloy Melt

A preliminary thermodynamic calculation indicates that the $AlZr_3$ phase can form in the melt at 660–630 °C and fcc_A1 phase starts to form below 640 °C. As a first step, the Zr was partially removed by keeping the melts at 650 °C and followed by filtering (assume a 100% removal efficiency). The Zr content decreases to 0.05105 wt% (see Table 1, De-Zr). And then, Zn can be removed from the remaining melt by evaporation. In the present work, Ar is selected. Ar bubbles can be introduced to the melts. Zn with a high vapor pressure can be take away from the melt. The vaporization temperature and pressure are low temperature (650 °C) and high pressure (1 bar). After 30,000 g of Ar gas equilibrium with the liquid melt for 750 times, the Zn concentration decreases to 0.21958 wt% (See Table 1, De-Zn).

Ti and Zr are removed by introduce B to form $(Ti, Zr)B_2$ precipitates. Here BCl_2 (B_2Cl_4) is used as B transformer. KCl (500 g), NaCl (500 g), and $MgCl_2$ (1500 g) mixture is added to cover the melt to prevent Mg consumption. The Zr content decreases to 150 ppm after adding 150 g BCl_2 for 8 times. Ti also apparently decreases. The current thermodynamic calculation indicates that further adding of BCl_2 will dramatically decrease Zr, but the B content in melt will increase. As a compromise, 150 ppm of Zr is kept in the melt. The liquid metal composition is shown in Table 1 as “De-TiZr”. Finally, 5500 g Cu, 1300 g Si, and 650 g of Ti are added to the melts to adjust the alloy composition. Fe is not added since it is always treated as impurity. The final alloy composition is presented in Table 1 as “new alloy”. All the alloying composition except for Fe is within the nominal composition of 2024 alloy. Similar recycling process can also be utilized to refine the alloy composition of “Mix 7xxx” alloy to 7075 alloy.

Summary

Thermodynamic calculations are carried out to study the recycling of aircraft Al alloys using FactSage software. The removal of the major alloy elements is discussed. The sorted 7xxx and 2xxx aircraft Al alloys can be recycled to the most used 7075 and 2024 alloys, respectively. The unsorted aircraft Al alloys can also be recycled into the 2024 alloy with more recycling steps.

References

1. E. A. Starke, Jr., and J. T. Staley, Application of modern aluminium alloys to aircraft, Fundam. Alum. Metall. 747–83 (2011)
2. T. Dursun, C. Soutis, Recent developments in advanced aircraft aluminium alloys. Mater. Des. **56**, 862–871 (2014)
3. http://www.factsage.com/
4. I.-H. Jung, S. Cui, J.-K. Lee, S.-M. Park, Thermodynamics of the Mg recycling process. JOM **65**, 1310–1316 (2013)
5. P. Ashtari, K. Tetley-Gerard, K. Sadayappan, Removal of iron from recycled aluminium alloys. Can. Metall. Q. **51**, 75–80 (2012)
6. H.L. de Moraes, J.R. de Oliveira, D.C.R. Espinosa, J.A.S. Tenorio, Removal of iron from molten recycled aluminum through intermediate phase filtration. Mater. Trans. **47**, 1731–1736 (2006)
7. P. Ashtari, B. Davis, K. Sadayappan, Removal of Si from recycled aluminium alloys. Int. J. Cast Met. Res. **25**, 100–102 (2012)

Lithium-Ion Battery Recycling Through Secondary Aluminum Production

Reza Beheshti, Ali Tabeshian and Ragnhild E. Aune

Abstract Today's lithium-ion batteries (LiBs) are found in more than half of the hybrid-electric and plug-in hybrid vehicles on the marked, all-electric vehicles, and nearly all consumer electronics. As a result, recycling of LiBs will be a strategic necessity in the near future. In the present study, the feasibility of application of conventional aluminum recycling process (secondary aluminum production) as a suitable process for treatment of LiBs was studied. The overall idea is to recover aluminum, copper and lithium from a mixed waste stream of LiBs and aluminum scrap. A two-stage process, consisting of a preheating step and a melting step, were tested on cell phone batteries mixed with pure aluminum in a ratio 1:10 and the distribution of elements between the metal phase and the slag phase evaluated. The obtained results are compared and discussed based on the thermodynamic calculated by the use of the FactSage™ software version 7.

Keywords Li-ion battery · Pyrometalurgy · Lithium · Aluminum · Copper · Recycling

Introduction

The popularity of the Lithium-ion batteries (LiBs) application in the field of electronic appliance such as cellphones and electrical vehicles (EVs) is increasing dramatically [1, 2]. The EVs have higher energy efficiency and less CO_2 emission than the traditional vehicles. In Scandinavian countries, the production and sale of EVs is widely promoted. Norway is fortunate enough to have close to 100% renewable and cheap hydro power production. The country (population: 5 million) is the world leader in electric cars per capita and has just become the fourth country in the world to have 100,000 of them on the roads, after the US (population: 320

R. Beheshti (✉) · A. Tabeshian · R.E. Aune
Department of Materials Science and Engineering, Norwegian University of Science and Technology (NTNU), 7491 Trondheim, Norway
e-mail: rezakb@kth.se

L. Zhang et al. (eds.), *Energy Technology 2017*,
The Minerals, Metals & Materials Series, DOI 10.1007/978-3-319-52192-3_26

million), Japan (population: 130 million) and China (population: 1.3 billion). In March 2016, 25% of sold cars in Norway were EVs [3–5]. So recycling of LiBs will be a strategic necessity for the world and especially for Norway in near future. The LiBs technology development is very active area and LiBs manufacturers are moving to cheaper cathode materials such as lithium manganese oxide and lithium iron phosphate to lower the price. On the other hand, current LiBs recycling processes are mainly focused on recovering cobalt and other valuable metals. Moreover, these processes are energy intensive and complicated to be applied for recycling automotive LIBs. As the result, they may not be profitable to recycle batteries with the new, cheaper, chemistries. So still researcher are looking for alternative hydrometallurgical, pyrometallurgical or mechanical processes for handling upcoming scraps [6–9]. One common practice to improve the economy of the processes is addition of spent batteries to existing large-scale processes, which are not dedicated to battery recycling, e.g. extractive cobalt or nickel metallurgy [10]. Since many LiBs including automotive batteries has pouch shape cells with aluminum casing, the aluminum recycling process is investigated for LiBs recycling. Recovery of all battery components, especially lithium, aluminum and copper were of interest. For the initial experiments, cellphone batteries, because of their smaller size, have been chosen (Fig. 2).

LiBs pouch cells consist of anode, cathode, separator, electrolyte, and polymer pouch, Fig. 1. The anode is graphite mixed with a binder deposited on a copper foil. The cathode is lithium metal oxide (e.g. $LiCoO_2$, $LiMnO_2$, $LiNiCoAlO_2$, $Li_3NiMnCoO_6$, and $LiFePO_4$) mixed with graphite and binder deposited on an aluminum foil. The electrolyte consists of a mixture of non-aqueous solvents and lithium salts. When charging, lithium de-intercalates from the metal oxide and intercalates in the graphite, where it is thermodynamically unstable. During discharging the process is reversed and lithium ions are intercalated back into the lithium metal oxide. Opening the batteries “as is” may results in an explosive reaction between lithium ions/lithium metal (if present) and oxygen in air [6, 8, 11].

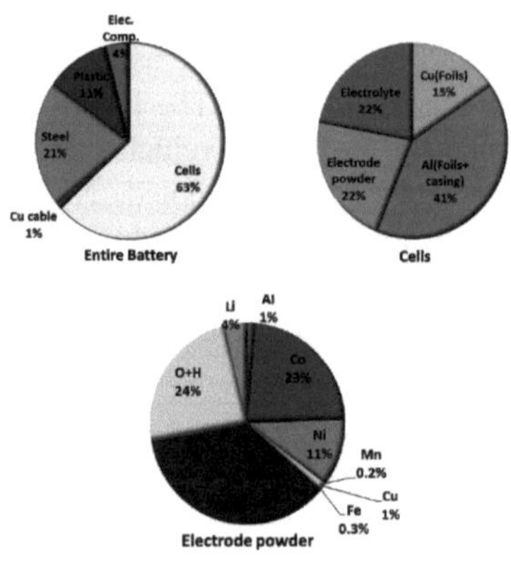

Fig. 1 Composition of an electrical car battery, cells and electrode powder [7]

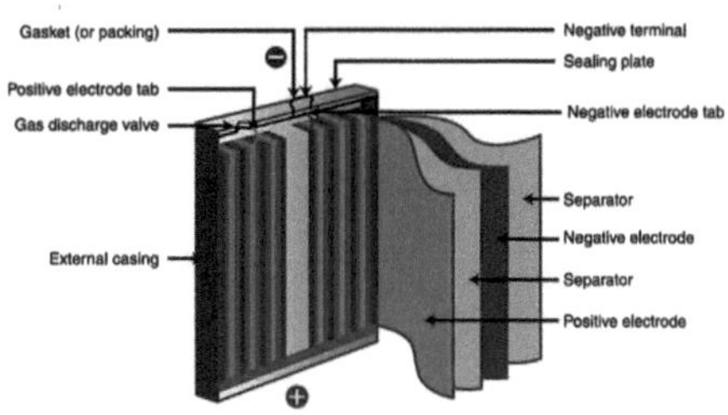

Fig. 2 A schematic drawing of a Li-ion battery cell [8]

Thermodynamic Evolution

FactSage™ software 7 [12] is used to establish the stable phases for a mixture of battery and pure aluminum, with ratio. This should be keep in mind that during battery production, some elements may form some other more stable phases and this thermodynamic calculation was done based on pure elements to get some cornerstone. The composition of lithium nickel cobalt aluminum oxide ($LiNiCoAlO_2$) batteries were used these evaluation [13]. The result of the thermodynamic evolution is summarized in Table 1. It can be seen presence of carbon and oxygen results some aluminum and lithium form oxide and carbide and be transfer to the slag phase [14]. Moreover, the thermodynamic evolution shows the presence of lithium in gas phase at the temperature of interest will be less 0.01% and for other elements less 10^{-8}%. Trager et al. [8] reported the evaporation of lithium will significant that temperature higher 1340 °C.

Table 2 shows all elements except lithium have the maximum presence in melt phase at ratio higher than 2. But there is a competition between aluminum and lithium to react with oxygen. Nevertheless the thermochemical calculations can only give an indication for the evaluation of the melt and slags since kinetic aspects cannot be considered.

Table 1 The FactSage calculation result for x mol Al and 1 mol battery (0.01 Co + 0.06 Ni + 0.26 Al + 0.09 Li + 0.07 Cu + 0.08 O2 + 0.43 °C) at 900 °C

	Battery	Battery + 1 Al	Battery + 2 Al	Battery + 3 Al	Battery + 6 Al	Battery + 9 Al
C (graphite)	0.34					
$LiAlO_2$	0.8	0.08	0.08	0.08	0.07	0.06
Al_4C_3	0.02	0.14	0.14	0.14	0.14	0.14
Al_3Ni_5	0.01					
AlCo	0.01					
Li_2C_2	0.004					
Al3Ni		0.05				
Al_9Co_2						
Al_4O_4C					0.008	0.008
Melt	0.10	0.53	1.76	2.75	5.76	8.76

Table 2 The FactSage calculation for slag and melt elemental destribution, 2 mol Al and 1 mol battery (0.01 Co + 0.06 Ni + 0.26 Al + 0.09 Li + 0.07 Cu + 0.08 O_2 + 0.43 °C) at 900 °C

	Slag		Melt						
Phase	Al_4C_3	$LiAlO_2$	Al	C	Co	Cu	Li	Ni	O_2
Mol%	0.14	0.08	1.6	0.0002	0.01	0.07	0.01	0.06	0.0004
Total	0.22		1.76						

Experimental Procedure

Three sets of experiments on LiBs cell phone (with and without pretreatment) were carried out, LiBs i.e.:

- As is (without pretreatment)
- After preheating at 300 °C for an hour in air to evaporate the electrolyte from the batteries.
- After preheating at 500 °C under vacuum to evaporate the electrolyte without oxidizing the aluminum for15 min.

The pretreated batteries after cooling cut to several pieces. In each experiment, a battery cell (23 ± 1 gr) is put together with 100 gr aluminum cubes (each 20 gr) placed in an alumina crucible and heated to 900 °C under argon atmosphere, kept for 2 h.

Experimental Results

As it can be seen from Fig. 3, in the case of LIB without pretreatment, the polymeric coating act as the boundary layer and keep the battery separate from aluminum melt and after cooling battery can be removed easily and the melt and battery neither shows considerable weight change.

In the case of LiB pretreatment in air, it can be seen from Fig. 4, battery parts and aluminum melt form two separate phases without any interaction. The copper foil can easily identify from rest of the sample.

In the case of batteries which heat treated in vacuum, same as previous experiment the battery parts does not dissolved in the aluminum melt. It was seen major part of the aluminum foil after the experiment was oxidize and the same result was observed in pervious experiments as well. It is well-known fact that recycling of the aluminum foil has lower yield compare to recycling of bulk aluminum scrap. The reason is a thin layer of oxide form on aluminum surface in room temperature and foils has high surface to volume ratio [15]. It can be conclude aluminum and lithium oxidize easily even under control atmosphere and this make recycling of aluminum portion of the batteries more challenging.

Fig. 3 Sample of battery as it is and pure aluminium after 2 h at 900 °C

Fig. 4 Sample of battery after preheating in air at 500 °C and then co-melting with pure aluminium

By considering low solubility of carbon in aluminum (4% mol) [12] at temperature lower than 1700 °C, it was concluded that presence of high amount of carbon in the samples makes a barrier between aluminum melt and battery components, to verify this theory; two extra sets of experiments were designed and carried out. After the pretreatment under vacuum, copper foil coated by carbon and aluminum foil with lithium oxide coating separated from each other. The carbon coating can scratch out by knife from the copper foil very easily, (see Fig. 5a and b). But the binding between aluminum and coat was strong. The weight of the different parts of the battery is reported in Table 3.

Copper foil (2.8 gr) put together with a 28 gr aluminum cube and heated to 900 °C under argon atmosphere and after 2 h cooled to room temperature. The copper foil was dissolved completely in aluminum melt. The same experiment is repeated for the aluminum foil with pure aluminum (28 gr cube + 20 gr granule), but not any mixing between the pure aluminum and aluminum foil happens nor any meaningful weight change observed. It is believed aluminum may reduce the other oxides and the formation of oxide layer hinder progress of the dissolution, see Fig. 6.

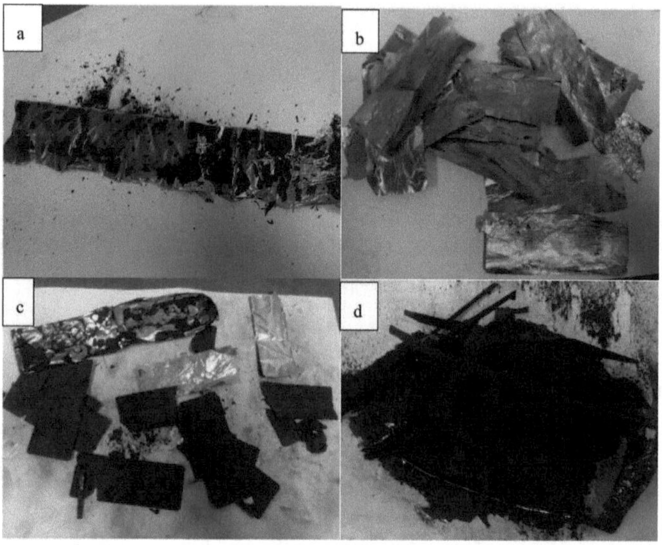

Fig. 5 Cell phone battery parts after pre-treatment at 500 °C for 15 min, **a** copper foil partially carbon removed, **b** copper foil after carbon removal **c** coated aluminium foil, **d** carbon and polymeric cover

Table 3 The weight of different parts of a cell phone battery after pre-treatment at 500 °C for 15 min

Parts of the battery	Copper foil	Coated aluminum foil and the casing	Carbon material
gr	2.8	4.8 + 0.6 (casing)	12.3

Fig. 6 Aluminium foil portion of the cell phone battery and pure aluminium after 2 h holding at 900 °C

Summary

The Li-ion batteries accumulator and its subsystems will dominate the future market. Therefore intelligent, material specific and economically optimized processing techniques have to be developed. Already today it is assumed that new chemical battery systems based on more cost effective electrode materials will be used for this application. The substitution of the valuable cobalt will challenge today's recycling processes with respect to cost effectiveness as well as recycling technology. The recycling of the Li-ion batteries metallic compound through aluminum recycling process has been studied. It was found out direct metal extraction (no pre-treatment) from Li-ion batteries is not possible. Battery cell has a very complex structure which consists of different layers. Batteries need to be preheated and divided to separate portions. Also It was seen high amount of carbon compounds hinder the dissolution of the aluminum and copper foils. However, copper foil can be separated from the rest and easily be de-coated. Except copper and aluminum, other metallic elements are present as oxide, so they need to first be reduced, which is not happen at 900 °C. Aluminum also competes with carbon to react with present oxide, as the result most of the aluminum portion of the battery and other oxides will end up in the slag phase. The recovery of the metallic oxides from the slag is needed to be further studied.

References

1. X. Wang, G. Gaustad, C.W. Babbitt, C. Bailey, M.J. Ganter, B.J. Landi, Economic and environmental characterization of an evolving Li-ion battery waste stream. J. Environ. Manage. **135**, 126–134 (2014)
2. K. Richa, C.W. Babbitt, G. Gaustad, X. Wang, A future perspective on Lithium-ion battery waste flows from electric vehicles. Resour. Conserv. Recycl. **83**, 63–76 (2014)
3. Anon: European Alternative Fuels Observatory, 2016 [viewed 2016 22 September]; Available from: http://www.eafo.eu/content/norway#country_ev_market_share_graph_anchor
4. J. Vidal, Norway has fallen in love with electric cars—but the affair is coming to an end, 2014 [viewed 2016 October]; Available from: https://www.theguardian.com/environment/2014/jan/29/norway-electric-cars-sale
5. C. Gibson, Why do they love electric cars in the Arctic Circle?, 2016 [viewed 2016 October]; Available from: http://www.bbc.com/news/business-36402942
6. A. Sonoc, J. Jeswiet, V.K. Soo, Opportunities to improve recycling of automotive lithium ion batteries. Procedia CIRP **29**, 752–757 (2015)
7. H. Wang, B. Friedrich, Development of a highly efficient hydrometallurgical recycling process for automotive Li–Ion batteries. J. Sustain. Metall. **1**(2), 168–178 (2015)
8. T. Träger, B. Friedrich, R. Weyhe, Recovery concept of value metals from automotive Lithium-Ion batteries. Chemie Ing. Tech. **87**(11), 1550–1557 (2015)
9. A.J. da Costa, J.F. Matos, A.M. Bernardes, I.L. Müller, Beneficiation of cobalt, copper and aluminum from wasted lithium-ion batteries by mechanical processing. Int. J. Min. Process **145**, 77–82 (2015)
10. T. Georgi-Maschler, B. Friedrich, R. Weyhe, H. Heegn, M. Rutz, Development of a recycling process for Li-ion batteries. J. Power Sources **207**, 173–182 (2012)

11. J.-K. Park, *Principles and Applications of Lithium Secondary Batteries* (Wiley, New York, 2012)
12. Anon: FactSage 7, 2016 [viewed 2016 October]; Available from: http://www.factsage.com/
13. L. Gaines, J. Sullivan, A. Burnham, and I. Belharouak, Life-cycle analysis for Lithium-ion battery production and recycling, Transportation Research Board 90th Annual Meeting, Washington, DC, 2011, 23–27
14. R. Beheshti, R. E. Aune, Automotive Lithium-Ion battery recycling: A Theoretical Evaluation, Rewas 2016: Towards Mater. Resour. Sustain. 65–71
15. M. E. Schlesinger, *Aluminum Recycling* (CRC Press, Boca Raton, 2013)

Part IX
Deriving Value from Challenging Waste Materials: Recycling and Sustainability Joint Session: Poster Session

Alternative Method for Materials Separation from Crystalline Silicon Photovoltaic Modules

Pedro F.A. Prado, Jorge A.S. Tenório and Denise C.R. Espinosa

Abstract There has been an increasing push to develop environmentally sound recycling processes of electronic waste (WEEE), including end-of-life photovoltaic modules, to reclaim materials such as silver which is both valuable and pollutant to human health and wildlife. However, process sustainability and economic feasibility of PV panels recycling is strongly tied to the materials separation step, which currently presents challenges due to use of high temperature treatments and strongly acidic solutions. A method using an easily accessible solvent—isopropanol—dissolved the silicone-based encapsulant of crystalline silicon PV modules in 2 days at room temperature, separating the module into semiconductor wafer, glass, ribbon and backsheet.

Keywords PV waste · WEEE · Recycling

Introduction

Even though renewable energy systems are not associated with waste generation, the photovoltaic (PV) waste is of increasing concern given the amount expected for the next decades, adding to staggering 78 million tons of PV waste by 2050 [1]. The ideal method for disposal of PV waste is recycling [2] however there are few industrial plants aimed at this type of waste, given its complexity (due to presence of polymers, metals and ceramics), the long lag time from production to disposal (around 25 years) and the lack of legislation in the majority of countries worldwide.

P.F.A. Prado (✉) · J.A.S. Tenório · D.C.R. Espinosa
LAREX—Department of Chemical Engineering, University of Sao Paulo, Rua Do Lago 250, 2nd Floor, 05508-080 Sao Paulo, Brazil
e-mail: pfaprado@usp.br

J.A.S. Tenório
e-mail: jtenorio@usp.br

D.C.R. Espinosa
e-mail: espinosa@usp.br

L. Zhang et al. (eds.), *Energy Technology 2017*,
The Minerals, Metals & Materials Series, DOI 10.1007/978-3-319-52192-3_27

Currently, most of the defected or broken panels are discarded and buried, without any previous treatment, triggering negative environmental impacts in the near future [3]. PV panels are not only an environmental concern for waste management but also an opportunity for materials reclaiming.

Crystalline silicon (c-Si) PV modules comprise 80% of all panels produced worldwide and its components are depicted in Fig. 1. The main components are: glass surface, encapsulant, c-Si wafers, polymer-based backsheet and metal ribbons [4] to conduct current to the junction box.

The importance of the encapsulant lies in protecting the cells from environmental stress and provide electrical isolation while maintaining acceptable optical properties, considering the light spectrum of absorption by the cells. As important as, the lifetime of PV panels is mainly determined by the encapsulant degradation, mostly by UV radiation [6]. The most commonly employed encapsulant is ethylene-vinyl acetate (EVA) and even though it does not have the best combination of properties for a PV application, it is the most economical option for mass production so far [7]. However due to struggles to fulfill 25-year warranty of PV modules, many manufacturers have been switching to silicone as encapsulant in view of its increased water resistance [8] and ultraviolet (UV) radiation resistance [9] besides more stable mechanical properties with respect to temperature changes when compared to EVA [8], providing extended lifetime to PV modules. Polydimethylsiloxane (PDMS) is the most commercially relevant silicone, characterized by its Si–O–Si and CH_3–Si–CH_3 bonds (Fig. 2). It is normally liquid at room temperature, but once cross-linked, it forms an elastomer [10].

Removal of the encapsulant and separation of materials from modules is the most challenging step in recycling crystalline silicon modules and hence should be more studied [11]. The most common separation processes are by immersing modules in a concentrated nitric acid solution, by thermal decomposition or by fluidized bed combustion [6]. Most of these processes are either energy-intensive, release toxic gases or use strong acidic media, not only degrading the semiconductor—the most valuable material in PV modules—but also allowing partial

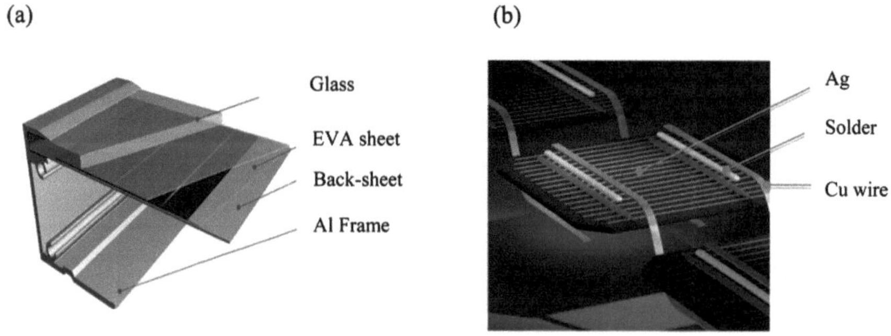

Fig. 1 Structure of a c-Si PV module. **a** The module has many layers of materials sandwiched by the encapsulant, showed here as EVA. **b** Metallic contacts on a PV cell [5]

Fig. 2 Chemical structure of the PDMS [7]

diffusion of its metals to the surroundings including to other components of the panels. This is not interesting when metals could be recovered.

The only study covering dissolution of encapsulants was developed by approaching the dissolution of EVA from working PV modules by many organic solvents, expressing that trichloroethylene is the most effective solvent for EVA [6]. The author dissolved a module completely at 80 °C after leaving it pressed by a binder clip for 7 to 10 days in trichloroethylene. For PV modules with PDMS-based encapsulant, however, there are no studies published so far.

Materials and Methods

Samples of broken polycrystalline silicon PV modules were cut into pieces of 5 cm × 5 cm and then weighed. The samples were immersed in isopropanol PA for 2 days at 25 °C, given a solid-liquid ratio S/L = 0.2 g/mL and the materials were recovered once semiconductor, glass and ribbon pieces were totally separated from the backsheet. The Becker with isopropanol was covered to avoid any losses of solvent to the environment. The recovered materials were then filtered with a NALGON 7.5 μm filter, then dried in a SOLAB SL 104/30 oven for 1 day at 60 ° C.

A sample of the encapsulant of the c Si module was mechanically extracted and characterized by FTIR-ATR, to identify the polymers present. A sample of the backsheet was analyzed as well, to verify if the backsheet would be dissolved by isopropanol and hence interfere in the encapsulant dissolution. The FTIR-ATR analyses were performed in a BRUKER Tensor 27, from 400 to 4000 cm^{-1} in 32 scans. All infrared spectra were identified with the aid of BIO-RAD KnowItAll software. A thermographic analysis (TGA) was also performed with a sample of the encapsulant. A NETZSCH STA 449 F1 Jupiter was used with rate of 20 °C/min from 25 to 1000 °C with 50 mL/min nitrogen gas, 6.2680 mg of polymer was used for the TGA.

Results

(a) FTIR-IR of backsheet

The backsheet was identified as a polyester, more specifically polyethylene terephthalate (PET) (Fig. 3). The characteristic peaks due to C = O bonds present in esters can be observed between 1740 and 1715 cm^{-1} and C–O bonds between 1300 and 1160 cm^{-1}. This compound is commonly used as backsheet for PV modules. Therefore the backsheet would not be dissolved by isopropanol and could be recovered in its entirety, besides avoiding interference in the dissolution of the encapsulant.

(b) FTIR-IR of encapsulant

The encapsulant sample from the broken module was identified as polydimethylsiloxane (PDMS), a commercial silicone, by FTIR-ATR (Fig. 4). The Si–O–Si bonds were represented by the strong peak between 1100 and 1000 cm^{-1}, hinting at silanol end groups.

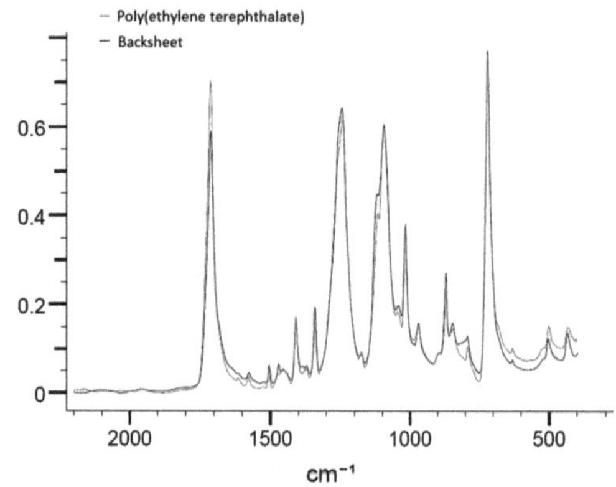

Fig. 3 Infrared spectrum for the backsheet

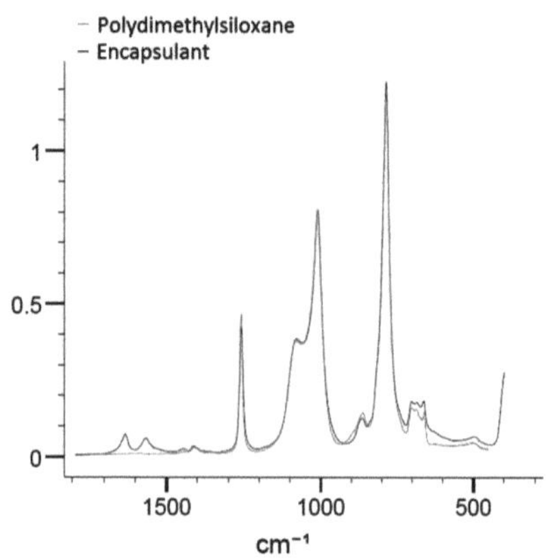

Fig. 4 Infrared spectrum of the encapsulant

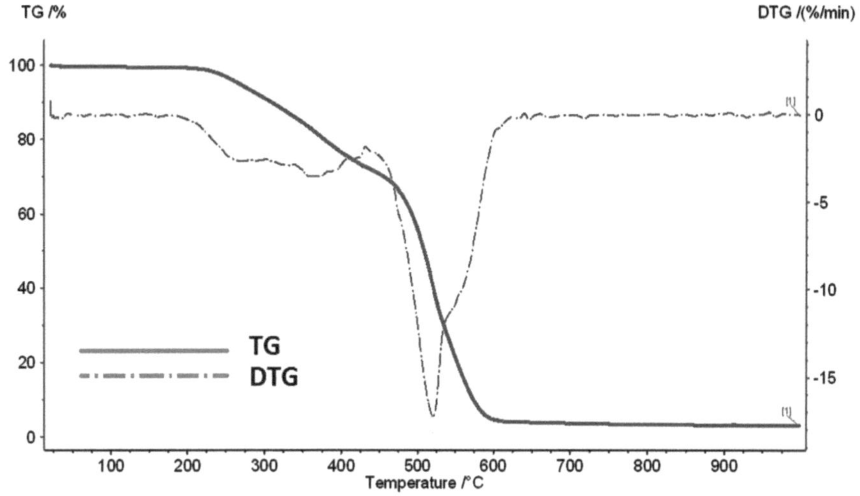

Fig. 5 TGA of encapsulant

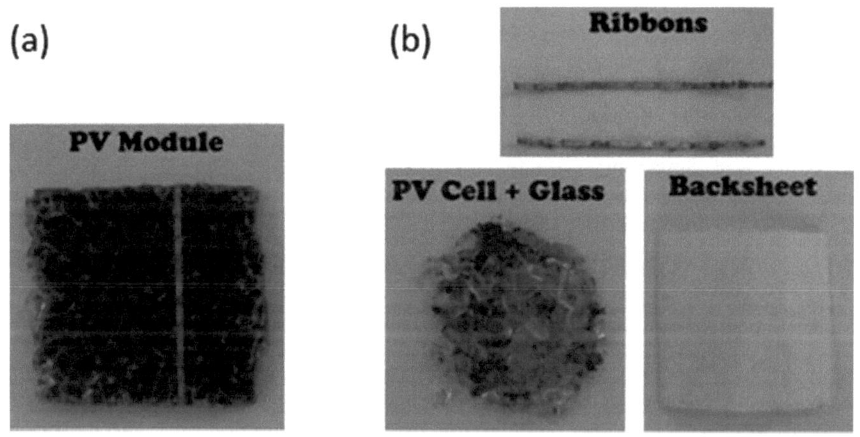

Fig. 6 **a** Broken PV module. **b** The materials out of the module after treatment with isopropanol

(c) TGΛ of encapsulant
The TGA for the encapsulant revealed that most of the weight loss was due to loss of solvents since little change in mass is seen below 200 °C (Fig. 5) and all encapsulant was lost by 600 °C. This is comparable to the TGA observed in the literature [12], confirming that the polymer present in the encapsulant was PDMS.

(d) Dissolution of PDMS with isopropanol
Once the module samples were left for 2 days in a Becker with isopropanol, they completely separated into PV cells and glass, ribbons and backsheet (Fig. 6a, b). The isopropanol was not consumed and it could be reused for a second time, dissolving another sample of panel in 2 days.

Conclusion

The PV module could be efficiently separated into PV cells, glass, ribbons and backsheet and the use of isopropanol to dissolve PDMS at room temperature was successful. This process can also be extended to monocrystalline panels which employ silicone as encapsulant as well, given the similarities of the other materials employed on those panels.

Acknowledgements We acknowledge the support of the Coordenação de Aperfeiçoamento de Pessoal de Nível Superior (CAPES) for the financial support through master grant and of the Fundação de Amparo à Pesquisa do Estado de São Paulo (FAPESP) for financial support through the research Project 2012/51871-9.

References

1. International Renewable Energy Agency, End-of-life management: Solar Photovoltaic Panels (Report on June, 2016)
2. F. Cucchiella, I. D'Adamo, P. Rosa, End-of-life of used photovoltaic modules: a financial analysis. Renew. Sustain. Energy Rev. **47**, 552–561 (2015)
3. S. Kang et al., Experimental investigations for recycling of silicon and glass from waste photovoltaic modules. Renew. Energy **47**, 152–159 (2012)
4. P.R. Dias, M.G. Benevit, H.M. Veit, Photovoltaic solar panels of crystalline silicon: characterization and separation. Waste Manage. Res. **34**(3), 235–245 (2016)
5. B. Jung et al., Sustainable system for raw-metal recovery from crystalline silicon solar panels: from noble-metal extraction to lead removal. ACS Sustain. Chem. Eng. **4**(8), 4079–4083 (2016)
6. T Doi et al., Experimental study on PV module recycling with organic solvent method. Sol. Energy Mater. Sol. Cells. **67**, 397–403 (2001)
7. M. Kempe, Overview of scientific issues involved in selection of polymers for PV applications (Presented at the 37th IEEE Photovoltaic Specialists Conference (PVSC 37), Seattle, Washington, 19–24 June 2011
8. B. Ketola et al., Silicones for photovoltaic encapsulation (Presented at the 23rd European Photovoltaic Solar Energy Conference, 2008)
9. F. Dross et al., Vacuum-free, cost-effective, developing-country-material-available solar cell encapsulation. Sol. Energy Mater. Sol. Cells **90**(14), 2159–2166 (2006)
10. J.P. Lewicki, J.J. Liggat, M. Patel, The thermal degradation behaviour of polydimethylsiloxane/montmorillonite nanocomposites. Polym. Degrad. Stab. **94**, 1548–1557 (2009)
11. E. Radziemska-Klugmann, P. Ostrowski, Chemical treatment of crystalline silicon solar cells as a method of recovering pure silicon from photovoltaic modules. Renew. Energy **35**, 1751–1759 (2010)
12. K. Jiao, C. Zhou, N. Becerra-Moura, J. Fiske, P. Kohli, Vapor-enhanced covalently bound ultra-thin films on oxidized surfaces for enhanced resolution imaging. J Mater. Chem. C. (2016)

Bioleaching Process for Metal Recovery from Waste Materials

Solange Kazue Utimura, Carlos Gonzalo Alvarez Rosario, Amilton Barbosa Botelho, Jorge Alberto Soares Tenório and Denise Crocce Romano Espinosa

Abstract In this study, the recycling of waste materials is interesting in the environmental aspect to reduce the amount of waste in landfill and attenuate the pollution of groundwater, air and soil due to the liberation of dangerous elements. Recycling materials is an important issue once it represents a potential source of valuable metals. Waste printed wired boards (PWBs) as a principal part of Waste Electrical and Electronic Equipment (WEEE) and it is a fast growing stream with complex composition. The composite materials from PWBs are ceramic, polymers and metals. Biohydrometallurgical route may offer possibilities to recycle PWBs from discarded computers with the order to recover copper. The extraction of copper from PWBs by bioleaching processes was studied using a mixed culture of acidophilic bacteria from acid mine drainage (AMD) and the bacteria *Acidithiobacillus ferrooxidans* (*A. ferrooxidans*). The bioleaching experiments were performed in shaker flasks at temperature 30 °C and 170 rpm with 10% (v/v) inoculum and pulp density of 35 g/L. The metals concentration was determined by energy dispersive X-ray Fluorescence (XRF). The results showed that the maximum copper recovery from PWBs using *A. ferrooxidans* was approximately 89% and with a mixed culture of bacteria from AMD the copper recovery was 92%.

Keywords Bioleaching · WEEE · Copper · PWB

Introduction

Material recycling will be a very important point in the future from environmental and economic aspects. Electronic waste (e-waste) consists of waste electric and electronic equipment (WEEE) and represents a fast growing disposal in a global

S.K. Utimura (✉) · C.G.A. Rosario · A.B. Botelho · J.A.S. Tenório · D.C.R. Espinosa
Department of Chemical Engineering, Universidade de São Paulo – Escola Politécnica, Rua Do Lago, 250 – Edifício Semi-Industrial, 05508-080 São Paulo, Brazil
e-mail: solange.utimura@usp.br

A.B. Botelho
e-mail: amilton.barbosa20@gmail.com

L. Zhang et al. (eds.), *Energy Technology 2017*,
The Minerals, Metals & Materials Series, DOI 10.1007/978-3-319-52192-3_28

dilemma [1]. According to the United Nations, the exponential amount of e-waste is a challenge to waste management in many countries. In 2014, the total e-waste generated was around 41.8 Mt and the global amount of electronic waste is forecasted 50 Mt in 2018 [2].

The e-waste management is important in view of the fact that it accumulates in disposal areas, contaminates the ground water and soil and generates toxic gas after incineration. Recycling WEEE proposes to transform it in sustainable resource and to reduce the disposal of waste [3].

The printed wired board (PWB) is an interesting part of WEEE that contains several elements of metals, nonmetals and hazardous substances. The composition of PWBs has been reported by percent of weight and could reach of 40% of metals, 30% of ceramic and 30% of plastics [4]. Recycling PWBs process of recovery metals involves the pretreatment to composition analysis [5, 6], comminution and separation [7, 8] and metallurgical processes [9, 10]. According to Takanori et al. [11], the most representative metal present on the PWBs by percent of weight was copper, which is applied in the circuitry.

Biohydrometallurgical processes are an alternative route in hydrometallurgical processes that have been used to mobilize metals especially copper from low-grade ores, concentrates and has been extended for metal recovery from waste PWBs by the microorganism cultures (Deveci et al. 2010). According to Brandl et al. [13] it was demonstrated that it is possible to recover metals from electronic waste by bacteria and fungi. The extraction of copper from PWBs using microorganism consortium from acid mine drainage (AMD) occurs via indirectly through oxidation by ferric ions (Fe^{3+}) generated from ferrous ions (Fe^{2+}) oxidation by bacteria [14, 15]. The copper recovery depends the pH, ferrous ions (Fe^{2+}) concentration and bio-oxidation of ferrous ions (Fe^{2+}). The reactions are [16]:

$$4Fe^{2+} + O_2 + 4H^+ \rightarrow 4Fe^{3+} + 2H_2O$$

$$Cu + 2Fe^{3+} \rightarrow Cu^{2+} + 2Fe^{2+}$$

This study evaluated the extraction of copper from PWBs using *Acidithiobacillus ferrooxidans* and a mixed culture of bacteria from acid mine drainage (AMD). The copper concentration was able by the technique of energy dispersive X-ray Fluorescence (XRF).

Material and Methods

PWBs Characterization

The printed wired boards (PWBs) samples from discarded computers were used in this experiment and were collected in a recycling unit in Brazil—Cedir (Re-used center of WEEE located at Universidade de São Paulo—USP).

The PWBs processing includes the physical route, which has been comminuted in cutter and hammer mill with 6 and 2 mm sizes grids [17]. The PWBs powder sample was digested in aqua regia (3 HCl: 1 HNO_3, v/v) mixture in a microwave digestion system (CEM, Mars 5). The solution was filtered and diluted to appropriate concentration range. Inductively coupled plasma-optical emission spectrometry (ICP-OES, Agilent Technologies, 700 Series) and energy dispersive X-ray Fluorescence (XRF, PANalytical, Epsilon 3^x) was applied to determine the metals concentrations present in the PWBs samples.

Metals concentration of PWBs samples after digestion in aqua regia revealed the presence of elements by percent of weight: Cu 24.92%; Pb 3.71%; Zn 2.35%; Al 2.22%; Fe 1.70%; Ni 0.66%; Ag 0.11%; Mn 0.06%.

Bacteria and Culture Conditions

The bacterial strain used in this study was *Acidithiobacillus ferrooxidans* and a mixed culture of bacteria from acid mine drainage (AMD) was collected in a mine located at south of Brazil. The bacteria were cultured in T&K growth media with a composition of 0.625 g/L $(NH_4)_2SO_4$; 0.625 g/L K_2HPO_4; 0.625 g/L $MgSO_4.7H_2O$; 166.5 g/L $FeSO_4.7H_2O$. The bacteria adaptation in pulp density of PWBs sample (35 g/L) and inoculation process were cultivated in a rotary shaker at 170 rpm, temperature at 30 °C, inoculum (10% v/v).

Bioleaching Experiments

Bioleaching tests were conducted in 500 mL Erlenmeyer shake flasks. For each flask, 270 mL of the growth media T&K was prepared followed by inoculation with 30 mL of adapted bacteria. The flasks were placed on a rotary shaker at 170 rpm and 30 °C. When the growth media and bacterial culture following the oxidation with redox potential reached approximately at 600 mV (E, mV vs Ag/AgCl), were added the amount of 10.5 g of PWBs sample. During the bioleaching test, the flasks were sampled at the predetermined intervals and these samples were used for the analysis of copper, redox potential (E, mV vs Ag/AgCl) and pH was controlled at pH 1.8 by the addition of sulphuric acid 5 M. Analysis of copper from the solution was performed using the technique of energy dispersive X-ray Fluorescence (XRF, PANalytical, Epsilon 3^x).

Results and Discussion

Effect of PH and Redox Potential

The effect of pH of each bioleaching sample was performed and is shown in Fig. 1. The results in Fig. 1 exhibited low pH at initial pH 1.80 and increase pH indicated in a lag phase. In this stage, the bio-oxidation of Fe^{2+} consumes H^+ in the solution and the oxidation process increase pH. The presence of a concentration of Fe^{2+} was determined by the color that was presented the green color. The characteristic of the presence of Fe^{3+} showed an orange-brown color. Other observations made from the data presented in Fig. 1 that were the high pH 3.2 by *Acidithiobacillus ferrooxidans* and pH 2.8 by the mixed culture of bacteria from AMD were resulted in the formation of a precipitate.

The rapid increase of pH was resulted the sedimentation of Fe^{3+} iron as many insoluble oxyhydroxides [18]. The pH has effect on the oxidation and hydrolysis reactions. As indicated by Jensen and Webb [18], the hydrolysis reactions were presented in precipitates of Fe^{3+} hydroxysulphates. These precipitates are known as jarosites complexes.

From the analysis of the redox potential presented in Fig. 2, it was observed evaluations by *Acidithiobacillus ferrooxidans* and a mixed culture of bacteria from AMD in a bioleaching experiments.

The redox potential (E, mV vs Ag/AgCl) evaluation was between 439 and 535 mV with *Acidithiobacillus ferrooxidans* and with a mixed culture of bacteria from AMD, the potential redox were between 390 and 545 mV. The potential redox with control sample was controlled in approximately 300 mV.

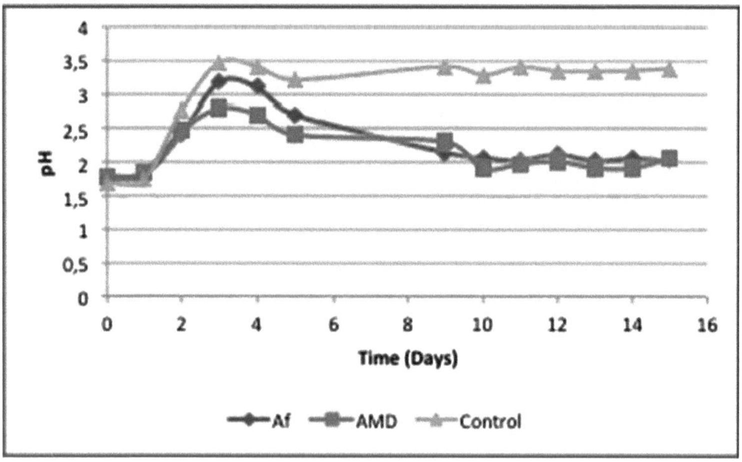

Fig. 1 pH during the bioleaching tests in the presence of *Acidithiobacillus ferrooxidans* and a mixed culture from AMD (3.5% w/v PWBs, 10% v/v inoculum, 30 °C, 170 rpm)

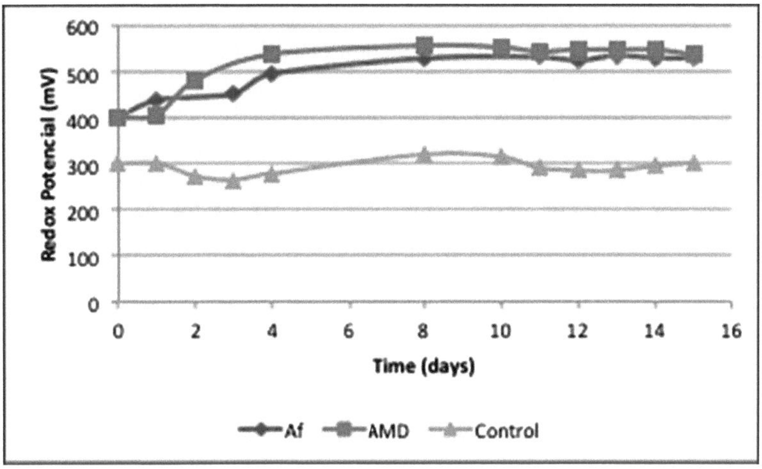

Fig. 2 Redox potential in bacterial leaching experiments with a mixture culture of microorganisms from AMD and in presence of *Acidithiobacillus ferrooxidans* (3.5% w/v PWBs, 10% v/v inoculum, 30 °C, 170 rpm)

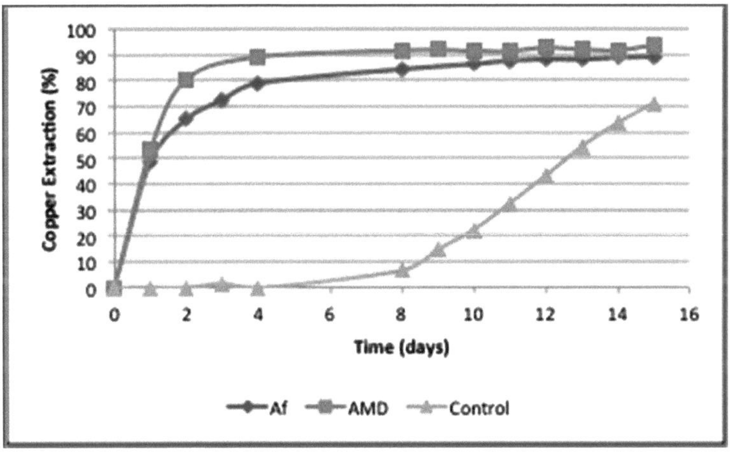

Fig. 3 Copper extraction during the bioleaching experiments with *Acidithiobacillus ferrooxidans* and a mixed culture of bacteria from AMD (3.5% w/v PWBs, 10% v/v inoculum, 30 °C, 170 rpm)

Copper Extraction

The results in Fig. 3 observed that the leaching rate with *Acidithiobacillus ferrooxidans* stabilized in 89% extraction of copper and with a mixed culture of bacteria from AMD the leaching rate stabilized in 92% extraction of copper.

The extraction of copper was achieved through the pH control and it can be observed in Fig. 3 that the copper extraction increased slowly (50%) at initial pH 1.8 due the inhibition effect on the bacterial metabolism [19]. Others observations from the Fig. 3 were at pH 1.9, in the first 3 days (lag phase) and a subsequent acceleration on the dissolution of copper were acquired. *Acidithiobacillus ferrooxidans* and a mixed culture of bacteria from AMD were approximately 80 and 90% of copper extraction, respectively. While at pH 2.8 by the mixed culture of bacteria from AMD and pH 3.2 by *Acidithiobacillus ferrooxidans* were produced the amount of precipitate that results in stabilization in the copper extraction (89% by *A. ferrooxidans* and 92% by the mixed culture of bacteria from AMD). The influence of the presence of precipitate on the surface of the substrate suggested a decrease rate during the bioleaching tests that can be evaluated in a deactivating the particle surface [20]. The pH control is necessary to reduce hydrolysis reactions and to prevent the amount of precipitates [21]. From the analysis of the fast increase of pH, has been reported that was accelerated the chemical oxidation of Fe^{2+} iron and resulted in the formation of Fe^{3+} iron precipitates [22, 23]. Other observations about the increase of pH in the bioleaching mechanism of copper by *A. ferrooxidans*, that was some components in PWBs contributed to the alkalinity in the culture media [13]. Consequently, the pH control and the effect on the dissolution of metals are an important parameters in the bioleaching process for the recovery copper from PWBs.

Conclusion

In the present work, the possibility of using the biohydrometallurgy technique for the extraction of metal from waste PWBs with the aim of recovering copper was evaluated. The results showed a good performance in bioleaching process with *Acidithiobacillus ferrooxidans* and a mixed culture of bacteria from AMD. The microbial leaching has the ability to oxidize ferrous iron to generate ferric ion and sulfuric acid, which dissolve metal ion from waste PWBs. In this view, the soluble iron controls the rate of extraction copper from waste PWBs. In the bioleaching process, the addition of ferrous iron was necessary for extraction copper because of the small quantity iron amount in waste PWBs. The performance of a mixture culture of bacteria from AMD in bioleaching system was better than the results of *Acidithiobacillus ferrooxidans.* The experimental conditions and adapted bacteria were the reasons that interfered in extraction copper from waste PWBs. Consequently, these findings suggested that the continuous bacteria adaptation process, the concentration of iron and the establishment of technical parameters were important for the development of an efficient bacterial leaching process for the recovery metals from waste PWBs.

Acknowledgements The authors would like to thank the financial support provided by CNPq for this work and the scientific research center—Larex (Laboratory of Recycling, Waste Treatment and Extraction) from the Department of Chemical Engineering of the Polytechnic School of the University of São Paulo for providing the analysis for this study

References

1. M. Kaya, Recovery of metals and nonmetals from electronic waste by physical and chemical recycling processes. Waste Manage. 2016
2. C.P. Baldé, F. Wang, R. Kuehr, J. Huisman, *The Global e-waste Monitor, United Nations University* (IAS—SCYCLE, Bonn, Germany, 2014)
3. A. Sepülveda, M. Schuep, et al., A review of the environmental fate and effects of hazardous substances released from electrical and electronic equipments during recycling: examples from China and India. Environ. Impact Assess. Rev. **30**(1), 28–41 (2010)
4. J.A.S. Tenório, R.P. Menetti, A.P. Chaves, Production of non-ferrous metallic concentrates from electronic scrap. (In: EPD Congress. TMS, Warrendale, PA, USA, 1997), 505–509
5. R.P. Menetti, A.P. Chaves, J.A.S. Tenório, Obtenção de concentrados metálicos não-ferrosos a partir de sucata eletrônica. (In: Anais do 51° Congresso Anual da ABM. Porto Alegre, RS. Associação Brasileira de Metalurgia e Materiais, v. 4, 1996)
6. H.M. Veit, T.R. Diehl, A.P. Salami, J.S. Rodrigues, A.M. Bernardes, J.A.S. Tenório, Utilization of magnetic and electrostatic separation in the recycling of printed circuit boards scrap. Waste Manage. **25**, 67–74 (2005)
7. J. Cui, E. Forssberg, Mechanical recycling of waste electric and electronic equipment: a review. J. Hazard. Mater. **99**(3), 243–263 (2003)
8. E.Y. Yazici, H. Deveci, Recovery of metals from e-waste. Madencilik. **48**(3), 3–18 (2009) (in Turkish)
9. G. Dorella, M.B. Mansur, A study of the separation of cobalt from spent Li-ion battery residues. J. Power Sources. **170**. 210–215 (2007). http://dx.doi.org/10.1016/j.jpowsour.2007.04.025
10. J. Cui, L. Zhang, Metallurgical recovery of metals from electronic waste: a review. J. Hazard. Mater. **158**, 228–256 (2008)
11. H. Takanori, A. Ryuichi, M. Youichi, N. Minoru, T. Yasuhiro, A. Takao, Techniques to separate metal from waste printed circuit boards from discarded personal computers. J. Mater. Cycles Waste Manage. 42–54 (2009)
12. A.D. Bas, H. Deveci, E.Y. Yazici, Bioleaching of copper from low grade scrap TV circuit boards using mesophilic bacteria. Hydrometallurgy. **138**, 65–70 (2013)
13. H. Brandl, R. Bosshard, M. Wegmann, Computer-munching microbes: metal leaching from electronic scrap by bacteria and fungi. Hydrometallurgy. **59**, 319–326 (2001)
14. P. Devasia, K.A. Natarajan, D.N. Sathyanarayana, G. Rao, Surface chemistry of *Thiobacillus ferrooxidans* relevant to adhesion on mineral surfaces. Appl. Environ. Microbiol. **59**(12), 4051–4055 (1993)
15. Y. Xiang, P. Wu, N. Zhu, T. Zhang, W. Liu, J. Wu, P. Li, Bioleaching of copper from waste printed circuit boards by bacterial consortium enriched from acid mine drainage. J. Hazard. Mater. **184**, 812–818 (2010)
16. A.C. Kasper, G.B.T. Berselli, B.D. Freitas, J.A.S. Tenório, A.M. Bernardes, H.M. Veit, Printed wiring boards for mobile phones: characterization and recycling of copper. Waste Manage. **31**. 2536–2545 (2011). http://dx.doi.org/10.1016/j.wasman.2011.08.013
17. F.P.C. Silvas, M.M.J. Correa, M.P.K. Caldes, V.T. Moraes, D.C.R. Espinosa, J.A.S. Tenorio (in press), Printed circuit board recycling; physical processing and copper extraction by selective leaching. Waste Manage. **46**, 503–510 (2015)

18. A.B. Jensen, C. Webb, Ferrous sulfate oxidation using *Thiobacillus ferrooxidans*: a review. Process Biochem. (Oxford). **30**(3), 225–236 (1995)
19. B.J. Anders, W. Colin, Ferric sulphate oxidation using Thiobacillus ferrooxidans: a review. Process Biochem. **30**, 225–236 (1995)
20. N. Zhu, Y. Xiang, T. Zhang, P. Wu, Z. Dang, P. Li, J. Wu, Bioleaching of metal concentrates of waste printed circuit boards by mixed culture of acidophilic bacteria. J. Hazard. Mater. **192**, 614–619 (2011)
21. J. Daoud, D. Karamanev, Formation of jarosite during Fe2+ oxidation by *Acidithiobacillus ferrooxidans*. Min. Eng. **19**(2006), 960–967 (2005)
22. N.S. Choi, K.S. Cho, D.S. Kim, D.J. Kim, Microbial recovery of copper from printed circuit boards of waste computer by *Acidithiobacillus ferrooxidans*. J. Environ. Sci. Health A. **39**, 2973–2982 (2004)
23. T. Yang, Z. Xu, J.K. Wen, L.M. Yang, Factors influencing bioleaching copper from waste printed circuit boards by *Acidithiobacillus ferrooxidans*. Hydrometallurgy. **97**, 29–32 (2009)

Chemical Analysis of Sludge Originating from Industrial Painting Performed in Brazil

Rita de Cássia S.S. Alvarenga, Henrique de Paula Santos, Beatryz C. Mendes, Maurício Paulo F. Fontes, Eduardo Antônio G. Marques and Kléos M.L. Cesar

Abstract Inadequate discard of industrial solid waste is a risk to human health and to the environment. Aiming at softening environmental damage caused by deposition on soil of an ink residue, from washing of printers used in carton box packages, it is intended to incorporate this residue into soil-cement bricks. The residue has other use potentialities such as in screeds, pavement parts, fertilizer, paints, etc. However, to explore these potentialities it is fundamental to assess the risks that the several forms of use of this residue can cause. For such, in this work the chemical analysis of the residue was performed in order to know its main chemical compounds, in addition to its characteristics of leaching and solubilization. The chemical composition assay showed traces of some heavy metals such as Pb, Cd, Cr, Zn, Al and Cu, although the residue-generating company uses ink labeled as heavy metals free. Among the elements analyzed in leaching and solubilization, some had concentration above the limit allowed by The Brazilian Codes. This code is based on *CFR—Title 40—Protection of Environment—Part 260–265—Harzardous waste management*. Therefore, it is suggested that the residue has a limited use to avoid contamination of ground water.

Keywords Waste · Paint · Soil-cement

Introduction

Economic activities have usually been generating sources of waste. Treatment and final discard of waste is more and more subject of great concern. According to Collatto and Bergmann [1], the main fear is that these residues eventually cause harm to human and environment, through soils and water contamination. The use of

R.C.S.S. Alvarenga (✉) · H. Paula Santos · B.C. Mendes · M.P.F. Fontes
E.A.G. Marques · K.M.L. Cesar
DEC—Civil Engineering Departament, UFV—Federal University of Vicosa,
Av. PH Rolfs, 36570-900 Vicosa, Minas Gerais, Brazil
e-mail: rcassia.alvarenga@gmail.com

L. Zhang et al. (eds.), *Energy Technology 2017*,
The Minerals, Metals & Materials Series, DOI 10.1007/978-3-319-52192-3_29

stabilization/solidification technique as pre-treatment of heavy metals residues and other toxic contaminants is common, aiming at discard on soil, as stated by Castro [2] and Lin [3].

The aim is to use a specific residue generated from effluent treatment, as aggregated in soil-cement bricks manufacturing. To determine the use potentialities of this residue, risks to human and environment should be assessed [4]. For such, it is necessary to know the main compounds of the residue, as well as its characteristics of leaching and solubilization.

Several industries generate solid waste in water treatment stations, industrial sewages or effluents, after going through flotation, decantation or filtration processes. This procedure type originates a sludge that has varied chemical composition due to the products used in the process and is constituted mostly of water, as stated by Sousa and Cordeiro [5]. Reuse of iodine is target of study by several researchers that seek to find an environmentally correct destination for such waste.

The residue under study has great use potential, especially in the construction industry. One of these applications concerns the possibility of incorporation of sludge into soil-cement bricks. There is also the possibility of use of this residue for other ends in constructions, such as in pavement parts and screeds. In this case, although there is direct contact with water, leaching will occur in lesser extent, due to the stabilization with cement and to the fact that the residue is encased in concrete. For being in direct contact with soil, the possibility of carrying of heavy metals is greater. Therefore, analysis of the possibility of leaching and solubilization of toxic agents in the parts with sludge incorporation is recommended.

For such, Dutch norm NEN 7345 (Determination of Leaching from Monolithic Construction Materials and Waste Materials by Means of a Diffusion Test), which uses monolithic samples with the end of assessing the potential and speed of compounds leaching throughout time, can be used. Other verification possibility is the use of Brazilian norms ABNT NBR 10006 [6] and ABNT NBR 10005 [7].

Other possibilities have been tested by some researchers for similar residues. Moreira et al. [8] studied sludge generated by the sewage treatment station of a textile industry in adding to clays for construction material manufacturing. Pietrobon et al. [9] studied sludge from textile laundry treatment for reuse as construction mortar. Silva [10] studied the potential of using sludge from textile industry sewage as agricultural fertilizer.

In the present work, chemical analysis of sludge resulting from washing treatment of machines that use pains was performed, assessing the following parameters: chemical composition, leaching and solubilization of sludge generated by the company Guarany Embalagens. The process of packages production includes printing of carton box surface in different colors, where a water based flexographic ink is used, which is free from heavy metals, according to the manufacturer. To change the ink color, printers are washed with water, which is directed to a processing machine. There, it goes through a flocculation process. From this process, sludge is obtained, which is removed in bags and dumped in an open external yard, where it waits for final discard while naturally dries. Wet sludge is shown in Fig. 1. In Fig. 2, sludge after dry is presented.

Fig. 1 Wet residue

Fig. 2 Dry residue

ABNT NBR 10004 [11] classifies the most common residues, which have known origin and can be identified in attachments A and B. Ink sludges from industrial painting are found in attachment A and are classified as toxic and identified by code F018. The following toxic agents are highlighted as dangerous constituents: cadmium, chromium, lead, cyanide, toluene and tetrachlorethylene. Despite this classification, that standard applies in statement that the generator of residues listed in attachments A and B can demonstrate by classification report that its residue specifically has none of the characteristics of hazardous specified in this norm. Relying on this statement and considering that company Guarany Embalagens uses an ink labeled as heavy metals free, analyses of chemical composition, leaching and solubilization became interesting, in order to study the possibilities of proper residue discard.

Concerning environmental problems, this research will provide a proper discard for industrial waste, thus avoiding the construction of industrial landfills and the indefinite accumulation in unfit areas.

It will also allow for obtaining new materials, by residues incorporation into soil-cement bricks, showing new alternatives that can contribute for decreasing the housing deficit in Brazil. Since the soil-cement is an adequate technique for construction on mutual aid, the disclosure of obtained results is expected to promote the aid of residue generator industries to organizations and families that seek solutions for their housing or community constructing needs.

Methodology

Chemical Analysis

Chemical analysis of the residue was performed in the Mineralogy Laboratory, of the Solid Department of Universidade Federal de Viçosa (UFV). This was performed by residue digestion in nitro-perchloric acid. From 0.5 g of the residue, which was transferred to an Erlenmeyer, 10 mL of nitric-perchloric solution 3:1 was added, in room temperature.

Following, the sample was transferred into a digester block at 180 °C where it was kept until total digestion (2:30 h). After cooling, hydrochloric acid was added to clean the Erlenmeyer walls and remove the elements attached. The sample was filtered in an assay tube that was completed with deionized water.

The reading of elements was performed by Plasma Emission Spectroscopy (ICP), which identified the concentrations of the following elements: magnesium (Mg), calcium (Ca), potassium (K), sodium (Na), phosphorus (P), manganese (Mn), chromium (Cr), lead (Pb), nickel (Ni), cadmium (Cd), copper (Cu), aluminum (Al), and iron (Fe).

Leaching

Leaching assay was performed according to the procedure of ABNT NBR 10004 [11]. Ink residue was dried in stove and milled before obtaining the leached extract, in order to obtain particles smaller than or equal to 9.5 mm.

A pre-test was performed, using 5 g of the sample that was transferred into a beaker where 96.5 mL of deionized water was added. The mix was vigorously stirred for 5 min. Next, pH was measured, finding 4.49. Since pH was inferior to 5, the following solution was used: 5.7 mL of glacial acetic acid and 64.3 mL of NaOH 1.0 N and adding distilled, deionized water, organic-free until the volume of 1L. The solution pH was measured, finding the value of 4.88. This value should be within 4.93 ± 0.05.

The leaching procedure for residues with solids assay equal to 100% was followed. For such, 2 g of the sample was weighed, which was transferred into leaching flasks and 40 mL of the previous solution was added, so that the solution quantity was 20 times the used mass. The flask was closed and kept under stirring for 20 h to complete (18 ± 2) h, at 20 °C of temperature with a rotation of (30 ± 2) rpm in the rotatory stirrer, as seen in Fig. 3.

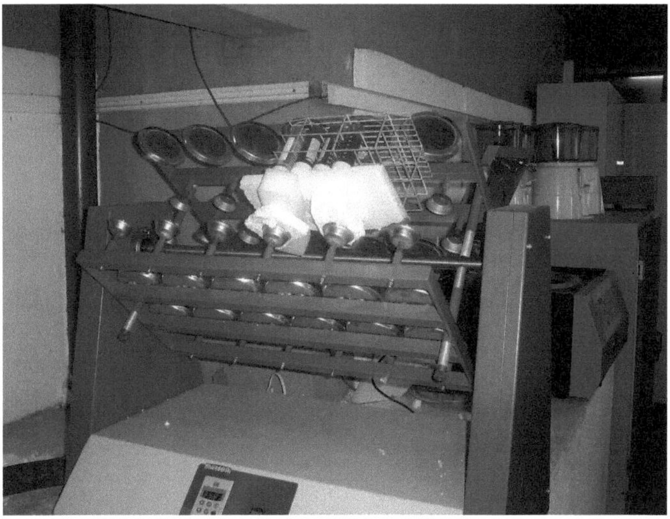

Fig. 3 Rotatory stirrer

After this period, the sample was filtered, obtaining the leached extract. The reading of elements was performed by ICP, which identified the concentrations of the searched elements (Mg, Ca, K, Na, P, Mn, Cr, Pb, Ni, Cd, Cu, Al, Fe).

Solubilization

Solubilization assay was performed according to the procedure of ABNT NBR 10006 [6]. To obtain the solubilized extract of ink residue, the sample was dried at temperature 42 °C, using a stove.

A dried sample of 250 g was placed in a 1500 mL flask and 1000 mL of distilled and deionized water was added. The flask was stirred in low speed for 5 min and, next, covered with plastic wrap (PVC) and left resting for 7 days at temperature up to 25 °C. The solution was filtered, obtaining the solubilized extract. The reading of elements was performed again by ICP.

Results and Discussion

Chemical Analysis

Chemical composition of the residue is presented in Table 1. It is important to stress that the assay method used verified only the existence of searched elements, not excluding the possibility of presence of other elements.

Table 1 Chemical composition of the residue

Elements	Mn	Cr	Zn	Pb	Ni	Cd	Cu
Concentration mg/L	118.08	10.75	771.19	6.90	0.00	2.05	817.29
Elements	Al	Fe	Mg	Ca	K	Na	P
Concentration mg/L	3849.99	62,483.04	2636.43	5531.39	61.87	378.12	0.00

Table 2 Leached elements of the residue

Elements	Mn	Cr	Zn	Pb	Ni	Cd	Cu
concentration mg/L	10.65	0.00	50.94	0.00	0.00	0.01	0.00
Maximum limit mg/L	–	5	–	1	–	0.5	–
Elements	Al	Fe	Mg	Ca	K	Na	P
Concentration mg/L	13.54	1.77	548.57	6088.36	23.46	0.00	4.73
Maximum limit mg/L	–	–	–	–	–	–	–

A high concentration of iron, calcium, aluminum and manganese was observed. The high concentration of iron can have its origin in the effluents treatment, in which a flocculant is used, whose chemical composition is unknown.

The ink used by the generator company of the residue is heavy metals free, according to the manufacturer. Still, traces of some heavy metals like Pb, Cd, Cr, Zn, Al and Cu were found.

Leaching

Leaching assay shows the elements that can be leached and contaminate ground water. It is important to be aware of the presence of heavy metals because they can cause adverse effects to human health.

Results shown in Table 2 are compared to the maximum limit allowed by attachment F of ABNT NBR 10004 [11]. Among the analyzed elements, none was superior to the limit. Therefore, concerning leaching, the residue can be used without restrictions.

Solubilization

Solubilization assay shows the elements that can be solubilized and contaminate ground water. Results shown in Table 3 are compared to the maximum limit allowed by attachment G of ABNT NBR 10004 [11].

Among the analyzed elements, concentrations of elements Mn, Zn and Fe can be noticed to be above maximum limit. Manganese must be stressed for having

Table 3 Solubilized elements of the residue

Elements	Mn	Cr	Zn	Pb	Ni	Cd	Cu
Concentration mg/L	6.60	0.00	15.59	0.00	0.00	0.00	0.00
Maximum limit mg/L	0.10	0.05	5.00	0.01	–	0.005	2.00
Elements	Al	Fe	Mg	Ca	K	Na	P
Concentration mg/L	0.00	1.94	352.33	2816.94	4.50	2.58	0.64
Maximum limit mg/L	0.20	0.3	–	–	–	200.0	–

concentration 66 times superior to the maximum limit allowed by the norm. Therefore, precautions must be taken for residue discard not to cause problems of water contamination and, consequently, harms to human health and to the environment.

Discussion

For the leaching extract, a residue is characterized as toxic when it has values superior to those of attachment F of ABNT NBR 10004 [11]. Since no element had values higher than the limit, the residue under analysis cannot be classified as toxic based on the leached extract.

Other possibility of the residue being classified as toxic is the presence of one or more substances present in attachment C and also presenting toxicity. Since lead and cadmium were detected in the sample, and they are listed in attachment C, some factors present in the same attachment should be checked to see the presence of toxicity.

Although the norm mentions factors that should be considered to verify the presence of toxicity, there is no information on how to proceed to verification. Therefore, it is hard to state that the residue is not toxic.

Considering that the residue is not toxic, the analysis of solubilized extract is useful for classification as inert or not inert. Since some concentrations superior to standards of drinking water present in attachment G were found, for some elements the residue has not inert characteristics.

Therefore, considering that the residue has no toxicity due to the presence of Lead and Cadmium, it should be classified as not harmful: Class II A—Not inert. If considered presence of toxicity, it should be classified as class I harmful residue.

Conclusions

By analyzing the leaching results, there are no restrictions concerning residue use according to recommendations of the ABNT NBR 10004 [11] and Wiemes [4]. However, by analyzing solubilized values, some of them are above the standard for drinking water. Therefore, it is interesting to take special measures concerning final discard of the residue, because it cannot be directly put on soil, due to the possibility of contamination of water that can eventually be ingested. It is valid to notice that performed classification is based on water for human consumption. Thus, detected contamination does not prohibit direct contact with the residue and, accordingly, the use of this residue as construction material causes no harm to human health. In this way, care to be taken concerning the use of residue are solely about water contamination and ingestion.

In this way, some sludge reuse technologies, such as fertilization, for the sludge in concern, demand higher attention because there are possibilities of ground water contamination. However, according to Santos [12] for the purpose intended for the residue, which is its use as alternative material for construction, there is no problem, as long as the possibility of leaching is avoided.

In case of incorporation of the studied residue into soil-cement bricks, these by principle should be protected from action of direct water. In this way, the end intended for this sludge has good potential from the environmental perspective.

Acknowledgements The authors thank the Brazilian agencies: National Counsel of Technological and Scientific Development (CNPq), Coordination for the Improvement of Higher Education Personnel (CAPES) and Research Support Foundation of Minas Gerais (FAPEMIG) for the support provided to this study.

References

1. D. Collatto, C.P. Bergmann, Emprego de resíduo de celulose da indústria de papel como matéria-prima para fabricação de material cerâmico. Cerâmica Ind. **14**, 30–35 (2009)
2. T.M. Castro, Solidificação/estabilização de lodo gerado no tratamento de efluente de lavanderia industrial têxtil em blocos cerâmicos acústicos e seu desempenho em câmara reverberante (Sc.M. thesis, State University of Maringá, 2010), p. 108
3. C.K. Lin, J.N. Chen, C.C. Lin, An NMR and XRD and EDS study of solidification/stabilization of chomium with Portland ciment and C_3S. J. Hazard. Mater. **56**, 21–34 (1997)
4. L. Wiemes, Utilização de lodos da indústria automobilística como componente de valor para fabricação de materiais cerâmicos (Ph.D. thesis, Federal University of Paraná, 2013), pp. 47–48
5. F.G.C. Sousa, J.S. Cordeiro, Resíduos gerados em estação de tratamento de água com descargas diárias. AIDIS. Forjando el Ambiente que Compartimos, 1–8 (2004)
6. Brazilian Association of Technical Norms, in *ABNT NBR 10006: Procedure for obtention of solubilized extraction of solid wastes* (Rio de Janeiro, 2004), p. 3
7. Brazilian Association of Technical Norms, in *ABNT NBR 10005: Procedure for obtention leaching extract of solid wastes* (Rio de Janeiro, 2004), p. 16
8. A.H. Moreira et al., Efeito da adição do lodo de águas residuais da indústria têxtil nas propriedades de materiais de construção. Cerâmica **47**, 158–162 (2001)
9. C.L.R. Pietrobon et al., Lixiviação de contaminantes de compostos de lodo-cimento. Acta Scientiarum **26**(1), 45–52 (2004)
10. E.P. Silva et al., Potencial de utilização do lodo de esgoto de indústria têxtil como fertilizante agrícola. Engenharia Sanitária e Ambiental **8**, 69–76 (2003)

11. Brazilian Association of Technical Norms, in *ABNT NBR 10004: solid waste—classification* (Rio de Janeiro, 2004), p. 71
12. H.P. Santos et al., Tijolos de solo-cimento com incorporação de resíduo de tinta: uma alternativa que envolve indústria e sociedade em busca de soluções para problemas ambientais e habitacionais, in IV Encontro Nacional e II Latino-Americano Edificações e Comunidades Sustentáveis. ANTAC, Campo Grande (2007), pp. 1–15

Preparing Ferrosilicon Alloy with Copper Slag

Ruirui Wei, Mingrui Yang and Xuewei Lv

Abstract Copper slag is a by-product obtained during smelting of copper from copper ores. In China, for one ton of metal production about 2.2 tons of copper slag was generated, and approximately 15 million tons slag is produced in 2013. Dumping and disposal of such huge quantities of slag will cause environmental and space problems. Most importantly, copper slag can not be made full use of to maximize its value. In this study, copper slag was used to take the place of scrap iron and part of silica in traditional process and to produce FeSi alloy as the main material. Utilizing silica and copper slag as the raw materials and coke as the reducing agent, FeSi45 was produced in the induction furnace. In addition, the effect of different n(C)/n(O) on FeSi45 has been investigated. Increasing n(C)/n(O) is beneficial to form the stable phase of FeSi. Meanwhile, n(C)/n(O) greatly affects the reduction and recovery of silicon than iron. When n(C)/n(O) equals 1.10, the recovery of silicon and iron is about 77 and 95%, respectively.

Keywords Copper slag · Ferrosilicon alloy · Carbothermic reduction

Introduction

Since the beginning of the industrial era, slags the glassy materials left over when metals are pyrometallurgically extracted from ores, have been considered waste. Copper slag is the one, which is mostly produced during matte smelting of pyrometallurgical production of copper. Production of one ton of copper generates, approximately 2.2 tons copper slag [1]. In recent years, world copper production rose rapidly, China has become the second-largest producer of copper in the world. Approximately 47.06 million tons slag is generated in 2013 and this figure will increase in the next few years. According to CRU consulting, Peru is still the world's third-largest producer of copper, headed by Chile and China in 2013.

R. Wei (✉) · M. Yang · X. Lv
College of Materials Science and Engineering, Chongqing University, Chongqing, China
e-mail: wrui1990@outlook.com

L. Zhang et al. (eds.), *Energy Technology 2017*,
The Minerals, Metals & Materials Series, DOI 10.1007/978-3-319-52192-3_30

The report points out that copper output is 5.79 million tons in Chile, 1.60 million tons in China, and 1.35 million tons in Peru in 2013. Therefore, the production of waste slag can be calculated in Chile, China and Peru, respectively about 12.74 million, 3.52 million, 2.97 million tons. Dumping and disposal of such huge quantities of slag will cause seriously environmental and space problems [1–3]. Consequently, how to make full use of the waste slag with an efficient and environmental way to maximize its value has been under the spotlight [4, 5].

Table 1 shows the chemical composition of copper slag produced by different smelting methods [6]. This table indicates Fe content accounts for 30–40% and SiO_2 content accounts for 25–40% in copper slag in despite of different smelting methods. Meanwhile, many valuable metals exist in the copper slag. Therefore, in order to utilize the waste slag with a better way, researchers worldwide carry out more works according to mechanical properties and components of copper slag. In the current research, copper slag always was used (1) to extract valuable metals (2) to recycle Fe from waste slag (3) in cement and construction industry [7]. Currently, copper slag has been widely used for abrasive tools, roofing granules, cutting tools, abrasive, tiles, glass, road-base construction, railroad ballast, asphalt pavements, cement and concrete industries [8–12].

However, the present ways utilizing waste copper slag can not make full use of copper slag's value. Many researchers are paying more attention to the efficient and environmental method to dispose copper slag.

Ferrosilicon alloy is widely used in manufacturing industry for its good performance as deoxidizer in steelmaking industry [13], reductant in ferroalloy industry [14–17], and main raw material in producing semiconductor.

Silica, coke and scrap iron are the main material to produce ferrosilicon alloy in the traditional process [18]. According to the main composition of copper slag in Table 1, silicon oxide and iron oxide accounting for about 90% in the slag, which are the main composition of silica and scrap iron. Therefore, the copper slag can be used to prepare ferrosilicon alloy.

Table 1 Typical chemical composition of copper smelting slag by mass,%

Melting method	Cu	Fe	Fe_3O_4	SiO_2	S	Al_2O_3	CaO	MgO
Closed blast furnace	0.42	29	–	38	–	7.5	11	0.74
Outokumpu flash furnace	1.5	44.4	11.8	26.6	1.6	–	–	–
Inco flash furnace	0.9	44	10.8	33.0	1.1	4.72	1.73	1.61
Noranda furnace	2.6	40.0	15.0	25.1	1.7	5.0	1.5	1.5
Vanjukoy furnace	0.5	40.0	5.0	34.0	–	4.2	2.6	1.4
Baiyin furnace	0.45	35.0	3.15	35.0	0.7	3.8	8.0	1.4
Teniente furnace	4.6	43.0	20.0	26.5	0.8	–	–	–
Isa furnace	0.7	36.61	6.55	31.48	0.8	3.64	4.37	1.98
Ausmelt furnace	0.65	34	7.5	31.0	2.8	7.5	5.0	–
Mitsubishi smelting furnace	0.60	38.2	–	32.2	0.6	2.9	5.9	–

In this study, copper slag, coke and silica were taken for the main raw materials to produce FeSi alloy, and the effect of reducntant on the ferrosilicon will be studied. This process can make full use of copper slag, and reduce the addition of iron scrap and silica, which can achieve the objective of energy saving and emission reduction.

Experimental

Experimental Materials

The copper slag in this paper is taken from Jinchuan corporation in China. Table 2 shows the chemical composition of copper slag. In this table, SiO_2 and Fe take up about 80% in the slag, and CaO, MgO and Al_2O_3 also exist in it. The chemical composition of coke and silica is shown in Tables 3 and 4. Figure 1 shows the XRD pattern of copper slag, the phase in this figure is very complicated, the spinel containing iron is the main phase. Figure 2 shows the XRD pattern of silica.

Experimental Method

1. TG-DSC experiments

The copper slag was ground into 200 mesh with mill machine and place it into oven at 120 °C for 2 h, and then 100 g copper slag was mixed with 24.10 g pure carbon as the raw materials. The thermal analysis measurements were carried out using a NETZSCH STA 449C TG-DSC instrument, at a heating rate of 10 °C·min^{-1}, to heat to 1500 °C in argon atmosphere.

Table 2 Chemical composition of copper slag, wt%

Composition	CaO	SiO_2	Al_2O_3	TFe	Cr_2O_3	MnO	MgO	ZnO	Others
Contents	6.00	34.92	3.86	31.49	0.09	0.09	1.44	3.03	5.58

Table 3 Chemical composition of coke, wt%

Composition	Fixed carbon	Moisture	Ash	Volatile
Contents	76.11	1.57	14.20	8.12

Table 4 Chemical composition of silica, wt%

Composition	SiO_2	CaO	Al_2O_3	MgO
Contents	97.55	1.12	0.56	0.77

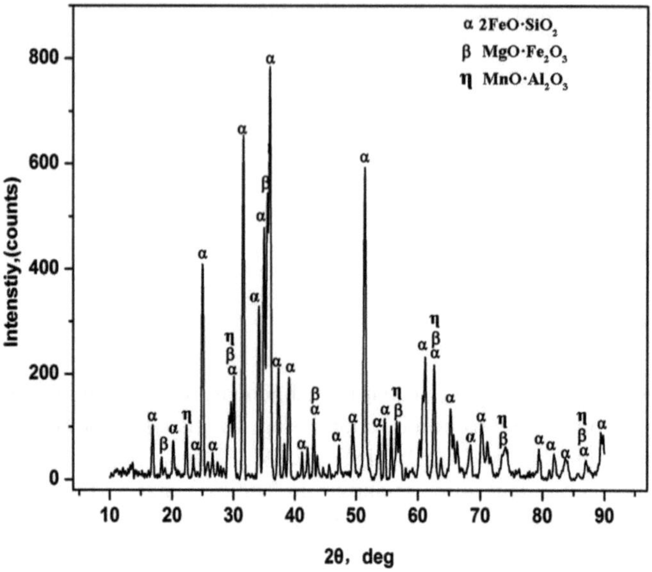

Fig. 1 XRD pattern of copper slag

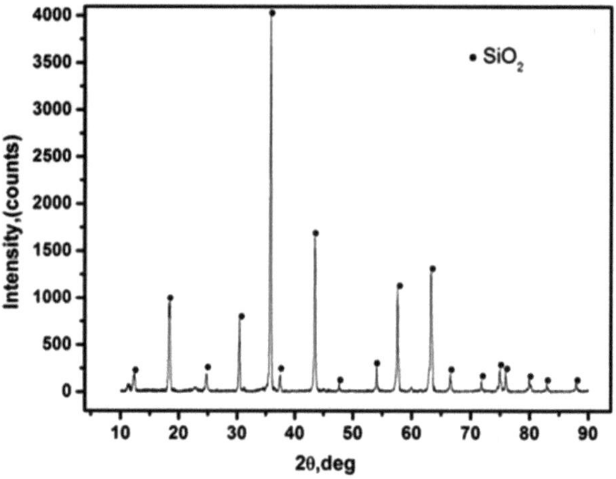

Fig. 2 XRD pattern of silica

2. Carbothermic reduction experiments

In the experiment, the copper slag, coke and silica were ground into 200 mesh and then mixed, and pressed into balls with weight of 50 g, diameter of 20 mm and height of 8 mm. The balls were put into drying oven for 3 h before starting the experiment every time. The induction furnace used in the experiment is showed in Fig. 3. And the experimentation process was as follows: (1) the balls were put into

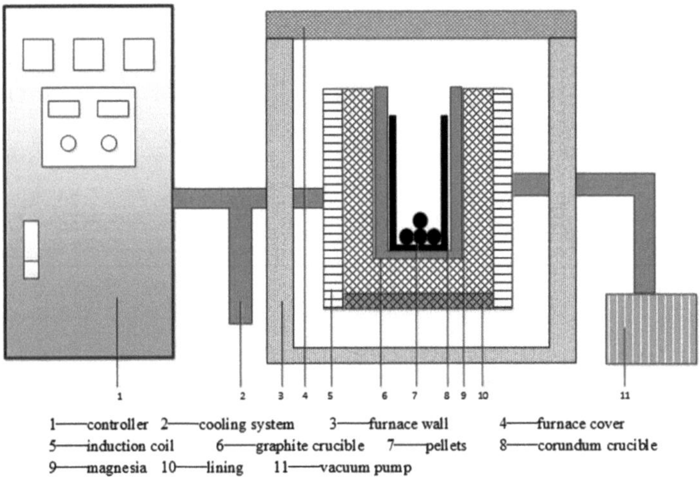

Fig. 3 Illustration of the devices used for the experiments

corundum crucible and then taken them into induction furnace; (2) sample was heated to 1650 °C within 0.5 h in argon atmosphere, and kept the temperature for 0.5 h, (3) the sample was cooled with the furnace. After the experiments, the sample was detected by X-ray diffraction and chemical analysis.

Results and Discussion

Thermal Analysis Experiments

See Fig. 4.

Carbothermic Reduction Experiments

Equations (1–9) show the detailed reaction process in the carbothermic reduction experiments [18].

$$FeO + C = Fe + CO \quad \Delta G = 1956.8 - 1216.8T \tag{1}$$

$$Fe_2O_3 + C = Fe + CO \quad \Delta G = 263060 - 175250T \tag{2}$$

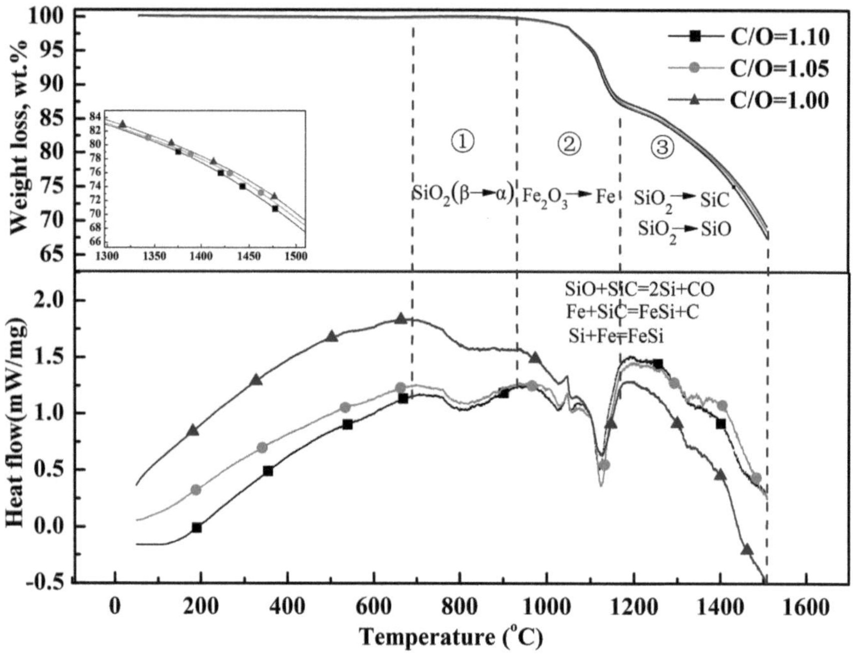

Fig. 4 TG-DSC curves of copper slag with carbon

$$SiO_2 + 3C = SiC + 2CO \quad \Delta G = 67035 - 43.89T \tag{3}$$

$$SiO_2 + C = SiO + CO \quad \Delta G = 159600 - 77.94T \tag{4}$$

$$2SiC + SiO_2 = 3Si + 2CO \quad \Delta G = -28500 - 0.64T \tag{5}$$

$$SiC + SiO = 2Si + CO \quad \Delta G = 49150 - 24.59T \tag{6}$$

$$SiO_2 + 2C = Si + 2CO \quad \Delta G = 167{,}400 - 86.40T \tag{7}$$

$$Fe + Si = FeSi \quad \Delta G = -28{,}500 - 0.64T \tag{8}$$

$$Fe + SiC = FeSi + C \quad \Delta G = 9900 - 9.14T \tag{9}$$

From the aforementioned Equations, iron oxide was reduced to Fe at lower temperature in the experiments, and silicon oxide was reduced to Si at high temperature, and FeSi can be produced through the reaction between Fe and Si, which is in accord with the results of TG-DSC.

Table 5 shows the chemical composition of samples after the experiments. The chemical composition is in according with the composition of FeSi45 in National Standard of China [19]. In this table, Si and Fe account for a large proportion in the samples, and with the increase of carbon addition, Fe content decreases and silicon

Table 5 Experimental scheme for preparing ferrosilicon

No.	Temperature, °C	Copper slag, g	Silica, g	Coke, g	n(C)/n(O)
1#	1650	100	27.62	43.13	1.00
2#				45.29	1.05
3#				47.45	1.10

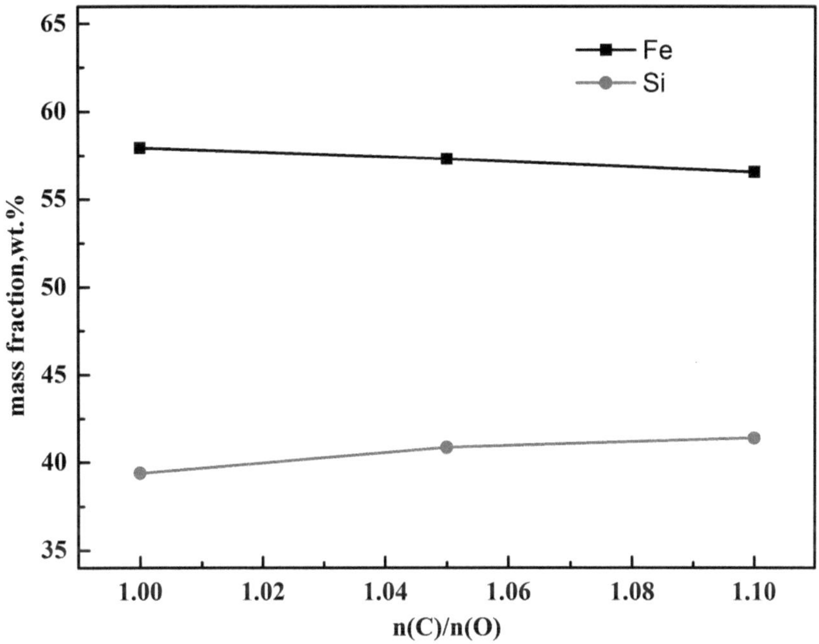

Fig. 5 Si and Fe contents in the final samples for different n(O)/n(O)

Table 6 Chemical composition of final samples(wt%)

No.	Fe	Si	Cu	Cr	P	S	Others
1#	57.92	39.38	0.43	0.16	0.09	0.13	0.89
2#	57.29	40.85	0.44	0.15	0.08	0.04	1.15
3#	56.55	41.39	0.44	0.13	0.06	0.02	1.41

content increases. It is inferred that high C/O mole ratio is benefit to the reduction of silica and preparation of FeSi45, which coincide with the results of TG-DSC (Fig. 5; Table 6).

Si content obviously increases whereas Fe content slightly decreases with the increase of n(C)/n(O), as observed in Fig. 6. It is demonstrated that n(C)/n(O)

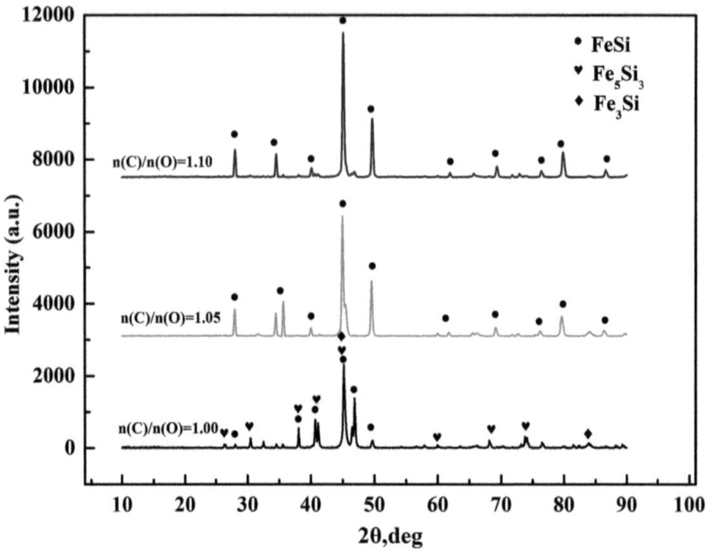

Fig. 6 XRD patterns of samples after experiments for different C/O mole ratio

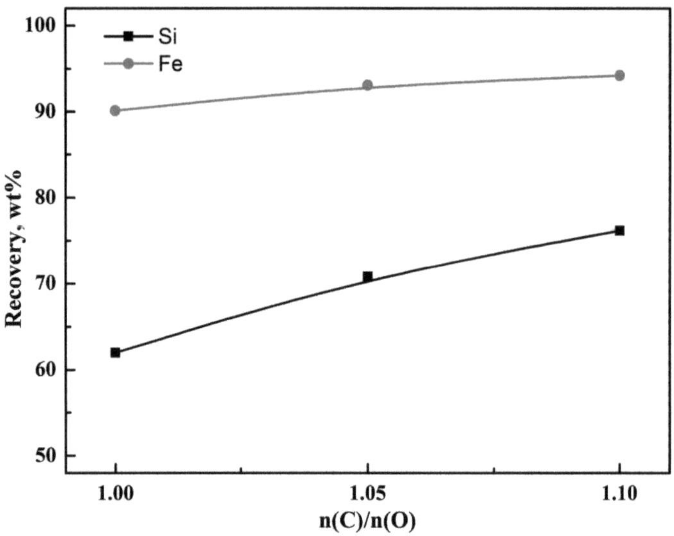

Fig. 7 Recovery of Si and Fe in the alloys

affects the reduction of silicon oxide than iron oxide, which is in according with the aforementioned TG-DSC results.

Figure 6 shows the XRD patterns of samples after reduction experiments for different C/O mole ratio. When the C/O ratio equals 1.00, FeSi, Fe3Si and Fe5Si3 are all exist in the samples. When the C/O increases to 1.10, FeSi is the unique phase in the sample. That is because silica has not been reduced enough with C/O ratio of 1.00 due to lack of reductant. Therefore, it can be indicated that high n(C)/n(O) is in favor of the stable phase's generation of FeSi.

Figure 7 shows the changes of recovery for different C/O mole ratio. With the increase of carbon addition, the recovery of Si and Fe increases. When the n(C)/n(O) reaches to 1.10, the recovery of Fe becomes constant and the recovery of Si still increases slowly. Therefore, high n(C)/n(O) is greatly benefit for the reduction of silicon oxide than iron oxide, when n(C)/n(O) equals 1.10, the recovery of Si and Fe reaches to 77% and 95%.

Conclusions

In order to take full advantage of the copper slag, the experiments of utilizing copper slag to prepare FeSi alloy has been carried out.

1. Recycling Fe and SiO_2 of copper slag to replace the scrap iron and part of silica in the traditional process of producing ferrosilicon alloy, FeSi45 alloy was obtained.
2. According to the thermal analysis experiments and XRD patterns of final samples, the unable phases of Fe_3Si and Fe_5Si_3 was observed because no enough reductant to reduce silicon oxide in the slag. With the increase of n(C)/n(O), FeSi became the unique phase. n(C)/n(O) was a main factor influencing generation of FeSi phase.
3. With the increase of n(C)/n(O), the recovery of silicon increases and iron increases then becomes constant, the recovery is about 77 and 95% with n(C)/n(O) of 1.10, respectively. n(C)/n(O) greatly affects the recovery of silicon than iron with the increase of n(C)/n(O).

Acknowledgements This study was performed with the financial support of the Natural Science Foundation of China (No. 51522403) and the Fundamental Research Funds for the Central Universities (No. CDJZR14135501).

References

1. C.C. Cheng, G.L. Guang, T.T. Xu, C.L. Li, Application of energy saving and emission reduction technology in ferrosilicon production. Mod. Chem. Ind. **9**, 115–117 (2013)
2. S.T. Chen, Recovery and utilization of waste heat of ferrosilicon furnace. Ferro-Alloys **2**,39–42 (2008)

3. H. Wan, in *Research of energy saving and consumption reduction and utilization of waste heat in ferrosilicon production* (Ph. D. thesis, Xi'an University of Architecture and Technology, 2009), pp. 7–13
4. Q.M. Xu, in *Study on ferrosilicon stove's dust control and recycling* (Ph. D. thesis, Wuhan University of Technology, 2006), pp. 7–19
5. Q. Chen, Y. Zheng, Y.T. Jia, Controlling impurity elements in ferrosilicon production. Ferro-Alloys **5**, 15–17 (2006)
6. B. Liu, H. Wang, J. H. Hu, L. Li, Progress in recovery technology of valuable metals from copper slag. Mining Metall. **1**, 44–48 (2009)
7. W.A. Moura, J.P. Goncalves, M.B.L. Lima, Copper slag waste as a supplementary cementing material to concrete. J. Mater. Sci. **42**(7), 2226–2230 (2007)
8. A.C. Raposeiras, A. Vargas-Ceron, D. Movilla-Quesada, D. Castro-Fresno, Effect of copper slag addition on mechanical behavior of asphalt mixes containing reclaimed asphalt pavement. Constr. Build. Mater. **119**, 268–276 (2016)
9. P.S. Prasad, G.V. Ramana, Feasibility study of copper slag as a structural fill in reinforced soil structures. Geotext. Geomembr. **44**(4), 623–640 (2016)
10. R.S. Edwin, M. De Schepper, E. Gruyaert, N. De Belie, Effect of secondary copper slag as cementitious material in ultra-high performance mortar. Constr. Build. Mater. **119**, 31–44 (2016)
11. C. Shi, C. Meyer, A. Behnood, Utilization of copper slag in cement and concrete. Resour. Conserv. Recycl. **52**(10), 1115–1120 (2008)
12. B. Gorai, R.K. Jana, Characteristics and utilisation of copper slag—a review. Resour. Conserv. Recycl. **39**(4), 299–313 (2003)
13. L.T. Khajavi, K. Morita, T. Yoshikawa, M. Barati, Removal of boron from silicon by solvent refining using ferrosilicon alloys. Metall. Mater. Trans. B-Process Metall. Mater. Process. Sci. **46**(2), 615–620 (2015)
14. T. Hu, X.W. Lv, C.G. Bai, Z.G. Lun, G.B. Qiu, Carbothermic reduction of titanomagnetite concentrates with ferrosilicon addition. ISIJ Int. **53**(4), 557–563 (2013)
15. Q.F. Tang, J.C. Gao, X.H. Chen, X.W. Wei, Smelting reduction of MgO in molten slag by liquid ferrosilicon. Revista De Metalurgia **52**(2) (2016)
16. R. Huang, X.W. Lv, C.G. Bai, K. Zhang, G.B. Qiu, Enhancement reduction of panzhihua ilmenite concentrate with coke and conglomeration of metal with ferrosilicon. Steel Res. Int. **84**(9), 892–899 (2013)
17. X. Wang, T. Van Gerven, B. Blanpain, M.X. Guo, In-situ investigation on the reduction of magnesiochromite with ferrosilicon between 1373-1573 K. ISIJ Int. **55**(11), 2289–2296 (2015)
18. N.C. Zhao, Q.X. Zhang, *Technical Manual of Ferroalloys Production* (Metallurgical Industry Press, Beijing, 1988), pp. 13–20
19. Ministry of Industry and Information Technology of the People's Republic of China, GB/T2272–2009, PRC National Standard

Research on Optimization of Sintering Mixture with Low-Grade Complex Ore

Yuchuan Ding, Zizong Zhu, Zhiqiang Zhou, Hao Xiong and Libin Zhu

Abstract In order to utilize the lean ore resources from XinJiang rationally and optimize mixture, laborarory measurements have been carried out under different basicity, various mixed carbon content as well as the properties (ratio) of fuel particle size by the sintering pot for further microstructure analyses. The results show that sinter strength falls by 5.0%, low temperature reduction disintegration index (RDI) drops off to the lowest in certain ranges. The natural degradation index of sinter is much higher when sinter basicity is between 1.65 and 1.75. The sinter strength increases when sinter basicity is between 1.75 and 2.0. Meanwhile, the sinter strength and $RDI_{+3.15}$ improve with increasing carbon content. And the sinter strength and $RDI_{+3.15}$ both decrease with the decrease of the distribution ratio of small fuel particle, thereby decreases the yield of sinter. Microstructural analyses have been investigated to figure out the mechanism of the influence of the three factors on sinter.

Keywords Sintering · Basicity · Particle powder · Metallurgical properties · Tumbler strength · Microstructure

Introduction

The low-grade complex ore has a very high comprehensive utilization value with the continuous exploitation of high-quality ore. In recent years, an increasingly high quality requirement for iron bearing charge has been called for along with the rapid

---

Y. Ding · Z. Zhu (✉) · Z. Zhou · H. Xiong
College of Material Science and Engineering, Chongqing University,
Chongqing 400044, China
e-mail: zhuzizong@163.com

Y. Ding
e-mail: dingyutsuan@163.com

L. Zhu
Xinxing Ductile Iron Pipes Xinjiang Co., Ltd., Hejing 841300, Xinjiang, China

L. Zhang et al. (eds.), *Energy Technology 2017*,
The Minerals, Metals & Materials Series, DOI 10.1007/978-3-319-52192-3_31

development of the production of steel. However, the easily accessible reserves are being depleted, and new resources of iron for sinter are required. It's essential to study the mechanical properties and metallurgical properties of sinter. While, the transportation cost of iron ore from the mainland to XinJiang area is considerably high due to the geographical location. Thus a promising direction is to develop a rational ore blending scheme by using the low-grade complex ore from XinJiang. Therefore, in order to meet the reduction of cost and achieve the rational utilization of the resources. Researches have investigated for several years on low-grade complex ores and have obtained some improvement in some regions like Akto, Hami [1], and Bachu [2]. However, some problems still remain. The operational conditions need to be improved and ore matching structure need to be optimized by using the low-grade complex ores from XinJiang during the sintering process reasonably.

At present, the researches on the improvement of sinter mainly are focused on the optimum of the sintering process, new method of sinter [3], energy consumption, sintering model [4, 5] and so on. Besides, Chen steel have conducted a research on the effect of carbon content on vanadium titanium sinter microstructure, which provided some theoretical basis on improving the quality of sinter. The effect principle of carbon content on quality of chrome-bearing vanadium-titanium sinter has been studied by Yang et al. [6] in sintering pot test. And JSW [7] company has conducted a research on the impact of fuel ratio of coke powder on the microstructure and the quality of sinter. All above researches are aimed at finding a better way to match and optimize sinter.

Experimental

Basicity, coke and mixed carbon are three important factors that affect the quality of sinter, especially for low grade ore. Therefore, a reasonable ore-blending proportion is of great significance to sintering process. The low-grade complex ore used in Xinxing Ductile Iron Pipes XinJiang Co., Ltd. is rich in impurity, resulting in bad variations in metallurgical properties. In this paper, several kinds of low grade and complex ore powders used in this study were supplied by Xinxing Ductile Iron Pipes XinJiang Co., Ltd. Sinter pot test researches on the impact of alkalinity, carbon content and ratio of coke powder (<1 mm fraction) have been investigated to provide references to the optimization of low grade complex ore sinter.

Analyses of the chemical composition of several kinds of ores finely ground to powder were taken by optical microscope, as shown in Table 1. It is obvious that the grades of Hehe fine ore and Beizhan fine ore were high, which reached more than 64%, while, Jintai ore and Jinte ore were relatively lower.

The contents of Pb and Zn of Jinte ore from XinJiang area are, respectively higher than those of other kinds of mineral powder. Such trace elements may cause the blast furnace get blocked or even form nodules. The chemical compositions of each mineral ore has great differences in composition. And especially, the SiO_2

Table 1 Chemical constituents of sintering materials (mass%)

No.	TFe	SiO_2	CaO	MgO	Al_2O_3	S	P	Zn	P
Beizhan	65.04	3.76	1.21	2.02	1.06	0.52	0.02	0.006	0.01
Kaihong	62.28	7.43	2.41	0.84	1.60	0.28	0.01	0.005	0.02
Jingang	62.26	7.90	0.99	0.96	0.98	0.32	0.05	0	0.02
Hehe	64.47	5.76	2.17	0.68	0.93	0.14	0.01	0.005	0.013
Dongyuan	62.34	6.28	1.33	0.97	1.60	0.09	0.02	0.001	0.02
Ruidi	58.90	9.11	2.23	2.30	0.58	0.06	0.01	0	0.04
Lianxu	48.43	3.89	12.9	4.59	0.98	0.07	0.02	0	0.12
Jintai	46.92	5.02	4.45	5.16	0.72	0.05	0.02	0.002	0.091
Jinte	48.56	15.8	1.38	0.77	2.26	0.60	0.06	0.345	0.30

Table 2 Scheme of sinter experiment

Group	1	2	3	4	5	6	7	8	9	10	11	12	13	14
Basicity (CaO/SiO_2)	1.65	1.75	1.80	1.85	1.90	2.0	1.75	1.75	1.75	1.75	1.75	1.75	1.75	1.75
Carbon content (%)	3.5	3.5	3.5	3.5	3.5	3.5	3.3	3.5	3.7	3.5	3.5	3.5	3.5	3.5
Fuel ratio of <1 mm particle (%)	40	40	40	40	40	40	40	40	40	20	30	40	50	60

content of Jinte ore reached above 15% and MgO content of Jintai ore was more than 5%. The experimental schemes of ore matching are shown in Table 2, in which 1–6 groups are schemes for alkalinity, 7–9 groups for carbon content and 10–14 groups for ratio of coke powder (<1 mm fraction) respectively.

The sintering pot used in this work is shown in Fig. 1, and it shows that the 60 kg ore in total granule sinter mixture was charged into it. The diameter of the sintering pot is 300 mm and the bed height maintains at 600 mm. Some major operating parameters of the sintering tests are listed as follows: granulating time (3 min), thickness of feed layer (500 mm), duration of ignition (2 min), thickness of bedding material (20 mm), ignition temperature (1100 ± 50 °C), ignition negative pressure(7–8 kPa), water content (7%) [17]. Experimental method: combination of quantitative water and empirical analysis. A series of comparative tests on the influence of the three factors were carried out based on Table 2; and the experimental scheme about sintering test is shown in Table 3, where, lime, lime activity, the ratio of Mg/Al and outside rejection ore, respectively, is 8.5, 300, 1.6, 20.0%. The tumbler strength tests of sinter are conducted according to YB/T5166 [8]. Below are five kinds of ores, which were picked out from the nine different ores for the combination of ore blending. And the balance is rejection ore, pellets powder and sintering dust.

Based on Tables 2 and 3, the chemical composition of sinter and the parameters of the metallurgical properties [9, 10] in each group are shown in Table 4.

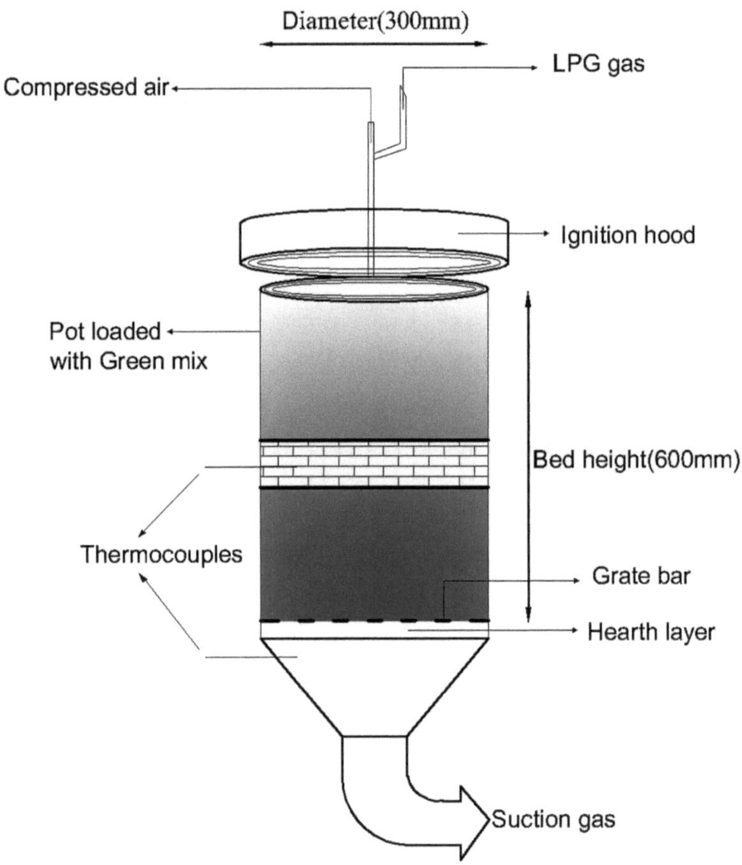

Fig. 1 Facility of sinter pot and parameters

Table 3 Scheme of experimental on sintering test (%)

Jingang	Hehe	Kaihong	Jintai	Jinte	Rejection ore	Pellets powder	Sintering dust
25	19	25	12	5	7	5	2

Results and Discussion

In order to find out the detailed influence of basicity on sintering strength and $RDI_{+3.15}$, mineral phase analyses on SFCA and glassy phases were taken by optical microscope, as shown in Fig. 3. The results about mineral composition and microstructure of sintering samples with different basicity are shown in Figs. 2 and 3. Figure 2 shows that both of the sinter strength and $RDI_{+3.15}$ decreases first and then increases with the basicity improved from 1.65 to 1.80. When the alkalinity is 1.75,

Table 4 Results of sinter experiment

No.	Chemical composition (%)				Sintering strength (%)	Yield (%)	$RDI_{+3.15}$ (%)
	TFe (mass%)	FeO (mass%)	SiO_2 (mass%)	CaO (mass%)			
1	52.20	9.36	6.71	12.49	67.67	71.25	84.51
2	51.73	9.10	7.61	12.85	62.34	72.58	72.53
3	51.48	9.04	6.78	13.34	70.00	73.03	87.37
4	49.87	8.72	7.49	14.46	70.47	72.02	88.56
5	49.56	7.72	7.05	14.97	71.33	74.58	90.84
6	49.40	7.34	6.75	14.80	71.51	76.87	91.72
7	51.61	9.40	7.26	12.37	61.43	67.08	84.71
8	52.45	9.54	6.81	12.10	63.39	66.27	86.57
9	51.97	11.10	6.67	12.08	64.67	65.56	89.40
10	52.16	10.90	6.47	11.89	71.91	75.19	90.33
11	52.75	11.73	6.33	11.56	70.49	74.41	88.71
12	52.68	10.89	7.24	12.40	68.91	72.33	85.79
13	52.43	10.41	6.48	11.84	66.52	71.22	84.13
14	52.83	10.25	6.83	11.88	62.32	68.21	83.52

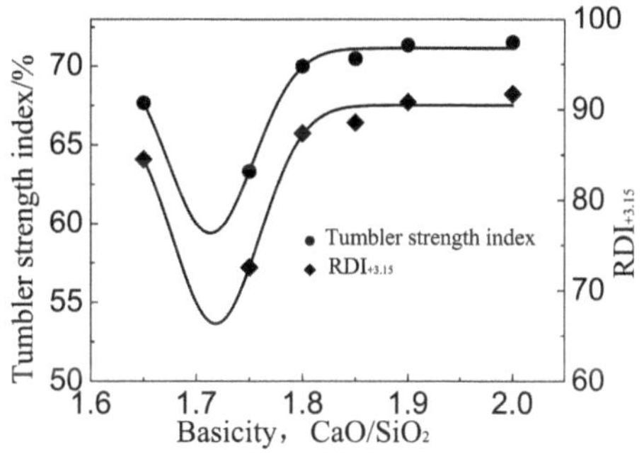

Fig. 2 Effect of basicity on sintering strength and $RDI_{+3.15}$

the strength of sinter reaches to 62.34% only. Analyses results of mineral composition of sintering samples in Fig. 3a, b show that the content of calcium ferrite is low and the distribution of that is not uniform either [11], while the hematite shows the opposite tendency. However, the tumbler strength enhances to 70% above as the alkalinity changes from 1.75 to 2.0, as well as the sinter $RDI_{+3.15}$. It is found that a large number of needle calcium ferrite is well-distributed (Fig. 3c, d) [12].

The reason for this phenomenon is that when alkalinity went from 1.65 to 1.75, the reaction between SiO_2 and CaO in the progress of sinter forms much more $2CaO{\cdot}SiO_2$, which would transform into another type of crystalline solid like $\alpha \rightarrow \gamma$ and $\beta \rightarrow \gamma$. However, the increase of the volume caused by the former was 12% and that caused by the latter was 10% only, which lead sinter to be fragile. As to $RDI_{+3.15}$, during this stage, it is observed that the contents of CaO and calcium

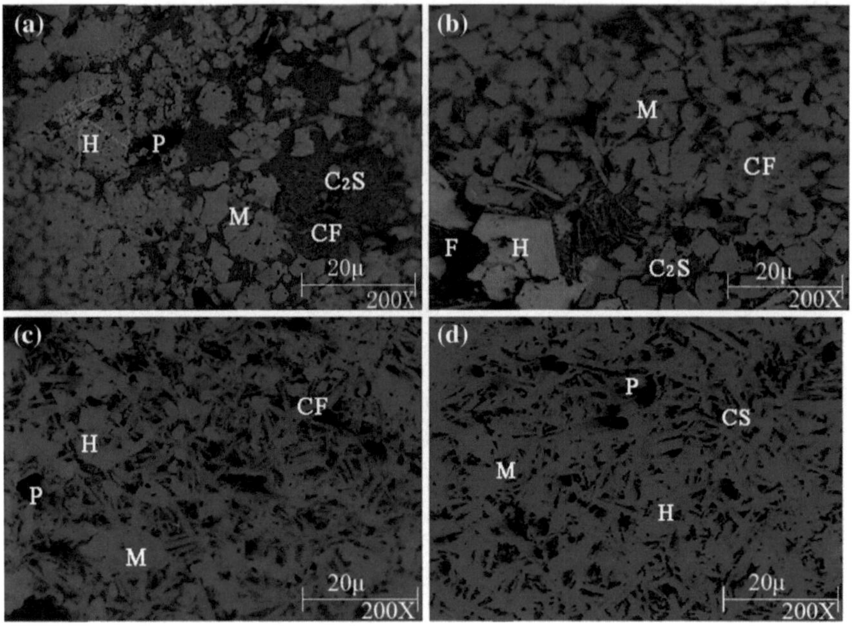

Fig. 3 Microstructure of sintering samples with different basicity

ferrite were low; on the contrary, that of FeO, iron olivine and silicate was high, respectively, resulting in reducing of $RDI_{+3.15}$ [13].

With the increase of alkalinity from 1.75 to 2.0, the non-equilibrium thermodynamics process, to a certain extent, promoted the rapid reaction between CaO and Fe_2O_3, thus the content of calcium ferrite increasingly augmented. The formation of holes between the skeletal crystal hematite leads to the increasing decomposition of calcium ferrite, and become bigger so that a plurality of small holes would combine into a large one. The density of sinter ore reduces with the results of increasing $RDI_{+3.15}$. Besides, the content of FeO shown in Table 4 descends and the mechanical strength improves with the decline content of low strength fayalite [11].

As we all know that sintering is a process of heating raw material on the condition of a reducing atmosphere [14]. And the fluctuation of reaction conditions will seriously affect the quality of sinter. As shown in Fig. 4, the tumbler strength of sinter increases with carbon content increasing, when the carbon content is 3.3, 3.5, 3.7%, respectively. Because of the increase of carbon content in sintering burden, amounts of heat would transfer to the sintering reaction, which causes temperature rising fiercely. Therefore, amount of liquid phase occurs with the increasing temperature of material layer, and the liquid phase contributes to the formation of the bonding phase [15], which helps to improve the sinter strength index.

While, the tumbler strength of sinter will decrease when the carbon content is more than 3.7%. As shown in Fig. 5a, the mineral compositions of sintering samples are simple and the major ferrous mineral is hematite while the bonding

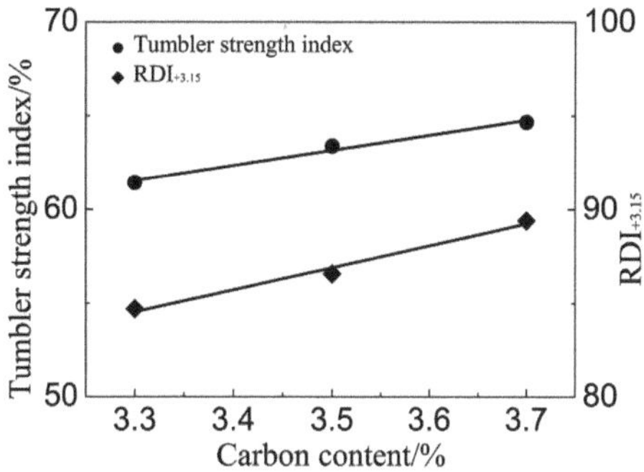

Fig. 4 Effect of different carbon content on sintering strength and $RDI_{+3.15}$

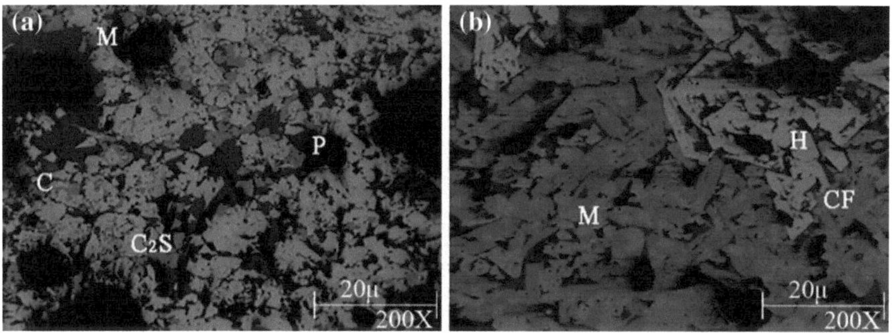

Fig. 5 Microstructure of sintering samples with different carbon content

phases of the sample A are SFCA and glassy phases [9, 15]. Many irregularly shaped holes appear in the uneven-distributed calcium ferrite, which results in sinter decreasing strength and bad $RDI_{+3.15}$. Figure 5 shows that a number of calcium ferrite with network structure is observed, which was distributed homogeneously around some magnetite phase. Figure 5b shows that the formation of some secondary pores around the SFCA will seriously affect the quality of sinter.

Thus, it can be concluded that the sinter produced by using low grade complex ore and 3.7% content of carbon, is beneficial for the sinter strength and yield of sinter. On the other hand, $RDI_{+3.15}$ of sinter also goes up with an increase of carbon content because of the strongly reducing atmosphere in sintering process. As it is known that Fe_2O_3 will be reduced to Fe_3O_4 and FeO, and a quantity of FeO emerged as well as the iron olivine phase. In addition, the reduction of Fe_2O_3 results in higher internal stress and volume expansion becoming smaller, thereby, which has a better effect on improving the $RDI_{+3.15}$ of the sinter [16]. Coke breeze or powder, the most commonly used fuel in the sintering process, has a significant influence on the physical and chemical properties and metallurgical properties of sinter [17]. The granulometric composition of coke powder is as follows.

Table 5 Distribution of coke breeze particle size (%)

Distribution	−1 mm	1–5 mm	10–5 mm	16–10 mm	25–16 mm	40–25 mm	+40 mm
1	20.13	10.10	16.87	13.32	13.43	16.31	10.03
2	29.86	8.64	16.95	11.96	12.27	13.28	6.90
3	40.00	7.12	15.55	10.13	10.57	12.13	4.50
4	50.63	7.77	13.01	8.07	8.53	7.49	5.13
5	60.71	6.49	11.30	5.19	6.17	8.13	2.72

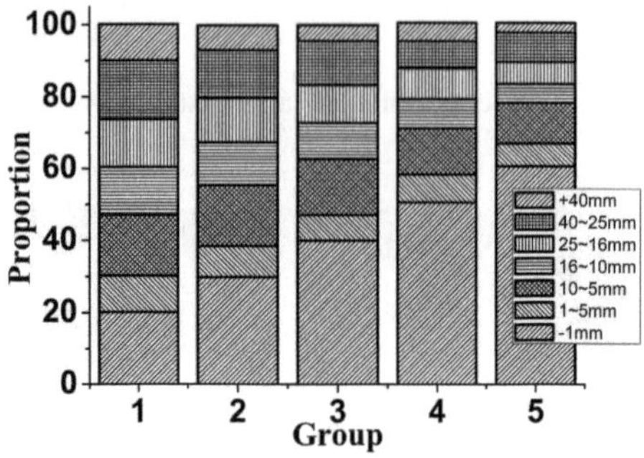

Fig. 6 Distribution of coke breeze particle size

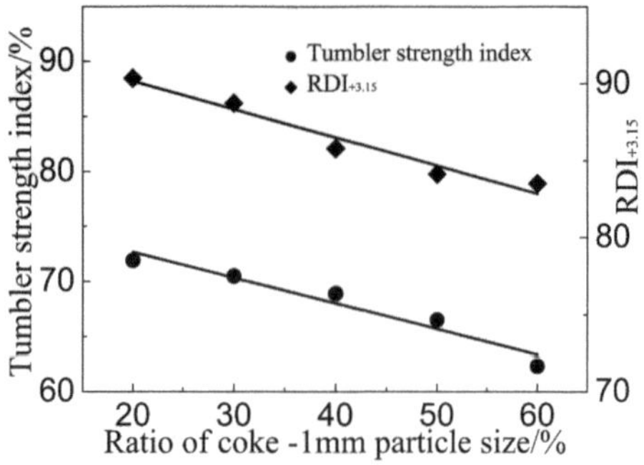

Fig. 7 Effect of different ratio of coke powder (<1 mm fraction) on sintering strength and $RDI_{+3.15}$

Especially, as the ratio of coke powder (<1 mm fraction) accounting for approximately 20, 30, 40, 50 and 60% shown in Table 5 and Fig. 6, the tumbler strength index of sinter and $RDI_{+3.15}$ tend to decrease. The distribution of coke breeze particle size shown in Fig. 6 is obvious that the smaller the distribution ratio of coke powder (<1 mm fraction) amounts for, the more homogeneous the overall coke powder will be. As a consequence, the coke combustion is more sufficient. The negative influences of tumbler strength and $RDI_{+3.15}$ are believed to be a result of the decrease in the proportion of coke powder (<1 mm fraction) [14]. And the productivity of the sinter varied from 75.19% down to 68.21%, as shown in Fig. 7.

Figure 8 shows the influence of ratio of coke powder (<1 mm fraction) on the yield of sinter. The yield of sinter decreases from 75.19 to 68.21% with the increase of the percentage the content of coke powder (<1 mm fraction), which is mainly

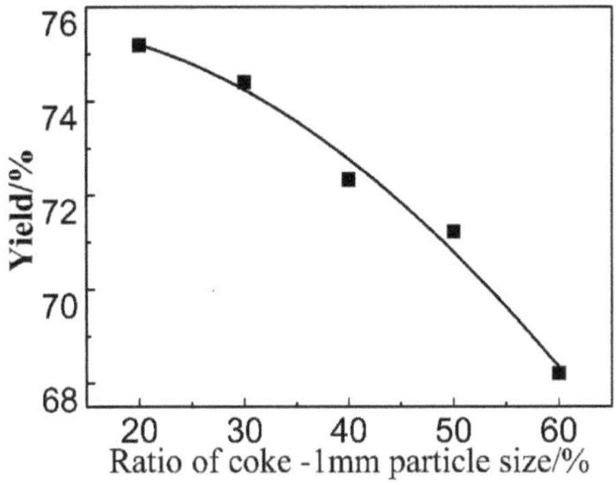

Fig. 8 Influence of the percentage ratio of coke powder (<1 mm fraction)

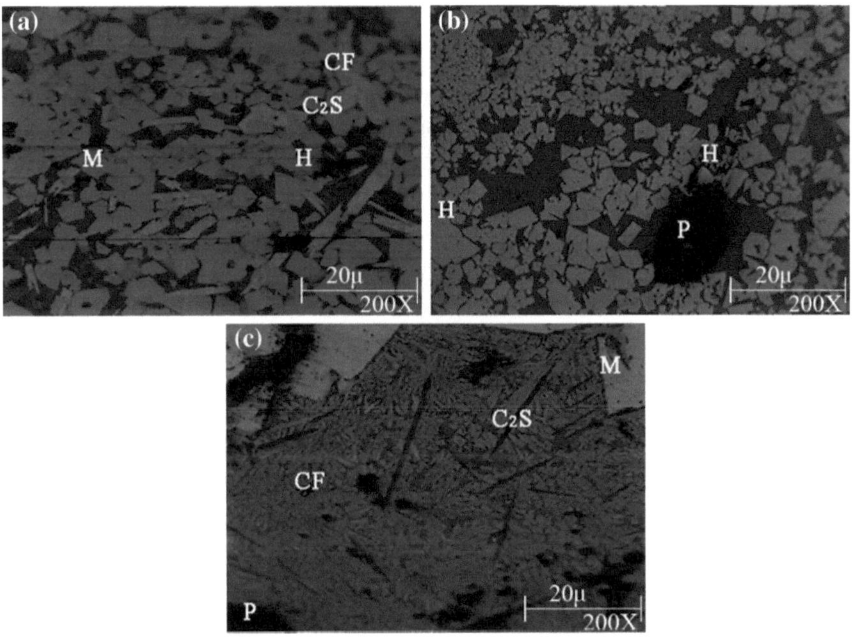

Fig. 9 Microstructure of sintering samples with different ratio of coke powder (<1 mm fraction)

due to a poor bed permeability and hence reduction of the amount of air passing through the bed, [14] leading to a poor tumbler strength of sinter [18].

Figure 9 shows that neither the distribution nor composition of the bonding phases and mineral exhibits well, resulting in the decrease of the amount of the overall quality of sinter. Obviously, the most difference is that the holes become much bigger from Fig. 9a–c. A large quantity of magnetite and a few of calcium ferrite are observed. The mineral composition of the binding phase is much complex, and $2CaO \cdot SiO_2$ shaped in fine granular or leaf is the main binding phase in the sintering process [19].

According to above analysis, it can be concluded that ratio of coke powder (<1 mm fraction) have a great effect on the composition of sinter and bonding phases. The explanation is that there exists a large number of magnetite, which was formed from calcium ferrite in heating and melting process. With the increase in the proportion of coke powder (<1 mm fraction), the content of magnetite and oxidized silicon get promoted. Based on the fundamental theory, the burning rate decreases when the ratio of coke powder (<1 mm fraction) is smaller, generating more magnetite and silicate melts. While, in practical, with the specific surface area becoming bigger, the reduction of iron within a short time was not sufficient. In such circumstances, the FeO decreased, resulting in the decrease of the reaction rate from 11.7 to 10.2%. While, the $RDI_{+3.15}$, declines when the alkalinity and lime activity are on the same conditions respectively. Besides, with the decrease of FeO content and the increase of calcium ferrite and hematite phase, the phase change in the process of expansion results in the reduction of the $RDI_{+3.15}$ [20, 21].

The air permeability of raw material layer turns worse because of the average particle size becoming smaller. A large amount of CO generated by the reaction of single particle of coke and O_2, on the condition of the rapid speed of burning and the lack of O_2, causes the reduction of the fuel utilization, and brings the elevated temperature in the process of sinter down, resulting in poorer strength and a lower yield.

Conclusions

1. When the alkalinity is between 1.65 and 1.75, it will result in increasingly serious behavior of natural degradation of sinter. Thereby, in the practical production, it is appropriate to avoid this range and adjust the basicity of sinter. By increasing the alkalinity from 1.8 to 1.85, tumbler strength, yield, and $RDI_{+3.15}$ all have increased significantly. The optimal alkalinity range is 1.8–1.85.
2. The strength and $RDI_{+3.15}$ of sinter are improved with the increase of carbon content; and the quality of sinter is the best when the carbon content accounts for 3.5%.
3. The strength and $RDI_{+3.15}$ of sinter have a tendency to become poor, as the ratio of coke powder (<1 mm fraction) increases. It is expedient to adopt 20% of ratio of coke powder (<1 mm fraction) as a reference point in subsequent industrial production.
4. The sinter productivity gradually decreased and reached minimum value with the increase of the proportion of coke powder (<1 mm fraction).

References

1. J. Shi, Study on mineral processing technology of a low grade iron ore in Hami. XinJiang, Mining Metall. Eng. **8**(34), 179–180 (2014)
2. Y. Long et al., Experimental study on the iron of a low-grade titanomagnetite in XinJiang Bachu, pp. 77–78 (2011)
3. J.P. Zhao et al., Modelling fuel combustion in iron ore sintering
4. C. Wu, *Metallurgical Engineering Experiment* (Chongqing University Press, Chongqing, 2005)
5. J. Zhang et al., Model of iron ore sintering based on melt and mineral formation. J. Iron Steel Res. Int. **22**(4), 288–296 (2015)
6. S. Yang et al., Effect of basicity on sintering behavior of low-titanium vanadium-titanium magnetite. Trans. Nonferrous Met. Soc. China **25**, 2087–2094 (2015)
7. L. Liu et al., Experimental study on sintering ore proportioning. Hebei Metall. **150**, 9–11 (2005)
8. C.S. Teo, Positioning coke particles in iron ore sintering. ISIJ Int. **32**, 1047 (1992)
9. L.H. Hsieh, Effect of raw material composition on the sintering properties. ISIJ Int. **45**, 551–559 (2005)
10. A. Selecka, A. Salak, H. Danninger, The effect of boron liquid phase sintering on properties of Ni-, Mo- and Cr-alloyed structural steels. J. Mater. Process Technol. 141–379 (2003)
11. G. Luo et al., Effects of compound silicate gangue on formation of complex calcium ferrite during sintering process. J. Iron Steel Res. Int. **20**(3), 18–23 (2013)
12. M. Zhou et al., Vanadium–titanium magnetite ore blend optimization for sinter strength based on iron ore basic sintering characteristics. Int. J. Miner. Process. **142**, 125–133 (2015)
13. G. Li et al., Effect of basicity on sintering behavior of saprolitic nickel laterite in air. Powder Technol. **249**, 212–219 (2013)
14. T. Umadevi et al., Optimisation of FeO in iron ore sinter at JSW Steel Limited. Ironmaking Steelmaking **39**(3), 180–189 (2012)
15. K Sheikhi Moghaddarn et al., Effect of sinter hardening on microstructure and mechanical properties of astaloy 85Mo. J. Iron Steel Res. Int. **19**(10), 43–46 (2012)
16. T. Umadevi, Effect of coke powder particle size on sinter quality. Iron making Steelmaking 8–12 (2008)
17. A. Formoso et al., Influence of nature and particle size distribution on granulation of iron ore mixtures used in a sinter strand. Ironmaking Steelmaking **30**, 447–460 (2003)
18. C.E. Loo, Role of coke size in sintering of a hematite ore blend. Ironmaking Steelmaking 18–33 (1991)
19. G. Fillllri, R. Causton, A. Lawley, Effect of cooling rates during sinter hardening, in II international conference on powder metallurgy and particulate materials, Las Vegas, MPIF, p. 174 (2003)
20. T. Hu et al., Reduction behavior of Panzhihua titanomagnetite concentrates with coal. Metall. Mater. Trans. B **44**(2), 252–260 (2013)
21. T. Umadevi et al., Influence of coke breeze particle size on quality of sinter. Ironmaking Steelmaking **35**(8), 567–574 (2008)

Part X
Solar Cell Silicon: Silicon Production, Crystallization, and Properties

Study on Producing Solar Grade Silicon by Carbothermal Reduction of Andalusite Ore

Shilai Yuan, Huimin Lu and Panpan Wang

Abstract A big andalusite ore bed was found in Dingxi District, Gansu, China. In this paper, the feasibility of producing solar grade silicon from aluminum-silicon alloys by carbothermal reduction of andalusite ore was studied. First, the carbothermal reduction experiments were conducted in a 100 kVA DC submerged arc furnace with reducing temperature 2000–2200 °C and reducing time 4 h, the hypereutectic aluminum-silicon alloys containing 50–55 mass% aluminum and 35–45 mass% silicon were obtained; then the investigation on solidification of the hypereutectic Al-Si melts under the temperature gradient as an intensified separation silicon way was carried out. According to previous investigation, the optimized technical process for solar grade silicon and eutectic Al-Si alloy from andalusite ore was proposed.

Keywords Carbothermal reduction · Andalusite ore · Al-Si melts · Solar grade silicon · Solidification

Introduction

Solar energy, due to its low green house gas emissions, has been paid great attention. Accordingly, the demand for solar grade silicon (SOG-Si) has been growing rapidly. Refining of metallurgical-grade silicon (MG-Si) attracts research interest due to its low material and energy costs, and more environmentally friendly technology. As such, impurity elements in MG-Si must be removed to low levels

S. Yuan · H. Lu (✉) · P. Wang
School of Materials Science and Engineering, Beihang University,
37 Xueyuan Road, Beijing 100191, China
e-mail: lhm0862002@aliyun.com

S. Yuan
e-mail: 13331151800@126.com

P. Wang
e-mail: 13910518379@126.com

L. Zhang et al. (eds.), *Energy Technology 2017*,
The Minerals, Metals & Materials Series, DOI 10.1007/978-3-319-52192-3_32

for the photovoltaics (PV) cell to operate at optimum efficiency. The main focus in refining MG-Si to SOG-Si is on boron and phosphorus which are the most difficult to remove by the conventional metal refining method. The refining efficiency depends on the segregation behavior of different elements in solid silicon and the liquid phase. The segregation ratios of all investigated impurities were found to be much smaller than the segregation coefficients between solid and liquid Al-Si melts [1]. For instance, the segregation coefficient of B in Si is 0.8 at the silicon melting point (1687 K), whereas it is around 0.2 when silicon is solidified from Si-Al melt at 1273 K. For phosphorus the segregation coefficient is 0.06 when silicon is solidified from Si-Al melt at 973 K, which is much less than 0.35 for recrystallization from liquid silicon. Since the segregation coefficient of boron in silicon is 0.8, it cannot be removed by directional solidification.

This article describes the feasibility of producing SOG-Si from aluminum-silicon alloys by carbothermal reduction of andalusite. By this way, we can get both the SOG-Si and the aluminum silicon alloys. Aluminum-silicon alloys are important aluminum alloys. It is widely applied as structural material, such as construction material or in the automotive industry and so on [2]. The market of Al-Si alloys is about 4 million tons per year in China. It is therefore of great interest to be able to produce an aluminum-silicon alloys directly from the widely available aluminum-silicon minerals, such as andalusite. The energy consumption in such a process will be considerably lower. This process also has many other advantages such as short production time, small investments for equipment, high production and utilization rate of energy [3, 4].

A big andalusite ore bed was found in Dingxi District, Gansu, China, and its reserves are the largest and its quality is the best in the world. The proved reserves of andalusite ore are 4 hundred million tons and the andalusite ore has little harmful impurities. In this work, first, the carbothermal reduction experiments were conducted in a 100kVA DC submerged arc furnace, the hypereutectic aluminum-silicon alloys containing 50–55mass% aluminum and 35–40 mass% silicon were obtained; then the investigation on solidification of the hypereutectic Al-Si melts under the temperature gradient as an intensified separation silicon way was carried out. The principle for the silicon growth in the Al-Si melts was studied. According to previous investigation, the optimized technical process for SOG-Si and eutectic Al-Si alloy from andalusite ore was proposed.

Experimental Procedure

Laboratory experiments were conducted in a 100 kVA DC submerged arc furnace, as pictured in Fig. 1. Process conditions were as follows: electrode diameter ϕ150 mm; furnace crucible diameter ϕ300 mm; depth of furnace 380 mm; transformer primary voltage 380 V, three-phase bridge rectifier DC voltages divided into four levels 22, 25, 28, 32 V; ratio of current to voltage 4; maximum current 4000A when voltage at 1000 V. The electrodes used in the furnace were graphite electrodes.

Fig. 1 100 kVA DC submerged arc furnace

Table 1 Analysis of andalusite ore used in this work

Na	Mg	Al	Si	K	Ca	Ti	Mn	Fe	O
0.27	0.37	28.01	19.69	1.15	0.29	0.06	0.13	1.49	48.54
MnO	MgO	Al_2O_3	SiO_2	K_2O	CaO	TiO_2	Na_2O	Fe_2O_3	
0.17	0.61	52.91	42.13	1.39	0.41	0.10	0.37	2.12	

Raw Materials

Andalusite. Andalusite ore powder was taken from Gansu Zixin Mineral Co. Ltd., the size was smaller than 0.5 mm. The main minerals in andalusite ore are 84.8 mass% andalusite, 5 mass% quartz, 3.6 mass% plagioclase, 4.6 mass% clay minerals. The assay of the andalusite ore is shown in Table 1.

Bitumenite. Bitumenite (BT) was from Shanxi Datong Washing Coal Plant. Prior to use it was ground to <0.5 mm. Its composition was 12.64 mass% ash content, 28.92 mass% volatile content, 3.70 mass% moisture content, 54.74 mass% solid carbon content. The chemical compositions of the ash content are as follows: 19.52 mass% Al_2O_3, 49.84 mass% SiO_2, 6.52 mass% Fe_2O_3, and 11.72 mass% CaO.

Petroleum coke. Petroleum coke (PC) was from Tianjin Dagang Petroleum Chemical Industrial Company. Before experiments it was ground to <0.5 mm. Its composition was 0.28 mass% ash content, 8.04 mass% volatile content, 0.39 mass % moisture content, 91.29 mass% solid carbon content.

Charcoal. Charcoal (CL) used in experiments was purchased from market, its sizes were smaller than 0.5 mm. Its composition was 1.03 mass% ash content, 2.99 mass% volatile content, 3.94 mass% moisture content, 92.04 mass% solid carbon content.

Adhesive paper industry wastewater. Adhesive paper industry wastewater was from Kuerle Paper Mill. Its composition was 26.08 mass% ash content, 47.74 mass % volatile content, 6.26 mass% moisture content, 19.92 mass% solid carbon content.

Test Methods

The raw materials were mixed uniformly in the given proportions in a stirring machine, briquetted into ovals with the major axis 60 mm and the minor axis 40 mm, and dried in the electrothermal drying box at 150 °C for 2 h until the moisture content was lower than 1%. The average density of briquettes was 1.52–1.56 g/cm^3, the porosity was 43–45%. The dried briquettes were charged into the 100 kVA DC submerged arc furnace with temperature 2000–2200 °C and time 4 h. The Al-Si alloy was discharged. The samples were analyzed by XRD, XRF analysis, SEM analysis and chemical analysis.

The experiment for preparing SOG-Si was done as follows. First, the Al-40%Si alloy from the 100 kVA DC submerged arc furnace was filled into a graphite crucible with an inner diameter of 30 mm and a height of 90 mm. The samples were directionally solidified in a vertical Bridgman-type furnace, and then cooled by water at (293 ± 5) K. Two WRR-130 thermocouples were used to determine the temperature gradient, which was measured to about 200 K/cm. When the vacuum of the furnace chamber was pumped up to 5×10^{-2} Pa, the samples were solidified in a high thermal gradient Bridgman directional solidification furnace, as shown in Fig. 2b, c. When an Al-40%Si alloy was solidified under the condition of uniformly cooled, the solidified structure was composed of uniformly distributed needle-like silicon crystals as shown in Fig. 2a. To reveal clearly the phase morphologies and some other growth features of the solidified Al-Si alloy, the ingots were sectioned, polished, and etched.

The morphologies of the solidified Al-Si alloy were observed using scanning electron microscope (SEM). The chemical composition of inclusions in SOG-Si and

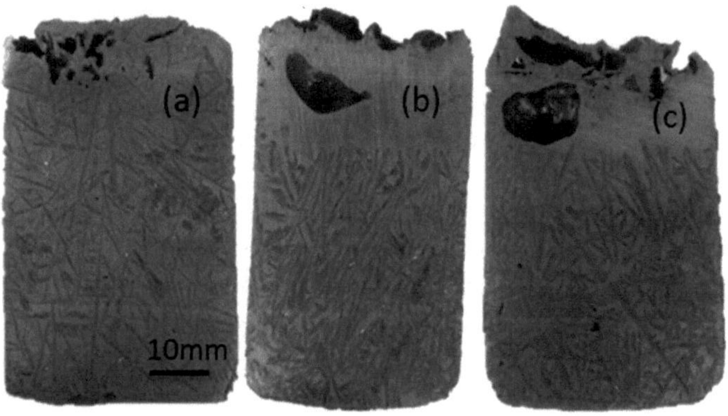

Fig. 2 Cross section of solidified Al-40% Si alloys at different cooling conditions: **a** uniformly cooled; **b** directional solidification at the pulling rates of 2.5 μm; **c** on one end of cooling, on the other end of insulation

Al-Si melt was analyzed by EPMA. The chemical composition of the solidified sample after acid leaching was analyzed by inductively coupled plasma atomic emission spectrometry (ICP-AES).

Results and Discussion

Comprehensive Experiment for Producing Al-Si Alloys

The content of the furnace charge is 60 mass% andalusite ore in which the impurities are quartz, calcium oxide and magnesia etc. and the total impurities amount is not in excess of 20 mass%; 30 mass% reducing agent (bitumenite and petroleum coke, mixing ratio of bitumenite and petroleum coke 70:30) and 6 mass % adhesive paper industry wastewater. First, all the raw materials are mixed uniformly, briquetted and dried, then the carbothermal reduction experiments are conducted in the 100kVA DC submerged arc furnace with reducing temperature 2200 °C and reducing time 4 h, the aluminum-silicon alloys containing 55 mass% aluminum, 40 mass% silicon are obtained with aluminum recovery rate 81% and silicon 72%. Table 2 lists the results of energy spectrum analysis for the Al-Si alloys and Table 3 lists the results of chemical analysis for the Al-Si alloys.

The analysis results of the slag by element fluorescence analysis were listed in Table 4. The slag can be used as a deoxydizing or dephosphorizing agent in steel smelting or as a reducing agent in magnesium smelting by thermal method or in smelting MnSi alloys. Therefore, no waste accumulates when using andalusite as raw material for carbothermal Al-Si alloy production.

Table 2 The results of energy spectrum analysis for the Al-Si alloys (mass%)

Sample number	Al	Si	Al/Al + Si
1	58.52	36.22	61.76
2	51.27	44.64	53.45
3	58.56	36.41	61.66
Average	56.11	39.09	58.93

Table 3 The results of chemical analysis for theAl-Si alloys (mass%)

Al	Si	Al/Al + Si	Fe	Cu	Ti
57.00	35.40	61.69	0.54	0.016	0.020
Mg	Zn	Mn	Ni	Pb	
0.014	0.11	–	0.015	0.02	

Table 4 The analysis results of the slag by element fluorescence analysis (mass %)

Al	Si	O	Ca	Fe	Sr	Mg	Cr
201.13	22.50	18.79	5.54	1.53	0.05	10.05	0.05
S	K	Co	Zr	Ni	Y	C	
0.04	0.02	0.11	0.01	0.01	0.01	31.24	

Table 5 The analysis of the smoke from the experiments

N_2/vol.%	O_2/vol.%	CO_2 vol.%/	Solid particles/g/Nm^3	
78	19	1.5	0.6	
SO_2 mg/Nm^3	Al_2O_3/mass%	SiO_2/mass%		
140	50	26		

Table 6 The impurity contents of purified silicon, [ppmw]

Silicon source	Fe	Ca	Al	B	P
MG-Si	4700	260	640	14	35
Si-2	15	3.1	15	6.8	4.5
Si-3	14	2.4	15	0.73	0.50

Si-2 MG-Si after acid leaching
Si-3 Solidified silicon from Al-Si melts after acid leaching

The analysis of the smoke from the experiments was listed in Table 5. Al-Si fine powder collected from the off-gas could be used as the raw material for smelting Al-Si alloy.

Silicon Solidification in Al-Si Melts

The mass fractions of impurities in primary silicon from Al-Si melts are less than those in MG-Si in Table 6. The primary Si crystal would initially precipitate from liquid alloy when the temperature of Al-Si melt decreases. The Si crystals solidified from the Si-Al melt tended to be uniformly distributed in the alloy melt, as shown in Fig. 2a. Under appropriate growth conditions, however, it was found that that bulk Si crystals can be obtained from the Si-Al melt. To intensify the separation, temperature gradient is introduced into the Al-Si refining process. The primary silicon is separated from the Al-Si melt under the constant temperature gradient and the refined silicon is obtained after acid leaching.

Al-40% Si (mass fraction) binary alloy was melted uniformly in a graphite crucible at 1300 K, using a directional induction furnace under the conditions of vacuum. After cooled to and holding at 1000 K under the constant temperature gradients, 90% silicon was solidified at the bottom of ingots.

Figure 2 shows the vertical cross sections of samples obtained at different cooling conditions and temperature gradients. When an Al-40%Si alloy is solidified without any temperature gradient, the solidified structure is composed of uniformly distributed needle-like silicon crystals in the solidified Al-Si alloy, as shown in Fig. 2a.

For directional solidification, the results show that the upper part is mainly uniform Al-22% Si alloy accompanying with a little primary silicon, and at the bottom of ingot needle-like silicon, as shown in Fig. 2b. It is due to the change of temperature gradients. The primary silicon at the bottom of the alloy is short dendrite-like silicon shown in Fig. 2c. It is under the condition of the constant temperature gradients. From the comparison of Fig. 2b, c, it is clear that morphology of solidified silicon is not due to the pulling rates, but due to the temperature gradient.

Optimized Technological Process of Producing SOG-Si

The optimized technical process of producing SOG-Si may be explained as follows: first the sample was kept in the coil and melted in 1300 K under the constant temperature gradient; after cooling and holding at 1000 K, the primary silicon was solidified in the Al-Si melts, needle-shaped silicon crystals were successfully agglomerated at the top part of the sample; Pouring the solidified silicon was to separate primary silicon from the Al-Si melts; then the separate solidified primary silicon were crashed into the particles, the refined silicon was collected after acid leaching and was subjected to the chemical analysis; additionally, through directional solidification most metallic impurities could be removed.

The segregation ratios of impurities between solid silicon and the Si-Al melt were much smaller, indicating the effective purification after the low-temperature solidification of silicon.

This solidification allowed us to collect silicon crystals selectively from the alloy and to reduce the loss of aluminum and silicon considerably during the acid leaching. Pouring the solidified silicon was the effective method to separate primary silicon from Al-Si melts, it could be further to avoid the waste of acid and the loss of aluminum and most of metal concentrations were left in the Al-Si melts. The concentrations of Fe, Ca can be significantly reduced.

The results show that the removal effect in the Al-Si melts is better than in other system. The boron content in MG-Si is successfully reduced from 34 ppm to less than 1 ppm during the process. Meanwhile, Al, Ca, and Fe elements in MG-Si are also well removed. Many metallic elements such as Fe, Al, Ti, Ca, Mn, V, Zn and Ni have high segregation coefficients in silicon. The unidirectional solidification method is effective to remove metallic impurities and a cost-effective technique for large-scale production of multi-crystalline silicon material. Finally, the low-cost overall process for producing SOG-Si was proposed, According to previous investigation, the optimized technical process for SOG-Si and eutectic Al-Si alloy from andalusite ore was proposed. The process is used in combination to convert MG-Si into SOG-Si.

Conclusions

1. It is feasible to produce aluminum-silicon alloys by carbothermal reduction of andalusite. The optimum mixing ratio of furnace charge was determined to be 50–80 mass% andalusite ore in which the impurities are quartz, calcium oxide and magnesia etc. and the total impurities amount is not in excess of 20 mass%; 20–40 mass% bitumenite and petroleum coke as reducing agent, the mix proportion of bitumenite and petroleum coke should be in the range of 8:2–6:4; 5–8 mass% adhesive paper industry wastewater. All these raw materials are mixed uniformly, briquetted and dried, the carbothermal reduction experiments are conducted in a 100kVA DC submerged arc furnace with reducing temperature 2000–2200 °C and reducing time 4 h, the aluminum-silicon alloys containing 50–55 mass% aluminum, 35–40 mass% silicon can be obtained with aluminum recovery rate 80–85% and silicon 70–75%.
2. Characteristics of silicon morphology after directional solidification process were considered. From these considerations, we further understand the impurities distribution and the principle for refining of silicon in Al-Si melts.
3. The principle for the silicon growth in the Al-Si melts was studied. It suggests that the solidified primary silicon distribution is due to the temperature gradients. It has been found that the temperature distribution and particularly changes in temperature gradient are the main factors governing the growth of silicon.
4. The optimized technical process for SOG-Si production from carbothermal reduction of andalusite ore was evaluated. The most suitable process to fabricate SOG-Si and will be the combination of carbothermal reduction of andalusite ore, solidification refining and acid leaching.

References

1. Kazuki Morita, Takeshi Yoshikawa, Thermodynamic evaluation of new metallurgical refining processes for SOG-silicon production. Trans. Nonferrous Met. Soc. China **21**, 685–690 (2011)
2. F. King, *Aluminum and Its Alloys* (Ellis Horwood Limited, Chichester, 1987), p. 23
3. Chongping He, Enhui Wang, *The Production Technology of Aluminum-Silicon Alloy by Electrothermal Method* (Science and Technology Press, Beijing, 1994), p. 45
4. Feng Zhang, Mengji Li, Technology of producing aluminum–silicon alloys (Al50%). Ferro-Alloys **171**(4), 17–21 (2003)

Phase Analysis of the Si-O_2 System

Shadia J. Ikhmayies

Abstract In this work the equilibrium temperature-composition phase diagrams of the Si-O_2 system were calculated using Thermo-Calc software in the temperature range 500–3000 K. Constituents of stable phases were obtained, and the amounts of certain phases where investigated with the variations in the composition of oxygen and temperature. It is found that quartz with silicon represent a stable phase in the temperature range 500–1150 K, while tridymite with silicon represent a stable phase in the temperature range 1150–1700 K. A liquid phase is found besides tridymite in the temperature range 1700–1750 K, and a stable phase of cristobalite with liquid appears in the temperature range 1750–2000. The maximum amounts of quartz, tridymite, and cristobalite are found at oxygen mole percent of 67.7%. These results are important to better understand the process steps in the production of silicon for different applications including silicon solar cells.

Keywords Quartz · Equilibrium · Phase diagrams · Silicon · Silicon solar cells

Introduction

Quartz is one of the most abundant minerals in the Earth's crust. Therefore, the knowledge of its properties is important for several branches of materials, minerals, and geological investigations. Moreover, the study of quartz has numerous practical applications in high-tech industries that include semiconductors, high temperature lamp tubing, telecommunications, optics, microelectronics, and solar silicon applications [1, 2]. In addition, Silica polymorphs are of great importance in solid state physics, technology, and Earth sciences [3]. However, only very few deposits are suitable in volume, quality and amenability to tailored refining methods for specialty high purity applications. As such high purity quartz has become one of today's key strategic minerals [4–9].

S.J. Ikhmayies (✉)
Faculty of SciencePhysics Department, Al Isra University, Amman 11622, Jordan
e-mail: shadia_ikhmayies@yahoo.com

L. Zhang et al. (eds.), *Energy Technology 2017*,
The Minerals, Metals & Materials Series, DOI 10.1007/978-3-319-52192-3_33

Equilibrium phase diagrams are important sources of information for materials development [10]. High-temperature phase diagrams of solid-gas-liquid phase transition for silica (SiO_2) are very important in many applications. In the present study the equilibrium temperature-composition phase diagrams of the Si-O_2 system were calculated using Thermo-Calc software in the temperature range 500–3000 K. Stable phases were predicted and analyzed, and the amounts of these phases were calculated as functions of temperature and oxygen composition.

Method of Calculations

The computational thermodynamic software Thermo-Calc utilizes a Gibbs free energy minimization procedure to calculate the thermodynamic properties and phase equilibria of the chosen system [11]. For a given set of conditions, the computer determines the free energy change for each possible combination of phases and phase compositions and selects the state that minimizes the total Gibbs free energy. Thermo-Calc is used in conjunction with different thermodynamic databases which contain the descriptions of the Gibbs free energies assessed using the CALPHAD approach, where CALPHAD is a phase-based approach to modeling the underlying thermodynamics and phase equilibria of a system through a self-consistent framework [12]. In this work, MAP calculations -a type of the calculations in the Thermo-Calc software- were used to calculate the equilibrium phase diagrams of the Si-O_2 system and to compute the amounts of the selected stable phases. The used database is the oxide database, and the mol percent of oxygen to silicon was chosen to be the same as that in silica 2:1. The equilibrium calculator was chosen with phase diagram as the calculation type, the pressure $P = 10^5$ Pa, and system size = 1 mol.

Results and Discussion

Figure 1 displays the equilibrium temperature- composition phase diagram of the Si-O_2 system. It is noticed that silicon and quartz can be found in a stable phase in the temperature range 500–1150 K. So, quartz with silicon is the lowest temperature stable phase in the Si-O_2 system. Tridymite can be found with silicon in a stable phase in the temperature range 1150–1700 K.

The silica starts to melt after 1700 K, and hence the stable phase in the short temperature range 1700–1750 K is tridymite with silica liquid. For T = 1750–2000 K, tridymite disappears and cristobalite with silica liquid are found in a stable phase. With the increase in temperature, cristobalite disappears and the whole silica is in the form of ionized liquid, which is the unique stable phase in the temperature range 2000–2150 K. For T = 2150–3000 K the ionic silica liquid dissociates to gases, so there is just one stable phase which is the ionic liquid with a gas. This gas

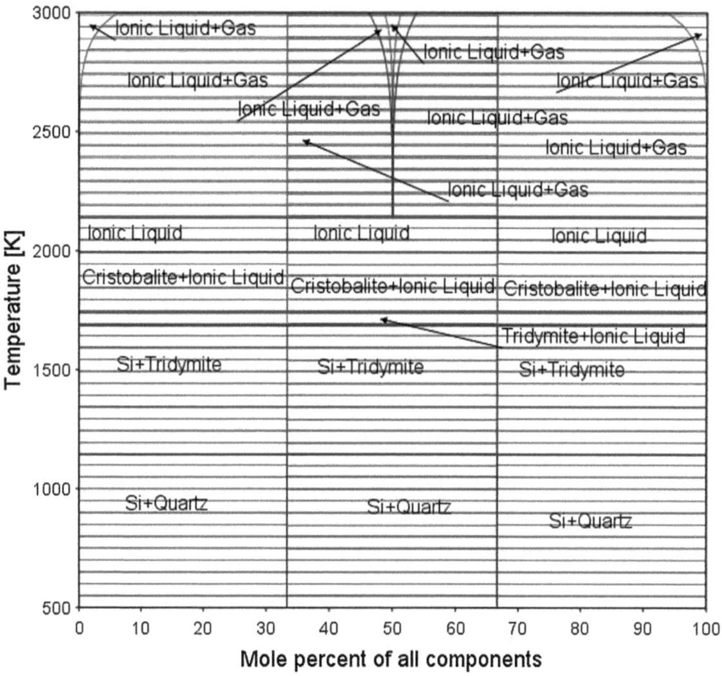

Fig. 1 Temperature- components mole fraction phase diagram of the Si-O_2 system, where the composition is that of all components (silicon and oxygen)

was found to be a mixture of O_2, O_3, SiO_2, Si_2, Si_3, SiO, Si_2O_2, and Si. From these results it is noticed that the dissociation of liquid SiO_2 to gases occurs before boiling temperature.

Figure 2 displays the equilibrium temperature- oxygen composition phase diagram of the Si-O_2 system in the temperature range 500–3000 K. Quartz with silicon as a stable phase can be found in the temperature range 500–1150 K, and oxygen mole percent 66.7%, which is the stoichiometric ratio of oxygen to silicon in silica. Also quartz can be found with gas in a stable phase in the same temperature range and oxygen mole percent 66.7–100%. Tridymite can be found with silicon as a stable phase in the temperature range 1150–1700 K and oxygen mole percent 66.7%, but mixed with gas in the temperature range 1150–1750 K and oxygen mole percent 66.7–100%. After the silica starts to melt for T > 1700 K, ionic liquid appears with tridymite as a stable phase in a short temperature range 1700–1750 K and oxygen mole percent 66.7%. Then, cristobalite appears with ionic liquid as a stable phase in the temperature range 1750–2000 K for oxygen mole percent 66.7%, and with gas as a stable phase in the same temperature range and oxygen mole percent 66.7–100%. For T = 2000–2150 K there is no cristobalite, but the ionic liquid is the stable phase in the range of oxygen mole percent 66.7%. For T = 2150–3000 K with oxygen mole percent 66.7%, and 2000–3000 K with oxygen mole percent 66.7–100%, the ionic liquid mixed with gas becomes the

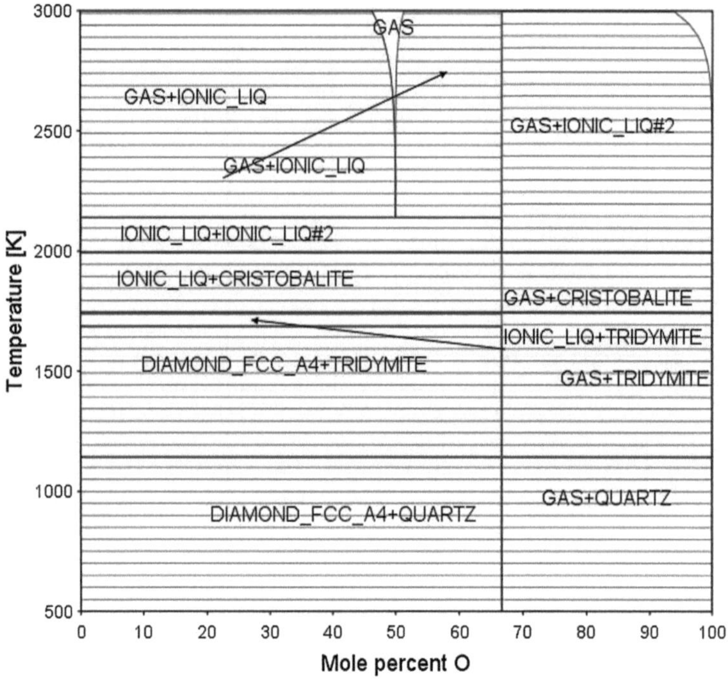

Fig. 2 Equilibrium temperature-oxygen composition phase diagram of Si-O_2 system

unique stable phase except in the upper zone in Fig. 1 at T = 2150 K and oxygen mole percent of 50% which extends to T = 3000 K and oxygen mole percent 46–52%, in which the stable phase is just a gas. The gas for all regions in Fig. 2 is the same which consists of O_2, O_3, SiO_2, SiO, Si_2, Si_3, Si_2O_2, and Si. It is noticed that these gases are formed before the melting of silica for oxygen mole percent 66.7–100%, but after it for oxygen mole percent of 66.7%.

The amounts of quartz were calculated (in mol percent) as functions of temperature and oxygen mole percent and displayed in Fig. 3, where Fig. 3a shows a heat map and Fig. 3b displays numerical values. From Fig. 3a it is apparent that the presence of quartz is restricted in the region T < 1150 K, and oxygen mole percent 5–95%, where the maximum amount of quartz is found in the interval of oxygen mole percent centered at 66.7%, then it decreases from both sides of the colored spectrum in an asymmetric way, and becomes 0 for oxygen mole percent <5%. From Fig. 3b the maximum amount of quartz is 0.819 which occurs at oxygen mole percent of 55% in the temperature range 500–1150 K. It decreases to 0.637 at oxygen mole percents of about 79 and 42%, then decreases to 0.091 at oxygen mole percents of about 5% and to 0.364 at oxygen mole percents of about 95% in the same temperature range.

Figure 4 displays the amounts of tridymite with temperature and oxygen mole percent, where Fig. 4a shows a heat map and Fig. 4b shows contours with

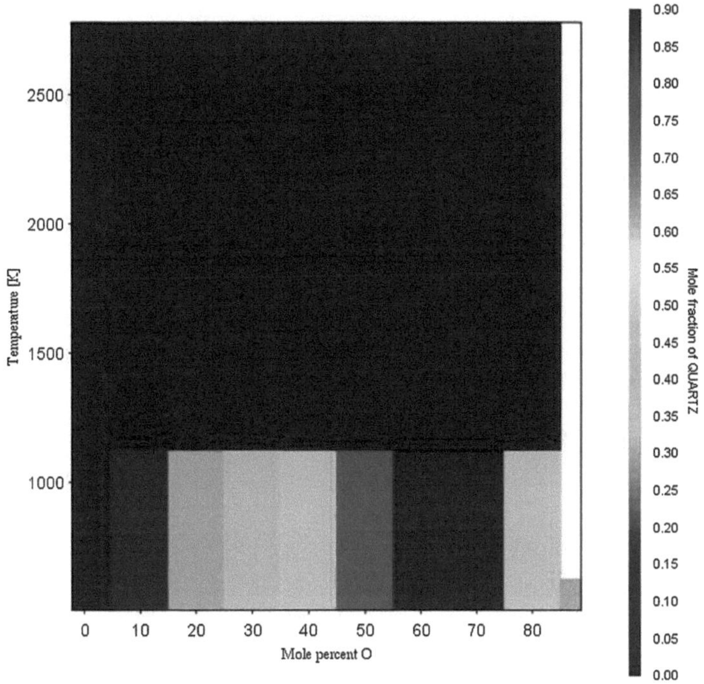

(a)

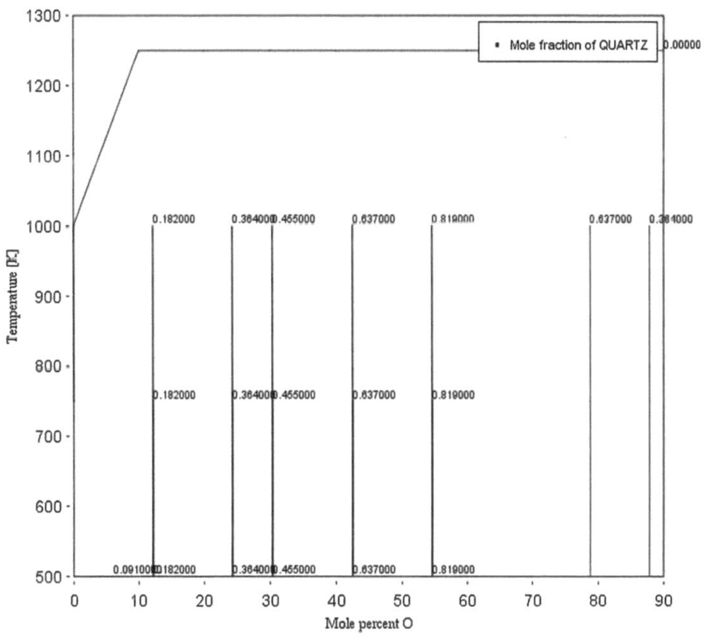

(b)

◂**Fig. 3** Mol percent of quartz with changes in oxygen mole percent and temperature. **a** Heat map. **b** Contour

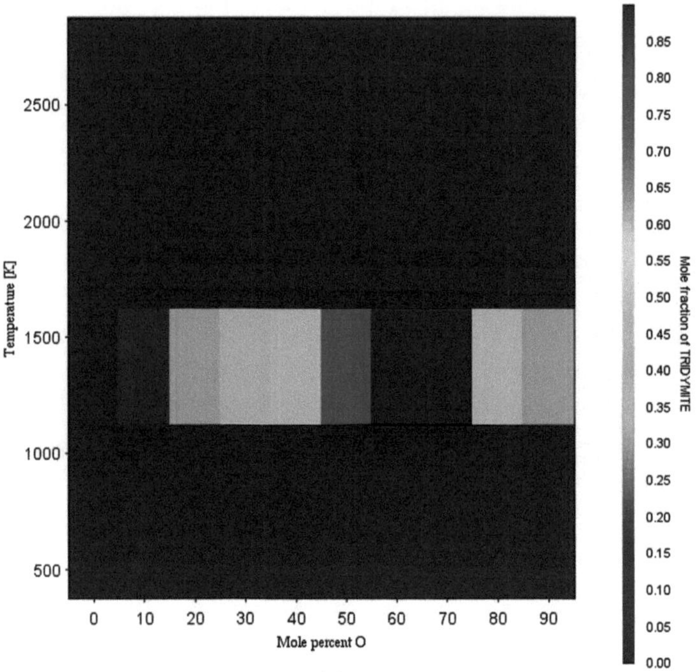

(a)

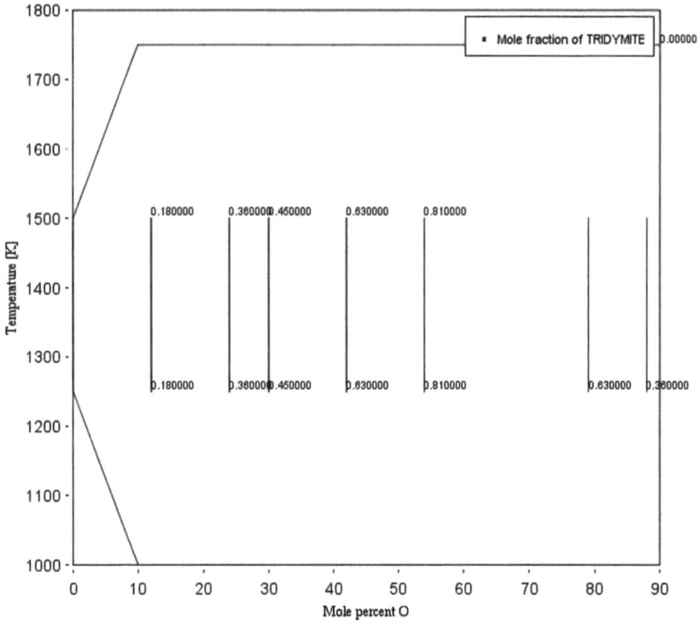

(b)

◂**Fig. 4** Amounts of tridymite with oxygen mass percent and temperature. **a** Hot map. **b** Contour

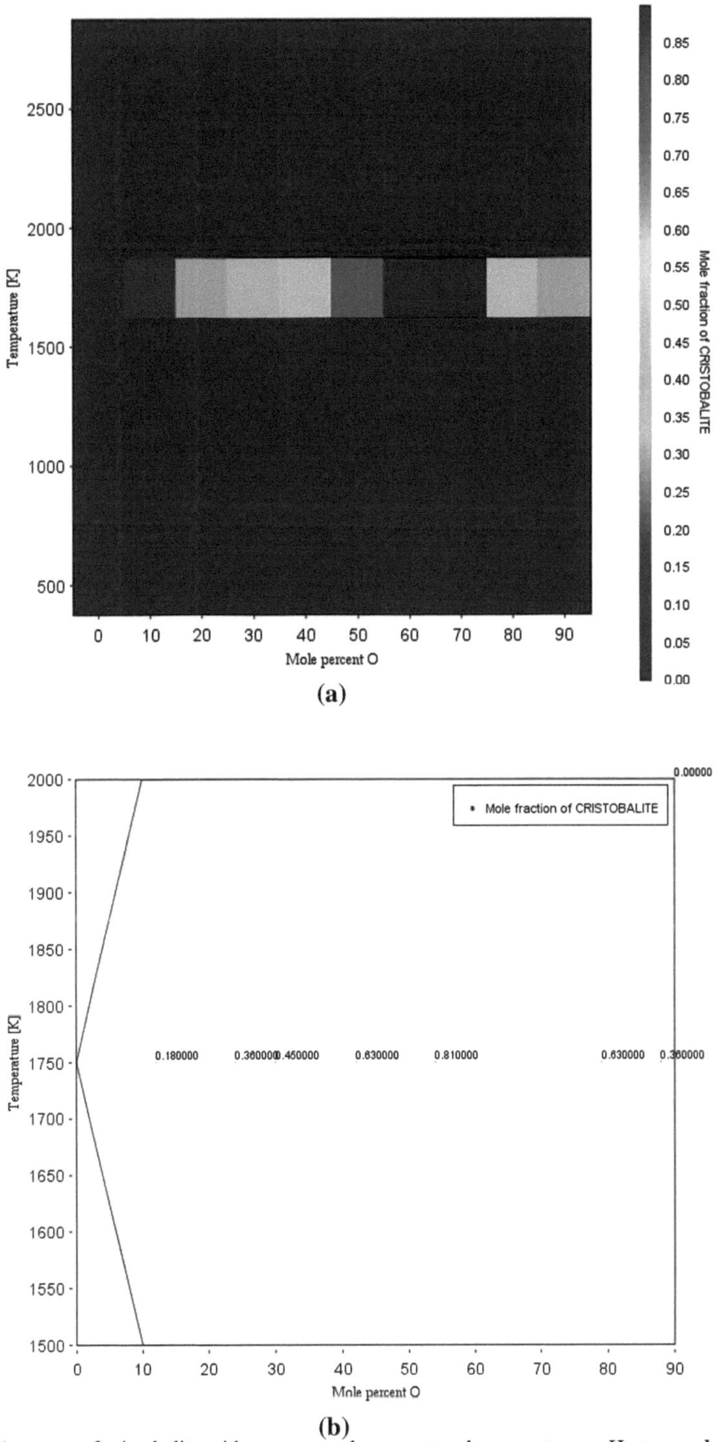

◀**Fig. 5** Amounts of cristobalite with oxygen mole percent and temperature. **a** Heat map. **b** Contour

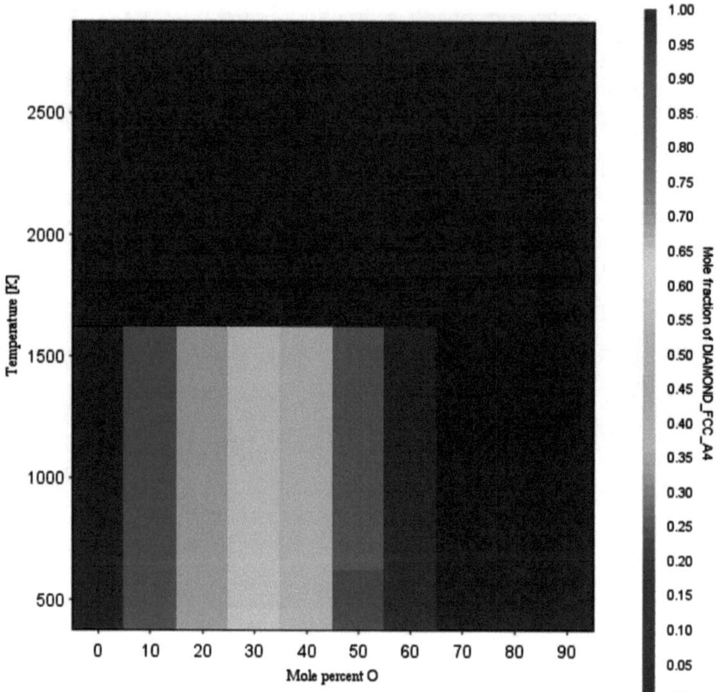

(a)

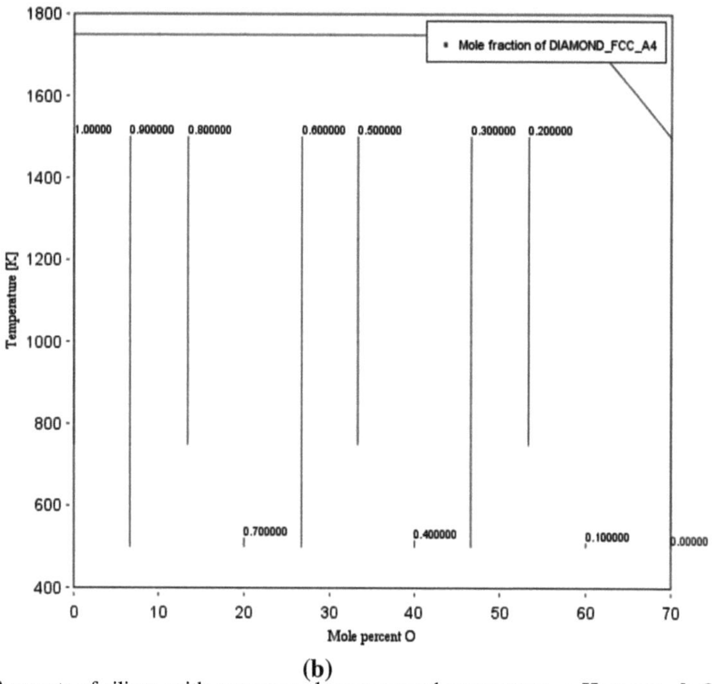

(b)

◂**Fig. 6** Amounts of silicon with oxygen mole percent and temperature. **a** Heat map. **b** Contour numerical values. Presence of tridymite is restricted in the temperature range T = 1150–1700 K, and oxygen mole percent range of 5–95%. Also the maximum amount of tridymite is found in the oxygen mole percent interval centered at 66.7%, and decreases in both directions to be zero for oxygen mole percent <5% and >95%. From Fig. 4b, the maximum mole fraction of tridymite is 0.81 in the oxygen mole percent interval centered at 67.7% in the temperature range 1250–1500 K, and at oxygen mole percent interval centered at 66.7%. It then decreases to 0.63 mol fraction in the oxygen mole percent 79–88%, and decreases to 0.19 mol fraction in the range 12–24% then to zero.

Figure 5 depicts the amounts of cristobalite with oxygen mole percent and temperature, where Fig. 5a shows a heat map and Fig. 4b shows contours with numerical values in mole fraction. From Fig. 5a, the maximum amount of cristobalite is also at the interval of oxygen mole percent centered at 66.7%. Presence of cristobalite is restricted in the region 1750–2000 K as seen in Fig. 5b. From Fig. 5b, the maximum mole fraction of cristobalite is 0.81 mol fraction at 1750 K in the same afore mentioned interval.

Figure 6 displays the same graphs for pure silicon, where silicon can be found from temperatures T = 500 to 1750 K, and from oxygen mole percent 0 to 66.7% as seen in Fig. 6a, b. The maximum amount of Si is when there is no oxygen, which is 1.0 mol fraction at T = 1500 K, and then it decreases gradually with oxygen mole percent until it becomes zero.

Conclusions

The equilibrium temperature-composition phase diagrams of the Si-O_2 system were calculated using Thermo-Calc software. From these phase diagrams the temperature intervals and oxygen mole percent intervals in which the stable phases exist were determined. Quartz can be found from 500 to 1150 K, while tridymite and cristobalite the polymorphs of silica can be found in the temperature intervals 1150–1700 K and 1750–2000 K respectively. The maximum amounts of these three minerals are found at oxygen mole percent of 66.7%. Pure silicon can be found in a high temperature range that extends from T = 500–1750 K. These results are important for determining the best conditions for producing silicon and silica for solar cells and other applications.

References

1. J. Götze, M. Plötze, D. Habermann, Origin, spectral characteristics and practical applications of the cathodoluminescence (CL) of quartz - a review. Mineral. Petrol. **71**, 225–250 (2001)
2. R. Haus, S. Prinz, C. Priess, Assessment of high purity quartz resources, in *Quartz: Deposits, Mineralogy and Analytics*, ed. by J. Götze, R. Möckel. Springer Geology (Springer, Berlin, 2012), pp. 29–51. doi:10.1007/978-3-642-22161-3_2

3. A.R. Oganov, M.J. Gillan, G. David Price, Structural stability of silica at high pressures and temperatures. Phys. Rev. B **71**, 064104 (2005)
4. H.-J. Blankenburg, J. Götze, H. Schulz, *Quarzrohstoffe* (Deutscher Verlag für Grundstoffindustrie, Leipzig-Stuttgart, 1994), p. 296
5. R. Haus, High demands on high purity. Ind. Miner. **10**, 62–67 (2005)
6. R. Haus, High-purity quartz resources, in *PHOTON's 8th Solar Silicon Conference*, Stuttgart (2010)
7. P. Moore, High purity quartz. Ind. Miner. **8**, 54–57 (2005)
8. E. Dal Martello, S. Bernardis, R.B Larsen, G. Tranell, M. Di Sabatino, L. Arnberg, Electrical fragmentation as a novel route for the refinement of quartz raw materials for trace mineral impurities, Powder Technol. **224**, 209–216 (2012)
9. E. Dal Martello, G. Tranell, S. Gaal, O. Raaness, K. Tang, L. Arnberg, Study of pellets and lumps as raw materials in silicon production. Metall. Mater. Trans. B **42**(5), 939–950 (2011)
10. T. Yamashita, K. Okuda, T. Obara, Application of Thermo-Calc to the development of high-performance steels. J. Phase Equilibria **20**(3), 231–237 (1999)
11. B.A. Boutwell, R.G. Thompson, N. Saunders, S.K. Mannan, J.J. deBarbadillo, Phase formation modeling of an alloy 706 casting using computational thermodynamics. Paper is presented at the 4th International symposium on superalloys 718, 625, 706 and derivatives, Pittsburgh, Pennsylvania, 15–18 June 1997
12. Introduction to Thermo-Calc, available on line at: http://www.thermocalc.com, 18 Sept 2016

Characterization of Composition, Morphology, and Structure of Disi Raw Sandstones in Jordan

Shadia J. Ikhmayies, Bothina M. Hamad, Abdulkader M. Abed, Belal S. Amireh and Yulia Valery Meteleva

Abstract This work is a first step on the way to characterize raw sandstone in Jordan, and to extract silica and silicon for the use in solar cells and other industries. Seven samples of raw sandstone were brought from Disi in Jordan. The samples were characterized by X-ray diffraction (XRD), scanning electron microscopy (SEM), energy dispersive X-ray spectroscopy (EDS), Fourier transform infrared spectroscopy (FTIR), and X-ray fluorescence (XRF). The XRD diffractograms showed that silica is the main constituent of the samples, and two samples contain Kaolinite $Al_2(Si_2O_5)(OH)_4$. EDS reports showed that all samples contain aluminum and carbon, but some of them contain calcium. XRF measurements revealed that some samples contain more than 97% silica, while one sample contains more than 90% silica, and the main impurity is Al_2O_3. FTIR revealed the known bands of silica, and this sandstone is free of organic compounds impurities.

Keywords Sandstone · Silica · Silicon · X-ray diffraction · X-ray fluorescence · Solar cells

---

S.J. Ikhmayies (✉)
Faculty of Science, Department of Physics, Al Isra University, Amman 11622, Jordan
e-mail: shadia_ikhmayies@yahoo.com

B.M. Hamad
Faculty of Science, Physics Department, University of Jordan, Amman 11942, Jordan
e-mail: b.hamad@ju.edu.jo

A.M. Abed · B.S. Amireh
Faculty of Science, Geology Department, University of Jordan, Amman 11942, Jordan
e-mail: aabed@ju.edu.jo

Y.V. Meteleva
Fokker Aerostructures B.V., Rotterdam, Netherlands
e-mail: yulia.meteleva.fischer@gmail.com

L. Zhang et al. (eds.), *Energy Technology 2017*,
The Minerals, Metals & Materials Series, DOI 10.1007/978-3-319-52192-3_34

Introduction

White silica sand deposits, from the Early Ordovician and Lower Cretaceous sandstone, are found exposed on the surface in southern Jordan. They were studied by several Jordanian geologists [1–4]. Geological maps at a scale of 1:50,000 covering most of the silica sand outcrops in Ras El Naqb, Qa' Ed Disi and Wadi Gharandal areas were published by the Jordanian Natural Resources Authority (NRA) [e.g. 5, 6]. The area of Qa' Ed Disa east of Quweira town, Wadi Ram, southern Jordan, comprises huge reserves of silica sand of Early Ordovician (50 km north-east of Aqaba city) (Fig. 1) [7, 8]. The exposures of the Disi sandstone formation are extended over an area of more than 150 km^2 with a thickness of about 265–350 m [4].

The white sandstones, cropping out in southern Jordan, have been technically studied starting in 1970 when Nimry and Haddadin studied the white sandstone outcrops of Ras El Naqb [4, 9]. In 1974, Swindell Dressler Company [10] carried out a study on the properties of these white sands. In 1984, Technostone [11] investigated the possibility of exploitation of glass sand deposits in the Ras En Naqb area. In 1995, Barjous [12] studied silica sand samples of Disi Sandstone formation from Ras En-Naqb and Wadi Es-Siq areas. There are seven mining

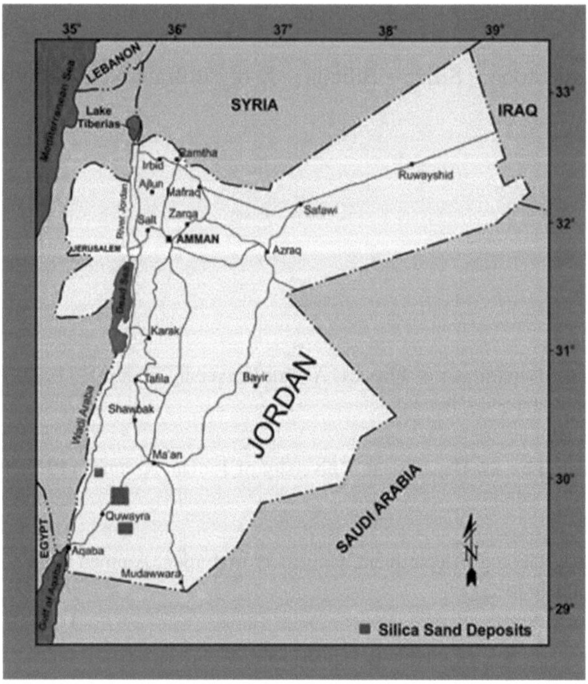

Fig. 1 Locations of some white sandstone in Jordan

companies in Jordan, who are currently working on sandstones, but only two are producing processed silica flour. Most of the products of these companies are size classification for sand blasting and construction material with little products for other industrial applications [1]. But there is no any company that produces silicon and/or silica for solar cell applications.

It is, thus, obvious that all studies on sandstones in Jordan are old and restricted in the geological part and interests of companies in some limited glass industries. That is, there is a lack in the scientific studies of the characteristics of sandstones as a resource of silica and silicon and for the use in solar cell applications. Comprehensive studies of sandstones of Jordan are necessary to start producing silica and silicon for the use in solar cells.

The objective of this work is to characterize raw sandstone of the Disi Formation in Jordan, as a first step in a comprehensive scientific research towards production of silica and silicon for the use in solar cells and different industries. It is expected that the Disi Formation outcrops in southern Jordan may be the preferable place for a silicon production plant in Jordan.

Materials and Methods

Seven sandstones samples were selected from the Disi Formation in southern Jordan. X-ray diffraction was used to explore the structure of the samples, using XRD-7000 X-Ray Diffractometer (SHIMADZU, Japan) in the 2θ mode, with X-ray Cu K_{α} line (λ = 1.54 A°). The morphology of the samples was determined by scanning electron microscopy (SEM) using a FEI scanning electron microscope (Inspect F 50), which was supplied by X-rays energy dispersive (EDS). The composition was analyzed by X-ray fluorescence (XRF) using a Panalytical Axios Max WD-XRF spectrometer, and data evaluation was done with SuperQ5.0i/Omnian software. Chemical bonding was examined by Fourier transform infrared (FTIR) measurements, using a Thermo Nicolet 380 FT-IR spectrometer (Thermo Electron Corp.).

Results and Discussion

The XRD diffractograms of the seven Disi raw sandstone samples are displayed in Fig. 2. The diffraction peaks of silica (quartz) are apparent, where the preferential orientation is the reflection from the (1 0 1) plane. All samples show the (0 0 1) diffraction peak of kaolinite $Al_2Si_2O_5(OH)_4$, where the intensity of this peak is the largest in sample (DS5). These diffractograms didn't show peaks of dickite as those obtained by Amireh [1].

The SEM photo micrographs of the seven samples, show at least two types of minerals (Fig. 3). Quartz particles are those that have distinct edges and corners, sometimes prismatic with two pyramids on both sides (Fig. 3b, f). The other type of

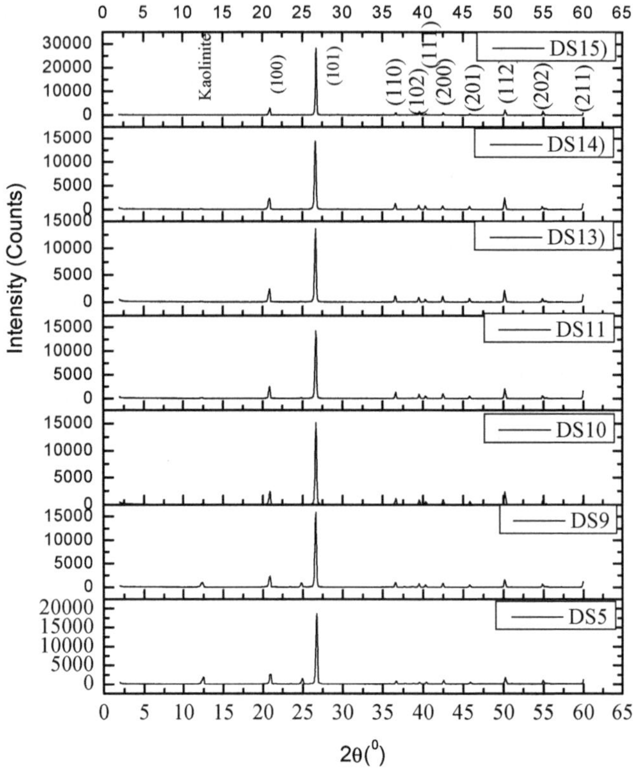

Fig. 2 XRD diffractograms of seven Disi raw sandstone samples, where D in the legend refers to Disi, S to sample, and the numbers are those of the samples

particles are those showing regions of booklets morphology, which are expected to be kaolinite (e.g. 3d). Rare dolomite rhombs are seen in Fig. 3g.

Figure 4 shows the EDS spectra of two of the samples. EDS analysis gave Si, O and Al along with trace amounts of C and Ca in the samples. Aluminum is incorporated into kaolinite ($Al_2Si_2O_5(OH)_4$), as seen in the XRD diffractograms.

Tables 1 and 2 list the XRF chemical analysis of two raw sandstone samples (DS9 and DS10). Table 1 shows that total silica in sample DS9 is 90.72%. If all the 8.4% alumina is assigned to kaolinite, then the free silica will be around 80%, while kaolinite is approaching the 20% of the sample. For sample DS10, Table 2 shows that total silica is 97.17%, with 2.22% alumina. Thus, there is about 93% free silica and around 6% kaolinite [13]. The other constituents form about 0.6% of the composition of this sample. TiO_2 and ZrO_2 are present mostly as the heavy minerals rutile and zircon respectively [14, 15]. Calcium and magnesium form part of the minor cementing material in the Disi sandstone [14, 15].

Both Tables 1 and 2 show Disi sandstones are free of boron, and it is well known that, for the use in solar cells, it is essential to have silica with low boron

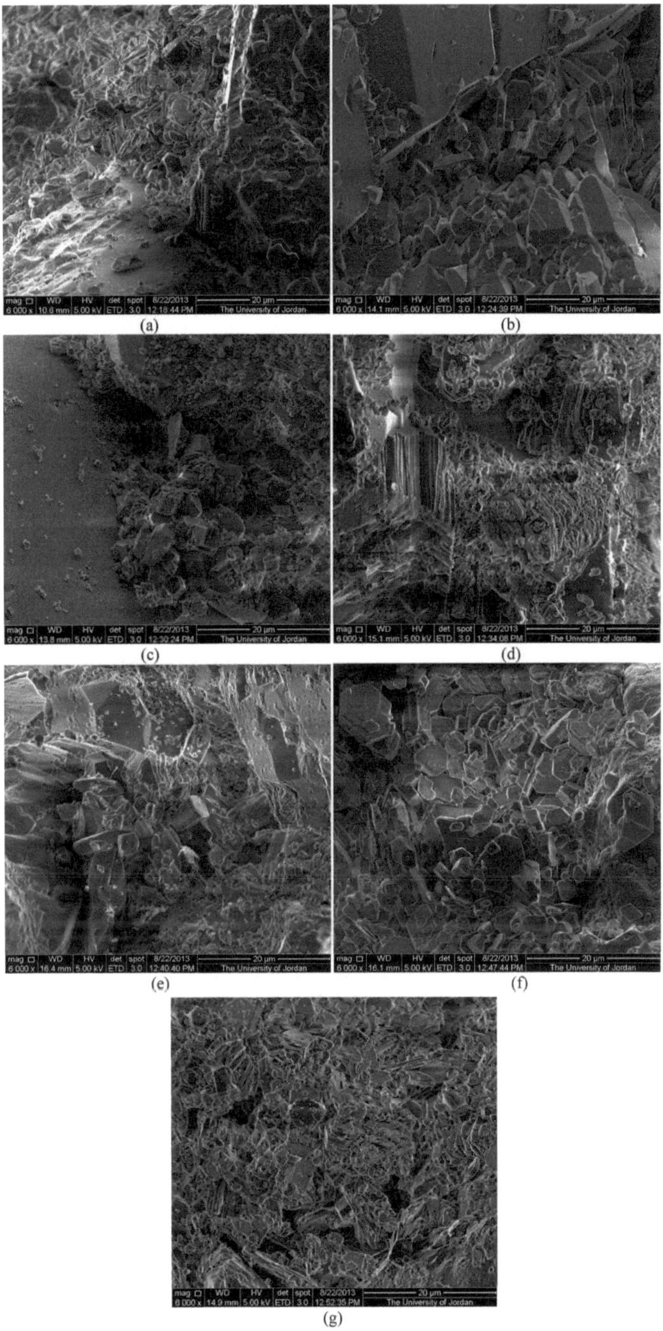

Fig. 3 SEM images of the Disi raw quartz. **a** DS5. **b** DS9. **c** DS10. **d** DS11. **e** DS13. **f** DS14. **g** DS15

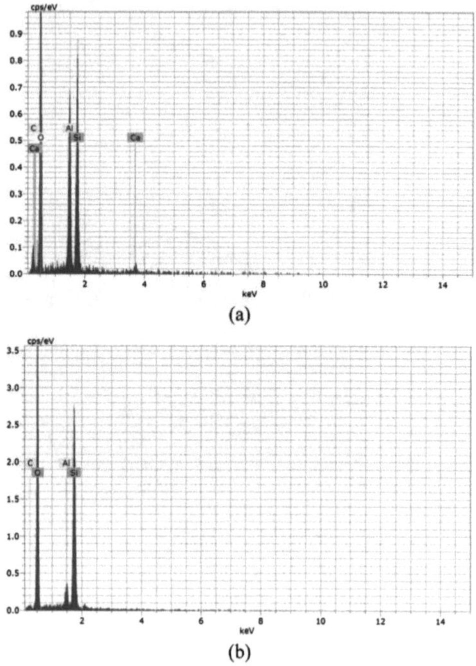

Fig. 4 EDS spectra for Disi raw sandstone samples. **a** Sample DS9. **b** Sample DS10

Table 1 XRF chemical analysis for sample DS9

Number	Compound name	Conc. wt (%)	Absolute error (%)
1	SiO_2	90.722	0.09
2	Al_2O_3	8.406	0.08
3	TiO_2	0.367	0.02
4	CaO	0.156	0.01
5	CeO_2	0.085	0.009
6	P_2O_5	0.065	0.008
7	Fe_2O_3	0.053	0.007
8	MgO	0.041	0.006
9	ZrO_2	0.037	0.006
10	SO_3	0.025	0.005
11	Cl	0.013	0.003
12	SrO	0.01	0.003
13	Na_2O	0.009	0.003
14	K_2O	0.008	0.003
15	Y_2O_3	0.002	0.001

Table 2 XRF chemical analysis for sample DS10

Number	Compound name	Conc. wt (%)	Absolute error (%)
1	SiO_2	97.172	0.08
2	Al_2O_3	2.215	0.04
3	TiO_2	0.215	0.01
4	CaO	0.149	0.01
5	CeO_2	0.036	0.006
6	P_2O_5	0.071	0.008
7	Fe_2O_3	0.039	0.006
8	MgO	0.018	0.004
9	ZrO_2	0.021	0.004
10	SO_3	0.028	0.005
11	Cl	0.009	0.003
12	SrO	0.013	0.003
13	Na_2O	0.008	0.003
14	K_2O	0.004	0.002

content, because it is difficult to remove the boron impurity from silicon. So, these results are encouraging for extracting solar grade silicon from Disi sandstone.

The chemical bonding of the samples was examined by Fourier Transform Infrared Spectroscopy (FTIR) in transmittance mode, and shown in Fig. 5. The vibrations at 1140, 1206 and 1231 cm^{-1} can be assigned to SiO_2 vibrations, and 768.7 can be assigned to Si–O stretching, and the peak at 626.7 can be assigned to Al_2O_3 vibrations. Si–O and Al–O bending modes dominate the 150–600 cm^{-1}. Lattice vibrational modes in the far-IR range (33–333 cm^{-1}) are related to the interlayer cations [16]. There are no interlayers in the kaolinites. Hence the main absorption band at 360 cm^{-1} can be assigned to Si–O bending, whereas the peaks at 536.7 and 476.7 cm^{-1} are due to O–Si–O stretching in silica and in kaolinite. Peaks at 231.3, and 265 cm^{-1} may represent other bending modes of bonds in kaolinite.

Conclusions

The number of samples investigated is too small to represent the around 300 m, Early Ordovician Disi Formation. However, this characterized is a first step towards producing silica and silicon for solar cell applications. XRD diffractograms showed that the samples are formed of silica and minor kaolinite. SEM images show the morphology of the samples, where the differences are due to impurities found in each sample. EDS analysis showed that the samples mainly contain silica, and impurities of carbon and/or calcium. XRF measurements revealed other impurities, which were not detected by EDS reports. FTIR spectra revealed the bonds in the

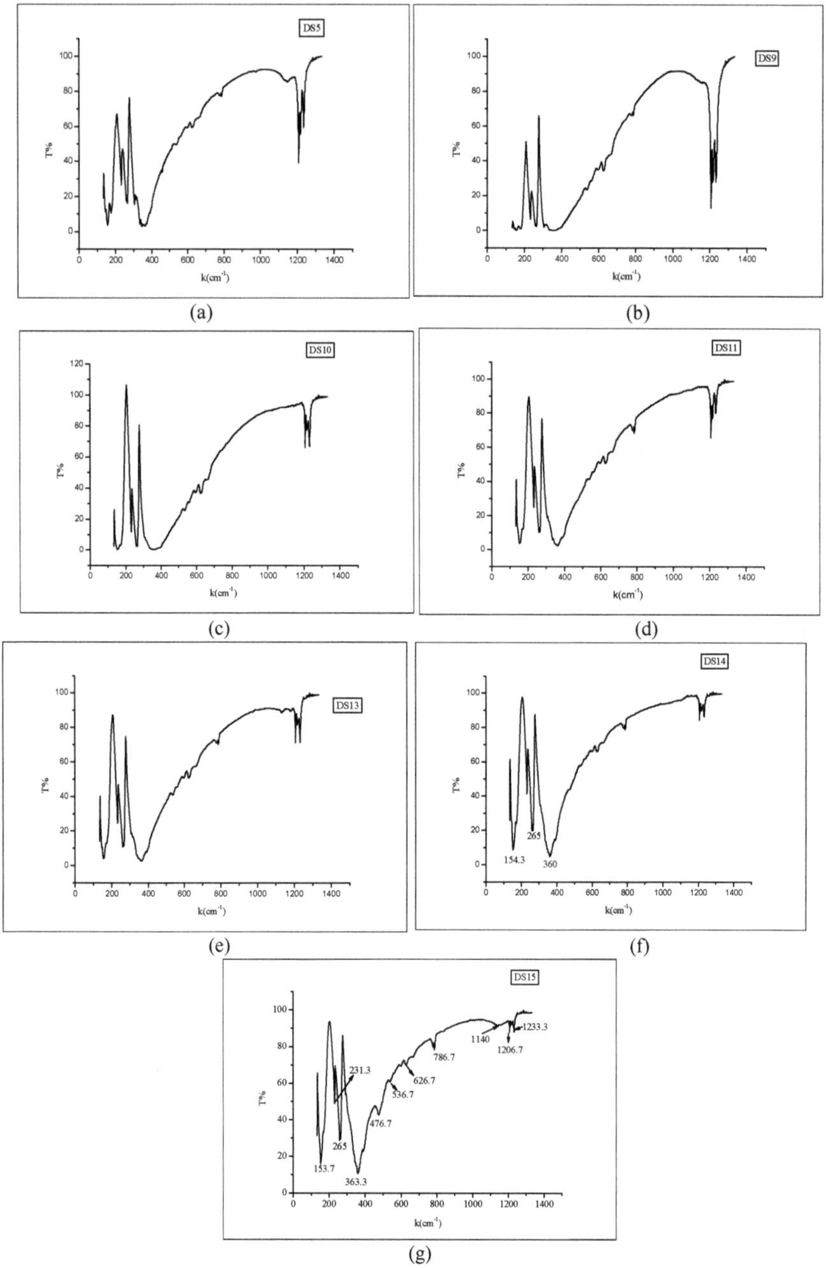

Fig. 5 The chemical bonding of the samples was examined by Fourier transform infrared spectroscopy (FTIR) in transmittance mode. **a** Sample 5. **b** Sample 9. **c** Sample 10. **d** Sample 11. **e** Sample 13. **f** Sample 14. **g** Sample 15

samples, where the spectra contain the main silica bands and some alumina bands. The samples are free of boron which is a good finding for solar cell silicon.

References

1. B.S. Amireh, Heavy and clay minerals as tools in solving stratigtaphic problems: a case study from the Disi sandstone (early ordovician) and the Kutnub sandstone (early cretaceous) of Jordan. Neues Jahrbuch fur Geologie und Palaontologie-Monatshefte **4**, 205–222 (1994)
2. M.M. Abu Ajamieh, *Mineral Resources of Jordan* (Natural Resources Authority, Amman, 1989), 149p
3. M. Abu Ajamieh, F. Bender, R. Eicher, *Natural Resources in Jordan* (Natural Resources Authority, Amman, 1988), 224p
4. Y. Nimry, M. Haddadin, Glass sand of Ras En Naqab. Report on Phase 1. NRA (1970)
5. B.M. Khalil, The geology of the Wadi Gharandal area, map sheet No. 3050 III. NRA. Mapping Project Bull. **24**, 80p (1993)
6. I. Rabba'a, *Al-Quwayra Sheet 3049 I. 1:50,000 Geological Map Series* (NRA, Amman, 1988)
7. F. Bender, *Geology of Jordan* (Gebrüder Borntraeger, Berlin, 1974), 196p
8. J. Powell, Stratigraphy and sedimentation of the Phanerozoic rocks in central and south Jordan, Part A: Ram and Khreim Groups. Geol. Map. Div. Bull. **11** (NRA, Amman, 1989), 72p
9. Silica Sand, Available at: http://emrc.gov.jo/images/silica_sand.pdf, 10 Sept 2016, 5p
10. Swindell Dressler Company, Glass, building material and component industry. Internal Report, NRA, Jordan (1974)
11. S.P.A. Technostone, *Final Report on Glass Sand Ras En Naqab Area* (Rep. Prepared for the Public Mining Co., Carrara, 1984)
12. M. Barjous, Mineralogy and processing of industrial minerals from Jordan. Ph.D. Thesis, Cardif University, U.K (1995)
13. C. Weaver, L. Pollard, *The Chemistry of Clay Minerals. Developments in Sedimentology*, vol. 15 (Elsevier, Amsterdam, 1975)
14. B.S. Amireh, New occurrences of the Disi sandstone formation (early ordovician) in central Jordan. Dirasat **20B**, 21–44 (1993)
15. B.S. Amireh, W. Schneider, A.M. Abed, Diagenesis and burial history of the Cambrian-Cretaceous sandstone series in Jordan. Neues Jb. Geol. Paläontol. Mh. **192**, 151–181 (1994)
16. L. Vaculíková, E. Plevová, Identification of clay minerals and micas in sedimentary rocks. Acta Geodyn. Geomater. **2**(2) (138), 167–175 (2005)

Part XI
Solar Cell Silicon: Silicon Impurity Removal and Refining

Effect of Magnesium Addition on Removal of Impurities from Silicon by Hydrometallurgical Treatment

Stine Espelien, Gabriella Tranell and Jafar Safarian

Abstract Hydrometallurgical treatment of silicon is an effective process for the removal of metallic impurities for producing solar grade silicon feedstock. In the present research, the removal of impurities from silicon is studied through the addition of Mg into a commercial silicon, its solidification, and acid leaching. The effect of concentration of HCl, temperature and leaching duration on the removal of impurities are studied. It is shown that the impurities such as P, Fe, Al, Ti, Ca,… are effectively removed through the applied method and Mg improves the segregation and the removal of the impurities. Scanning electron microscopy examination of the samples indicates the disintegration and dissolution of Mg_2Si phase in acid leaching which causes the chemical and physical removal of the impurities existing in the phases between the silicon grains. It is shown that the leaching process rate can be optimized through the control of the acid leaching conditions.

Keywords Silicon · Magnesium · Solar grade silicon · Acid leaching · Magnesium silicide

Introduction

The demand for solar energy has risen quickly in the last decades, and it will continue to rise due to the greenhouse warming and energy crisis. To produce solar cells, the main substrate materials is silicon, and the costs and environmental factors for the production of the silicon has an impact on making solar cells competitive against non renewable resourccs dominating the marked today. Today, the silicon for the photovoltaic (PV) industry mainly comes from the processes applied for production of electronic grade silicon (EG-Si) and the semiconductor industry. EG-Si is produced from metallurgical grade silicon (MG-Si) through the Siemens

S. Espelien · G. Tranell · J. Safarian (✉)
Norwegian University of Science and Technology, Alfred Getz Vei 2, 7491, Trondheim, Norway
e-mail: Jafar.Safarian@ntnu.no

L. Zhang et al. (eds.), *Energy Technology 2017*,
The Minerals, Metals & Materials Series, DOI 10.1007/978-3-319-52192-3_35

process developed in the 1950s, or the newly developed Fluid Bed Reactor (FBR) technology [1]. EG-Si is of ultra-high purity, usually above 9N (+99.9999999%), which is much higher purity than solar grade silicon (SoG-Si) of 6N purity, required in PV-devices. In the Siemens process, silicon is transformed to a silicon-containing gas, and the gas is then purified and reduced by hydrogen, where the silicon is deposited on silicon rods in bell jars. This is a very slow batch process, and it is very energy consuming to keep the bell jar hot. The production of toxic chlorine gases in the process, in combination with high-energy consumption and rather low yield; this process is not environmentally friendly nor energy efficient for producing SoG-Si [1, 2]. It would therefore be more convenient to use a metallurgical refining process, which has shown good results in different studies. Several metallurgical refining techniques can be used to refine silicon, and a combination of them yields SoG-Si. One of the key techniques in all integrated metallurgical processes is *directional solidification,* where many impurities are effectively removed from silicon except B (boron) and phosphorous (P). These two elements show small segregation between solid and liquid silicon during solidification and therefore other specific processes must be employed for the removal of these impurities. The main techniques are slag refining, vacuum refining, plasma refining, solvent refining and acid leaching.

Acid leaching as a method for silicon refining, like directional solidification, is based on the segregation of impurities between a solid silicon and a molten rich silicon phase. This leads to that the major portion of the impurities will precipitate at grain boundaries during solidification of a molten silicon doped by a refiner metal. In acid leaching the impurities located between the silicon grains and inside a silicide phase will be dissolve or physically separated, while the silicon matrix will not dissolve [2]. For production of SoG-Si it is important that the levels of the dopants B and P are not higher than 0.3 and 0.6 ppmw, respectively. For the metals the level should be 1–10 ppmw. As seen in Fig. 1, the segregation coefficients of B and P is relatively high compared to the metallic impurities. This leads to a lower concentration/less quantities of these impurities form at the grain boundaries, and cause small removal efficiency of these impurities by acid leaching compared to other elements. To lower the segregation coefficient, there is possibility to alloy the silicon with reactive alkaline metals like calcium [3–10], magnesium [2] or iron compounds [11]. Inoue [10] alloyed with calcium and leached with aqua regia, and the experiment showed a removal of boron up to 40% when 6.2% of calcium was contained in silicon. Recently, Safarian and Tranell [2] added Mg to molten silicon, and found that the best removal degree for boron was with a content greater than 2.1% Mg, and leaching with 15% HCl was effective for dissolving of the Mg_2Si particles. In other studies, where Ca was added into silicon as the refiner and further acid leaching was employed, it has been observed that leaching with HCl showed significantly better results than HNO_3, H_2SO_4 and their mixtures, while addition of HF, or another leaching step, before or after, with HF will improve the leaching effect on Fe–Si and Fe–Si–Ti phases [6, 8].

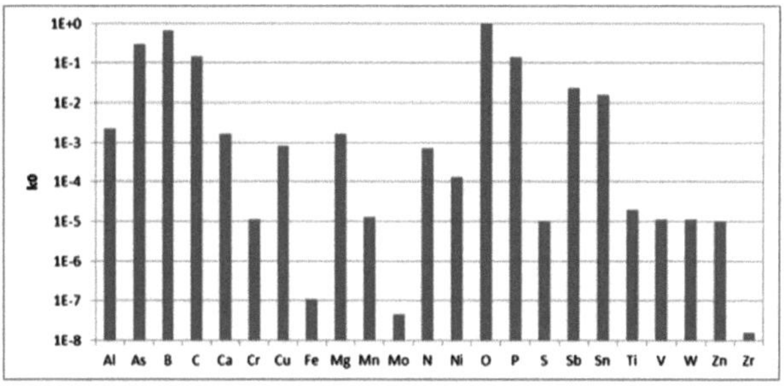

Fig. 1 Segregation coefficient of different elements between solid and liquid silicon

Table 1 The chemical composition (ppmw) of the Mg-doped silicon

Impurity	Mg	Fe	Al	Ca	Ti	B	P	Zr
Content	19424.9	1619.3	259.8	252.2	138.4	44.9	12.2	5.5

In this study magnesium is added into molten silicon as the refiner metal, and the effect of this element addition, and different leaching parameters for acid leaching purification of silicon is investigated. Leaching by different concentrations of HCl, using different temperatures, acid concentrations, and leaching durations are studied. The Mg-doped silicon is examined through an innovative electron microscopy-leaching study, and employing leaching of silicon particles.

Experimental Procedure

Experimental procedure employed in the present study is described as follows.

Silicon Material

A commercial metallurgical grade silicon (MG-Si) was doped by Mg through the addition of magnesium silicide (Mg_2Si) with purity above 99% in order to reach 2.0–3.0 wt% Mg in silicon. The mixture was then heated and smelted in a high-density graphite crucible at 1500 °C in an induction furnace for 30 min duration at this temperature. The molten Si–Mg alloy was then cooled down and solidified slowly. The solidified silicon was crushed, and the particles in size range 0.315–1.0 mm were separated. The average chemical compositions of the particles with this size are presented in Table 1.

Table 2 Applied conditions for the leaching trials, and the obtained silicon purities

Sample (#)	Temperature (°C)	Concentration (%)	Time (h)	Si concentration after leaching (%)	Overall purification — including Mg (%)	Overall purification — excluding Mg (%)
1	60	15	1	99.72	87.01	63.04
2	60	15	3	99.85	93.28	63.33
3	60	15	5	99.87	94.22	68.20
4	22	15	5	99.66	84.48	27.31
5	40	15	5	99.85	93.21	59.80
6	60	15	5	99.88	94.48	70.26
7	80	15	5	99.92	96.23	72.16
8	60	5	1	99.33	69.10	21.62
9	60	10	1	99.75	88.35	56.26
10	60	20	1	99.68	85.53	50.89
11	60	5	3	99.84	92.53	53.38
12	60	10	3	99.88	94.33	67.95
13	60	20	3	99.84	92.56	61.86
14	60	5	5	99.86	93.71	60.86
15	60	10	5	99.89	95.02	69.99
16	60	20	5	99.79	90.28	55.33

Acid Leaching

Around 2 g silicon samples were leached with hydrochloric acid, with different concentrations (5, 10, 15, 20% HCl), temperatures (22, 40, 60, and 80 °C) and leaching times (1, 3 and 5 h). The details of the leaching conditions are given in Table 2. Each leaching trial was performed by using maximum 20 ml of solution and mixing by a magnet stirrer with 60 rpm. The temperature was controlled through inserting a thermometer from the top into the solution with accuracy ±1 °C, while the required heat was supported from the container bottom. The samples before and after leaching were analyzed by high-resolution Inductively Coupled Plasma-Mass Spectroscopy (ICP-MS) technique. The analysis was done through dissolution of 20–40 mg silicon in HF + HNO_3 solution. The calibration of the results was done using blank samples and measuring the chemical composition of a reference standard silicon sample.

Leaching-SEM Study

Metallographic samples were prepared from Mg-doped silicon sample through mounting particles in resin and following a general metallographic procedure without etching. The applied method for studying the dissolution of the phases in leaching [2] was used in this study in which the combination of etching-SEM analysis was employed. In this case, the metallographic samples were studied by SEM, then they were suspended in acid (15% HCl) for one hour and the interacted silicon surface with acid was again studied by SEM. The leaching conditions were similar to the conditions used for the particles: both temperature and also magnetic stirring. It is worth noting that the sample was carbon coated to make a conductive surface for SEM study.

Results and Discussion

The results of leaching-microscopic studies of the bulk silicon samples and also the leaching of silicon particles are presented and discussed as follows.

Microstructural Analysis, Magnesium Addition Effect

Microscopic examination of the commercial received silicon indicated that it contains impurities and they are distributed non-continuously in grain boundaries of silicon grains. The SEM image of the MG-Si (before addition of Mg) is shown in Fig. 2a and for an enlarged area in Fig. 2b. The X-ray map of this area shows the distribution of the main impurities: Al, Ca, Fe, and Ti. The Fe and Al containing phase, which is the largest portion of the secondary phase, contains also Si and it shows that the main metallic impurities are appearing in silicide.

The microstructure of Mg-doped silicon is shown for comparison with the original silicon sample in Fig. 3. As we see the introduction of Mg yields a

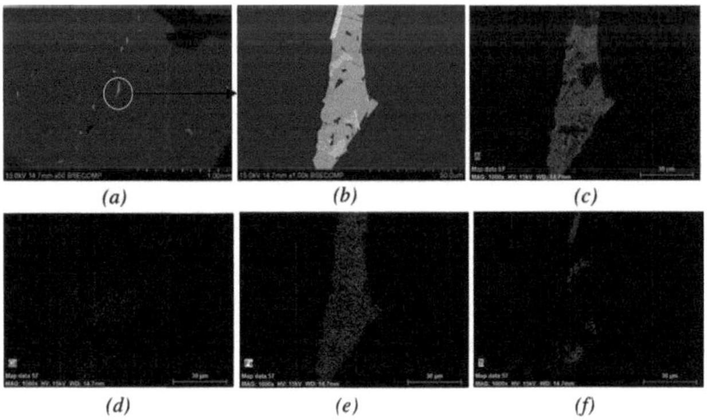

Fig. 2 **a**, **b** are BSE-images of MG-Si sample, and **c**–**f** are the corresponding X-ray mapping of the main impurities in the sample

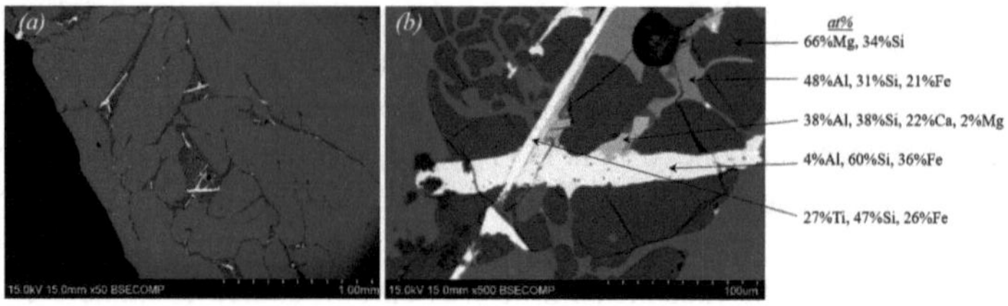

Fig. 3 The microstructure of MG–Si doped by 2 wt% Mg, the BSE images at two magnifications, and the EDS analysis results of the observed phases

microstructure in which the impurities appear in more continues areas between the silicon grains and a new secondary phase is appearing in the microstructure. According to the EDS analysis in Fig. 3, the Mg addition yield Mg_2Si phase, which is a little darker than silicon matrix in BSE images. Figure 3 indicates the existence of other small particles in the microstructure, which are mostly distributed next to or inside Mg_2Si phase. The EDS analysis of these phases shows Fe_xSi_y, $Al_xCa_ySi_z$, $Fe_xSi_yTi_z$ and $Al_xSi_yFe_z$-phases, which are obviously silicides of the existing impurities in silicon. The approximate quantitative EDS-analysis of these phases is shown in Fig. 3.

Leaching-SEM Study

The microstructural analysis of the Mg-doped silicon indicated that the Mg_2Si phase existing between the silicon grains is dissolved by HCl and after 1 h leaching, it is completely removed. This is clearly observed through comparing the SEM images before and after leaching in a specific area as illustrated in Fig. 4. This figure shows also that the iron and titanium containing silicide phases are still present, and there is just a small change between the images before and after etching. This observation can also be seen from the mapping in Fig. 5. In comparison between Fig. 5a and b it can be seen that after leaching the Mg_2Si-phase is completely removed, except from some spots that probably are contamination. The Ca and Al-phases are also well removed, but there are some spots that probably contain some iron. As it can be seen, both Fe and Ti-containing silicide phases are removed in small quantities. Obviously, these phases are not dissolved by HCl, and therefore their removal is most likely be physical removal. These silicide particles are well locked between two silicon grains as seen in images in Fig. 4 and their removal may need crushing of silicon sample, which is a case for the silicon subjects to leaching. We also may come to the conclusion that if more Mg is added into the silicon, it causes the creation of larger Mg_2Si phase and therefore the Fe–Ti silicide particles may be more likely formed in the Mg_2Si phase and therefore their separation due to Mg_2Si solution will be better.

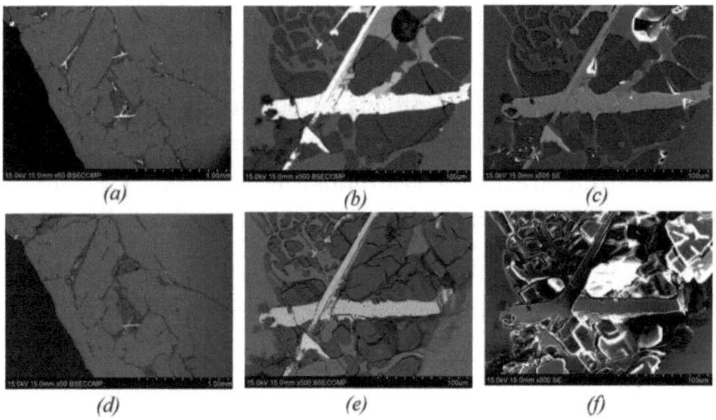

Fig. 4 Images **a–c** show microstructure in BSE and SE modes before etching, while images **d–f** show the corresponding images after etching with 15% HCl for 60 min, respectively

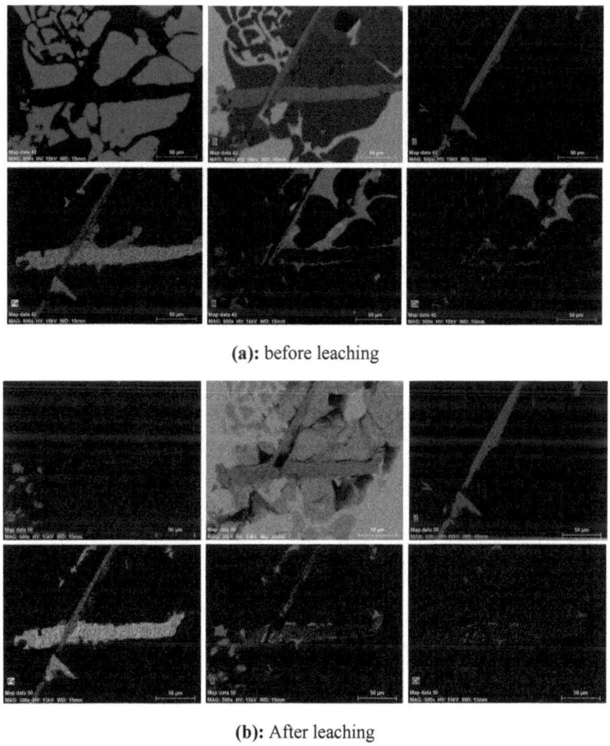

Fig. 5 X-ray mapping of area in images b and c in Fig. 4, before leaching and after leaching

Effect of Leaching Conditions on Silicon Purification

In the present work we studied the effects of acid concentration, temperature, and leaching time on purification of silicon. The calcium phases are also well removed, and there are just some traces in combination with Al. For simplification, the removal degree of each component was calculated by the following equation:

$$Removal\ degree\ (\%) = \frac{C_i - C_f}{C_i} \times 100\% \tag{1}$$

where C_i and C_f are the concentrations of impurity i in silicon before and after leaching, respectively. The effect of leaching parameters on silicon purification is presented as follows.

Effect of Acid Concentration

As mentioned before four different acid concentrations were tested in the present work: 5, 10, 15 and 20% HCl. The best removal results were obtained with the use of 10 and 15% HCl for almost all the impurities as seen in Fig. 6 for a given leaching time. As we observe, all the impurities are better removed when a 10% HCl is used, except for boron and phosphorous. With use of a concentration of 15% the results are slightly worse than for 10%, but using 15% HCl is significantly better than for 5%, and some better than for 20%. For 1 and 3 h leaching time, the results

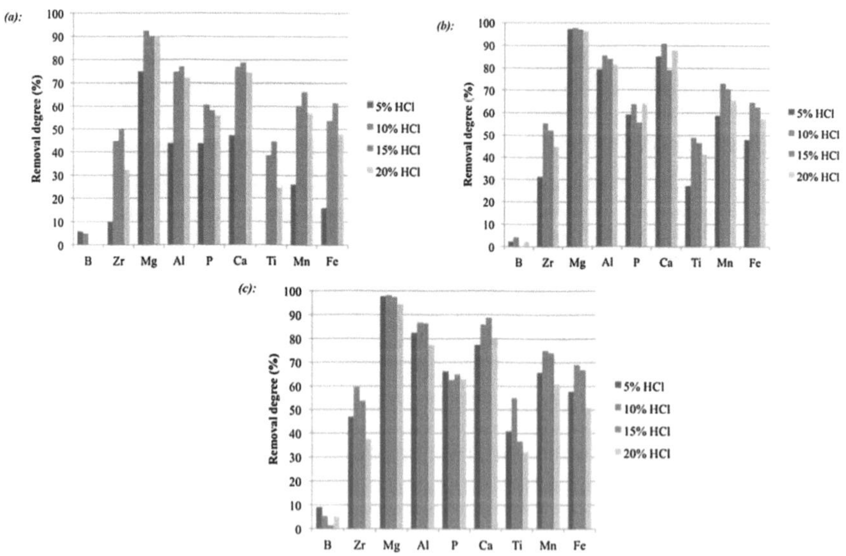

Fig. 6 The removal degree of impurities after **a** 1, **b** 3 and **c** 5 h leaching time

were better for 20% than for 5%, while for 5 h leaching time, the 5% was better in all elements, however, there are some uncertainties about Ca.

Effect of Temperature

Experiments 4–7 were run at different temperatures (Table 2) using 15% HCl and leaching for 5 h. The obtained results about the removal degree of the impurities are shown in Fig. 8. As we see the purification is better at higher temperatures, however, the removal degree is minimally increased with increasing temperatures above 40 °C. The difference between 40 and 22 °C is significantly higher, where it can be seen from Fig. 7 that the plots are quite steep in this area for all impurities. Overall, the best results are obtained for 80 °C, except for titanium, where 60 °C is slightly better (3%), but this may be related to errors in the analysis.

Effect of Leaching Time

There is a noticeable improvement when the leaching time is increased (Fig. 8). The improvement is significantly better when increasing from 1 to 3 h, than for 3 to 5 h: 6 and 1%, respectively, when looking at the overall purification in Table 2. For 20% HCl solution, the removal efficiency was actually observed to go down when leaching for 5 h compared to 3 h.

Overall Purification

The concentration of silicon and the overall purification of the samples after the different leaching trials are shown in Table 2 in Sect. 2.2. Here it can be seen that the highest silicon concentration and the best overall purification degree is obtained for sample 7. This sample was leached with 15% HCl, at 80 °C and for 5 h, and yielded a concentration of 99.92% Si, an overall purification degree including Mg

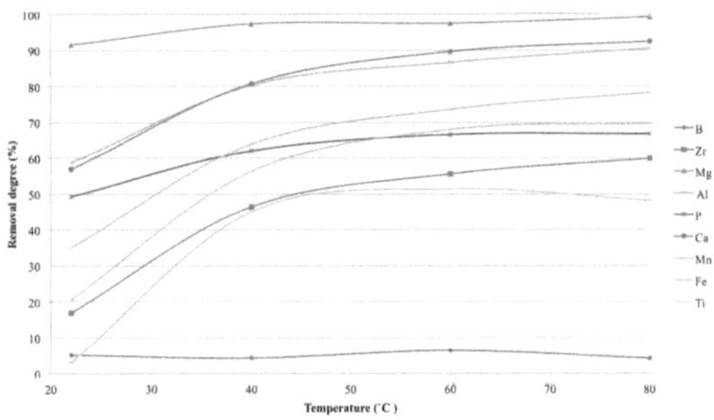

Fig. 7 The removal degree of impurities at different leaching temperatures

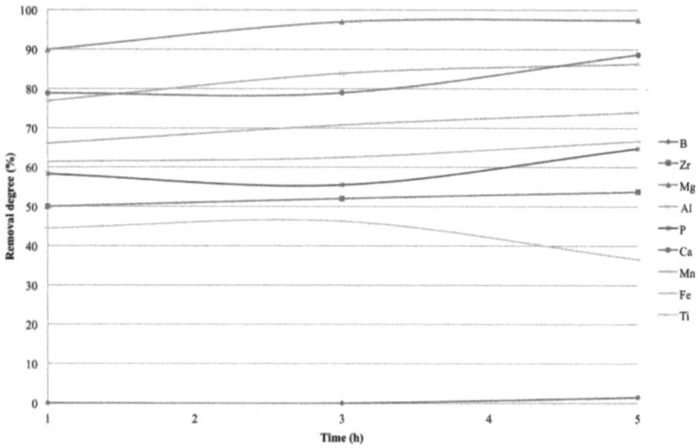

Fig. 8 The removal degree of elements in different leaching times

of 96.23%, and an overall purification degree excluding Mg of 72.16%. The overall purification was calculated from the equation below:

$$Overall\ purification\ (\%) = \frac{\sum C_i - \sum C_f}{\sum C_i} \times 100\% \qquad (2)$$

where C_i and C_f are the concentrations of impurity i in silicon before and after leaching, respectively. Since the removal degree for Mg is quite high, the overall purification without Mg is significantly lower. It can also be seen that the removal degree of the other impurities has some higher temperature dependence above 40 °C than Mg. If we look at the overall purification including Mg, there is a minimal increasing in removal above 40 °C, while excluding Mg the removal degree is also increasing some more up to 60 °C. In comparison, the difference in overall purification from 40 to 60 °C is 1% including Mg, and 10% excluding Mg. Regardless the difference is still much higher from 22 to 40 °C. The effect of concentration and time is quite the same for both overall purification calculations.

Conclusions

A commercial silicon was doped by around 2% Mg and its behaviour was studied through employing leaching-SEM analysis, and leaching of small particles under different conditions in a laboratory leaching bath. The following conclusions have been obtained in this study:

1. The impurities existing in silicon are distributed between the silicon grains in the forms of complex silicides such as Fe_xSi_y, $Al_xCa_ySi_z$, $Fe_xSi_yTi_z$ and $Al_xSi_yFe_z$. After doping with Mg, the complex silicides were observed being distributed next to or inside the formed Mg_2Si phase.

2. The Mg_2Si phase distributed between the Si grains is completely removed by leaching, and it causes removal of $Al_xCa_ySi_z$ intermetallic phases, and also physical separation of other intermetallic phases.
3. The silicon purification degree is increased with increasing temperature up to 40 °C, and it is increased minimally with applying higher temperatures.
4. The 10% HCl solution provided better refining than the lower and higher concentrations.
5. The purification degree is increased with the leaching time, however, the purification was not significantly higher in longer times than 1 h.
6. The best result was obtained for the sample leached with a concentration of 15% HCl, a temperature of 80 °C and a leaching time of 5 h. This sample yielded a Si concentration of 99.92% and an overall purification degree of 96.23% including Mg, and an overall purification degree of 72.16% excluding Mg.

Acknowledgements The authors appreciate the financial support of this study through the Research Domain 3-refining and recycling of SFI-metal production (Norwegian acronym for Centre for Research driven Innovation (CRI)).

References

1. J. Safarian, G. Tranell, M. Tangstad, Processes for upgrading metallurgical grade silicon to solar grade silicon. Energy Procedia **20**, 88–97 (2012)
2. J. Safarian, G. Tranell, Silicon purification through magnesium addition and acid leaching, 1011–1014 (2016)
3. F. He, S. Zheng, C. Chen, The effect of calcium oxide addition on the removal of metal impurities from metallurgical-grade silicon by acid leaching. Metall. Mater. Trans. B, **43b**, 1011–1018 (2012)
4. T. Shimpo, T. Yoshikawa, K. Morita, Thermodynamic study of the effect of calcium on removal of phosphorus from molten silicon by acid leaching treatment. Metall. Mater. Trans. B **35B**, 277–284 (2004)
5. Y. Zhan-liang, M. Wen-hiu, D. Yong-nian et al., Removal of iron and aluminium impurities from metallurgical grade-silicon with hydrometallurgical route. Trans. Nonferrous Met. Soc. China **17**, 1030–1033 (2007)
6. I.C. Santos, A.P. Gonzales, C.S. Santos et al., Purification of metallurgical grade silicon by acid leaching. Hydrometallurgy **23**, 237–246 (1990)
7. J. Dietl, Hydrometallurgical purification of metallurgical grade silicon. Solar cells **10**, 145–154 (1983)
8. Z. Jian, L. Tingju, M. Xiaodong, et.al., Optimization of the acid leaching process by using ultrasonic field for metallurgical grade silicon, **30**, 053002-1 (2009)
9. J.M. Juneja, T.K. Mukherjee, A study of the purification of metallurgical grade silicon. Hydrometallurgy **16**, 65–75 (1986)
10. G. Inoue, Effect of calcium on thermodynamic properties of boron in molten silicon. High Temp. Mater. Processes (London) **22**, 21–26 (2003)
11. S. Esfahani, Solvent refining of metallurgical grade silicon using iron. Master thesis, University of Toronto (2010), pp. 13–14, 26–43

Evaporation Removal of Boron in Molten Silicon Using Reactive Fluxes

Ye Wang and Kazuki Morita

Abstract In order to optimize the B removal during slag-refining process for solar-grade silicon production, a new process for the removal of B in molten Si was proposed based on the principle of oxidized chlorination and evaporation. B can be generated and evaporated in the form of B-containing gas (BOCl) from molten Si using $CaO–SiO_2–CaCl_2$ slag system. Compared with the binary systems of $CaO–CaCl_2$ and $SiO_2–CaCl_2$ slag, the ternary slag system $CaO–SiO_2–CaCl_2$ showed a better potential (90 pct) for B removal at 1723 K for 2 h. Moreover, the diffusion coefficient (D) of B in the slag was also determined using tube-molten pool method. Through this, it was found that the rate-limiting step of B-removal in the slag refining process is controlled by B transfer in the slag interface and surface.

Keywords Molten silicon · Slag refining · Boron removal · Diffusion coefficient

Introduction

In the photovoltaic industry, the purification of silicon through metallurgical treatments may be a promising way to produce solar grade silicon (6N) from metallurgical grade silicon (3–4N) [1]. Among the metallurgical treatments, the slag method [2] was used to remove boron, which is difficult to eliminate due to its high segregation coefficient ($L_B = 0.8$) and low partial pressure [3]. It has been proved that the molten slag based on the $CaO–SiO_2$ binary system is an efficient agent for boron removal from molten metallurgical grade silicon (MG-Si). Several slag systems including $CaO–SiO_2$ [4–6], $CaO–SiO_2–CaF_2$ [7], $CaO–SiO_2–Al_2O_3$ [8],

Y. Wang (✉)
Department of Metallurgical Engineering, School of Chemical Engineering,
Sichuan University, Chengdu, China
e-mail: wangye@scu.edu.cn

K. Morita
Department of Materials Engineering, The University of Tokyo, Bunkyo-ku, Hongo,
Tokyo City, Japan

L. Zhang et al. (eds.), *Energy Technology 2017*,
The Minerals, Metals & Materials Series, DOI 10.1007/978-3-319-52192-3_36

$CaO–SiO_2–MgO$, $CaO–Li_2O–SiO_2$ [9], $CaO–SiO_2–Na_2O$ [6, 10] and $CaO–SiO_2–Al_2O_3–CaF_2$ [11] have been investigated; however, they all exhibit limiting boron removal efficiency.

In this study, for the purpose of removing B from metallurgical grade Si by reactive fluxes, refining experiments of Si by chlorination with a $CaCl_2–CaO–SiO_2$ slag were carried out, and expanded to large-scale tests of Si refining in a vacuum furnace. Furthermore, the kinetics mechanism and rate-limiting step was also investigated by measuring the diffusion coefficient of B in molten $CaO–SiO_2–CaCl_2$ slag at 1723 K.

Experiment

B Removal Experiment from Molten Si Using $CaO–SiO_2–CaCl_2$ Slag System

5 g of pure Si (10N) containing 100 ppmw of B and various mole ratios of $CaO–SiO_2–CaCl_2$ slags were pre-melted at 1723 K under an Ar atmosphere in an induction furnace. Second, the pre-melted Si (B $\approx$ 100 ppmw) and 10 g of slag were placed in a graphite crucible and held for 2 h at 1723 K in a SiC resistance furnace under an Ar atmosphere (Fig. 1). In order to ascertain whether this performance held true at various scales, 500 g of silicon refining tests using 1 kg of $CaO–SiO_2–CaCl_2$ slag were carried out in the vacuum resistance furnace (VRF) under low vacuum, the photo of sample after refining was shown in Fig. 2. The real metallurgical-grade Si with 20 ppmw B was used, rather than doped pure Si. The B concentration of the slag and Si were then measured using inductively coupled plasma atomic emission spectroscopy (ICP-AES).

Fig. 1 Experimental set-up

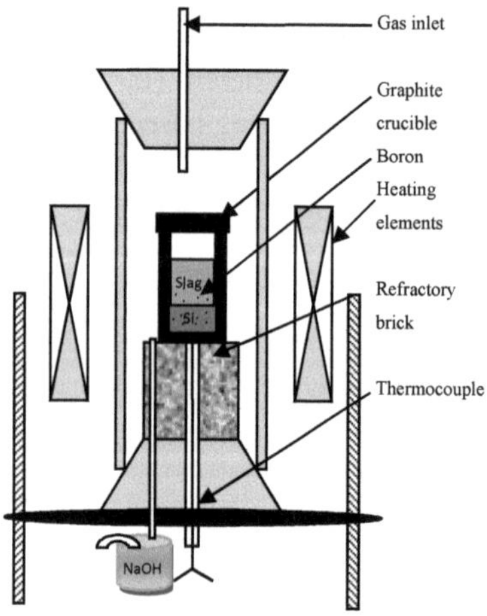

Fig. 2 Photo of large sample

Measuring the Diffusion Coefficient of B in the Slag

To measure the diffusion coefficient of B in molten slag, a graphite tube (length = 20 mm, width = 15 mm, height = 12 mm) was prepared with holes (diameter = 1.5 mm). Next, 20 g of 30 mol% CaO-23 mol% SiO_2-47 mol% $CaCl_2$ slag system compositions were mixed in a graphite crucible and pre-melted in an induction furnace under vacuum, into which the graphite tube was immersed for 10 min. A flow of Ar gas was then introduced to force the slag into the vacuum tube, which was finally lifted up and furnace cooled to produce a slag-filled tube.

Diffusion coefficient experiments were carried out by first melting 30 g of the same slag composition with 100 ppmw of B in a resistance furnace at 1723 K for 5 min, into which the slag-filled tube was immersed for 20 min. After the tube was removed, it was cut into portions (Fig. 3), and the slag was collected from the hole. The B concentration (C_B) of each portion (l = 2 mm) of the tube was analyzed by ICP-AES.

Results and Discussion

B Removal Experiment from Molten Si Using CaO–SiO_2–$CaCl_2$ Slag System

Since boron oxychlorides with a high vapor pressure, such as BOCl, will tend to evaporate from a molten slag, the B in molten Si can be removed by an oxidation, chlorination and evaporation process. The mechanism of B removal was shown in Fig. 4. The overall reaction for this B removal by slag treatment can be represented as follows [12]:

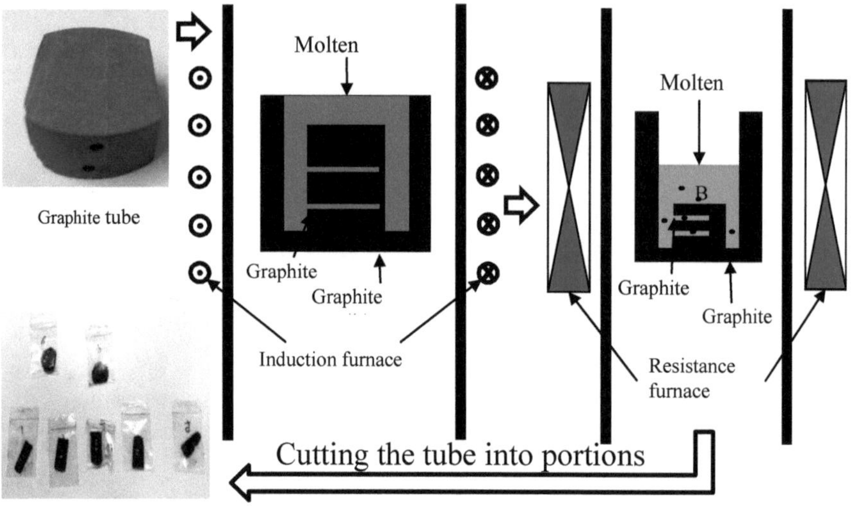

Fig. 3 Schematic of the diffusion coefficient experiment

Fig. 4 Concept of B removal with reactive fluxes

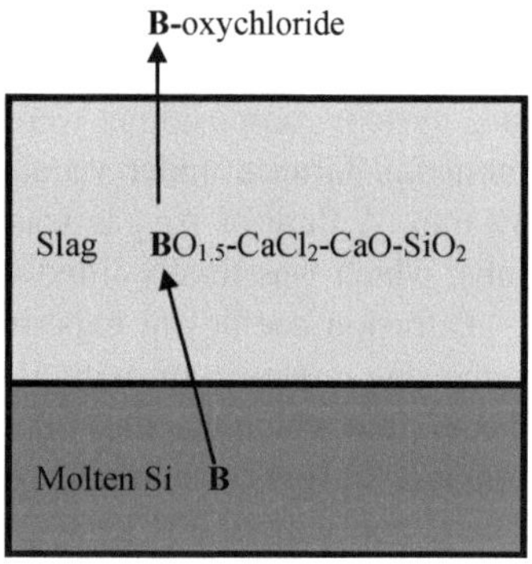

$$\begin{aligned} &B(1) + 0.75O_2 + 0.5CaCl_2(1) \rightarrow BOCl(g) + 0.5CaO(s) \\ &\Delta G^0 = -272 + 0.029T\,kJ/mol \end{aligned} \quad (1)$$

From the melting of various slags with B-containing Si, the B removal efficiency (R.E.) was evaluated as $1 - m_{B\ \mathrm{final\ in\ Si}}/m_{B\ \mathrm{initial}}$, and the evaporative efficiency (E.E.) of B was estimated as $1 - (m_{B\ \mathrm{final\ in\ Si}} + m_{B\ \mathrm{final\ in\ slag}})/m_{B\ \mathrm{initial}}$. The basicity of each slag was defined as the ratio by weight between CaO and SiO_2 (w_{CaO}/w_{SiO_2}). L_B was defined as the mass ratio between B in the slag and that in the Si.

Figures 5 and 6 showed the relationship between L_B, removal efficiency (*R.E.*) and evaporation efficiency (*E.E*). It indicated that, as the L_B values of CaO–SiO_2–$CaCl_2$ slag increased, the *R.E.* also increased to 90%. Furthermore, in the L_B values, the R.E. in the larger scale was larger than that in the Lab scale due to convection

Fig. 5 Relationship between L_B and basicity

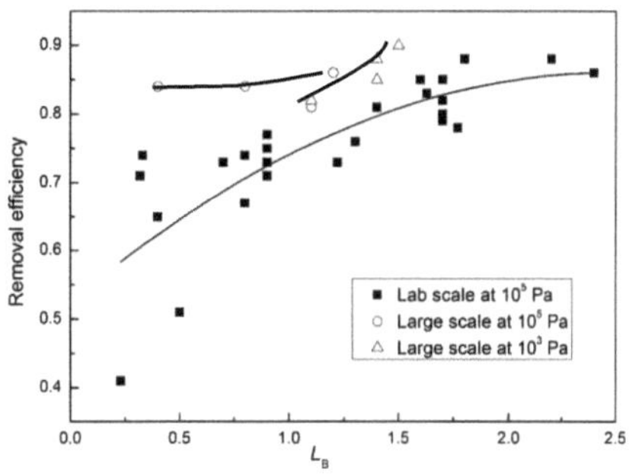

Fig. 6 Relationship between E.E. and R.E

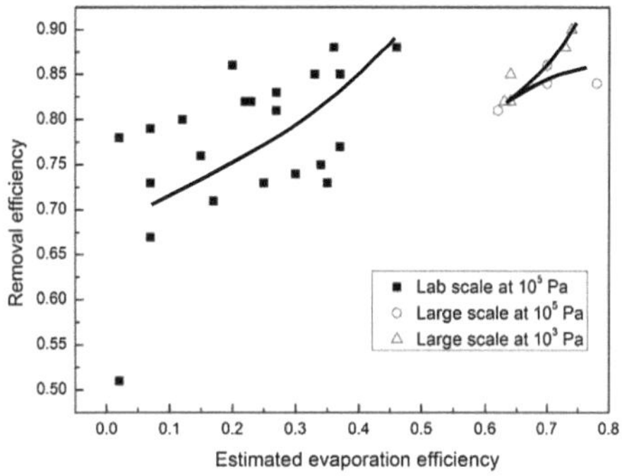

Fig. 7 Relationship between L_B and basicity [4]

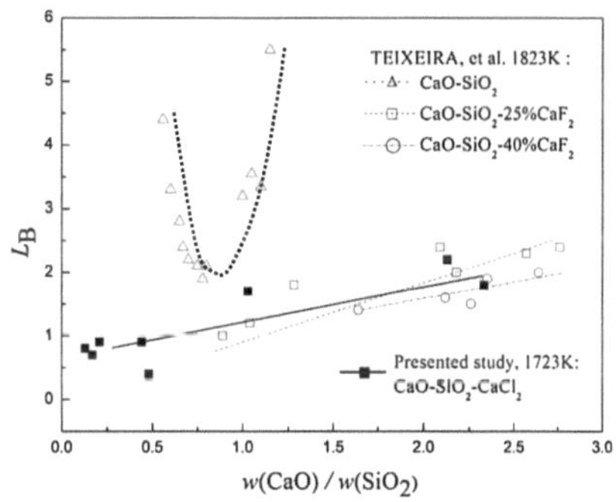

and stirring in the induction furnace. In the same pressure (10^5 Pa), the E.E. of large scale was larger than that of in the Lab scale due to large surface area. These two figures also showed that at 10^3 Pa, the E.E of large scale was higher than that at 10^5 Pa, as it was easier to vaporize BOCl gas. Figure 7 showed the relationship between L_B and the basicity (w_{CaO}/w_{SiO_2}) of a $CaO–SiO_2–CaCl_2$ slag, with the results of other slag systems shown for comparison. This indicated that with an increase in the basicity value of a $CaO–SiO_2–CaCl_2$ slag, the L_B value was also increased.

Measuring the Diffusion Coefficient of B in the Slag

Figure 8 showed the B distributions obtained with the four different slag compositions after 20 min at 1723 K, in which it could be clearly seen that the concentration of B (C_B) decreased from 100 to 0 ppmw with increasing distance from the end of the tube (x).

If it was assumed that diffusion in the slag is dimensionally sufficiently long, then diffusion could be considered to begin at one edge of the slag, with the concentration remaining 0 at the other side. This type of diffusion can be expressed using the semi-infinite one-dimensional model, wherein the range of the diffusion distance, x, is [0, ∞]. At time t = 0, it can be expressed as follows:

$$\frac{\partial C}{\partial t} = \frac{\partial}{\partial x}\left(D\frac{\partial C}{\partial x}\right),\ C_B(x,t) = C_B^0 - \left(C_B^0 - C_0\right) erf\left(\frac{x}{2\sqrt{D_s t}}\right) \quad (2)$$

$$\text{Initial condition } \mathrm{C}(x,0) = \mathrm{C}_0(t=0) \quad (3)$$

$$\text{Boundary condition} \begin{cases} C(0,t) = C_B^0 \\ C(\infty,t) = C_0 \end{cases} \quad (t > 0) \quad (4)$$

Here, C_B represents the concentration of B in the slag at various positions, while C_B^0 is the initial concentration at the edge and is regarded as being equal to the concentration of the solute diffusion source (100 ppmw); C_0 is the concentration of slag in the tube, which is 0; D_s is the diffusion coefficient; and t is the diffusion time, t = 1200 s. Erf (y) is the error function.

After taking the definite integral, Eq. (2) can be expressed as Eq. (5). The right-hand side of Eq. (5) is the error function, which is erf(y) and C_0 = 0. Thus, the equation can be simplified as Eq. (6). The value of D can therefore be calculated from C_B, C_B^0 and the table of error for erf(y).

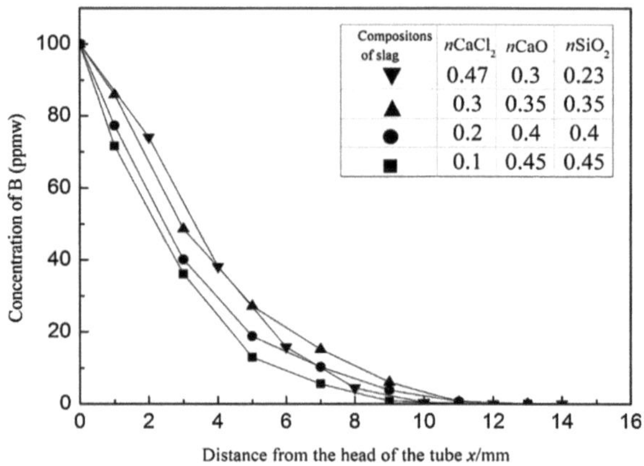

Fig. 8 Relationship between slag composition and the concentration of B at various positions

$$\frac{c_B - c_B^0}{c_0 - c_B^0} = \frac{2}{\sqrt{\pi}} \int_0^{x/(2\sqrt{D_s t})} e^{-\lambda^2} d\lambda \tag{5}$$

$$C_B = C_B^0 \left[1 - erf\left(\frac{x}{2\sqrt{D_s t}}\right)\right] \tag{6}$$

The diffusion coefficients obtained using Eq. (6) were shown in Fig. 9, in which the slopes of x^2 and $4y^2t$ were the diffusion coefficients of the four different slag compositions.

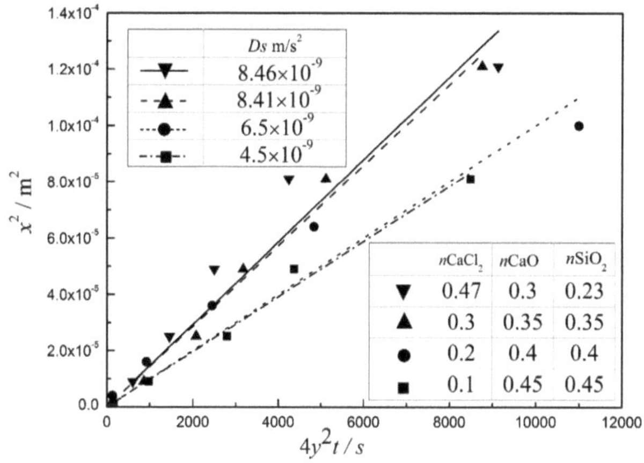

Fig. 9 Diffusion coefficients of four different slags

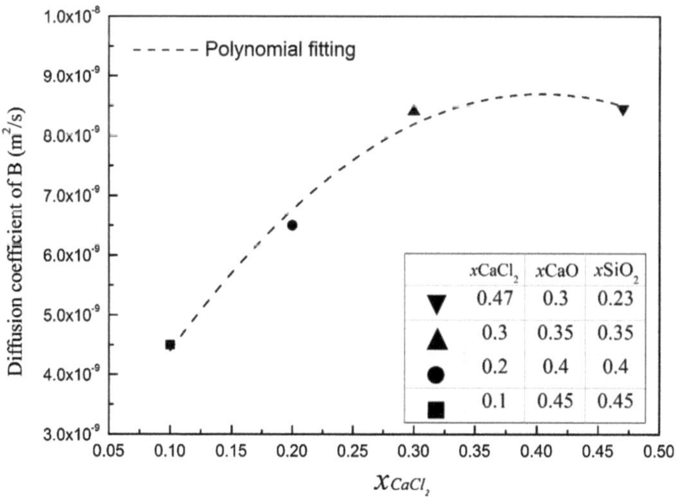

Fig. 10 Change in the diffusion coefficient of B with slag composition

The value of *Ds* was found to increase from 4.5×10^{-9} to 8.46×10^{-9} m/s^2 when the $CaCl_2$ concentration was increased from 10 to 47 mol%, as shown in Fig. 10. Given that $CaCl_2$ can decrease the viscosity of the slag, the maximum value shows that the solubility of $CaCl_2$ in a $CaCl_2$–CaO–SiO_2 slag is approximately 50%.

Compared with the diffusion coefficient of B in the molten Si ($D = (2.4 \pm 0.7) \times 10^{-8}$ m^2/s) [12], the diffusion of B in a CaO–SiO_2–$CaCl_2$ slag is slower at 1723 K, suggesting that the rate-limiting step in the B removal process may be controlled by the transportation of B in the slag.

Conclusions

The ternary CaO–SiO_2–$CaCl_2$ slag system demonstrates good potential for B removal due to the dual action of O and Cl. Among the tests of ternary slag systems, 30 mol% CaO-23 mol% SiO_2-47 mol% $CaCl_2$ system shows the best potential to achieve B-removal (90%) efficiency at 1723 K for 2 h. The diffusion coefficient of B in slag (D_s) is 8.46×10^{-9} m/s^2, which means the rate of B diffusion in slag is slower than that of B diffusion in Si ($(2.4 \pm 0.7) \times 10^{-8}$ m/s^2). Thus, the transfer of B in the slag controls the whole kinetic steps of B removal from Si by slag refining.

References

1. I. Kero et al., Refining kinetics of selected elements in the industrial silicon process. Met. Mat. Trans. B **46**(3), 1186–1194 (2015)
2. K. Suzuki et al., Gaseous removal of phosphorus and boron from molten silicon. J. Jpn. Inst. Met. **54**(2), 161–167 (1990)
3. F.A. Trumbore, Solid solubilities of impurity elements in germanium and silicon. Bell Sys. Tech. J. **39**(1), 205–233 (1960)
4. L.A.V. Teixeira, K. Morita, Removal of boron from molten silicon using CaO–SiO_2 based slags. ISIJ Int. **49**(6), 783–787 (2009)
5. K.X. Wei, et al., Boron removal from metallurgical-grade silicon by CaO–SiO_2 slag refining. Rare Met. **34**(7), 522–526 (2015)
6. J.J. Wu, et al., Boron removal in purifying metallurgical grade silicon by CaO–SiO_2 slag refining. T. Nonferr. Metal. Soc. **24**(4), 1231–1236 (2014)
7. J. Cai et al., Boron removal from metallurgical silicon using CaO–SiO_2–CaF_2 slags. T. Nonferr. Metal. Soc. **21**(6), 1402–1406 (2011)
8. M. Li, T. Utigard, M. Barati, Removal of boron and phosphorus from silicon using CaO–SiO_2–Na_2O–Al_2O_3 flux. Met. Mat. Trans. B **45**(1), 221–228 (2014)
9. JJ. Wu et al. "Boron removal from metallurgical grade silicon using a CaO-Li_2O-SiO_2 molten slag refining technique," *J. Non-Cryst. Solids,* 358 (23) (2012), 3079–3083

10. L. Zhang et al., Removal of boron from molten silicon using Na_2O–CaO–SiO_2 slags. Separ. Sci. Technol. **48**(7), 1140–1144 (2013)
11. J.Y. Li, et al., Research of boron removal from polysilicon using CaO–Al_2O_3–SiO_2–CaF_2 slags. Vacuum **103**, 33–37 (2014)
12. Y. Wang, X.D. Ma, K. Morita, Evaporation removal of boron from metallurgical-grade silicon using CaO–$CaCl_2$–SiO_2 slag. Met. Mat. Trans. B **45**(2), 334–337 (2014)

Part XII
Solar Cell Silicon: Silicon Photovoltaics

Electrodynamic Eddy Current Separation of End-of-Life PV Materials

York R. Smith, James R. Nagel and Raj K. Rajamani

Abstract In this work, we examine the efficacy of Electrodynamic Eddy Current Separation (EECS) to recover valuable materials from end-of-life solar panels. Traditional rotary-based eddy current separators are capable of excitation frequencies of ≈1 kHz or less and struggle to economically separate particles smaller than ≈1 cm. A new design of eddy current separators has been developed at the University of Utah which has no mechanically moving parts. The design is capable of excitation frequencies up to 50 kHz, allowing sorting of particles as small as 1.0 mm. Recently, we have been successful in separating mixtures of Si/Al and CdTe/Al particles (1–3 mm) with recovery and grades greater than 85%, an energy demand of 68 kWh/short ton of sorted material, and throughput of roughly 10 kg/h. Current and future challenges utilizing this method for valuable material recovery from end-of-life solar panels are discussed.

Keywords Eddy current · Recycling · PV materials

Introduction

Photovoltaic materials have been the focus of recent work on resource scarcity and material criticality [1–3]. Overburden of end-of-life (EOL) solar panel materials is currently not an issue, as many of the panels in place have yet to reach their maturity (15–20 years). However, a large influx of EOL PV modules is anticipated in the next 10–15 years. The annual waste from the photovoltaic industry from EOL panels is expected to be ~250,000 tons by 2016, and according to early-loss scenarios, this value could reach up to 78 Mt by 2050 [4].

Some EOL PV modules are considered hazardous waste and require special disposal methods with additional cost (e.g. CdTe and CIGS). Current recycling methods utilizing chemical treatment are more expensive than hazardous waste

Y.R. Smith (✉) · J.R. Nagel · R.K. Rajamani
Metallurgical Engineering Department, University of Utah, Salt Lake City, UT, USA
e-mail: york.smith@utah.edu

L. Zhang et al. (eds.), *Energy Technology 2017*,
The Minerals, Metals & Materials Series, DOI 10.1007/978-3-319-52192-3_37

disposal costs. A report by Bio Intelligence Services [5] highlighted the potential environmental problems related to the improper disposal of waste PV panels as: leaching of hazardous substances (i.e., Pb and Cd), losses of conventional material resources (i.e., Al and glass), and losses of precious and scarce metals (e.g., Ag, Ga, In, Ge). About 10–15% of the weight of a Si solar module consists of Al, Sn, Ag, and In. Recycling of such materials has relevance, not only from the point of waste management, but also from the recovery aspect of valuable materials. Several major benefits are achieved when recycled materials are used over virgin materials, namely natural resource preservation and energy conservation.

A first study on the technical and economic feasibility of the recycling crystalline silicon PV modules was presented in a photovoltaic technology conference in the 1990s [6]. However, the interest on PV recycling only started to rise around one decade later. For example, a study by Fthenakis [7] identified the challenges and the possible approaches for PV recycling in the USA. The study concluded that recycling was technologically and economically feasible, but not without careful forethought. The methods adopted so far for the recycling of silicon PV panels have been based on physical treatments, chemical treatments, or a combination of both. A description of these methods is reviewed [8]. However, a detailed analysis of the impacts related to such treatment in a lifecycle perspective is still missing from the literature.

To date, under most scenarios, the best option is landfill disposal [1] of EOL PV materials. However, this presents itself as only a short-term solution and is due to a lack of current innovative recycling methods/technical processes that are economically viable. The concept of recycling EOL PV materials is relatively new ($\sim$20 years), and only in the last few years have innovative commercial treatments been developed by *Deutsche Solar* for the recycling of crystal silicon panels, and by *First Solar* for the recycling of CdTe panels [5, 8].

As the market continues to expand and innovation cycles become even shorter, the replacement of equipment accelerates, making electrical and electronic equipment a quickly growing source of waste. One effective and established method of removing non-ferrous metals from such streams of industrial or municipal waste is by eddy current separation techniques [9]. The process is used to separate aluminum and copper from automobile scrap and to remove metals from recycled glass. The separation technique is based on the premise that conductive materials resist being moved in a magnetic field and, vice versa, will accelerate in a moving field [10]. The design and performance of rotating belted-drum type eddy current separators have been improved substantially in the last ten years, primarily due to advances in magnet materials and magnet configurations, as well as a better understanding of the separation mechanisms [11, 12]. Yet, problems associated with separation of small particles ($\leq \sim$5 mm), remains an arduous challenge to date.

In Electrodynamic Eddy Current Separation (EECS), particles are allowed to fall freely through an alternating magnetic field with adjustable frequency. This applied magnetic field induces eddy currents but only in the electrically conductive particles. The movement of the eddy currents generated in the conducting particles leads to produce a Lorentz forces on the nonferrous metallic particles. The generated Lorentz forces results in expelling them out of the main stream. The use of EECS for recycling EOL solar panels represents a potential low-cost and low-environmental impact technology.

Theory of Eddy Current Separation

Eddy current separation is a developed technology with many practical applications in the recycling industry. The basic principle is derived from the fact that when a permanent magnet passes over a conductive metal object, electric charges within the metal tend to experience a net magnetic force [13]. This causes distinct swirling flow patterns of charges, commonly referred to as eddy currents or Foucault currents. With a strong enough magnetic field with relative quick motion, enough force can be generated to significantly accelerate the entire metal particle. Modern eddy current separators are generally a large, rotary drum implanted with a series of permanent magnets. When the drum is rotated at a high velocity, nearby metal particles tend to deflect along distinct kinematic trajectories away from nonmetallic particles/material. Two distinct material streams are then isolated by a physical barrier placed between their respective trajectories.

For our case of electrodynamic sorting design, spinning permanent magnets are replaced with a single fixed electromagnet excited by an alternating electric current. This significantly changes the mathematical nature of the physics involved. In such a configuration, the relative velocity between magnet and particle is no longer a significant factor. We may therefore begin with Faraday's law of electromagnetic induction, which states that a time-varying magnetic field **B** will give rise to an electric field **E** in accordance with [14]

$$\nabla \times \mathbf{E} = -\frac{\partial \mathbf{B}}{\partial t}. \tag{1}$$

If we next assume sinusoidal steady-state operation at angular frequency ω, all time derivatives can be replaced with $\partial/\partial t = -j\omega$. The phasor form of Faraday's law is therefore written as

$$\nabla \times \mathbf{E} = j\omega\mathbf{B}. \tag{2}$$

Now let us now consider a metal particle placed somewhere in the magnetic field **B**. Because of the induced electric field **E**, electric charges within the metal are accelerated accordingly, thus giving rise to an eddy current density **J**. The relation between **E** and **J** is given by the point form of Ohm's law, written as

$$\mathbf{J} = \sigma \mathbf{E}, \tag{3}$$

where σ is the electrical conductivity of the metal. Changing magnetic fields therefore give rise to electric fields, which in turn give rise to eddy currents in metallic objects.

Next, we note that electrical currents also give rise to magnetic fields of their own in accordance with Ampere's law,

$$\nabla \times \mathbf{B}_e = \mu_0 \mathbf{J}, \tag{4}$$

where $\boldsymbol{B}_e$ indicates the magnetic fields produced just by the eddy current density $\mathbf{J}$. This implies more changing magnetic fields that must be accounted for in Eq. (1), which in turn cause changes in the resultant eddy current $\mathbf{J}$. Thus, a complete solution for $\mathbf{J}$, $\mathbf{E}$, and $\mathbf{B}$ requires us to solve Eqs. (1)–(4) simultaneously. While such a process extends well beyond the scope of this article, we can at least note that once a solution for $\mathbf{J}$ is finally obtained, it becomes possible to calculate the net Lorentz force $\mathbf{F}$ acting on the metal particle. This is given by the magnetic force law,

$$\mathbf{F} = \iiint \mathbf{J} \times \mathbf{B} \, dV, \tag{5}$$

where the integral is carried out over the volume defined by the space within the particle. Thus, if the excitation field $\mathbf{B}$ is strong enough and the frequency ω great enough, the net force $\mathbf{F}$ will be great enough to significantly deflect arbitrarily large particles of metal.

Description of Experimental Setup

The eddy current separator designed and used in this study overcomes some of the limitations of belted-drum type eddy current separators. One advantage is there are minimal mechanically moving parts, which enables the system to operate at higher excitation frequencies. The principle of this design was originally pioneered by Saveliev to extract gold particles from ore [15], but the original design has remained relatively unrefined. Previously, Dholu et al. [16] demonstrated the sorting of 6.0 mm sphere mixtures of Al/Cu, Cu/Brass, Al/Brass with nearly perfect recovery and grade. The same configuration also demonstrated promising sorting of Al alloys from Al alloys. For example, the separation of Al-110/Al-2024, Al-110/Al-6061, and Al6061/Al-2024 series alloy mixtures using 12 mm size spheres, recoveries and grades of 85–95% can be realized. Operating at a higher frequency improves separation of smaller conducting particles. In this experimental campaign, 1–3 mm particles of Si, CdTe, and Al where examined.

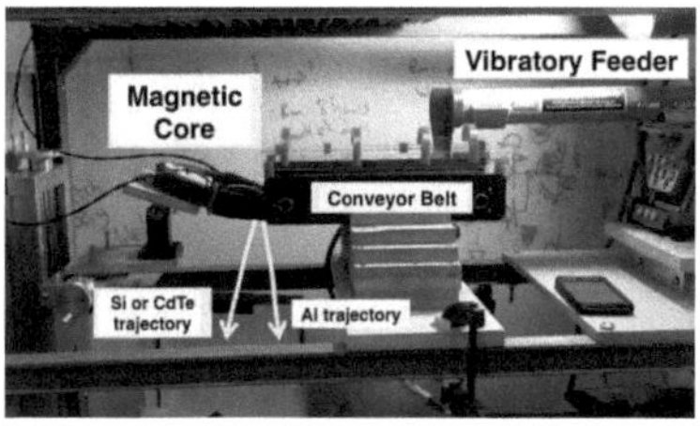

Fig. 1 A picture of the experimental setup utilized for separating Si/Al and CdTe/Al 1–3 mm particles

Table 1 Summary of experimental results for sorting 1–3 mm particles mixtures of Si/Al and CdTe/Al

	Si	Al	CdTe	Al
Recovery, %	97.5	98.2	95.1	98.8
Grade, %	99.5	98.3	85.7	99.5
	Si/Al		CdTe/Al	
Feed rate, kg/h	5.6		12.4	
Energy consumption, kWh/short ton	68.0		68.0	

Particles 1–3 mm in diameter of Si (American Elements, 99.9%) and Al (American Elements, 99.9%) was carried out using an experimental setup depicted Fig. 1 and described in further detail elsewhere [16]. The ferrite core (CMD5005 NiZn ferrite) has an outer diameter of 160 mm and an inner diameter of 120 mm. A specialized gap of approximately 1.1 cm in width was cut on the ferrite core, which is used to funnel flux into a tight volume of space. The core is wound (N = 173) with two windings in parallel on either side of the gap. The magnet was powered by a 400 W power supply, operating at 52.8 V and 6.60 Å with at a resonance frequency of 21.4 kHz. Under these operating conditions a $\boldsymbol{A}$ field of 60–80 mT is achieved in the gap. In each of the experiments, approximately 200 g of each sorting material was used for a total of ~400 g sorted per each experimental run. An equivalent circuit model for the drive electronics is detailed in a previous publication [16].

Results of Sorting Demonstration

The results from this experimental campaign demonstrate the capability of this technology to separate Si/Al and CdTe/Al particle mixtures of 1–3 mm with recoveries and grades >85% under the given conditions. The results for Si/Al and CdTe/Al mixtures after one pass are given in Table 1. To define quality metrics of the separation, such as recovery (R) and grade (G), we consider two initial masses, m_0 and M_0, of differing materials mixed together before the sort. The collection bins consist of two bins, labeled bin 'A' and bin 'B'. The desired outcome of the experiment is for all of the mass m_0 to collect in bin 'A' and all of mass M_0 to collect in bin 'B'. We can define M_A and M_B as the total masses from M_0 that fall

Fig. 2 An example of a sorting result separating 1–3 mm particle mixtures of Si/Al or CdTe/Al

into their respective bins. Common industry terms for these bins are "throw" and "drop." For our case, bin 'A' represents the "drop" bin (Si or CdTe particles) and the "throw" bin is bin 'B' (Al particles).

The recovery of bin 'A' and 'B' as the fraction of the initial material can be defined as:

$$R_A = \frac{m_A}{m_0} \tag{6}$$

$$R_B = \frac{M_A}{M_0} \tag{7}$$

The grade of the separation is then defined by the following:

$$G_A = \frac{m_A}{(m_A + M_A)} \tag{8}$$

$$G_B = \frac{m_B}{(m_B + M_B)} \tag{9}$$

The feed rate was determined by dividing the total mass sorted over the time duration of the sort. The energy consumption was determined using a P3 International Kill A Watt® real-time power monitoring device, which monitored the energy consumption of the whole operating unit. A typical sorting outcome is depicted in Fig. 2, which shows particle mixtures after one sorting pass.

Discussion and Concluding Remarks

Under the given conditions, the application of EECS to separate simple mixtures of semiconductor/metal particle mixtures is demonstrated. The particles used in this demonstration were not spherical, but rather granular chunks of material. Using such particles contributes to deviation of a perfect separation. Perfect spherical particles would yield higher consistency due to symmetry of the particle. For example, the majority of Al particles that were found in the Si or CdTe bin were thin sheet-like particles. The particle experiences a torque as well as a net force. Due to the geometry, (i.e. sheet-like particles) the torque tries to orientate the particle along a direction that minimizes the cross-sectional area exposed to the magnetic field. Since these thin particles have small cross-sectional area, they are exposed to less magnetic flux and thus do not kick as strong out of the stream.

Another cause for error in the sort has to do with how the particles are fed to the magnetic core gap, which has its limitation of throughput. Particle trajectories can become altered or screened by too much of material. For example, as a material is ejected from the gap, its trajectory is altered by a collision with another particle and ends up in the wrong bin. For our configuration, the particle mixtures are fed to a conveyor belt by a vibratory feeder. By using a relatively fast enough belt velocity and channels above the conveyor belt, the particles are fed to the core gap in such a way as to not fill the gap with too much material. Techniques and methods to deliver material to the gap for optimal separation remain an area of active investigation.

The current configuration is only able to effectively sort $\sim$10 kg/h of material while maintaining high selectivity and grade. Increasing the throughput by an order of magnitude while maintaining high selectivity and grade (>95%) is necessary if commercial application is to be realized. Although the system configuration has a low throughput, the energy consumption for one unit is rather low. It should be noted this does not take into account any active cooling of the system. Since the energy demand is relatively low, several unit configurations of parallel and series can be utilized to achieve desired throughput while maintaining high enough selectivity and grade.

Since particle size plays a large role in application of this technology, detailed characterization on the comminution of PV panels and materials is of great necessity. Moreover, how various parts of the panel break up under milling conditions will be valuable information. This technology demonstrates an ability to effectively separate particles of 1–3 mm under these set conditions, whereas in real-world applications the particles will likely be of much smaller size. Separation of particles smaller in size has yet to be realized and warrants further investigation. These are currently active areas of investigation.

Separation based on materials that have relatively large differences in conductivity appears to be easily achievable with this technique. On the other hand, separation of materials with relatively close conductivities presents a challenge. Moreover, it is not just also the differences in conductivity, but the absolute magnitude of the conductivity. A higher excitation frequency of the magnet will be required to achieve effective separation of less conducting particles (i.e., Si, CdTe), which faces physical limitations of the system. One approach to potentially increase the conductivity of Si and CdTe particles during separation is exposure to light irradiation, thereby increasing the conductivity of the semiconductor particles.

Acknowledgements YRS would like to acknowledge support provided though the U.S. Department of Energy, Office of Energy Efficiency & Renewable Energy, Sunshot Initiative Postdoctoral Research Award. RKR and JRN acknowledge support through the United States, Advanced Research Project Agency-Energy (ARPA-E) METALS Program under cooperative agreement grant DE-AR0000411. The authors would also like to acknowledge technical contributions by Dave Cohrs and Jacob Salgado.

References

1. M. Goe, G. Gaustad, Estimating direct climate impacts of end-of-life solar photovoltaic recovery. Sol. Energy Mater. Sol. Cells **156**, 27–36 (2016)
2. M. Goe, G. Gaustad, Identifying critical materials for photovoltaics in the US: a multi-metric approach. Appl. Energy **123**, 387–396 (2014)
3. N.T. Nassar et al., Criticality of the geological copper family. Environ. Sci. Technol. **46**, 1071–1078 (2012)
4. S. Weckend, A. Wade, G. Heath, End-of-life management: solar photovoltaic panels. Int. Renew. Energy Agency, 100 (2016)
5. V. Monier, M. Hestin, Study on photovoltaic panels supplementing the impact assesment for a Recasr of the WEEE directive, Bio Intelligence Service, 2011, pp. 86
6. T. Doi et al., Experimental study on PV module recycling with organic solvent method. Sol. Energy Mater. Sol. Cells **67**, 397–403 (2001)
7. V.M. Fthenakis, End-of-life management and recycling of PV modules. Energy Policy **28**, 1051–1058 (2000)
8. C.E.L. Latunussa et al., Life cycle assessment of an innovative recycling process for crystalline silicon photovoltaic panels. Sol. Energy Mater. Sol. Cells **156**, 101–111 (2016)
9. W.L. Dalmijn, Practical applications of eddy current separators in the scrap recycling industry, 2nd international symposium recycling of metals and engineered materials, 1990, pp. 303–314
10. T.A. Edison, Ore Separator, U.S. Patent 400,317, in: U.S. Patent (Ed.) USA, 1889
11. S. Zhang, P.C. Rem, E. Forssberg, The investigation of separability of particles smaller than 5 mm by eddy current separation technology. Part I: rotating type eddy current separators. Magn. Electr. Sep. **9**, 233–251 (1999)
12. S. Zhang, P.C. Rem, E. Forssberg, Particle trajectory simulation of two-drum eddy current separators. Resour. Conserv. Recycl. **26**, 71–90 (1999)
13. P.C. Rem, P.A. Leest, A.J.V.D. Akker, A model for eddy current separation. Int. J. Min. Process. **49,** 193–200 (1997)
14. J.D. Jackson, *Classical Electrodynamics* (Wiley, New York, 1999)
15. V. Saveliev, System and method for separating electrically conductive particles, in: U.S. Patent (Ed.)USA, 1998
16. N. Dholu et al., Eddy current separation of nonferrous metals using a variable-frequency electromagnet. KONA Powder Part. J. 1–7 (2016). http://doi.org/10.14356/kona.2017012

Investigation on Quartz Crucibles for Monocrystalline Silicon Ingots for Solar Cells

M. Di Sabatino, F.W. Thorsen, A. Lanterne, Y. Hu, J.A. Bones and E. Øvrelid

Abstract This study presents a new testing method to analyze the bubble content and distribution in quartz crucibles for monocrystalline silicon ingots. Two different types of silica (SiO_2) crucibles have been investigated, before and after use during Czochralski (Cz) silicon ingot production for solar cells. Samples have been cut from three different positions along the crucible wall and then investigated by X-ray tomography, optical microscopy and scanning electron microscopy. The unused samples were heat-treated at 1400 °C for 4 h in Ar at 15 mbar. It is observed that the crucible bubbles grow in size and density during the Cz process. More bubbles are detected closer to the inner part of the crucible wall after use in the Cz process. The results also indicate that there are some differences in quality between different crucible producers and that bubble formation and growth are significantly affected by the Cz process parameters, such as temperature and time.

Keywords Silicon · Quartz · Crucible · Solar cells

Introduction

Silicon-based solar cells are the most used types of solar cells in the market [1]. The highest solar cell efficiency is obtained by using monocrystalline silicon wafers [2] and wafers are cut from silicon ingots grown by the so-called *Czochralski* (Cz) method. In this process, silicon feedstock is first stacked in a quartz crucible and melted, and a monocrystalline silicon ingot is then pulled out of the melt. The crucibles used in this process are made from high purity amorphous quartz (SiO_2),

M. Di Sabatino (✉) · F.W. Thorsen · A. Lanterne
Department of Materials Science and Engineering, NTNU, Trondheim, Norway
e-mail: marisa.di.sabatino@material.ntnu.no

Y. Hu
NorSun AS, Årdalstangen, Norway

J.A. Bones · E. Øvrelid
Materials and Chemistry, SINTEF, Trondheim, Norway

L. Zhang et al. (eds.), *Energy Technology 2017*,
The Minerals, Metals & Materials Series, DOI 10.1007/978-3-319-52192-3_38

and should prevent unwanted impurities from entering the silicon melt during the crystal pulling. However, crucible materials are the main sources for oxygen contamination in monocrystalline silicon ingots and oxygen-related defects are among the most detrimental factors for the minority carries lifetime, and thus solar cells efficiency [3–6].

Furthermore, the quartz crucible contains many bubbles. Most of these bubbles form during crucible manufacturing while some also grow during the crystallization process. These bubbles may become possible cause for impurity contamination and dislocation formation in the monocrystalline silicon ingot. Indeed, the bubble can be released into the melt during the Cz process as the crucible is slowly consumed, along with the possible detachment of small pieces of quartz or cristobalite. If the bubbles or quartz particles reach the solidification front, they can interfere with the crystal growth and may lead to, e.g., structure loss [7] and, thus, it is an important issue for the photovoltaic (PV) industry.

In general, crucibles consist of two different layers: one bubble free layer (BF) and one bubble-containing layer (BC). The bubble free layer is facing the inside of the crucible. As this side is in direct contact with the silicon melt, it is important to keep this layer free from bubbles through the entire process, to ensure that no bubbles are released into the melt and subsequently into the silicon ingot. It is also important to understand whether the bubbles are formed in the bubble free layer during the pulling process, or if they grow from very small, invisible bubbles to larger visible bubbles. They might also move from the bubble-containing layer to the bubble free layer. The goals of this work are, therefore, (i) to improve the understanding of the mechanisms occurring inside the crucible during use, (ii) to compare the impact of two different types of crucibles, and finally (iii) to discuss about the possible impact on the silicon ingot quality and yield.

Experimental Procedure

Two types of crucible materials, named A and B, from two different manufacturers were investigated in this work. They are made from high purity fused silica, which has an amorphous structure, low thermal conductivity and high heat tolerance. Three positions for each used and unused crucibles were cut as shown in Fig. 1; each sample being $10 \times 10 \times 10\ mm^3$. The unused samples were heat-treated at 1400 °C for 4 h in argon atmosphere and 15 mbar (which is a similar pressure used during industrial Cz process). This heat-treatment is referred to as HT4 in this manuscript.

The selected samples were investigated by X-ray tomography to measure the bubble density and distribution. The X-ray tomography tool is a Nikon XT-H225-ST with a resolution of 9 μm. Each sample was scanned with X-ray at a voltage of 130 kV and 100 mA, while the sample was rotated 360°.

Fig. 1 Each crucible investigated (*A* and *B*) both in unused and used form, was cut in several cubic pieces. For this study positions 1, 7 and 10 were selected

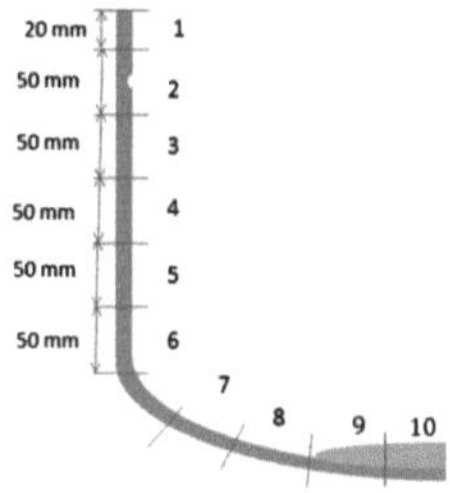

Fig. 2 Optical micrograph from the wall of a used crucible, showing the BF layer to the left and the BC layer to the right (50× magnification)

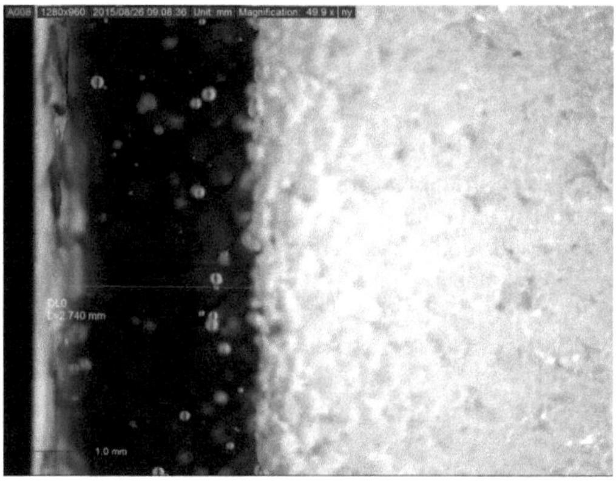

The samples were polished and then analysed by optical microscopy and scanning electron microscopy (SEM), which allows for better resolution and thus to detect bubbles that are smaller than 9 μm. Furthermore, in order to predict the temperature profile on the crucible during the process, CGSim simulation software from SINTEF was used.

Results and Discussion

The optical microscopy investigation indicates that both used and unused crucibles have two well defined layers inside the wall; one opaque bubble containing (BC) layer, and one transparent bubble free (BF) layer. Figure 2 shows an optical micrograph from one of the samples investigated. A thick layer of cristobalite (approximately 0.5 mm) was observed on the surface of the used crucibles.

Figure 3 shows a simulation of the temperature distribution of the crucible with the silicon melt while the silicon crystal is pulled. The simulation indicates that the corner and bottom parts of the crucible will maintain a much higher temperature than the top, with a longer holding time, as the crucible is raised to let the melt level stay constant relative to the crystal puller. During the ingot body growth, the difference in temperature between position 1 and 7 in the crucible was measured to

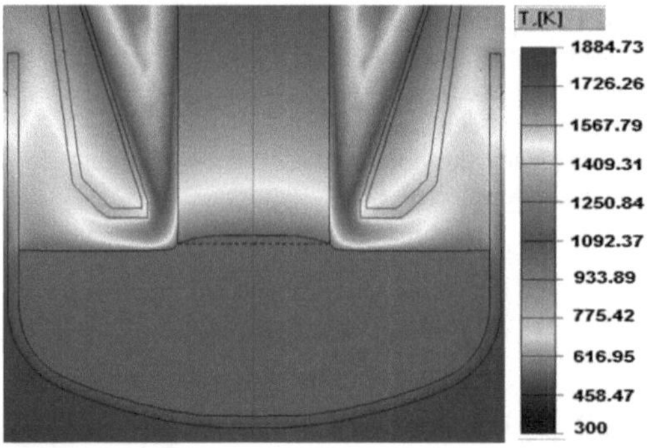

Fig. 3 CGSim simulation of the temperature distribution of the crucible with the silicon melt while the silicon crystal is pulled

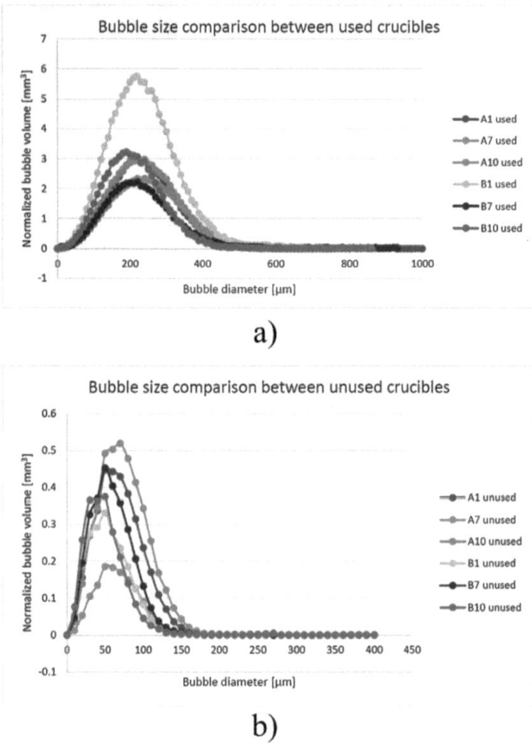

Fig. 4 Bubble size distribution in the two crucible types at three different position **a** used, after Cz process and **b** unused

be around 300 °C. This has a significant impact on the viscosity properties of the quartz, and hence its bubble growth [7]. The temperature difference in position 7 and 10 is only around 20 °C, which will not contribute to a significant change in viscosity. It is, therefore, assumed that the bubbles would grow more in positions 7 and 10 than in position 1. However, the results of the X-ray tomography (see Fig. 4) show a different trend.

Figure 4 presents the bubbles size distribution in the two crucible types at three different positions before and after use. It reveals that the highest growth in bubble volume is found in position 1 from crucible B. The bubble content increased from 0.22% before use to 12.84% after use in the Cz pulling process. Position 7 from crucible B has the lowest bubble volume growth, with an increase in bubble content from 0.29 to 4.97%. The samples from the different positions in crucible A have a more similar bubble growth, and crucible A has a lower total bubble volume growth than crucible B. Crucible A has a total bubble content of 0.30% before use, and 6.43% after use in the Cz pulling process. Crucible B has a total bubble content of 0.25% before use and 8.11% after use in the Cz process.

After the heat-treatment, HT4, of the unused crucible samples of both types A and B (see Fig. 5), the bubble content looks very similar to the results from the unused samples. The highest and lowest bubble content is still found in A10 (position 10 of crucible A) and A7 (position 7 of crucible A), respectively. In all the samples except from B1, the bubble content has decreased after the heat-treatment. In B1, the increase is from 0.22 to 0.23%, which is considered to be not significant.

The bubble distribution in the crucible wall from the used and unused samples as well as the samples from the HT4 are shown in Fig. 6a–c, respectively. In Fig. 6a, the different crucibles have a various distribution of bubbles both between the positions and between the crucible types. However, a lower bubble content is always measured close to the inner wall. A7 and A10 have higher bubble content in the middle of the crucible than in the outside, the other samples have a decreasing bubble content close to the inner wall. In Fig. 6b, B7 has a high bubble content in the inner part, close to the melt. The other samples have close to no bubbles in this region. For most of the samples, the highest concentration of bubbles is around 4 mm into the sample (from the outer side). Sample A7 has the lowest bubble content, while A10 has the highest. The distribution of bubbles in the HT4 samples, shown in Fig. 6c, look very similar to the unused. In this case, all the samples have no bubbles in the inner part of the crucible wall.

The growth of the bubbles is expected to rely on three main parameters: (i) temperature, (ii) pressure, and (iii) holding time. The comparison of the size distribution indicates that the samples from the used crucible have a much higher bubble content than the other samples, consisting of bigger bubbles. The bubbles in the used samples are up to 700–800 μm in diameter, while the bubbles in the

Fig. 5 Bubble size distribution for the crucibles after HT4 heat-treatment (1400 °C for 4 h in Ar)

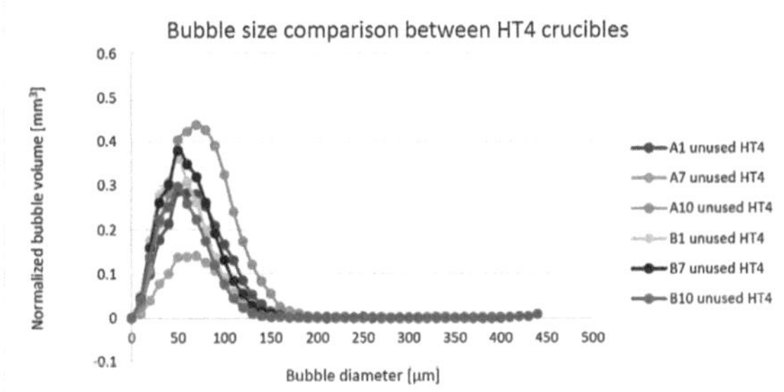

Fig. 6 Bubble distribution in the crucible wall for the **a** used, **b** unused and **c** after HT4 conditions

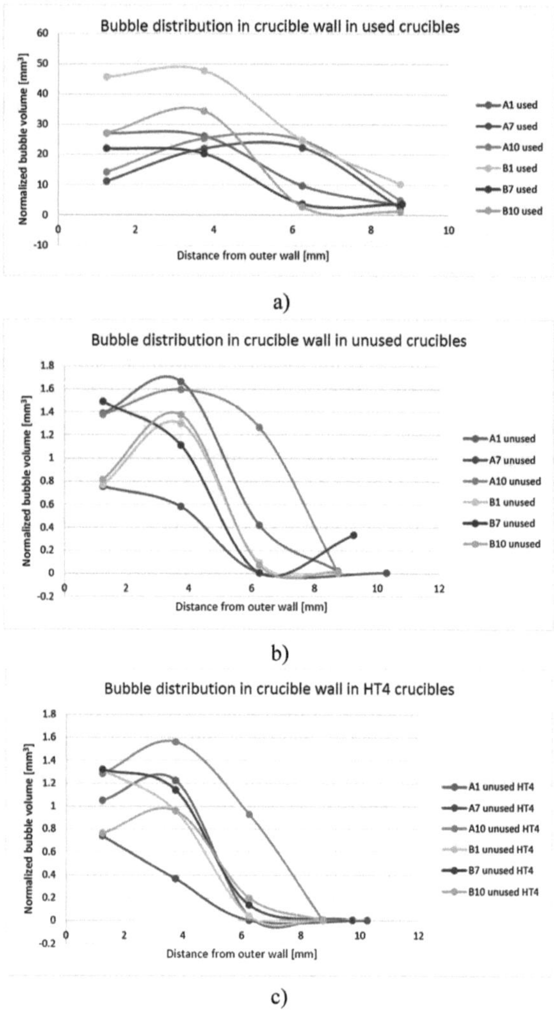

unused and heat-treated crucibles are up to approximately 200 μm in diameter. The sample from position 1, in the used crucible, has 12.84% bubble content, which is the highest measured in this work. This is significantly higher than position 7 and 10, which had 4.97 and 6.53%, respectively. Sample B7 has a high bubble content in the BF layer before use, but one of the lowest bubble contents after use.

Before use, the crucible B has the most equal distribution of bubbles throughout the different positions, that is to say, the distribution profiles are similar. After use, however, the situation is the opposite, the crucible B has the least similar bubble distribution profiles and crucible A has a very even bubble content in the three positions, with 6.60, 5.78 and 6.90% for A1, A7 and A10, respectively.

Both crucibles A and B, except from position B1 which is close to stationary, have lower bubble content after HT4 than before. These changes in bubble content are between one and two orders of magnitude lower than between the unused and used samples. Small variations in bubble content between the samples could be explained by the deviation in the cubic shape of the samples. However, this seems unlikely, since all samples except from B1 experienced lower bubble content after HT4. It is worth noting that B1 also has the largest growth in bubble content when comparing used and unused crucibles.

There are big differences between the two crucible types in how the bubbles are distributed through the crucible wall. Nevertheless, the bubbles have similar size. Both crucibles have a very low bubble content close to the melt in the bottom position (position 10) before use; however, only crucible B keep a BF layer free of bubbles after use. The bubbles in crucible A grow less than the bubbles in crucible B, when compared before and after the crystal pulling process. Crucible A goes from a total bubble content of 0.30–6.43%, while crucible B goes from 0.25 to 8.11%. This means that crucible A had the lowest growth in bubble content, while crucible B had the highest growth in bubble content. This might indicate that crucible B has a higher carbon content inside the bubbles [8], which can be due to insufficient quality control in the manufacturing. This, in turn, may affect the quality of the final monocrystalline silicon ingot grown with this type of crucibles.

The SEM investigation (images not shown here but from Ref. [8]) revealed bubbles smaller than the detection limit of the X-ray tomography, down to approximately 1 μm in diameter on unused sample. These small bubbles could explain the increase in bubble density observed in the used samples by the X-ray analysis. The observed bubbles in the used samples could be already present in the material before use but not visible in the X-ray analysis.

Conclusions

This study presents a new testing method to analyze the bubble content and distribution in quartz crucibles for Cz silicon by X-ray tomography measurements. It is believed that this technique can help understanding the bubble growth in crucibles during crystal pulling, which is of great interest to the PV industry.

The bubbles inside the crucible grow significantly in size during the crystal pulling process. The bubbles in the used crucibles were around 200 μm in diameter, with bubbles up to 1 mm diameter in some samples, while the bubbles were closer to 50 μm in diameter before the Cz pulling with very few bubbles measured over 200 μm in diameter.

Crucibles from supplier A and B differ in bubble volume and distribution in unused crucible and show different evolution during the Cz process. For example, position 1 in crucible B shows one of the lowest bubble content before use while it shows the highest bubble volume after use. The knowledge of the size and number of bubble before use is then not enough to predict the evolution during the Cz process.

The carbon and other impurities content in the quartz may have an important role in the bubble evolution.

The formation of bubble in the bubble free layer during the Cz process could be due to the presence of bubbles in the unused samples with diameter smaller than the detection limit of the X-ray tomography. These small bubbles with diameter down to 1 μm were observed by SEM.

Finally, a very different evolution of the bubbles was observed after the heat-treatment, where a small decrease of the bubble content was observed. Concerning the variation between various crucible position, position 1 with the lowest thermal budget (up to 300 °C less than in position 10) shows the maximum increase of bubble volume in crucible B, contrary to what could be expected. Considering all the samples investigated, the variation in thermal budget between position 1, 7 and 10 does not seems to have a significant impact on the bubble formation as well as the contact with the silicon melt in position 7 and 10. The holding time may then have an important impact as no bubble increase was observed for a shorter holding-time in the heat-treated samples.

Acknowledgements The authors thank the FME Centre for Solar Cell Technology and the Research Council of Norway for financial support. Scientific discussions with Dr Ove Paulsen and Mari Dr Mari Juel at SINTEF are gratefully acknowledged.

References

1. Fraunhofer Institute for Solar Energy (ISE), *Photovoltaics Report* (Freiburg, 2016)
2. A. Willoughby, Solar cell materials: developing techniques. Wiley Series (2014)
3. L. Arnberg, M. Di Sabatino, E.J. Øvrelid, State-of-the-art growth of silicon for PV applications. J. Cryst. Growth **360**, 56–60 (2012)
4. M.P. Knudson, M. Juel, E.J. Øvrelid, M. Di Sabatino, Effect of impurities in mono-crystalline silicon for solar cells. TMS Annu. Meet. 219–226 (2013)
5. M. Di Sabatino, S. Binetti, J. Libal, M. Acciarri, H. Nordmark, E.J. Øvrelid, Oxygen distribution on a multicrystalline silicon ingot grown from upgraded metallurgical silicon. Sol. Energy Mater. Sol. Cells **95**, 529–533 (2011)
6. G. Gaspar, G. Coletti, M. Juel, S. Würzner, R. Søndenå, M. Di Sabatino, L. Arnberg, E.J. Øvrelid, Identification of defects causing performance degradation of high temperature n-type Czochralski silicon bifacial solar cells. Sol. Energy Mater. Sol. Cells **153**, 31–43 (2016)
7. A. Lanterne, G. Gaspar, Y. Hu, E.J. Øvrelid, M. Di Sabatino, Investigation of different cases of dislocation generation during industrial Cz silicon pulling. Phys. Status Solidi. (2016) (Current Topics in Solid State Physics)
8. F.W. Thorsen, Master Thesis, NTNU, 2016

Influence of Oxygen Content on the Wettability of Silicon on Graphite

Zineb Benouahmane, Lifeng Zhang and Yaqiong Li

Abstract The wettability behavior of graphite with silicon is a key factor for its application as a crucible in silicon making industry. A previous study was carried out to determine the correlation between the physico-chemical behaviors of different types of graphite and the final contact angle for the graphite/silicon system. In this study, experimental vacuum annealing treatment by varying the exposure time at high temperature of 1000 °C processed by X-ray photoelectron spectroscopy and SEM analyses enabled to found a correlation between the oxygen content in the graphite and the final contact angle for the graphite/silicon system. Finally, the experimental results with and without vacuum annealing treatment have shown that a deep infiltration of silicon inside the graphite occurs in case of high level of oxygen content inside the graphite's pores.

Keywords Vacuum treatment · Wettability · Silicon · Graphite · Oxygen content

Introduction

World energy needs, the lessening of the limited amount of fossil fuels and their significant health and environmental impacts have leaded to show a deep interest in the development of solar energy technology. Improving the solar cells efficiency by producing a high purity solar grade silicon material with low level of impurities without the contamination of silicon remains on the conception and design of new crucibles and dies with high hydrophobic surfaces wetting properties. In this study, the one of particular interest is the graphite crucible, a refractory material that features high-temperature strength, machinability, and its purification process offers a low investment cost. Considering the interest of this material as a refractory material, there has been considerable interest in understanding the wetting

Z. Benouahmane · L. Zhang (✉) · Y. Li
School of Metallurgical and Ecological Engineering, University of Science and Technology Beijing, 100083 Beijing, China
e-mail: zhanglifeng@ustb.edu.cn

L. Zhang et al. (eds.), *Energy Technology 2017*,
The Minerals, Metals & Materials Series, DOI 10.1007/978-3-319-52192-3_39

dynamics, infiltration and reactivity of the system Si/C [1–4]. Previous studies show that, the wettability of a material is tailored by its surface chemistry modification; mainly the surface roughness variation and the heterogeneity tend to dictate the spreading behavior of the droplet over the refractory material [5]. Although, the wettability behavior can be influenced by the amount of oxygen absorbed by the graphite, a key elements in rendering a surface hydrophilic, but there is a lack of a rigid quantitative correlation between this factor, surface wettability and the depth of infiltration of silicon in graphite's pores [6]. This paper aims to explore the possibility of using vacuum annealing to modify the surface chemistry of graphite in order to control its wettability.

Experimental Procedure

For this study, several graphite samples were respectively prepared and placed in the chamber vacuum oven. The experiments were carried out at temperature of 1000 °C exposing them to vacuum annealing for several holding times (0, 6, 8, 12, and 24 h). The total exposure annealing time depend on degree of wettability that needs to be accomplished. The samples were weighted before and after the treatment to quantify the mass loss, which were further analyzed by X-ray photoelectron spectroscopy to measure the oxygen to carbon atomic ratio (O/C ratio). The graphite sample has been preserved in an argon environment to prevent a recombination of carbon with the oxygen in the air. The vacuum annealing treatment was followed by sessile drop wetting experiments under Ar atmosphere. A piece of silicon (approximately 0.07 g and $\Phi 2.9 \times 4.90$ mm) with a purity of 99 wt% was placed on the annealed graphite. The argon was purified from water, oxygen and organics elements by passing over magnesium turnings at 500 °C before to be injected in the furnace. The sample was then introduced into the furnace chamber; the heating and cooling rate was the same in every experiment. The heating rate was approximately 10 °C/min until a temperature of 1000 °C and reduced to 5 °C/min, then the furnace was cooled down to the room temperature. A video camera was used to record a video from the furnace. The samples then were prepared for a SEM analysis to quantify the silicon infiltration into the graphite.

Vacuum Annealing Treatment of Graphite

Vacuum annealing treatment was performed on the graphite samples with the same routine for different exposure time, a weight loss and oxygen concentration versus holding times curve is shown in Fig. 1. The mass decreased approximately by 5 wt% after 24 h due to evaporation of the following elements: O_2, H_2O, CO_2, and CO [7]. A decrease in weight could be observed as the exposure time was further increased.

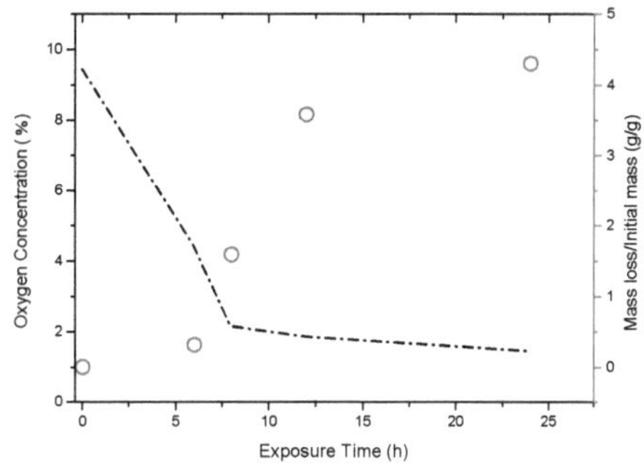

Fig. 1 Development of mass loss and oxygen concentration as a function of vacuum annealing exposure time at 1000 °C

Table 1 Oxygen concentration and O/C ratio results for 1000 °C temperature annealing

No.Experiment	1 (0 h)	2 (6 h)	3 (8 h)	4 (12 h)	5 (24 h)
Oxygen concentration (%)	9.43	4.4	2.1	1.8	1.4
O/C ratio at%	0.0943	0.044	0.021	0.018	0.014

The X-ray photoelectron spectroscopy showed a decrease of oxygen concentration in Fig. 1 with increasing annealing time. Annealing at 1000 °C resulted in less O/C ratio at%, which decreased furthermore with increasing exposure time, resulting in a reduction of surface oxygen content by 8% into the graphite.

Graphite captures easily oxygen from air to form various oxygen functional groups. This procedure helps to reduce the atomic oxygen concentration by desorption process without using any harsh reducing agents. We believe also that the vacuum annealing helped softening the graphite surface, inducing a decrease of the surface roughness, which is a vital criteria controlling the wetting behavior of the refractory material [1]. Softening of the graphite is due to the thermal expansion at high temperature [8].

The oxygen concentration in the annealed graphite does not go lower despite furthermore exposition to the vacuum treatment and stagnates around a value of 1.8% in Table 1.

Wettability and Reactive Infiltration Behavior

Wetting angles between silicon and NO-annealed and annealed graphite were measured at temperature of 1460 °C above the melting point of silicon 1414 °C under Ar atmosphere. As shown in Fig. 2, all wettability experiments with graphite annealed at temperature of 1000 °C, exposing them to vacuum annealing for several holding times (6, 8, 12, 24 h) resulted in a higher final contact angle than the NO-annealed one. Complete melting is achieved after 60 s, and the contact angle takes generally the high values at the moment of complete melting of the metal, but decreases rapidly with time to stabilize at relatively low values after the highest

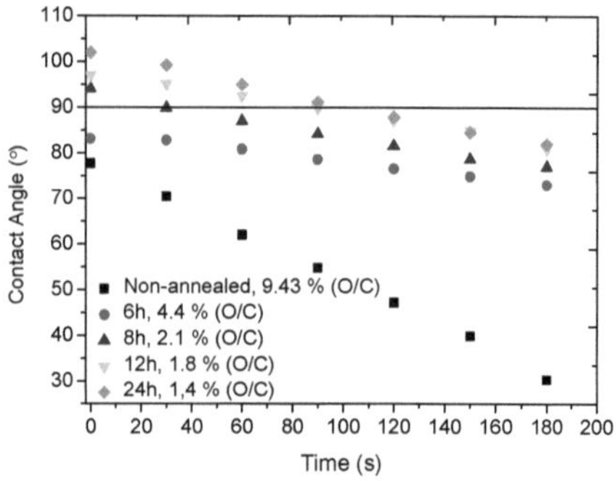

Fig. 2 Plot of contact angle as a function of vacuum annealing exposure time at 1000 °C

temperature of experiment is reached. All the wettability experiments for (8, 12, 24 h) with an oxygen content less than 2% resulted in an initial contact angle above 90° in Fig. 2. The final results displayed a quite large decrease; it takes only 3 min for the contact angle to drop rapidly from a wetting to no-wetting angle (102°–82.27°).

The spreading behavior was dependent of the temperature, resulting in a decrease of the wetting angle due to the reaction between silicon and graphite which lead to the formation of SiC in Fig. 3. For all the experiments, the silicon evaporated and condensed leading to a layer on the top of the substrate. The annealed graphite at 1000 °C in Fig. 2 exhibited a slow spreading velocity. The wetting experiments showed a clear correlation between wetting angle, surface tension, and droplet volume versus the temperature, annealing time, and oxygen concentration in the graphite as shown in Fig. 4.

Despite increasing the vacuum annealing time from 12 to 24 h at 1000 °C, the graphite demonstrated slightly the same wetting ability. This restraint is owing to the fact that O/C ratio of the vacuum annealed graphite reaches a certain limit around 2%, although after being treated by vacuum annealing for an extended time. When the contact angle is below 90°, the infiltration happens in the porous media of

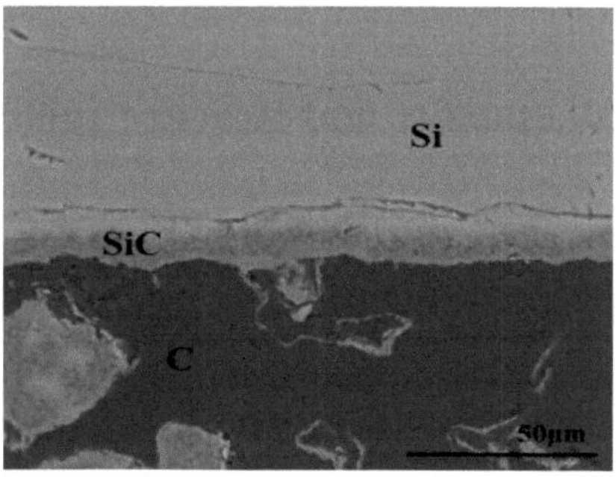

Fig. 3 SEM of the sample consisting of silicon carbide

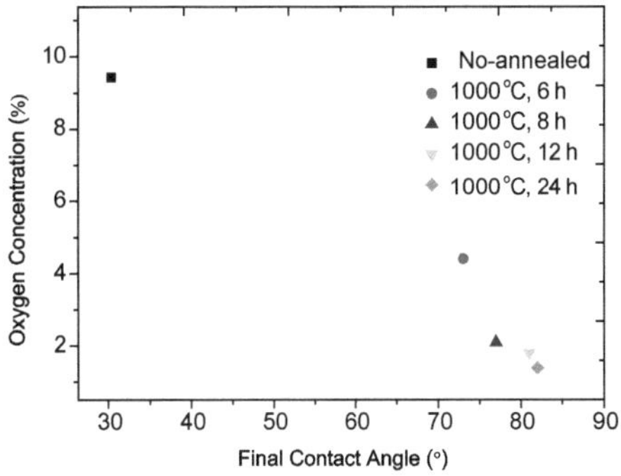

Fig. 4 Final contact angle versus oxygen concentration for graphite annealed at 1000 °C

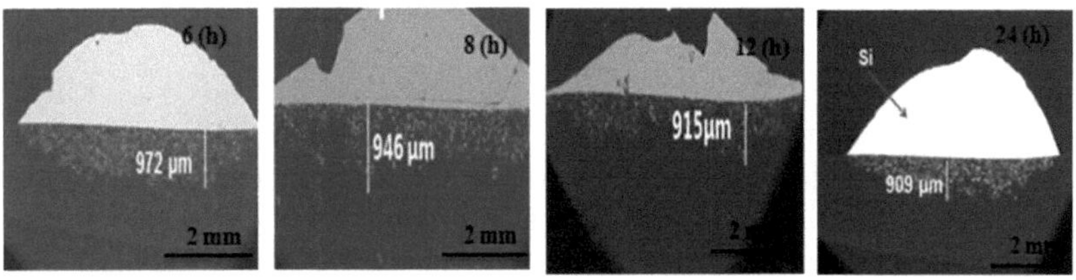

Fig. 5 Depth of infiltration for the graphite annealed at 1000 °C (SEM)

graphite, the rate of infiltration is function of the evacuation of the CO and CO_2. The spreading rate was faster for the non-annealed graphite compared to the annealed one. Compared to other results [3–8], this work exhibited a higher contact angle. This could be explained by the oxygen concentration contained in the graphite.

Samples were then prepared for a SEM analysis to evaluate the depth of infiltration as shown in Fig. 5, it showed that a reaction occurred between silicon droplet and the graphite substrate leading to the formation of silicon carbide, the infiltration of silicon in the pores, decrease in size of the silicon droplet.

The depth of silicon infiltration was different for the no-annealed and annealed graphite. Without reduction of graphite, the depth of infiltration increased due to the presence of more oxygen functional groups on the surface and inside the porous media of the refractory material in Fig. 6. Molten silicon will react with C to form SiC and therefore infiltrate the refractory material. Wetting is associated by a chemical reaction between the silicon droplet and carbon with an accumulation of solid silicon carbide at the interface [3, 9–11].

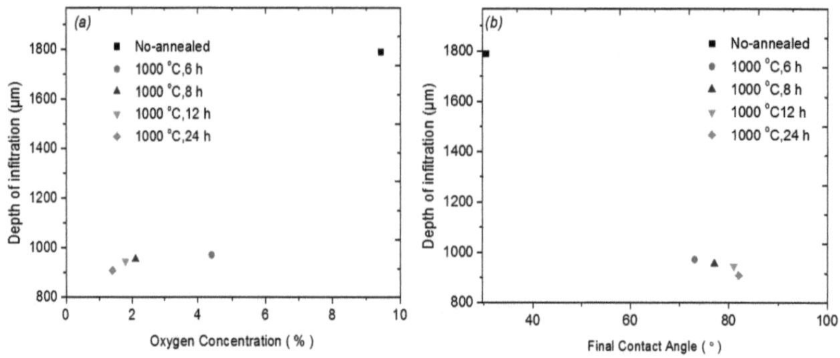

Fig. 6 Infiltration depth of molten silicon versus oxygen concentration (**a**) and final contact angle (**b**)

Summary

The vacuum annealing treatment can be used to control the O/C ratio inside the pores of graphite which plays a major role in the wetting behavior of the system silicon graphite. A high ratio of oxygenated groups exhibit a final contact angle lower than the one with a low ratio of O/C. The management of the contact angle of a graphite thermally reduced leads to a chemical reaction over the triple line and the formation of a new solid compound at the interface at lower spreading rate and dissolution depth.

Acknowledgements The authors are grateful for support from the National Science Foundation China (Grant No. 51274034, No. 51334002, No. 51604023, No. 51504020, and No. 51404019), Beijing Key Laboratory of Green Recycling and Extraction of Metals (GREM), the Laboratory of Green Process Metallurgy and Modeling (GPM2) and the High Quality steel Consortium (HQSC) at the School of Metallurgical and Ecological Engineering at University of Science and Technology Beijing (USTB), China.

References

1. T. Ciftja, Abel Engh, M. Tangstad, Wetting properties of molten silicon with graphite materials. Metall. Mater. Trans. A **41**(12), 3183–3195 (2010)
2. P. Sangsuwan, S.N. Tewari, J.E. Gatica, M. Singh, R. Dickerson, Reactive infiltration of silicon melt through microporous amorphous carbon preforms. Metall. Mater Trans. B **30**(5), 933–944 (1999)
3. O. Dezellus et al., Wetting and infiltration of carbon by liquid silicon. J. Mater. Sci. **40**(10), 2307–2311 (2005)
4. A. Ciftja, *Solar Silicon Refining, Inclusions, Settling, Filtration Wetting,* vol. 11. (Lambert Academic Publishing, New York, NY, 2010), pp. 29, 47–48

5. T.T. Chau, W.J. Bruckard, P.T.L. Koh, A.V. Nguyen, A review of factors that affect contact angle and implications for flotation practice. Adv. Colloid Interface Sci. **150**(2), 106–115 (2009)
6. M. Acik, G. Lee, C. Mattevi, A. Pirkle, R.M. Wallace, M. Chhowalla, K. Cho, Y. Chabal, The role of oxygen during thermal reduction of graphene oxide studied by infrared absorption spectroscopy. J. Phys. Chem. C **115**(40), 19761–19781 (2011)
7. Z.H. Ni, H.M. Wang, Z.Q. Luo, Y.Y. Wang, T. Yu, Y.H. Wu, Z.X. Shen, The effect of vacuum annealing on graphene. J. Raman Spectrosc. **41**(5), 479–483 (2010)
8. Z.H. Ni, X.F. Fan, H.M. Wang, Z. Zheng, Y.P. Feng, Y.H. Wu, Z.X. Shen, High temperature Raman spectroscopy studies of carbon nanowalls. J. Raman Spectrosc. **38**(11), 1449–1453 (2007)
9. R. Deike, K. Schwerdtfeger, J. Electrochem, Reactions between liquid silicon and different refractory materials. J. Electrochem. Soc. **142**(5), 609–614 (1995)
10. H. Zhou, R.N. Singh, Kinetics model for the growth of silicon carbide by the reaction of liquid silicon with carbon. J. Am. Ceram. Soc. **78**(9), 2456–2462 (1995)
11. J.G. Li, H. Hausner, Reactive wetting in the liquid-silicon/solid-carbon system. J. Am. Ceram. Soc. **79**(4), 873–880 (1996)

Particle Separation in Silicon Ingot Casting Using AC Magnetic Field

V. Bojarevics, G. Djambazov and K. Pericleous

Abstract The AC electromagnetic (EM) field effects are investigated for the possibility to extract particles (impurities, precipitates, oxides, bubbles, etc.) from molten silicon during the directional solidification of silicon ingot. The dynamics of particles are affected by the hydrodynamic drag, buoyancy, turbulent fluctuations and the EM force via the local pressure distribution. The EM force leads to high mixing rates, transitional flow structures and turbulence of the melt, contributing to the particle dispersion, transport and separation to a desired location. Application to silicon kerf loss recycling by melting and solidifying in the presence of AC EM field is presented for a variety of cases using 3D models for general flow structure and axisymmetric models for moving solidification front long time tracking. This paper demonstrates the effects of flow on the time dependent particle 'swarm' transport and the eventual modified distribution in the final ingot.

Keywords Particle separation · Silicon solidification · AC magnetic field · Lagrangian tracking

Introduction

Attempts to purify silicon of lower grades to a pure solar or electronic grade silicon material are well known for some time [1]. Recycling of the solar silicon lost during the wafering process (40–50% of initial weight) [2] adds to the challenge of removing unwanted particulate contaminants from the silicon melt. Typical size of the most undesirable (SiC, Si_3N_4) impurity particles is between 20–100 μm. EU sponsored project SIKELOR [3] aims to find a solution suitable for industrial scale implementation by applying EM fields to separate the particles from silicon melt. The principle of separation is based on the so called electro-magneto-phoretic forces [4], which are well known for a variety of MHD applications, such as

V. Bojarevics (✉) · G. Djambazov · K. Pericleous
University of Greenwich, London SE10 9LS, UK
e-mail: v.bojarevics@gre.ac.uk

L. Zhang et al. (eds.), *Energy Technology 2017*,
The Minerals, Metals & Materials Series, DOI 10.1007/978-3-319-52192-3_40

removal of inclusions from steel melt [5, 6], concentrating bubbles [7, 8], or insulating particles collected near the surface [9], and for many other purposes of separation or mixing [10]. The principle has been demonstrated in laboratory scale experiments, particularly for the case of silicon melt [11] where the liquid zone had 0.04 m diameter. Typical AC EM field frequencies used in these experiments are relatively high (10–100 kHz), ensuring the high EM force density in the thin skin-layer where the separated particles are collected. Application to large scale silicon melt purification requires to create similar conditions in thin layers and additional means to transport the particles to these regions.

The large scale electromagnetically driven flow circulation exerts a drag force which contributes to the particle transport. The particles travel due to the hydrodynamic drag, buoyancy, turbulent fluctuations [12] and the electromagnetic force action at the particle. The electromagnetic force acts directly onto electrically conducting inclusions, however the electromagnetic force in the surrounding fluid creates the pressure distribution gradient leading to the integral force on the non-conducting inclusions. Often the electromagnetic force expression is used as derived by Leenov and Kolin [4] for the case of particles positioned in a constant crossed electric and magnetic fields. An improved theoretical derivation accounting for the electric current perturbation by the particle is available [13]. Experimental observations of the particles of various shapes and electrical conductivities in the liquid metal carrying electrical current are described in [14], where the expression similar to Leenov-Kolin's was derived, but for the condition of a gradient magnetic field, which results in the two-fold increase of the effective force on the particles. In the AC magnetic field case the time average force effects are often treated using the same formulae as for the constant uniform fields. The latter contradicts the skin-layer distribution of the force in the AC field [15]. Moreover, the total force in AC field contains also the pulsating part contributing to the particle drag by the 'history' and 'added mass' contributions [16]. The high mixing rates ensure intense turbulence in the bulk of molten metal, additionally affecting the particle dispersion. This paper attempts to demonstrate the effects leading to organized particle transport and the eventual modified distribution in the final ingot.

Mathematical Models

The mathematical modelling of the time dependent problem describing the EM field, heat-mass transfer, the solidification process, and the fluid flow evolution is accomplished using the SPHINX code developed at the University of Greenwich [17]. The time dependent cooling rate imposed by external graphite heaters and supplemented by the internal Joule heating, eventually leads to solidification at the bottom part of the melt. When the solidus temperature is reached at the bottom part of the melt, the model accounts for the phase transition and the liquid volume shape change. The fluid flow is solved using the Reynolds averaged turbulent flow model for an incompressible fluid with the effective viscosity ν_e and the turbulent thermal diffusivity α_e. The modified heat transfer equations are given in detail in previous publications [17, 18]. The method is based on the continuous coordinate transformation adapting to the surface and the containing vessel shape. The dynamic Lagrangian passive particle tracking algorithm was implemented in [18]. The position **R**(x, y, z, t) of an individual particle can be determined following its path in the fluid volume according to the variable total force **F**(x, y, z, t) acting on the particle [12, 15]:

$$\partial_t \mathbf{R} = \mathbf{u}_p, \quad m_p \partial_t \mathbf{u}_p = \mathbf{F}, \tag{1}$$

where $\mathbf{u}_p$ is the particle's velocity and m_p its mass. The force **F** acting locally on a spherical particle can be decomposed into the fluid drag force $\mathbf{F}_d$, the buoyancy force $\mathbf{F}_g$ and the effective electromagnetic force $\mathbf{F}_e$. The drag force depends on the particle Reynolds number

$$Re_p = r_p |\Delta \mathbf{u}| / \nu, \tag{2}$$

where r_p is the particle's radius, $\Delta\mathbf{u} = \mathbf{u} - \mathbf{u}_p$ is the (slip) velocity relative to the fluid velocity **u**, ν—the kinematic viscosity. For small particles (1 nm–100 µm) the Reynolds number Re_p is rather small, of the order 0.1–10, therefore the drag force can be approximated as instantaneous modified Stokes formula [12]. The buoyancy force due to the gravity **g** action on the particle and the surrounding fluid is

$$\mathbf{F}_g = (\rho_p - \rho) V_p \mathbf{g}, \tag{3}$$

ρ is the fluid density, ρ_p—the particle density, V_p—the particle volume.

Generally, the electromagnetic force acts directly on the electric current passing in the particle, however the EM force distribution in the surrounding fluid leads to an additional pressure redistribution on the surface of the particle, which results in a total force on the particle. For the case of electrically non-conducting particles (oxides, carbides, bubbles etc.) the effective AC EM force (from the pressure redistribution) is concentrated in the fluid EM skin-layer. According to our derivation, the effective resultant force on the particle can be represented by the original EM force in the fluid at the location of the particle, given as the sum of time average and oscillating components

$$\mathbf{F}_e = -V_p \left(\frac{3}{2} \widehat{\mathbf{f}}_e + \mathbf{f}'(t) \right). \tag{4}$$

Note, that Leenov-Kolin [4] derived a similar expression for DC constant fields

$$\mathbf{F}_{eDC} = -V_p \frac{3}{4} \mathbf{f}_e, \tag{5}$$

which gives two times smaller force acting on a non-conducting particle. The presence of the oscillating force component in (4) requires the modified drag expression for an oscillating spherical particle [16]:

$$\mathbf{F_d} = 6\pi\nu\rho r_p\left[(1+\varepsilon)\Delta\mathbf{u} + \frac{\varepsilon}{\omega}\left(1+\frac{2}{9}\varepsilon\right)\frac{d\Delta\mathbf{u}}{dt}\right], \quad \varepsilon = \sqrt{\frac{\omega r_p^2}{2\nu}}, \tag{6}$$

where the Stokes number ε can reach an order 1 in an AC field oscillating at typical frequencies $\omega \sim 10^3$–10^4 Hz. The Eq. (6) contains the instantaneous Stokes drag, the memory term (Basset force) and the added mass force.

The typical mixing flow is turbulent ($Re \sim 10^3$–10^4), which further requires a modification due to the stochastic part of fluid velocity in accordance to the resolved turbulent kinetic energy k and the local eddy life time (or the particle transit time in that eddy, whichever is the shortest) [12]. The use of k-ω turbulence model in the SPHINX code facilitates to obtain these quantities, which are locally interpolated to the particle position at each time step. The numerical integration of Eq. (1) is done for each individual particle of various properties depending on the initial seeding locations. The time dependent forces $\mathbf{F}$ are sensitive to the location and the instantaneous update of local slip velocity $\Delta\mathbf{u}$ (including the stochastic contribution). Therefore, the stability of long term integration along the particle tracks can be adversely affected by the choice of numerical integration scheme. The classical explicit integration schemes are limited to extremely small time steps, which could make the numerical solution difficult for the dynamic flow conditions. The exponential numerical scheme, proposed in [18], permits stable time integration of the particle tracks with the time steps of the order 0.1–10 ms (adjustable) used for the unsteady fluid flow solution in the examples considered below.

Numerical Solution Examples

Silicon G2.5 Experiment

The first experiment to apply the numerical model is the G2.5 size silicon melting furnace installed at the Padua university [20]. The numerical model is constructed for an axisymmetric equivalent setup to the experimental rectangular liquid zone. The axisymmetric version permits to compute in a reasonable time the required large number of time steps of the turbulent flow development while continuously updating the Lagrangian particle positions for the 3 sizes (1, 20 and 50 μm) of 200 initially random seeded SiC particles in the r, z plane. Figure 1 shows the model setup including the side coil, quartz crucible, graphite enclosure and the fluid flow for the case of downward travelling magnetic field (TF). The force density is relatively low in this experiment (max 800 N/m^3). Figure 1 (right) represents the computed fluid flow for the downward travelling magnetic field (TF). The dominating single toroidal vortex transports the particles upwards in the axial region and downwards along the wall, recirculating periodically the particles along the streamlines. This flow leads to a gradual collection of the particles in the near to wall region where they could be potentially retained if the force density is sufficient

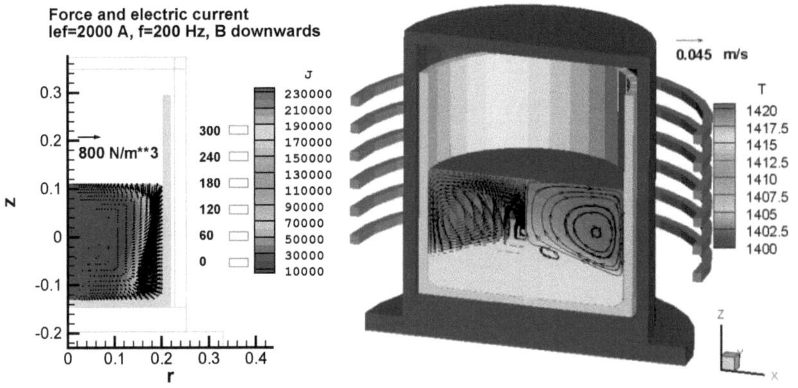

Fig. 1 The EM field distribution for downward TF in the G2.5 crucible (*left*) and the flow, temperature and solidification front at intermediate stage (*right*)

Fig. 2 The instantaneous streamlines and the Lagrangian particles for the downward TF at the intermediate solidification development stage

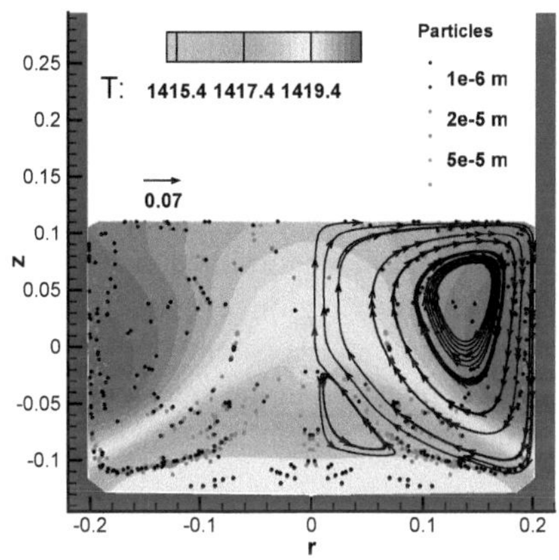

to overcome the drag and the sedimentation force (ρ_{SiC} = 3185, ρ_{Si} = 2560 kg/m^3) (see Fig. 2). The initially randomly generated 200 SiC particles tend to be organized in 'swarm' like structures aligned along the flow streamlines. Since the flow itself is transitional and oscillating, these particle clouds are periodically reorganized, being affected by inertia and the EM interaction when temporarily located in the AC force skin-layer. Clearly, the EM interaction is too weak to achieve any noticeable particle separation and retention for the field magnitudes used in this experiment. The dynamic tracking procedure used in this work is extending the previously used passive contaminate tracking in a stationary flow of silicon melts as presented, for instance, in [19].

The second case uses a single phase current in the same side coil, as shown in the Fig. 3. The EM force density is considerably higher (10,000 N/m^3) than in the previous example, however localized in a thin skin layer region. The fluid flow was calculated for the full 3D case of the experimental rectangular crucible, shown in

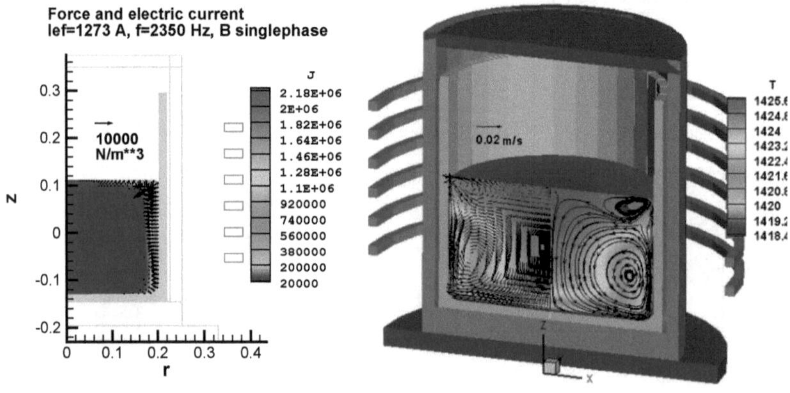

Fig. 3 The EM field distribution for single phase current in the G2.5 crucible (*left*) and the instantaneous flow and temperature at the initial stage (*right*)

the Fig. 4 (left). This compares rather favourably with the axisymmetric flow calculated with SPHINX for a long term solidification process (Figs. 3 and 4). The particle tracks were followed for a total time of 380 s. The cooling rate was set at 0.1 K/s, which is much higher than in a real experiment, but permits to compute the required time stepping in about 5 days computational time. The particles were recirculated in the toroidal loops many times, yet largely failed to be captured in the EM skin-layer region, eventually accumulating in the solidified material mostly in the bottom part. Apparently, a higher current magnitude will be required to ensure particle retention in the side skin-layer, or a combination of the TF and the higher frequency AC single frequency EM field.

Silicon G1 Experiment

The numerical experiments indicated that a smaller size crucible could benefit from higher EM force concentration, as for example in the G1 size crucible constructed at the Padua university. This new crucible uses a coil consisting of 3 sections to permit various combinations of AC EM fields (TF and/or singular phase). The Fig. 5 demonstrates a single phase current produced EM field distribution in the liquid zone and the corresponding fluid flow at the initial quasi-stationary stage. Following the similar random seeding of the particles and the long term particle tracking while the solidification sets in gradually at the bottom part, the flow and particle positions are shown in the Fig. 6. This case indicates a significant increase in the particle retention in the side layer where the high density EM force leads to the particle separation to the side and bottom layers. Further work is in progress to find the optimum combination and configuration of the EM field to achieve the goals of the project.

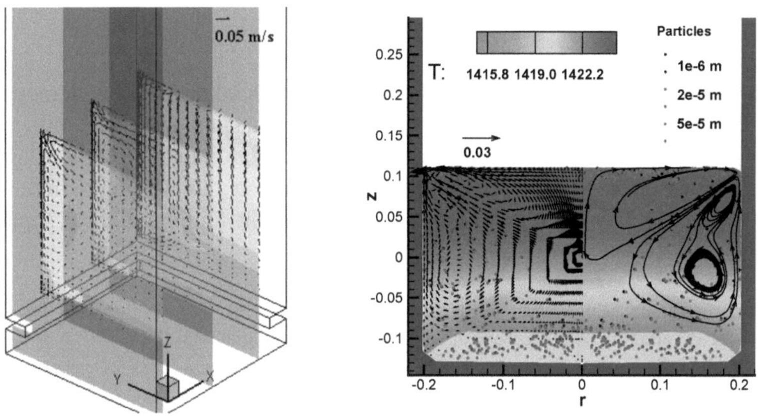

Fig. 4 The full 3D velocity field in the G2.5 crucible (*left*) and the Lagrangian particles for the single phase AC field at the initial solidification stage (*right*)

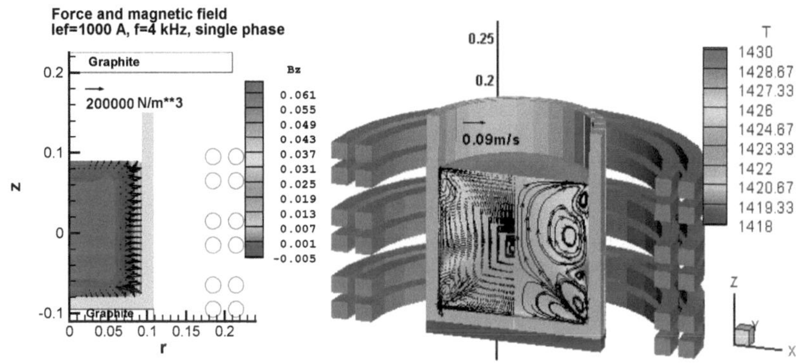

Fig. 5 The EM field distribution for single phase current in the G1 crucible (*left*) and the instantaneous flow and temperature at the initial stage (*right*)

Fig. 6 The particle distribution for the single phase AC field at the solidification stage in the G1 crucible

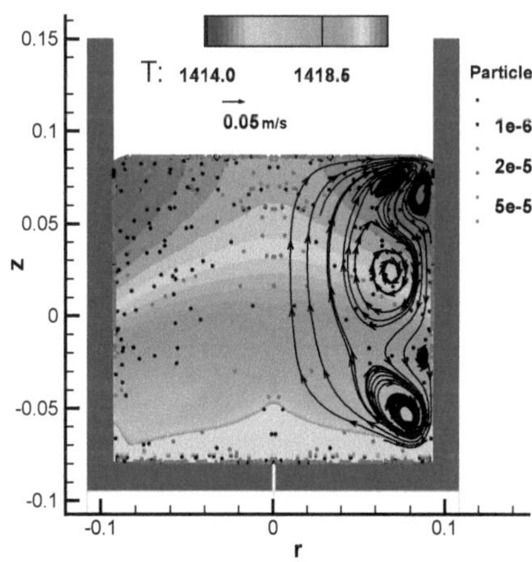

Conclusions

The dynamic particle tracking algorithm for AC EM field driven flow demonstrates the difference between the particle paths and the instantaneous streamlines. The fully developed turbulent flow affects the particle retention in the EM skin-layer via the competing effects of EM force, turbulent fluctuation and the drag variation. The smaller size G1 furnace type shows more promise in finding suitable separation conditions.

Acknowledgements This work is financed by SIKELOR [2]—a project funded by the 7th Framework Programme of the European Commission, sub-programme: ENV.2013.6.3–1, project reference 603718.

References

1. A. Ciftja et al., J. Metals **61**(11), 56–61 (2009)
2. I. Lombardi et al. (2011). *Proceeding of 26th European Photovoltaic Conference*, Hamburg. http://www.sikelor.eu/app/download/10479104399/Poster_Bamberg.pdf?t=1434606285
3. SIKELOR project of the 7th Framework Programme of the European Commission. Silicon kerf loss recycling. www.sikelor.eu
4. D. Leenov, A. Kolin, Journ. Chem. Phys. **22**(4), 683–688 (1954)
5. T. Toh, et al. *ISIJ International,* **47**(11), 1625–1632 (2007)
6. L. Barnard et al., Ironmaking and Steelmaking **20**(5), 344–349 (1993)
7. J.W. Haverkort, T. Peeters, Metall. Mater. Trans. **41B**, 1240–1246 (2010)
8. V. Bojarevics, A. Roy, *Magnetohydrodynamics*, **48**(1), 125–136 (2012)
9. S. Taniguchi et al., *The 6th PAMIR Conference on Fundamental and Applied MHD*, Riga, Latvia, 55 (2005)
10. L. Zhang et al., Metall. Materials Trans. **45B**, 2153–2185 (2014)
11. M. Kadhodabeigi et al., Trans. Nonferr. Met. China **22**, 2813–2821 (2012)
12. P.G. Tucker, J. Fluids Eng. **123**, 372–381 (2001)
13. H.K. Moffatt, A. Sellier, J. Fluid Mech. **464**, 279–286 (2002)
14. V. Bojarevics, J. Freibergs, E. Shilova, E. Shcherbinin, *Electrically Induced Vortical Flows* (Kluwer, Dordrecht, 1989)
15. V. Bojarevics, K. Pericleous, in *Proceeding of 8th International Conference Electromagnetic Processing of Materials*, Cannes, France (2015)
16. L.D. Landau, E.M. Lifshitz, *Fluid Mechanics* (Pergamon Press, 1987)
17. V. Bojarevics, K. Pericleous, M. Cross, Metall. Materials Trans. **31B**, 179–189 (2000)
18. V. Bojarevics, K. Pericleous, R. Brooks, Metall. Materials Trans. **40B**, 328–336 (2009)
19. N. Dropka, Ch. Frank-Rotsch, P. Rudolph, J. Crystal Growth **365**, 64–72 (2013)
20. F. Dughiero et al., in *2011 37th IEEE Photovoltaic Specialists Conference (PVSC),* 002151–002156 (2011)

Part XIII

Advances in Environmental Technologies: Recycling and Sustainability Joint Session: Advances in Environmental Technologies: Characterization and Uncertainty

Characteristics of Municipal Solid Waste Incineration Bottom Ash With Particulate Matters PM 2.5–PM 10

T. Thriveni, Ch. Ramakrishna and Ahn Ji Whan

Abstract Numerous reports and researches address the various environmental issues about the municipal solid waste incineration waste management of heavy metals and other particulate matters with the range of 10–2.5. Although in many developing and industrialization countries landfill with the disposal of municipal solid waste, open incineration has become a common practice. Bottom ash contains heavy metals including particulate matters. The investigation of size distribution for the heavy metals and particulate matters in the bottom ash is essential. However, ultra-fine particle size and heavy metal stabilization are the most important requirements for ecofriendly by products. There are wide variety commercial technologies available for monitoring PM 10, PM 2.5 and ultra-fine particles. In this paper, we reported the studies on physical and chemical characteristics of municipal solid waste incineration (MSWI) bottom ash containing particulate matter whose particulate sizes are lower than PM10, PM2.5 and heavy metal stabilization method by carbonation were investigated.

Keywords Municipal Solid Waste · PM 10–PM 2.5 · Heavy metals · Carbonation · Stabilization

Introduction

Currently, developing countries were faces problems associated with poorly managed solid waste management system. There are several reports published on the low performance of MSW accessibility of the city due to the lack of properly designed collection route system and poor condition of the final dump site, littering

T. Thriveni · A.J. Whan (✉)
Mineral Resources Research Division, Korea Institute of Geosciences and Mineral Resources, 124 Gwahagno, Gajeong-dong, Yuseong-gu Daejeon 305-350, Republic of Korea
e-mail: ahnjw@kigam.re.kr

Ch. Ramakrishna
Department of R&D Team, Hanil Cement Corporation, Danyang 395-903, Korea

L. Zhang et al. (eds.), *Energy Technology 2017*,
The Minerals, Metals & Materials Series, DOI 10.1007/978-3-319-52192-3_41

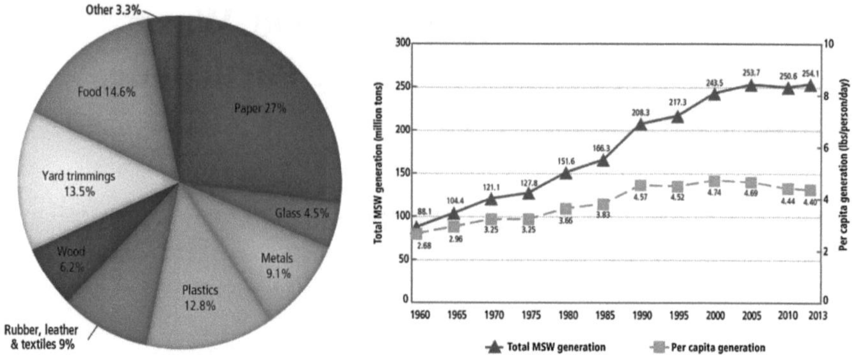

Fig. 1 Total municipal solid waste generation, 2013

of the corner around the skips which encouraged illegal dumping which leads the environmental and health risks. Large amounts of residues are generated from the various industries, which are very toxic nature and thus not been utilized beneficially. This effluents are being directly discharged to the nearby water bodies causing severe ailments like overexcitement, skin irritation, stomach pain, nausea and breathing problem. Adequate planning of waste management is essential if communities and regions are to successfully address the challenge of a sustainable development, including resource conservation, climate protection, and pollution prevention. Municipal solid waste management plays a significant role in climate change mitigation [1]. The global waste generation trend presented in the Fig. 1 [2].

Currently, air pollution is a major emerging global environmental problem. Particulate matter (PM) pollution in air affects people's living quality tremendously, and it poses a serious health threat to the public health as well as influencing visibility, direct and indirect radiative forcing, climate, and ecosystems. Particulate matter (PM) is a complex pollutant and it contains a mixture of both organic and inorganic particles, such as dust, pollen, soot, smoke, and liquid droplets found in the air which are toxic and hazardous. PM air pollution is often discussed in terms of particle size because of the distinct characteristics (origin, chemical species and atmospheric behavior) associated with different particle size classes. EPA and the states have established a network to monitor, assessment and record PM 2.5 and PM 10 concentrations throughout the U.S. for controlling the particulate matter emissions. In many developing countries, greenhouse gases (GHGs), reactive trace gases, particulate matter (PM), and toxic compounds emissions from waste combustors are more challenging to national and global inventories.

Atmospheric particulate matter (PM) is one of the main environmental pollutants in China and South Korea also gardually increase the atmospheric pollution due to rapid industrialization and urbanization have resulted in an accelerated rate of waste generation, and to the environmental challenges on the relatively small land mass of the Republic of Korea. Air pollutants emissions by various sectors presented in Fig. 2 [3].

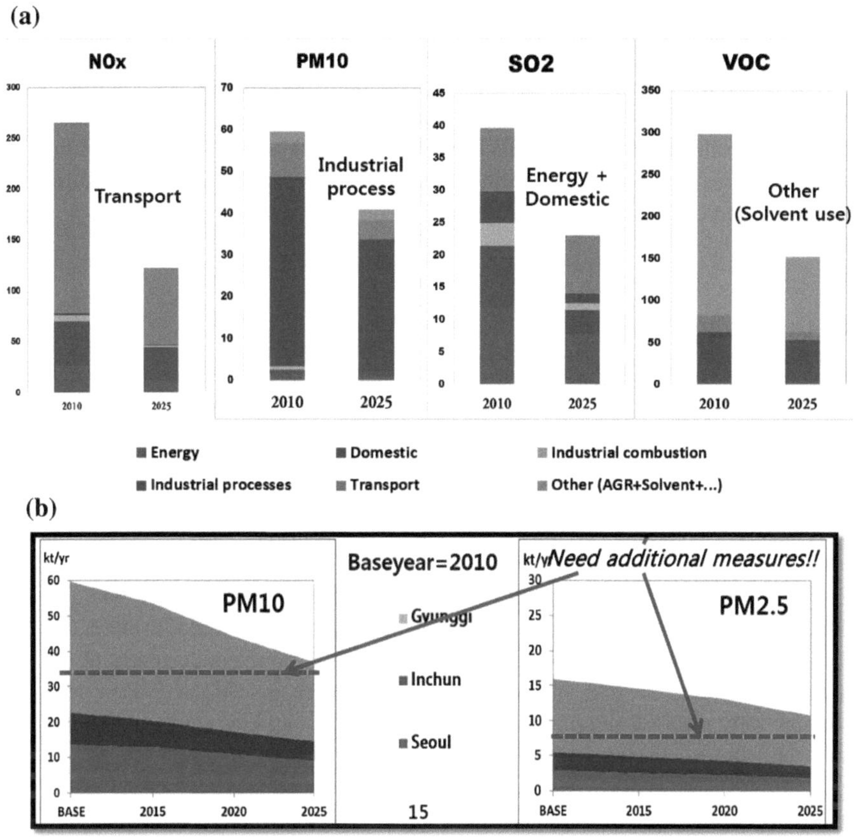

Fig. 2 **a** Change of emissions by sector (SAQMP)(adopted from Ref. 3) **b** Future emissions for SAQMP scenario (adopted from Ref. 3)

World Health Organization (WHO) set air quality standards in ambient air in 2005 and 2006 [4, 5]. In 2009, the Japanese government also set PM 2.5 environmental standards. The environmental standard states that PM 2.5 concentrations should be less than 15 lg/m^3 and annual average is less than 35 lg/m^3 as a daily average [6].

Several studies have hypothesized that atmospheric PM of smaller sizes (i.e., PM 2.5), due to their larger surface area to mass ratio, can be more potent than larger ones (i.e., PM 10) in inducing cytotoxic and/or inflammatory responses in various lung models [7, 8]. Moreover, PM 2.5 has been listed as an important air pollutant due to its potential effects of bioaccumulation, oxidation and inflammation in the human body [9–11]. Recent research has found that the ultrafine particles only account for a small proportion of PM, however, they contain many soluble metallic elements that can induce macrophage reactive oxygen species (ROS) activity [12].

Therefore, the relationships between specific PM and health conditions still remains as an active and important topic of aerosol research.

The purpose of this study was to determine the metallic elements of ambient PM samples surrounding MSWI and investigate their properties. Size-segregated particles were collected around a MSWI in South Korea. The mass size distributions of metallic elements in the PM (ultrafine, fine and coarse particles) were measured and to discuss the characteristics of PM 2.5.

Methodology

The aim of the study was to provide the spatial variation and characteristics of ambient PM 2.5 in South Korea and to determine the contributions of various pollution sources. To understand the spatial variation of respirable particulates, PM2.5 was sampled at five areas, representing the different areas of South Korea. To provide information on source profiles, fly ash of an MSW incinerator, and bottom ash of an MSW incinerator.

Sampling of PM 2.5: Ambient PM 2.5. The measurements were taken at five sites in South Korea, representing the ambient air quality in the Southeast Korea. At each site, samples were collected. The sampling was done concurrently at the five sites from Korea. So that spatial variations could be investigated. Five identical samples were used for sampling ambient PM2.5. The particles were collected on the poly tetra fluoro ethylene (PTFE) filters.

Chemical Analysis

The PTFE filters were stored in a drying cabinet and balanced in relative humidity of 40–45% at room temperature for 24 h for weighing both before and after sampling. Then PTFE filters were weighed using an electronic balance (resolution ± 1 μg, Crystal 250, Gibertini, Italy).

Result and discussion

Characterization of Particulate Matters

The basic composition of MSWI bottom ash presented in the Table 1. The flow sheet of experimental procedure presented in Fig. 3. We followed the two process (1) dry sieving method (2) wet sieving method (Fig. 4).

Table 1 Particle size distribution ratio (Boryeong B/A)

Sieve	1 time		2 time		3 time		Average (%)
16 mm over	199	12.04%	161.4	9.42%	167.8	10.22%	10.56
16–9.50	199	12.04%	238.3	13.91%	177.4	10.80%	12.25
9.50–4.75	178.2	10.78%	223.3	13.03%	226.4	13.79%	12.53
4.75–2.36	191.9	11.61%	207.2	12.09%	211.0	12.85%	12.18
2.36–1.18	189.6	11.47%	211.7	12.35%	219.4	13.36%	12.39
1.18–0.6	132.4	8.01%	150.7	8.79%	162.6	9.90%	8.90
0.6–0.3	99.9	6.04%	110.8	6.47%	122.0	7.43%	6.65
0.3–0.15	115.5	6.99%	99.6	5.81%	92.2	5.62%	6.14
0.15 under	347.2	21.01%	310.5	18.12%	263.1	16.02%	18.38
	1652.7	99.99%	1713.5	99.99%	1641.9	99.99%	99.98
Total weight	1667.2	−14.5	1728.6	−15.1	1654.3	−12.4	

Fig. 3 Flow sheet of dry sieving method experiments

(a)

(b)

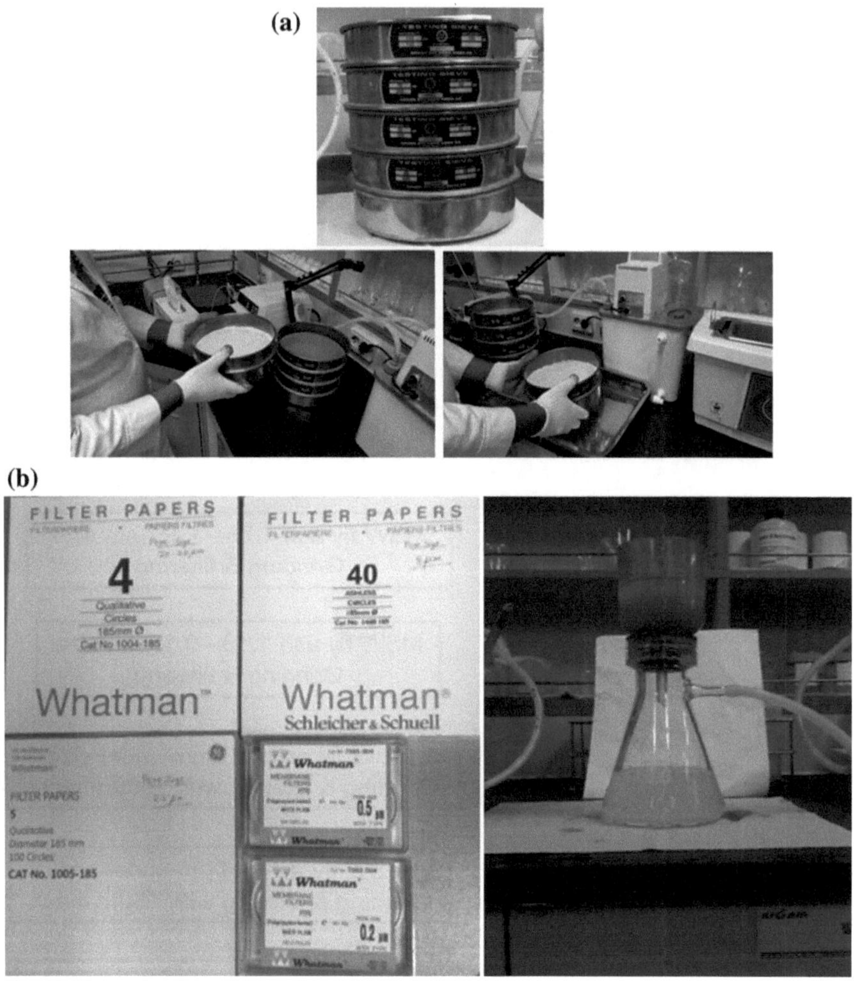

Fig. 4 **a** Dry sieving method experiments **b** wet sieving method experiments

Conclusions

In this study, we collected MSW incineration samples from five different areas of Seoul, South Korea. We investigated the primary characteristics of PM 2.5–10 PM, and we concluded our preliminary data suggested that MSW having PM 2.5–10 μm size. Our future aim is to solidify these particulate matter including heavy metals by carbonation process, these process is eco-friendly and cost effective process.

Acknowledgements The authors are very grateful to the Korea Institute of Energy Technology Evaluation and Planning (KETEP) through the ETI program, Ministry of Trade, Industry and Energy (MOTIE, Project No. 2013T100100021) for financial support of this research.

References

1. G.W. Asrat, W. Belay, S.C. Bhagwan, Wet coffee processing waste management prctice in Ethiopia. Asian J. Sci. Technol. **6**(5), 1467–1471 (2015)
2. Environmental Protection Agency (EPA), Advancing sustainable materials management: facts and figures 2013, in Assessing Trends in Material Generation, Recycling and Disposal in the United States, pp. 1–186, June 2015
3. J.H. Woo, Y. Kim, K.C. Choi, D.K. Lee, J.H. Park, T.Y. Jung, S.C. Hong, S. Kim, J.H. Hong, Future emissions of air pollutants and greenhouse gases over Korea and East Asia, in IHTAP Emission Scenario Workshop, 12 Feb 2015
4. EN14907, Ambient air quality standard gravimetric measurement method for the determination of the PM2.5 mass fraction of suspended particulate matter, p. 1 (2005)
5. World Health organization (WHO) air quality guidelines for particulate matter, ozone, nitrogen dioxide and sulphur dioxide. Global update 2005. Summary of risk assessment, WHO Geneva (2006)
6. M. Takaoka, K. Shiota, G. Imai, K. Oshita, Emission of particulate matter 2.5 (PM2.5) and elements from municipal solid waste incinerators. J. Mater. Cycles. Waste Manag. **18**, 72–80 (2016)
7. A.R. Osornio-Vargas, J.C. Bonner, E. Alfaro-Moreno, L. Martinez, C. Garcia-Cuellar, S.P.D. Rosales, J. Miranda, I. Rosas, Proinflammatory and cytotoxic effects of Mexico City air pollution particulate matter in vitro are dependent on particle size and composition. Environ. Health Perspect **111**, 1289–1293 (2003)
8. P.E. Schwarze, J. Ovrevik, R.B. Hetland, R. Becher, F.R. Cassee, M. Lag, M. Lovik, E. Dybing, M. Refsnes, Importance of size and composition of particles for effects on cells in vitro. Inhal. Toxicol. **19**, 17–22 (2007)
9. X. Deng, W. Rui, F. Zhang, W. Ding, PM2.5 induces Nrf2-mediated defense mechanisms against oxidative stress by activating PIK3/AKT signaling pathway in human lung alveolar epithelial A549 cells. Cell Biol. Toxicol. **2**, 143–157 (2013)
10. X. Deng, F. Zhang, W. Rui, F. Long, L. Wang, Z. Feng, D. Chen, W. Ding, PM2.5-induced oxidative stress triggers autophagy in human lung epithelial A549 cells. Toxicol. Vitro **27**, 1762–1770 (2013)
11. D. Wang, P. Pakbin, M.M. Shafer, D. Antkiewicz, J.J. Schauer, C. Sioutas, Macrophage reactive oxygen species activity of water-soluble and water-insoluble fractions of ambient coarse, PM2.5 and ultrafine particulate matter (PM) in Los Angeles. Atmos. Environ **77**, 301–310 (2013)
12. J.G. Watson, N.F. Robinson, C. Lewis, T. Coulter, Chemical mass balance receptor model–version 8 (CMB8) User's Manual; Document No. 1808.1D1, Desert Research Institute: Reno, NV (1997)

Part XIV
Advances in Environmental Technologies: Recycling and Sustainability Joint Session: New Areas of Value Recovery

Accelerating Life-Cycle Management Protocols for New Generation Batteries

Timothy W. Ellis and John A. Howes

Abstract The production of lead-acid batteries is defined by a closed-loop product life-cycle. Active material and grid alloys converge on an adaptable specification that improves both performance and recyclability. This mapping of product requirements for lead-acid batteries has taken over 80 years to develop. But, as the use of batteries made with other chemistries continues to grow, the need for a comparable life-cycle management process becomes increasingly apparent given societal needs to manage raw materials, e.g. the European Union Battery Directive recycling targets. What has taken more than 80 years for lead-acid to evolve in a closed-loop paradigm must now be accomplished in less time for batteries using other chemistries and technologies. End-of-life costs are embedded in the product cost of lead-acid batteries. However, end-of-life costs are not reflected in the product cost of batteries made with other chemistries. Those costs are borne by other constituencies with often inefficient and environmentally damaging results. Lead-acid is a materials technology management model for other products.

Keywords Circular economy · Recycling · Batteries · Lithium

Introduction

Lithium-ion batteries have become the dominant power supply source for portable consumer electronic products and their presence is increasing in larger applications such as automotive, utilities, military and aerospace. Where the global market for lithium-based batteries was approximately \$7.9 billion in 2008, it is estimated to

T.W. Ellis (✉)
SAE Battery Recycling Committee, & Chairman, Advanced Lead-Acid Battery Consortium, RSR Technologies, Inc, 4828 Calvert St, Dallas, TX 75247, USA
e-mail: tellis@rsrtechnologies.com

J.A. Howes
Principal, Vice-Chairman, American Bar Association Energy, Infrastructure and Reliability Committee, Redland Energy Group, 1875 Eye St., NW, Washington, DC 20006, USA

L. Zhang et al. (eds.), *Energy Technology 2017*,
The Minerals, Metals & Materials Series, DOI 10.1007/978-3-319-52192-3_42

grow to about $46.2 billion by 2022 [1]. This represents an estimated compound annual growth rate of 10.8% projected to occur between 2016 and 2022. The reason for this impressive growth is the lithium-ion's superior energy density compared to lead-acid, nickel-cadmium and nickel metal hydride batteries.

However, with this market growth comes certain risks in environmental protection, health, safety and other considerations. Efforts to mitigate these risks have not kept pace with the technological development of lithium-based batteries. Nor does a regulatory framework exist in the U.S. for lithium battery recycling targets. This is in contrast to lead-acid batteries, which have a closed-loop profile and a 99% recycling rate. This superior resource management profile of lead-acid batteries has resulted in high safety and environmental protection standards for more than 80 years. Moreover, where U.S. laws and regulations governing the manufacture, use and disposal of lead-acid batteries are among the world's most strict, the level of regulations applied to lithium batteries is considerably less developed [2]. Discarded lithium batteries have a recycling rate of less than 5%, meaning the majority of batteries end up in landfills or other waste streams.

Lithium-Ion Battery Safety Discussion

Batteries in the lithium family contain potentially toxic materials such as copper, nickel, lead, and fluoride compounds. They also have organic electrolytes that present fire hazards [3]. Another complexity is thermodynamic. Materials used in Li-ion cathodes and some anodes require greater energy input for separation process, DG = $LiFePO_4$ (−1819 kJ/mol) [4], $PbSO_4$(−811.3 kJ/mol) [5], PbO (−188 kJ/mol) [5], than lead-acid batteries, which have fewer materials in a more standardized design, thereby making lead-acid battery recycling much easier.

Automotive manufacturers and other industries are increasingly reliant on advanced energy storage technologies to meet higher electrification demands. Where flooded and valve-regulated lead-acid batteries have dominated the rechargeable battery market for more than 150 years, newer battery designs, particularly lithium-based, are becoming more popular.

As lithium-ion battery use increases, however, safety incidents are occurring, including those at secondary lead smelters because of the entrance (both intentional and unintentional) of lithium-ion batteries into the waste streams of the smelters. Explosions and fires have resulted from the combustion of organic electrolytes in lithium batteries as they are crushed at the start of the recycling process. These challenges require increased attention by industry, consumers, environmental activists and government. Several major lead processing and lead-based battery organizations [4] have proposed plans to prevent the improper flow of non-lead-acid batteries into secondary lead smelters. The plans include better on-site sorting, education of upstream manufacturers, distributors and collectors, along with global standards for energy storage identification and government policies to ensure enforcement.

The fact that recycling costs are not embedded in the retail price of lithium-ion and other non-lead-acid batteries creates an economic motive to put discarded lithium-ion batteries into lead-acid recycling streams.

Lithium batteries, especially those in large multi-cell packs, present problems to recyclers because of complex assemblies requiring special dismantling procedures. Some even contain multiple technologies/chemistries within a single package. This requires development of universal mechanisms for sorting to prevent safety and environmental hazards, cross contamination, and negative economic impact to manufacturers and recyclers.

The Society of Automotive Engineers (SAE) is working with battery standards organizations to develop an energy storage identification system for systems with maximum voltages >12 V and <60 V, and 6 V SLI (starting, lighting and ignition), to improve the safety and efficiency of dismantling, sorting and recycling operations at secondary smelters [5]. This identification system is designed to support the proper and efficient recycling of rechargeable battery systems.

Battery End-of-Life Discussion

Throughout the industrial age, many products have been designed, manufactured, used and discarded in a linear manner with little regard to the consequences of how waste would impact the environment. This linear perspective has been challenged in the past 30 years because of rising concern about waste pollution, giving rise to concepts such as resource efficiency and the circular economy.

Increased attention to concepts like the circular economy are, in turn, resulting in discussions on whether worldwide government policies are keeping pace with growing environmental concerns and whether incentives for economic development are adequately considering the impact of such incentives on the environment, worker safety, etc. There is growing recognition of a need to place environmental protection, public health, worker safety and other considerations on equal footing with economic development and investment decision making.

Investment decisions on manufacturing technologies and processes will in future years be made in one of two ways, either through the traditional linear method or through a more integrated life-cycle management method. As concerns about product end-of-life increase, the need for total life-cycle analysis will require significant international cooperation among governments, industries, the investment community and environmental advocates on the issues of product reuse, remanufacturing and recycling. In the metallurgical industry, lessons learned in the lead processing industry are becoming more apparent and relevant to industries involved in the processing of other materials.

In the case of batteries, the advisability of government intervention will be discussed as a way to incentivize the development and deployment of optimal end-of-life technologies for batteries using lithium, nickel, cobalt and other materials. This will require, among other things:

- Design of batteries with materials more suitable for recycling
- Development and expansion of recycling facilities
- Development of infrastructure to collect, transport and deliver batteries to recyclers
- Sufficient volumes of waste that will be collected and sorted in a manger best suited to metallurgical processing.

The U.S. lags behind many European countries in putting rechargeable battery take-back laws into effect. The European Union (EU) has mandated a 25% collection rate for lithium-ion batteries and laws in the EU are designed to ensure that end-of-life management for rechargeable batteries will be accomplished in an efficient and environmentally friendly manner [6]. However, in the U.S., there is no federal law in place. Only New York and California have laws that ban the disposal of rechargeable batteries in landfills and require battery dealers to take back batteries [7].

The current collection rate of spent lithium-ion batteries is less than 10 per cent and the recycling rate is less than five per cent. Lithium-ion batteries from the first generation of hybrid and electric vehicles have begun reaching their end-of-life and will grow to about 25,000 tons by 2025 [8]. This, however, is dwarfed by the ~10 million tons of lead-acid batteries presently recycled worldwide and remains insufficient to justify large scale investment in lithium recycling. Several factors could improve this rate and make recycling more economically viable. Among them is designing batteries so they can be readily placed into recycling infrastructure. Lithium-ion battery cell design continues to require attention to minimize its environmental footprint. Lithium-ion battery cells have a wide variety of materials, powders, metal foils, etc. in a number of designs that lack common definition [9].

Designing lithium-ion batteries so they can be more readily recycled would help improve investor confidence in lithium-ion battery production and could also result in a larger supply of "feedstock" for recyclers. It also would help stabilize the price of cobalt, copper, lead and lithium itself, thereby making elemental recovery a more attractive investment for battery manufacturers and recyclers. This increase in lithium-ion battery sales for vehicles parallels the increasing battery sales for consumer electronic products such as cellular phones, handheld appliances, etc.

Until these safety and environmental issues are proactively addressed, the risk of developing a recycling infrastructure is likely to remain too high for lithium battery manufacturers. Recycling of lithium-ion batteries is presently being undertaken by third parties under contract to manufacturers and collection companies. (An exception is Tesla, which is developing a large lithium-ion battery production facility and a companion recycling unit in Nevada.)

In the meantime, recycling is considered to be one of three options for handling lithium-ion batteries at their end of life. Other options are remanufacturing and repurposing [10]. However, recycling remains the optimal path for lithium-ion batteries when the other options become impractical once the batteries lose energy performance potential for any purpose.

Life-Cycle Analysis Considerations

Presently, most lithium batteries are produced on a cradle-to-grave basis, which does not include cost considerations for after life management. Nevertheless, cradle-to-cradle considerations are being carried out by various academic and policy advisory bodies, which have done valuable research and developed frameworks and indices that ultimately must be adopted by industry and government.

For metals lacking the sustainability profile of lead to reach a comparable level will require evaluation of a broad cross-section of issues in an integrated—rather than isolated—manner. This will become important for metals processors as they represent to customers how their materials conform to accepted industry standards. One such approach [11] is to evaluate how products can be ranked on a scale comprised of the following indicator groups:

- *Climate change*: The metallurgical industry is typically very energy intensive. When less raw materials and water are used, less energy is required to produce and process them.
- *Water*: To minimize water intensity, the most water-efficient and suitable raw materials need to be chosen and all processes selected so that they use as little water as possible.
- *Energy*: It is important to minimize energy consumption by using energy-efficient process technologies and renewable energy sources that decrease environmental impacts.
- *Land use*: Since land use cannot be avoided (except for recycled materials), the impacts must be minimized, mitigated, and if necessary, compensated.
- *Chemical risks*: The use of environmentally friendly chemicals or safety measures taken for hazardous chemicals may affect resource use.
- *Resource depletion*: Resource depletion is a globally recognized challenge and the metallurgical industry is an intensive consumer of nonrenewable resources.
- *Material efficiency*: Raw material efficiency can be improved by choosing suitable materials, choosing suitable processing methods, and using all fractions of the materials.
- *Urecovered materials*: The unrecovered materials indicators look at how much is lost to which phase, so that recovery could be improved.
- *Emissions*: Acidification and eutrophication potentials are important indicators because the environmental burdens related to these may not only be due to the energy use during the production process but also other factors typical of mining and the metal industry.
- *End use and end of life*: Recyclability has to be considered at the product design phase. The evaluation incorporates the reusability of a product, recycling infrastructure for a product, and also the possible contamination and material losses in the recycling phase.

Eco-Cost Model

An example of how such indicators can be utilized in an integrated framework is the Eco-cost model, which is based on the sum of marginal costs for the prevention of environmental and safety burdens arising from the production of batteries and the products in which they are used [12]. These costs of prevention (which are included in lead-acid batteries) are not included in the cost of production of other batteries. Including them in the cost of production, in theory, will create a "no-effect-level" on harmful emissions to air, land and water arising from the use of the project.

A calculation structure of the Eco-cost model has been created [13] by the Delft University of Technology, utilizing Simapro software:

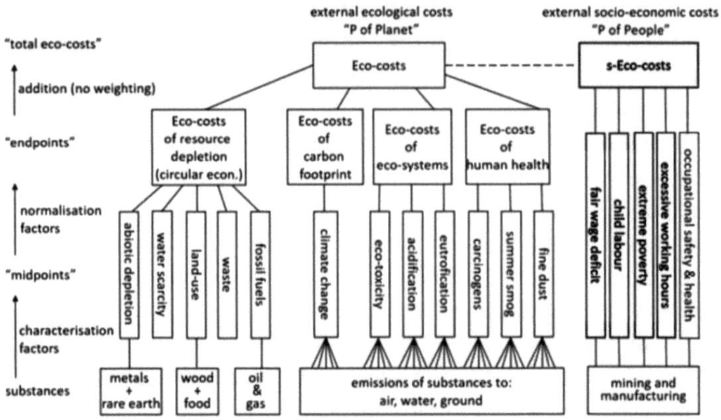

The Eco-cost system was originally designed to reflect the costs of preventing global warming (primarily CO_2) into the costs of products. Eco-costs were calculated to reflect the utilization of "non-carbon" sources of energy to displace "carbon-based" energy sources. The cost effects of toxicity on human health can also be calculated, thereby providing insight into the true costs of products and services entering into a worldwide economy that is becoming increasingly aware of the need for life-cycle "cradle-to-cradle" analysis.

In the case of non-ferrous metals, an eco-cost model [14] has been created to show how consideration of human health, toxicity, resource depletion and carbon footprint contributes to an assessment of how various metals compare on a life-cycle basis. It is important to note that many of these elements are co-produced from more than one ore body, i.e. lead from Zinc and/or Copper deposits, Li from salt brines containing Na, K and Cl, Co from Copper ores, etc. The possibility of extra costs for disposal of unwanted byproducts will impact life-cycle value calculations.

Selected materials, metals, non-ferrous

Unit	Name	Total Eco-Cost €	Human Health Eco-Cost €	Exo-toxicity Eco-Cost €	Resource depletion Eco-Cost €	Carbon footprint Eco-Cost €	Carbon footprint kg CO_2 equivalent
kg	2016 Aluminum (primary)	3.97	0.06	0.68	1.96	1.26	9.35
kg	2016 Aluminum (secondary)	0.28	0.00	0.02	0.01	0.25	1.86
kg	2016 Cobalt	45.65	0.01	0.80	43.80	1.04	7.72
kg	2016 Copper (primary)	3.08	0.00	0.21	2.44	0.43	3.19
kg	2016 Copper (secondary)	0.34	0.00	0.03	0.01	0.30	2.23
kg	2016 Lead (primary)	1.71	0.05	0.36	1.04	0.25	1.89
kg	2016 Lead (secondary)	0.12	0.00	0.01	0.01	0.10	0.77
kg	2016 Lithium	106.04	0.03	0.54	103.00	2.48	18.34
kg	2016 Nickel (primary)	14.92	0.11	1.46	9.02	4.33	32.10
kg	2016 Nickel (secondary)	0.34	0.00	0.03	0.01	0.30	2.23
kg	2016 Tungsten	5.41	0.79	4.15	0.08	0.39	2.87
kg	2016 Vanadium	39.10	0.42	3.97	28.30	6.41	47.45
kg	2016 Zinc (primary)	2.58	0.13	0.21	1.78	0.44	3.29
kg	2016 Zinc (secondary)	0.13	0.00	0.01	0.02	0.10	0.75

The Eco-cost data in the table above can be considered in parallel with data from the United Nations Environmental Program (UNEP) on the recycling rates of various metals:

Global Average Recycled Content

Recycled range	
• 50%	Lead
• 25%	Aluminum, Nickel, Tungsten
• 10%	Copper, Zinc
• 1%	Lithium
Data unavailable	Cobalt

It is apparent that metals processed in the secondary (recycled materials) market have considerably lower eco-costs than those processed in the primary (virgin materials) market. It also is apparent that the superior eco-cost profile of lead is directly related to its status as the most highly recycled of non-ferrous materials. The characteristics of lead can serve as a model for other metals as the development of advanced batteries continues.

Cost of Inaction

Finally, constituencies in the lithium battery recycling discussion will be required to consider the potential cost of inaction. These constituencies include battery manufacturers, recyclers, environmentalists, consumer advocates, industrial consumers, government agencies, non-government oversight organizations and research institutions. If the constituencies cannot agree on protocols, guidelines or standards for total life-cycle management, then the consequences will become readily apparent. These consequences ultimately include costs associated with the remediation of landfills where batteries are being dumped without an adequate end-of-life system.

Average cleanup cost for a hazardous waste landfill cleanup can be as much as $27 million per site [15]. Long term remediation cost constitutes the bulk of the overall cost and is highly dependent on the degree of permanence. These remediation steps range from soil capping ($79 per cubic yard); soil treatment ($231 per cubic yard) and excavation/removal ($1428 per cubic yard).

To the extent the lithium battery constituencies can agree on end-of-life protocols, liability burdens such as these may be avoided. Resource recovery and reuse can then proceed efficiently with greater safety and environmental protection benefits.

References

1. Allied Market Research titled, World Lithium-Ion Battery Market: Opportunities and Forecasts, 2015–2022
2. X. Wang, G. Gaustad, *Economies of Scale For Future Lithium-Ion Battery Recycling Infrastructure* (Rochester Institute of Technology, New York, 2013)
3. K. Chen, et al., *Potential Environmental and Human Health Impacts of Rechargeable Lithium Batteries in Electronic Waste* (University of California, Irvine, 2013)
4. S. Ping Ong et al., Li–Fe–P–O_2 Phase diagram from first principles calculations. Chem. Mater. **20**(5), 1798–1807 (2008)
5. CRC Handbook of Chemistry and Physics, 97th Edition, W. M. Haynes Editor, CRC Press, Boca Raton
6. Battery Council International, Association of Battery Recyclers, International Lead Association, and EUROBAT, statement, Marcy, 2016, http://www.altenergymag.com/article/2016/03/lead-recycling-threatened-by-lithium-ion/23120
7. Lead Recycling Threatened By Lithium-ion, altenergymag.com, 17 Mar 2016
8. European Union Battery Directive—http://ec.europa.eu/environment/waste/batteries/pdf/faq.pdf
9. X. Wang, Gabrielle Gaustad, op. cit
10. K. Richa, C.W. Babbitt, G. Gaustad, X. Wang, A future perspective on lithium-ion battery waste flows from electric vehicles. Resour. Conserv. Recycl. 2013
11. Gaines, *The Future of Automotive Lithium-Ion Battery Recycling: Charting a Sustainable Course* (Argonne National Laboratory, Lemont, 2014)
12. Standridge, *Remanufacturing, Repurposing, and Recycling of Post-Vehicle-Application Lithium-ion Batteries* (Mineta National Transit Research Consortium, 2014)
13. R. Reuter et al., Eco-efficiency Indicator Framework Implemented In The Metallurgical Market (Springer, Berlin, 2016)
14. The Eco-cost system was first introduced and published in 2004 in the International Journal of LCA
15. Delft University of Technology, http://ecocostsvalue.com/EVR/model/theory/subject/1-general.html
16. Excerpted from Idematapp 2016 with eco-costs and carbon footprint.xlsx excel file, http://ecocostsvalue.com/EVR/model/theory/subject/5-data.html
17. G. Cairns, Cost-benefit analysis of the clean-up of hazardous waste sites, in *Integrated Waste Management*, vol. I (Intech, 2011)

Recovery of Metals and Nonmetals from Waste Printed Circuit Boards (PCBs) by Physical Recycling Techniques

Muammer Kaya

Abstract This paper reviews the existing and state of art knowledge for electronic waste (e-waste) recycling. Electrical/electronic equipments (EEEs) which are unwanted, broken or discarded by their original users are known as e-waste. The main purpose of this study is to provide a review of e-waste problem, give e-waste management strategies and introduce various physical e-waste recycling processes with special attention towards extraction of both metallic values and nonmetallic substances. The hazards arise from the presence of heavy metals (e.g. Hg, Cd, Pb etc.), brominated flame retardants (BFRs) and other potentially harmful substances in e-waste. Due to the presence of these substances, e-waste is generally considered as hazardous waste and may pose significant human and environmental health risks if improperly managed. This review paper describes the potential hazards and economic opportunities of e-waste. First of all, an overview of e-waste/printed circuit board (PCB) components and substances are given. Current status and future perspectives of e-waste and waste PCB recycling techniques are described. E-waste characterization, dismantling, liberation, classification and separation processes, are also covered. Manual selective dismantling after desoldering and metal-nonmetal liberation at −150 μm with two step crushing are seen to be the best techniques. After size reduction, mainly physical separation processes employing gravity, electrostatic, magnetic separators etc., which are commonly used in mineral processing, have been critically reviewed here for separation of metals and nonmetals. The recovery of pure metals from e-waste after physical separation can be performed by pyrometallurgical, hydrometallurgical or biohydro metallurgical routes, which are not covered here. Suitable PCB recycling flowsheets for industrial applications are also given. E-waste recycling will be a very important sector in the near future from economic and environmental perspectives for both developing and developed countries. Recycling technology aims to take today's waste and turn it into conflict-free, sustainable polymetallic secondary resources (i.e. Urban Mining) for tomorrow. Recycling technology must ensure that e-waste is processed in an environmentally friendly manner, with high metal/nonmetal recovery efficiency and

M. Kaya (✉)
Mining Engineering Department, Eskişehir Osmangazi University, 26480 Eskişehir, Turkey
e-mail: mkaya@ogu.edu.tr

L. Zhang et al. (eds.), *Energy Technology 2017*,
The Minerals, Metals & Materials Series, DOI 10.1007/978-3-319-52192-3_43

lowered carbon footprint. Taking into consideration our depleting natural resources, this Urban Mining approach offers quite a few benefits; it provides a substantial secondary resource, while also ensuring that the environment is kept free from harmful toxins in e-waste. This results in increased energy efficiency and lowers demand for mining of new raw materials.

Keywords E-waste · PCB · Dismantling · Recycling · Metal recovery · Nonmetal recovery

Introduction

E-waste comprises of waste electric and electronic equipments (WEEE) or goods which are not fit for their originally intended use. Half of the e-waste comes from electrical appliances and the half from electronic goods. EEEs can become e-waste due to rapid advancement in technology; development in society; change in style, fashion and status; greater demands on EEE; nearing the end of their useful life and not taking precaution while handling them. This makes the replacement of EEE become more and more frequent, which results in large quantities of e-waste need to be disposed [1, 2].

E-waste is one of the fastest growing waste streams and it has been estimated that these items already constitute about 8% of municipal waste [3]. Therefore, recycle of valuable metallic and/or nonmetallic materials from them are necessary and compulsory in many developed/developing countries. The United Nations (UNs) estimate of the global WEEE production was 49 million tons in 2012 and more than 50 million tons today. The rate of e-waste generation is increasing by 10% every year [4]. In 2012, the USA, China, Japan, Germany and Russia were the biggest e-waste generating countries. The USA, Australia and the UK had the biggest e-waste production per capita at the same year. E-waste amount is 5–30 kg per person per year and grows at 3 times faster than the municipal waste [1, 5, 6].

There are hazardous substance occurrences in WEEE and possible adverse affects to the human health. Hg is used in relays, switches, batteries, liquid crystal displays (LCDs) and gas discharge lamps (i.e. fluorescent tubes in scanners and photocopiers). Yearly, about 22% of the Hg produced in the world is used in electronics industry. Rechargeable batteries contain Pb, Cd, Li and Ni. Old TVs, Personal Computers (PCs) and Cathode Ray Tubes (CRTs) contain Pb in cone glass, Ba in electron gun getter and Cd in phosphors. PCBs have Pb, Sn and Sb in solder, Cd and Be are found in switches. Polyvinyl chloride (PVC) and BFRs are main components of plastics. Cr^{+6} are found in data types and floppy discs. Condensers and transformers contain polychlorinated biphenyls. Chlorofluorocarbon (CFC) can be found in cooling units and insulation foams. Americium (Am) can be found in smoke detectors. LCDs include liquid crystals embedded between thin layers of glass and electrical control elements. Liquid crystals are mixture of 10–20 substances which belong to the groups of substituted phenylcyclohexanes, alkylbenzens and cyclohexyl benzenes. These substances contain O, F, H and C and they are suspected to be hazardous.

Proper e-waste management for all countries is necessary; because, e-waste pollutes the ground water, acidifies the soil, generates toxic fume and gas after burning, accumulates fastest in municipal disposal areas and releases carcinogenic substances into the air. For a proper e-waste management; waste prevention conserves scarce resources; minimization reduces material usage and reuse uses materials again. Burning (i.e. incineration or pyrolysis for energy recovery prior to disposal) and disposal by landfilling are the least favored options in the e-waste management. Disposal does not conserve any resources. Recycling e-waste is an intermediate polymetallic secondary resource recovery option. Open dumping is the most common form of e-waste disposal in the most developing countries. Burial or landfill disposal allows heavy metals to be leached into the ground water or methane off gassing, combustion of organic substances in waste by incineration makes hazardous material airborne, generates ashes and heat. Leaching of the ashes may cause water and soil contamination. E-waste constitutes 40% of Pb and 70% of heavy metals in landfills [7].

BFRs are used in both PVC and in other types of plastics to reduce the flammability of PCBs, cables and plastic covers of WEEEs. Incineration has a risk of generating and dispersing contaminants and toxic substances. The gases released during the burning and residue ash is often toxic and requires expensive flue gas purification systems. Studies have shown that Cu in PCBs and cables acts as catalyst for dioxin formation when BFRs are incinerated. These BFRs when exposed to low temperature (600–800 °C) uncontrolled burning can lead to the generation of extremely toxic polybrominated/polychlorinated dioxins and furans (Fs). PVC which can be found in e-waste in significant amount is highly corrosive when burned and also induces the formation of dioxins. Concerns have also been raised about the use of stabilizer Cd metals and phthalate plasticizers in PVC. Phenolic BFRs and glass fiber are generally used in PCBs [8].

In landfill, e-waste is placed in a hole, compacted and covered with soil. A double liner system (i.e. compacted plastic clay and plastic jeomembran liner) at the bottom prevents liquid waste from seeping into the ground water and collects leachate to seep through the solid waste. Current improper e-waste handling methods in developing countries include: open burning of circuit boards and cables for metals, acid/cyanide stripping of valuable metals and CRT cracking and dumping.

E-waste includes at least 57 valuable elements found in periodic table. E-waste contains valuable resources which offer opportunities for Urban Mining and job creation. The USA Environmental Protection Agency (EPA) and UNs estimate that only 15–20% of e-waste is recycled, the rest of these consumer electronics go directly to landfills and incineration [9]. In the EU, 9.3 million tons of e-waste were collected and only 35% of them are recycled in 2012. 50–80% e-waste collected in the USA and other developed countries exported to third world countries. PCBs represent the most economically attractive portion of WEEE and account for the weight for about 3–5% [2, 10]. Waste PCBs constitute a heterogeneous mixture of metals, nonmetals and some toxic substances. By containing many electronic components (ECs), such as resistors, relays, capacitors, and integrated circuits (ICs), Waste PCBs have a metal content of nearly 30% (copper (Cu); 10–20%, solder lead (Pb); 1–5%, nickel (Ni); 1–3%, iron (Fe); 1–3%, silver (Ag); 0.05%, gold (Au) 0.03%; and palladium (Pd) 0.01%), especially the purity of precious metals in PCBs is more than 10 times that of rich minerals [1]. It can be seen clearly that except the hazardous substances, a lot of valuable materials contained in PCBs make them worth being recycled. Therefore, developing a non-polluting, efficient and low cost processing technology for recycling of PCBs can not only avoid environmental pollution, but also help to recycle valuable resources, which have a great significance for continuous improvement of the human living environment, standards and resources recycling.

In the USA, e-waste recycling ratio for computers in 2010 was 40%, monitors 33%, mobile phones 11%, keyboards and mice 10% and TVs 17%. In the USA, about 40,000 mobile phones discarded every day [11]. A personal computer material composition includes about 26% silica/glass, 23% plastics, 20% ferrous metal, 14% Al and 17% other metals (such as Pb, Cu, Zn, Hg, Cd etc.). A CRT panel contains 0–2% Pb, frit 65–75% Pb, funnel glass 22–25% Pb and neck 28–30% Pb. A 43 cm CRT monitor contains about 950 g Pb. 41 smart phones contain about 1 g of Au. Every year 1 million smart phones recycled and 16 tons Cu, 350 kg Ag, 34 kg Au and 1.5 kg Pd can be recovered.

Besides all the hazards originating from e-waste, manufacturing mobile phones and PCs consumes considerable fractions of the Au, Ag and Pd mined annually worldwide. 43% of total production of Au in the world is used in electronics. A large fraction of the WEEE precious metals is found on the PCBs. Since PCBs as becoming more complex and smaller, the amount of materials is constantly changing. 1 ton of circuit boards can contain 80–1500 g of Au and 160–210 kg of Cu. Globally, 267.3 tons of Au and 7275 tons of Ag are consumed annually by electronic industry [12]. According to 2014 UNs' e-waste report, yearly about 300 tons of Au were recovered and the e-waste market size was 52 billion dollars.

There are several factors to consider in developing a new recycling technology for waste PCBs driven by innovations, social and environmental impact, an integrated waste management policy and economy of the process. Some of the key factors are:

- The waste PCBs are diverse and complex in terms of type, size, shape, components and composition. With time, the composition of PCBs is continuously changing, making it more difficult to obtain a stable material composition.
- The presence of plastics, ceramics and metals in PCBs in a complex manner leads to great difficulty in liberation and separation of each fraction.
- Presence of numerous metallic elements leads to a very complex recovery process. The recovery process becomes more complicated when the elements are available in ppm concentration.

- The driving force for recycling is the recovery of metal values, which is nearly 30% of the total weight of waste PCBs. The nonmetallic materials (~70%) have rather less economic values (i.e. filler material).
- The objective of most recycling processes is to recover maximum metallic values from waste PCBs but sometimes these processes are not very environment-friendly.

Printed Circuit Boards (PCBs)

PCBs are used to mechanically support and electrically connect electronic components using conductive pathways, tracks or signal traces etched from Cu sheets laminated onto a non-conductive substrate. PCBs constitute at least 3–5% of the total electronic scraps by weight. Most recycling approaches practiced can only recover metal contents of PCB scraps to an extent 30% of the total weight. More than 70% of PCB scraps cannot be efficiently recycled and recovered and have to be incinerated or land filled [13]. PCB assemblies contain a green/yellow board and ECs [i.e. resistors, relays, capacitors and integrated circuits (chips)] attached to it. PCBs are a mixture of phenolic/cellulose paper (yellow and low grade) or epoxy (green and high grade) resins, woven glass-fiber and multiple kinds of metals (Cu, Sn, Pb etc.). The basic structure of the PCBs is the Cu-clad laminate consisting of fiber glass-reinforced epoxy resin and a number of metallic materials including precious metals. The concentration of precious metals especially Au, Ag, Pd and Pt is much higher than their respective primary resources, making waste PCBs an economically attractive Urban Mining source for recycling. Additionally, PCBs also contain different hazardous elements including heavy metals (Cr, Hg, Cd etc.), rare earth elements (Ta, Ga etc.) and flame-retardants (Br and Cl) that pose grave danger to the eco-system during conventional waste treatment of landfilling and incineration [13].

Many research works have revealed that the composition of metals, ceramic and plastics in PCBs with ECs could reach 40, 30 and 30%, respectively [2, 6, 7, 14]. Meanwhile, the concentrations of precious metals such as Au and Pd in waste PCBs are richer than in natural ores, which makes their recycling important from both economic and environmental perspectives. Therefore, factors affecting the extraction of metals are economic feasibility, recovery efficiency and environmental impact. PCB recycling process with the aim of highest recovery of metallic fraction usually includes three stages:

- pretreatment (i.e. manual/automatic/semiautomatic selective disassembly of the reusable and toxic electronic parts by thermal/chemical desoldering) [14],
- size reduction and separation including comminution and separation of materials using mechanical/physical processing (such as shredding, crushing, pulverizing (100–300 μm), screening, upgrading etc.) [15–22] and

- metallurgical/chemical (pyro/hydro/electro/biohydro metallurgical processes) refining/purification [23–26].

PCBs without ECs contain about 30% metals and 70% nonmetals. Nonmetallic fractions consist of cured thermosetting resins, glass fiber, ceramics, BFRs, residual metals (Cu and solder) and other additives. Nonmetallic fractions composition contains 65% glass fiber, 32% epoxy resin and impurities (Cu: <3%, solder <0.1%) by weight [27]. Resins are organic plastic polymers, high cost and low quality products. If resins are land filled or incinerated like in the past, they create potential environmental problems. Glass fibers are reinforcing material in PCBs. They are about 50–70% of PCBs. Material content of the nonmetallic recycled from waste PCBs by air classification showed that residual Cu and glass fiber were in fine-size class and resins were in coarse-size class [1, 28–31].

PCB Assembly Structure

All PCBs essentially consist of three basic parts: a non-conducting substrate or laminate, conducting Cu substrate printed on or inside the laminate and the components attached to the substrate (chips (Ga, In, Ti, Si, Ge, As, Sb, Se and Te), connectors (Au, Ag), capacitors (Ta, Al), etc.). Depending on the structure and alignment, PCBs can be classified as single-sided, double-sided or multilayered. PCB assembly weights of some consumer electronics are: CRT TVs 7%, computers (PC) 18.8%, mobile phones 21.3% and LCD screens 11.9% [32]. Waste PCBs contain 33% semiconductors, 24% capacitors, 23% unpopulated circuit boards, 12% resistances and 8% switches and other materials by weight [33]. Many components are still functional and valuable. According to Takanori et al. [34] the PCBs contain about one-third metallic materials such as Cu and Fe, approximately one-fourth organic resin materials containing elements such as C and H, and approximately one-third glass-materials used as resin reinforcing fibers. In terms of metal composition, the highest content was Cu, which is used in the circuitry, followed by Sn, Fe and Pb, which is used in the soldering and lead frames. In terms of the precious metal composition, Au, Ag and Pd are found in ICs as contact materials or as plating layers due to their high conductivity and chemical stability [34, 35]. Table 1 shows representative material compositions of PCBs by percent of weight used in 11 previous studies [32]. The material present in PCBs can be categorized in three groups: organic materials, metals and ceramics. Organic materials in PCB are mainly composed of plastics with contents of BFRs, resins and paper. The type of plastics is predominantly C–H–O and halogenated polymers.

Metals in PCBs consist of a large amount of base metals (BMs); such as Cu, Fe, Al and Sn; rare metals (RMs) like Ta, Ga and other rare platinum groups metals (PGMs); noble/precious metals (PMs) such as Au, Ag, and Pd. Hazardous metals (HMs) such as Cr, Pb, Be, Hg, Cd, Zn, Ni are also present. Ceramics present in the PCB are primarily silica and alumina. Other ceramic materials include alkaline

Table 1 PCB material composition

Metals about 40%	(%)	Ceramics about 30%	(%)	Plastics about 30%	(%)
Cu	6–27	SiO_2	15–30	PE	10–16
Fe	1.2–8.0	Al_2O_3	6.0–9.4	PP	4.8
Al	2.0–7.2	Alkali-Earth oxides	6.0	PS	4.8
Sn	1.0–5.6	Titanates-micas	3.0	Epoxy	4.8
Pb	1.0–4.2			PVC	2.4
Ni	0.3–5.4			PTPE	2.4
Zn	0.2–2.2			Nylon	0.9
Sb	0.1–0.4				
Au (ppm)	250–2050				
Ag (ppm)	110–4500				
Pd (ppm)	50–4000				
Pt (ppm)	5–30				
Co	1–4000				

earth oxides, mica and barium titanate. WEEE plastics contain BFRs and the combustion of these produces the formation of highly toxic gases. Thermoset resins cannot be reformed or reshaped because of their net structure; thus, they are regarded as nonrecyclable. PC scrap primarily contains ABS, PS and PVC plastics with the density range of +1.0–1.5 g/cm^3, whereas PCB scrap mainly contains glass fiber reinforced epoxy resins plastics with the density range of +1.5–2.0 g/cm^3 [36].

Methods of Joining Components in PCBs

Electronic components are mounted on PCB assemblies using various types of connections. These connections are socket pedestal, through-hole device (THD), surface mounted device (SMD), screw joint, and rivet. There are several methods by which these connections can be broken. For example, components with socket pedestal connection can be disassembled directly by nondestructive force, but the method used to disassemble components with SMD or THD connections are always destructive, involving removal of solder or pins [14]. Solder welding is a process through which chemically and mechanically two metals are joined at a low melting point. Welding occurs at a temperature of 40 °C above the melting point of the solder alloy, and is valid for any type of solder, including electronic welding. It has a relatively low melting point (183 °C), good wettability, good mechanical and electrical properties and low cost [37]. The oldest and most common Pb-based solders are 63Sn-37Pb. Melting points of solder alloys change between 176 and 228 °C. Special liquor, methylphenyl silicon oil, kerosene or solder can be used as the medium to transmit heat to melt solder from the PCB assemblies in desoldering pretreatment process [38].

Physical/Mechanical Recycling Techniques

Physical processes are usually employed during the upgrading stage when various metals and nonmetals contained in e-waste are liberated and separated by some kinds of shredding and crushing processes. The drive to recover the valuable metals in particular Au, Ag, Pd and Cu has received tremendous attention in recent years using extraction processes such as physical, chemical and hydro/pyro metallurgical leaching separation techniques. Methodology generally includes PCB assembly desoldering pretreatment for ECs removal, size reduction for liberation and a combined route of physical and metallurgical processing. Physical recycling techniques include crushers, pulverizers, classifiers and separators. The outcome of a recycling process can be evaluated from two aspects: the material (i.e. metal and nonmetal) recovering efficiency and the environmental impact of the processing. Physical separation processes benefit from low capital and operating costs and suffer from a high valuable metal loss (10–35%) due to insufficient metal liberation.

Physical processes include dry crushing and pulverizing and then high voltage electrostatic separation (ES) to obtain a mixture of metal powder (Cu, Pb, Zn, Al, Sn, Au, Ag etc.) which is conductor and nonmetallic resin powder material which is not conductor. To avoid dust pollution three-stage dust removal equipments (i.e. cyclone, bag and air cleaner) are used. The separation efficiency of this type of physical PCB recycling systems is up to 99%. A physical dry PCB recycling production line layout is given in Fig. 1. This physical separation plant separates mixed metal powders from plastic resins using a shredder, pulverizer, cyclone classifier, air cleaning equipments, vibrating screens and ES. This type of dry separation plants must be dust and noise free and can process waste PCBs with or without ECs, Cu-clad plate and scrap circuit boards. Chinese PCB recycling machine manufacturers produce dry and wet working machines from a capacity between 1–10 t/d and 1500–10,000 t/y. This type of PCB recycling plants has no waste water and no smoke; requires only one or two workers to operate and is highly automated. This plant requires small space. Cable wire recycling can be achieved by dry shredding and air classification to separate Cu from plastic material. 80–800 kg/h capacity machines are available with 99% separation efficiency. These type of machines accept cables with a diameter between 0.5 and 25 mm. Cables contain about 40% Cu and 60% PVC plastic. Cu and PVC powders are obtained from these machines separately.

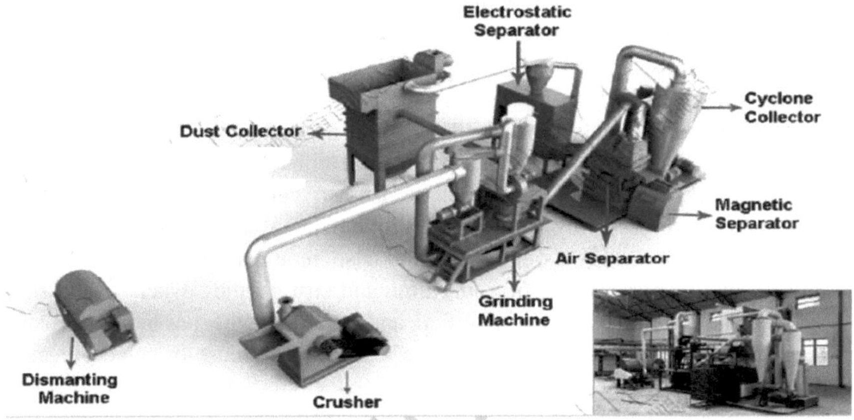

Fig. 1 Waste PCB recycling production line layout

Dismantling (Manual, Automated and Semi Automated)

Dismantling is the systematic removal of components and parts from e-waste. Attachment regimes of materials encountered in e-waste are fastening, inserting, welding, binding, wrapping, coating and plating. Basically, it is not energy intensive to unlock the associated materials such as ceramics, glasses and metals with different mechanical properties, depending on the locking regimes of materials involved. In general, materials locked through fastening by screws, clinks and rivets etc., inserting, wrapping and packaging can readily be detached. Materials locked by means of coating, binding, welding and encapsulating are relatively difficult to be unlocked and materials locked by alloying and filling cannot be liberated by mechanical means [35].

There are two forms of dismantling: selective and simultaneous disassembly. Selective dismantling can also be performed manually or mechanically. Currently, manual informal dismantling is common in developing countries and mechanical dismantling is common in developed countries [39]. Mechanical dismantling can be done by either semi-automatic or automatic way. Informal manual dismantling aims to heat the solder above melting point (i.e. desoldering) and resell the reusable ECs. Informal manual dismantling uses chisels, hammers and cutting torches to open solder connections and separate various types of metals and ECs. PCBs are cooked over a coal-heated plate in order to melt and resell the chips and other recovered components to acid strippers for further processing [38]. The components are heated using electric blower and then separated by dint of pliers. Disadvantages of this process are bad smell, black fumes and being a banned process. On the basis of assembly of PCBs, two methods can be used for automatic disassembly: selective disassembly and simultaneous disassembly:

- In selective disassembly, some specific components are located and removed. The connection type and coordinates of connection are determined and components are disassembled individually. This principle is called "look and pick".
- Under simultaneous disassembly, the whole flat board is heated in a tin furnace to desolder components from the board and all components are "wiped off" simultaneously. These components are then identified and sorted by geometrical and physical criterions. This principle is also called "evacuate and sort". Simultaneous disassembly method has high efficiency, but also a higher risk of damaging the components. This method also requires an additional sorting process that increases the processing time and cost. In the "evacuate and sort" system, the components are identified and sorted on the basis of geometry, density, or using magnetic effect. The components reuse-oriented selective disassembly is an indispensable process, since the reuse of valuable components has first priority, dismantling the hazardous components is essential, and it is also common to dismantle highly valuable components and highly graded materials such as battery in order to simplify the subsequent recovery of materials.

Mechanical dismantling can be performed by semi automatic or automatic approach. In semi automatic dismantling, the ECs on the PCBs are removed by a combination of heating them above melting point of solder and applying external forces such as impact, shearing, and vibration in part removal unit. Infrared (IR) heaters at 240–250 °C are required. These types of machines include exhausted gas controlling unit, solder removal unit and bare board collection units. Park et al. [40] developed an apparatus for ECs disassembly from PCB assembly in e-waste. 94% maximum disassembly ratio was obtained at a feeding speed of 0.33 cm/s and a heating temperature of 250 °C by them. Daily 800 kg capacity machines are available today [32]. Solder baths and liquids as a heating medium at 215–230 °C for desoldering pretreatment and mechanical vibration can be used. Daily 500 kg capacity equipments are available today. The use of liquids has two disadvantages: disposing the used fluid and cleaning the bare boards and components for further processing [32]. Haruta et al. [41] reported on using two (mild and strong) oxidation capability reagents in two steps to remove a layer of Sn and Sn–Pb alloy from PCBs for desoldering.

In modular semi-automatic systems; disassembly cells consist of a transportation system, a vision system and heating-desoldering stations. In this mechanical approach, PCBs are fixed on frames and fed into the dismantling cell. Vision system identifies reusable parts and toxic components by image processing. Using 3D pictures, applying hot air type of heaters and vacuum grippers, reusable and toxic parts are removed. SMD/THD components are removed by hot liquid and parallel jaw grippers [32, 42, 43]. Japanese NEC corporation system comprises two IR heating units and two removal units. The first removal unit is equipped with impacting brush and the second shearing propellers. Automatic disassemblers are expensive and ideally suited for high-metal content waste [32, 40].

Size Reduction for Liberation of Metallic and Nonmetallic Fractions (Shredding/Crushing)

An effective liberation of various materials like metals and plastics is a crucial step towards mechanical separation. In addition, classification of e-waste is also important to be able to provide an appropriate feed material for the subsequent physical/metallurgical separation process [36]. The purpose of shredding/crushing is to strip metals from PCBs. Crushing technology is intimately related to not only energy consumption, but also further selective efficiency. Waste PCBs are comprised of reinforced resin and glassfibers. Comminution of waste PCBs and high effective liberation of the metals from nonmetals is the prerequisite of the following sequence separation for better recovery of waste PCBs. Jiang et al. [10] claim that shredding or grinding may lead to 40% loss of precious metal and/or to the formation of dangerous metal fines, dust containing BFRs and dioxins.

A potential process flowsheet for metals recovery from e-waste with two step crushing, gravity, magnetic, electrostatic physical separations and metallurgical separations is given in Fig. 2 [2]. Much dust and harmful gas can be generated during the crushing process because of the strength and tenacity of the PCBs. Therefore, a very good dust collection system must be used. After removal of the hazardous components, different mineral processing operations such as shredding, crushing and grinding can be used to liberate metals from cladding materials such as resin and fiberglass. Various types of hammer, rotary and disc crushers; shredders; cutters equipped with a bottom sieve are used for liberation. Ball and disc milling are also reported for pulverizing the PCBs after cutting into small sizes [14]. Low speed high torque shear shredders (−10 mm) are basically ideal for the primary crushing, while a number of mills have been reported for the finer comminution [36, 44, 45], with the swing hammer types appearing to have become the industry standard [46–48]. Schubert and Bernotat (2004) gave a good review of the different modifications of the swing hammer mills; the comminution mechanism is essentially the same. It breaks mostly by impact, with shear, compression and abrasion; a combination of stress modes appropriate for the diverse material types, from brittle to ductile, present on PCBs. The product discharges through a screen in closed circuit grinding. Discharge size below 500 μm is necessary for liberation [49].

At −75 + 38 μm, very few composite particles remain. Beneficiation operation therefore cannot attempt to separate solder alloy particles into constituent elements; realistic schemes can aim only at achieving some bulk collection of metallic values into a concentrate. Cu is found between layers of resin. Ferromagnetics and Cu liberate at −2 mm and Al at +7 mm [36]. The PCB scraps having small size particles can be separated by various separation methods like: size, density, magnetic, electrostatic, eddy current and gravity separation. Zhang and Forssberg (1997) studied the liberation characteristics of PCBs and the effect of shape and size on liberation. It was concluded that below 6 mm size, ferromagnetic and Cu are completely liberated and at the same time, Al is found to be liberated in much

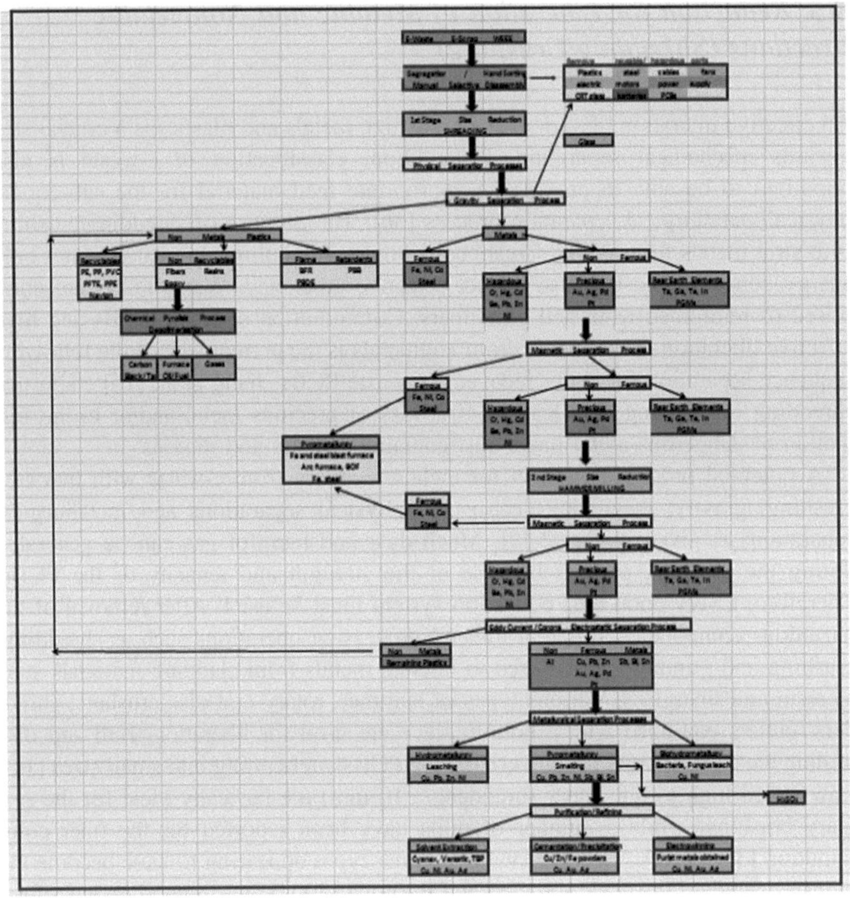

Fig. 2 E-waste recycling flow sheet with a combination of physical, chemical and metallurgical separation techniques [2, 6, 7]

coarser fraction. Koyànaka et al. (1999) reported that glass fiber-reinforced epoxy resin undergoes brittle fracturing more readily than metallic materials and concentrates in the finer fraction during impact milling of PCB scraps. Yoo et al. [50] demonstrated the use of stamp mill for the liberation of various metallic components. Vidyadhar and Das (2012) also reported that after milling to below 150 μm size, no interlocking of metallic and nonmetallic particles is observed. Wang et al. (2015) investigated the dust as the byproduct originating from mechanical recycling process of waste PCBs from the viewpoints of resource reuse and environmental protection. Their mineralogical analysis on the dust collected from a typical crushing and separating recycling line demonstrated 73.1% organic matters, 4.65% Al, 4.55% Fe, 2.67% Cu and 1.06% Pb. Most of the metals liberated at −0.75 mm size fraction and as the particle size decreases, the content of liberated elemental metals and magnetic materials increase.

Shredder is used to reduce the dimension of the material through cutting, tearing and extruding. Double shaft shredder is widely used in e-waste recycling. Diameter of shredder blades changes between 200 and 400 mm, capacity between 200 and 2500 kg/h, quantity of blade between 16 and 42 and feed inlet size between 700 and 1090 mm.

Screening for Classification

In size separation, a sieve is used to classify the different sized particles. The size of sieve apertures can be varied according to the desired particulate size for separation. Screening has not only been utilized to prepare a uniformly sized feed to a certain mechanical process, but also to upgrade metals contents. Screening is necessary because the particle size and shape properties of metals are different from that of plastics and ceramics. After primary shredding and vibrating screening, secondary stage crushing/pulverizing with dry air cyclone classification are used with dust collection systems.

Gravity Separation (GS)

Gravity concentration methods separate materials of different specific gravity and size by their relative movement in response to gravity. Besides gravity one or more of the other forces exerted by the viscous liquid (i.e. water or air) can serve as the separation medium. By using different heavy liquids (i.e. TBE), the metals can be separated from the plastics or ceramics. Water or airflow tables, heavy media separation and sifting are common gravity separators used in e-waste recycling. Different metal particles can be further separated. For this purpose, the PCB material is processed on shaking tables. The principle of the air classification technique is based on the suspension of the particles in a flowing air stream and the separation of the particles based on their density (d) difference. Density separation techniques have found their way into e-waste recycling based on the fact that electronic scrap consists essentially of plastics, $d < 2.0$ g/cm^3; light metal (Al and glass) $d = 2.7$ g/cm^3; and heavy metals (Cu and ferromagnetic) $d > 7$ g/cm^3. In sink-float separation, both PC and PCB scraps $\sim$50% (weight) of floats which is primarily plastics can be separated out at the specific density of 2.0 g/cm^3 [36]. Sarvar et al. [51] used a jig for +0.59 and 1.68 mm PCB particles.

Magnetic Separation (MS)

Magnetic separation can be used to separate magnetic metallic particles from non-magnetic particles (i.e. plastics, glass, Al etc.) Magnetic separators, in particular low-intensity drum separators are widely used for the recovery of ferromagnetic metals from nonferrous metals and other nonmagnetic wastes. Over the past decade, there have been many advances in the design and operation of high-intensity magnetic separators (HIMS), mainly as a result of the introduction of rare earth alloy permanent magnets capable of providing very high field strengths and gradients. There are some problems associated with this method. One of the major issues is the agglomeration of the particles which results in the attraction of some nonferrous fraction attached to the ferrous fraction [7]. This will lead to the low efficiency of this method. Through the process of MS, it is possible to obtain two fractions: magnetic fraction, which includes Fe, steel, Ni etc. and non-magnetic fraction, which includes Cu [30]. Veit et al. (2005) employed a magnetic field of 6000–6500 G to separate the ferromagnetic elements, such as Fe and Ni. However, there was a considerable amount of Cu impurity in the magnetic fraction as well. Yoo et al. (2009) used a two-stage MS to milled PCBs. The milled PCBs of particle size >5.0 mm and the heavy fraction separated from the <5.0 mm PCBs particles by gravity separation. In the first stage, a low magnetic field of 700 G was applied which led to the separation of 83% of Ni and Fe in the magnetic fraction and 92% of Cu in the non-magnetic fraction. The second MS stage was conducted at 3000 G which resulted in a reduction in the grade of the Ni–Fe concentrate and an increase in the Cu concentrate grade.

Electrostatic Separation (ES)

Electrostatic separation separates materials of different electrical conductivity. There are three typical separation techniques: Corona electrostatic separation (CES), Triboelectric separation (TES) and Eddy current separation (ECS). The ES capability depends on the difference in polarity and the amount of charge acquired by particles to be separated. Induction or Corona charging can successfully separate the mixed particles that have large difference in conductivities. Tribo-electricity or contact charging is useful for charging and separating materials that have similar conductivities.

In the separation zone of ECS, gravitational, centrifugal, frictional, and magnetic deflection forces influence the falling particles; but, only magnetic force deflects the ferrous particles to a higher degree [13]. ECS technology is used to recover Al [52]. Eddy current separators can remove Cu, Al, nonferrous metal scrap, broken glass etc. Al fraction is sent to Al smelter for further treatment.

The CES method is perhaps the most effective separation technology for the metallic and non-metallic fractions at present. The method has the advantage of being environmentally friendly, producing no wastewater and no gaseous emissions. In the CES, electrode system, rotor speed, moisture content and particle size have the greatest effect in separation. The PCBs with the metallic components removed must be reduced to less than 0.6 mm and then passed along a vibratory feeder to a rotating roll to which a high voltage electrostatic field is applied using a Corona and an electrostatic electrode [53, 54]. The nonmetallic particles that become charged and attached to the drum eventually falling off into storage bins; whereas the metallic particles discharge rapidly in the direction of an earthed electrode.

It has been found that particle sizes of 0.6–1.2 mm are the most suitable size for separation in industrial applications. Li et al. [55] found that as the angle of the static electrode reduced and the Corona electrode angle was increased, the separation efficiency was enhanced. Considerable work is continuing in this area with particular focus on the electrostatic behavior of the system and the field intensity. Corona electrostatic methods are now capable of producing two streams of a metallic and a nonmetallic portion with little cross-contamination; the method is dry at room temperature and almost zero polluting. The residual Cu metal in nonmetallic fraction, is in fine size range (7–9 μm) [52]. Industrial ESs adopt physical high voltage to separate conductor metals from nonconductor nonmetals with a separation efficiency of 95–99% purity. ES power changes from 1.5 to 3 kW, rotational speed from 20 to 300 rpm, static electricity from 50 to 150 kV adjustable [2].

Conclusions

E-waste is one of the fastest growing solid waste streams worldwide today. WEEEs are one of the largest known sources of heavy metals without effective collection, reuse and recycling systems. They are dangerous to environment. Recycling of WEEE and reuse of some electrical/electronic parts are a beneficial alternative to disposal. Waste PCBs account for about 3–5% of nearly 50 mt/year global E-waste generations. The 70% PCBs nonmetallic fraction has been traditionally discarded to landfill or used as very low cost fillers in the construction industry. A significant amount of Cu and Au in e-waste attracts illegal recyclers in some parts of Asia and Africa leading to substantial environmental and health problems. PCBs assay far above many base and precious metals natural deposits. The primary objective of recycling PCB scrap is to minimize the harmful environmental impact and ensure maximum material recovery.

Thermal dismantling of PCBs is the most important first step in the recycling chain. The removal of ECs from PCBs includes a combination of heating to a temperature 40–50 °C higher than the molten point of the solder, ranging 200–230 °C and applying a shear force. The major common point of these disassembling technologies is the recovery of the solder remaining on the board.

Physical recycling methods have been the traditional techniques for PCBs. Selective disassembly targets and singles out hazardous and valuable components for special treatment. Following the dismantling, good liberation of nonmetals from metals is required for effective separation. Upgrading increase the content of desired materials. In general, milling to −150 μm ensures no interlocking of metallic and nonmetallic particles. Many researchers have used various mechanical methods (multi-stage crushing, pulverizing, screening, GS, MS and ES) to separates metals from nonmetals. For physical separation the density-based separators are quite popular. The latest developments include CES, which is based on the difference in density and electrical conductivity between plastics, metals and ceramics with little cross-contamination.

The basic topic in nonmetallic fraction recovery by physical separation is how to use these recovered materials as effective, cheap and safe filler for different materials. Nonmetallic fraction materials are diverse, complex and toxic. Physical recycling methods are relatively simple, convenient and environmentally sound, the initial equipment investment and energy cost are low and potential application of products is diversified. But, there are still some challenges in physical separation between the metallic and nonmetallic fractions from waste PCBs. Significant dust generation and metal loss during shredding and grinding are some important drawbacks of the process. Today, Chinese and Japanese recycling facilities generally use dismantling and mechanical processing of e-waste for the recovery of raw materials. Process flowsheets include manual selective disassembly, shredding, MS and ECS [8]. The major hazards during e-waste recycling are associated with the size reduction and separation. Shredding leads to the formation of fine dust composed of plastics, metals, ceramic and silica. Additive chemicals like BFRs which are embedded in WEEE are also released during shredding. Some of the base and valuable metals are lost during size reduction as well. The author hopes that a cleaner, safer and economical recycling process can be achieved soon. Traditional technologies may not meet the future requirements of industry because of environmental contamination, high cost and low efficiency. Evidently, more studies are needed in the area of metal and nonmetal separation and recovery of valuable metals from PCB.

References

1. Y. Zhou, K. Qiu, A new technology for recycling materials from waste printed circuit boards. J. Hazard. Mater. **175**, 823–828 (2010)
2. M. Kaya, Recovery of metals and nonmetals from electronic waste by physical and chemical recycling processes. Waste Manag. **57**, 64–90 (2016)
3. R. Widmer, K.H. Oswald, D. Sinha-Kheeetriwal, M. Schnelmann, H. Boni, Global perspectives on e-waste. Environ. Impact Assess. Rev. **25**, 436–458 (2005)
4. P. Sakunda, Strategy of e-waste management (2013), http://www.slideshare.net/ketanwadodkar/e-waste-tce-r2?related=2

5. F. Wang, Y. Zhao, T. Zhang, C. Duan, L. Wang, Mineralogical analysis of dust collected from typical recycling line of waste printed circuit boards. Waste Manag. **43**, 434–441 (2015)
6. M. Kaya, Recovery of metals from electronic waste by physical and chemical recycling processes. Int. J. Chem. Nucl. Mater. Metall. Eng. **10**(2), 246–257 (2016)
7. M. Kaya, Recovery of metals from electronic waste by physical and chemical recycling processes, in *Proceeding Part VII, 18th International Conference on Waste Management, Recycling and Environment (ICWMRE 2016)*, Barcelona, Spain (2016c), pp. 939–950
8. O. Tsydenova, M. Bengtsson, Chemical hazards associated with treatment of waste electrical and electronic equipment. Waste Manag. **31**, 45–58 (2011)
9. United Nations University, Set world standards for electronics recycling, reuse to curb E-waste exports to developing countries, experts urge. *Science Daily* (2009), www.sciencedaily.com/releases/2009/09/090915140919.htm. Last accessed on 21 Dec 14
10. P. Jiang, M. Harney, Y. Song, B. Chen, B. Chen, Q. Chen, T. Chen, G. Lazarus, H.D. Lawrence, M.B. Korzenski, Improving the end-of-life for electronic materials via sustainable recycling methods. Procedia Environ. Sci. **16**, 485–490 (2012)
11. http://www.powershow.com/view0/660a6c-MjlmM/Lets_Not_Waste_E-Waste_powerpoint_ppt_presentation. Accessed 1 June 2016
12. M.C. Vats, S.K. Singh, Assessment of gold and silver in assorted mobile phone printed circuit boards (PCBs). Waste Manag. **45**, 280–288 (2015)
13. J. Li, P. Shrivastava, Z. Gao, H.C. Zhang, Printed circuit board recycling: a state of the art survey. IEEE Trans. Electron. Packag. Manuf. **27**(1), 33–42 (2004)
14. B. Ghosh, M.K. Ghosh, P. Parhi, P.S. Mukherjee, B.K. Mishra, Waste printed circuit boards recycling: an extensive assessment of current status. J. Clean. Prod. **94**, 5–19 (2015)
15. J. Cui, E. Forssberg, Mechanical recycling of waste electric and electronic equipment: a review. J. Hazard. Mater. **99**(3), 243–263 (2003)
16. H.M. Veit, T.R. Diehl, A.P. Salami, J.S. Rodrigues, A.M. Bernardes, J.A.S. Tenório, Utilization of magnetic and electrostatic separation in the recycling of printed circuit boards scrap. Waste Manag. **25**, 67–74 (2005)
17. J. Cui, L. Zhang, Metallurgical recovery of metals from electronic waste: a review. J. Hazard. Mater. **158**, 228–256 (2008). doi:10.1016/j.jhazmat.2008.02.001
18. A. Tilmatine, K. Medles, S.E. Bendimerad, F. Boukholda, L. Dascalescu, Electrostatic separators of particles: application to plastic/metal, metal/metal and plastic/plastic mixtures. Waste Manag. **29**, 228–232 (2009). doi:10.1016/j.wasman.2008.06.008
19. C. Guo, H. Wang, W. Liang, F. Jiangang, X. Yi, Liberation characteristic and physical separation of printed circuit board (PCB). Waste Manag. **31**, 2161–2166 (2011). doi:10.1016/j.wasman.2011.05.011
20. Z. Tan, Y. He, W. Xie, E. Zhou, Y. Zheng, Size distribution of wet crushed waste printed circuit boards. Mining Sci. Technol. (China) **21**(359–363), 2011 (2011). doi:10.1016/j.mstc.05.004
21. A.C. Kasper, G.B.T. Berselli, B.D. Freitas, J.A.S. Tenório, A.M. Bernardes, H.M. Veit, Printed wiring boards for mobile phones: characterization and recycling of copper. Waste Manag. **31**, 2536–2545 (2011). doi:10.1016/j.wasman.2011.08.013
22. A. Tuncuk, V. Atazi, A. Akcil, E.Y. Yazici, H. Deveci, Aqueous metal recovery techniques from e-scrap: hydrometallurgy in recycling. Miner. Eng. **25**, 28–37 (2012). doi:10.1016/j.mineng.2011.09.019
23. G. Dorella, M.B. Mansur, A study of the separation of cobalt from spent Li-ion battery residues. J. Power Sources **170**, 210–215 (2007). doi:10.1016/j.jpowsour.2007.04.025
24. Y. Pranolo, W. Zhang, C.Y. Cheng, Recovery of metals from spent lithium-ion battery leach solutions with a mixed solvent extractant system. Hydrometallurgy **102**, 37–42 (2010). doi:10.1016/j.hydromet.2010.01.007
25. K. Provazi, B.A. Campos, D.C.R. Espinosa, J.A.S. Tenório, Metal separation from mixed types of batteries using selective precipitation and liquid–liquid extraction techniques. Waste Manag. **31**, 59–64 (2011). doi:10.1016/j.wasman.2010.08.021

26. T.E. Lister, P. Wang, A. Anderko, Recovery of critical and value metals from mobile electronics enabled by electrochemical processing. Hydrometallurgy **149**, 228–237 (2014). doi:10.1016/j.hydromet.2014.08.011
27. S. Yokoyama, M. Iji., Recycling of printed wiring boards with mounted electronic parts, in *Proceedings of the 1997 IEEE International on Symposium* (1997), pp. 109–114
28. S. Koyanaka, H. Ohya, J.C. Lee, H. Iwata, S. Endoh, Impact milling of printed circuit board wastes for resources recycling and evaluation of the liberation using heavy medium separation. J. Soc. Powder Technol. Jpn. **36**, 479–483 (1999)
29. A. Vidyadhar, A. Das, Kinetics and efficacy of froth flotation for the recovery of metal values from pulverized printed circuit boards, in *XXVI International Mineral Processing Congress (IMPC)*, New Delhi (2012), pp. 236–243
30. L.H. Yamane, V.T. Moraes, D.C.R. Espinosa, Recycling of WEEE: characterization of spent printed circuit boards from mobile phones and computers. Waste Manag. **31**, 2553–2558 (2011)
31. E.Y.L. Sum, The recovery of metals from electronic scraps. JOM **43**(4), 53–61 (1991)
32. H. Duan, K. Hou, J. Li, X. Zhu, Examining the technology acceptance for dismantling of waste printed circuit boards in light of recycling and environmental concerns. J. Environ. Manag. **92**, 392–399 (2011)
33. A. Tohka, H. Lehto, *Mechanical and Thermal Recycling of Waste from Electric and Electronic Equipment* (Helsinki University of Technology, Department of Mechanical Eng. Energy Engineering and Environmental Protection Publications, Espoo, 2005)
34. H. Takanori, A. Ryuichi, M. Youichi, N. Minoru, T. Yasuhiro, A. Takao, Techniques to separate metal from waste printed circuit boards from discarded personal computers. J. Mater. Cycles Waste Manag. **11**, 42–54 (2009)
35. L.B. Jung, J.T. Bartel, Computer take-back and recycling, an economic analysis for used consumer equipment. J. Electron. Manufact. **9**, 67–77 (1999)
36. S. Zhang, E. Forssberg, Mechanical separation-oriented characterization of electronic scrap. Resour. Conserv. Recycl. **21**, 247–269 (1997)
37. A.C. Marques, J.M. Cabrera, C.F. Malfatt, Printed circuit boards: a review on the perspective of sustainability. J. Environ. Manag. **131**, 298–306 (2013)
38. K. Huang, J. Guo, Z. Xu, Recycling of waste printed circuit boards: a review of current technologies and treatment status in China. J. Hazard. Mater. **164**, 399–406 (2009)
39. X. Chi, M. Streicher-Porte, M.Y.L. Wang, M.A. Reuter, Informal electronic waste recycling: a sector review with special focus on China. Waste Manag. **31**, 731–742 (2011)
40. S. Park, S. Kim, Y. Han, J. Park, Apparatus for electronic component disassembly from printed circuit board assembly in e-waste. Int. J. Miner. Process. **144**, 11–15 (2015)
41. T. Harruta, T. Nagano, T. Kishimoto, Y. Yamada, T. Yuna, Process for removing tin and tin-lead alloy from copper substrates, US Patent issued on 30 July 1991
42. P. Kopacek, *IFAC-16th Triennial World Congress* (Czech Republic, Prague, 2005)
43. B. Kopacek, P. Kopacek, Extracting rare materials from electr(on)ic scrap. IFAC-PaperOnLine **48–24**, 157–161 (2015)
44. M. Iji, S. Yokoyama, Recycling of printed circuit board with mounted electronic components. Circuit World **23**(3), 10–15 (1997)
45. P. Mou, L. Wa, D. Xiang, J. Gao, G. Duan, A physical process for recycling and reusing waste printed circuit boards, in *IEEE Conference Record, International Symposium on Electronic and the Environment* (2004), pp. 237–242. doi:10.1109/ISEE.2004.1299722
46. M. Goosey, R. Kellner, Recycling technologies for the treatment of end of life printed circuit boards (PCBs). Circuit World **29**, 33–37 (2003)
47. S. Sander, G. Schubert, H.-G. Jackel, The fundamentals of the comminution of metals in shredders of the swing-hammer type. Int. J. Miner. Process. **74S**, 385–393 (2004)
48. G. Schubert, S. Bernotat, Comminution of non-brittle materials. Int. J. Miner. Process. **74S**, S19–S30 (2004)

49. I.O. Ogunniyi, M.K.G. Vermaak, D.R. Groot, Chemical composition and liberation characterization of printed circuit board comminution fines for beneficiation investigations. Waste Manag. **29**, 2140–2146 (2009)
50. J.M. Yoo, J. Jeong, K. Yoo, J.C. Lee, W. Kim, Enrichment of the metallic components from waste printed circuit boards by a mechanical separation using a stamp mill. Waste Manag. **29**, 1132–1137 (2009)
51. M. Sarvar, M.M. Salarirad, M.A. Shabani, Characterization of mechanical separation of metals from computer printed circuit boards (PCBs) based on mineral processing methods. Waste Manag. **45**, 246–257 (2015)
52. J. Guo, B. Ca, J. Guo, Z. Xu, A plate produced by nonmagnetic materials of pulverized waste printed circuit boards. Environ. Sci. Technol. **42**(14), 5267–5271 (2008)
53. H. Lu, J. Li, J. Guo, Z. Xu, Movement behavior in electrostatic separation: recycling of metal materials from waste printed circuit board. J. Mater. Process. Technol. **197**, 101–108 (2008)
54. L. Jianzhi, P. Shrivastava, G. Zong, Z. Hong-Chao, Printed circuit board recycling: a state-of-the-art survey. IEEE Trans. Electron. Packag. Manufact. **27**, 33–42 (2004). doi:10.1109/TEPM.2004.830501
55. J. Li, H. Lu, S. Liu, Z. Xu, Optimizing the operating parameters of corona electrostatic separation for recycling waste scraped printed circuit boards by computer simulation of electric field. J. Hazard. Mater. **153**, 269–275 (2008)
56. http://unu.edu/news/news/ewaste-2014-unu-report.html. Accessed 1 June 2016

The Use of Rice Husk Ash as an Aggregate for Foundry Sand Mould Production

A.O. Apata and F.V. Adams

Abstract The use of rice husk ash as aggregate for foundry sand mould production has been studied. 5–25wt% rice husk ash was added to the sand mixture. The chemical, physical and mechanical properties measured include: chemical analysis of the rice husk ash, particle size, density, permeability, compatibility mouldability, moisture content, green compression and shear strength, dry compression and shear strength. The result revealed that, addition of rice husk ash to River Niger Silica sand, increased the mouldability, compatibility, grain fineness number, both green and dry strength, slightly decreased the density, permeability and moisture content. The result shows that better properties is achievable by the addition of rice husk ash to River Niger silica sand and can be used to enhance the mould properties of foundry sand.

Keywords Foundry sand · Rice husk ash · Strength · Mouldability · Compatibility and particle size

Introduction

The urgent need to develop the foundry industries in Nigeria in order to meet the technological development of the country has generated great interest in the characterization of the locally available materials. Silica sand is an extremely good material for casting moulds because it has the ability to withstand the temperature of the molten metal, can absorb and transmit heat, and has sufficient permeability to allow gasses generated during casting to pass between the particles without causing casting defects [1, 2]. Foundry sands are produced within strict particle size distributions to tailor the

A.O. Apata (✉)
Department of Metallurgical and Materials Engineering, Federal Polytechnic Idah, Idah, Nigeria
e-mail: toyinben.toyinben@gmail.com

F.V. Adams
School of Arts and Sciences, American University of Nigeria, Yola, Nigeria

L. Zhang et al. (eds.), *Energy Technology 2017*,
The Minerals, Metals & Materials Series, DOI 10.1007/978-3-319-52192-3_44

properties of the material to the intended casting process [2–4]. There are many deposits of silica sand usable in foundry applications scattered all over the country [1, 4]. The local foundry men have been using this without the knowledge of its physical, chemical and foundry (moulding) properties. Except when these characteristics of the sand are known, it will be very difficult to use the sand and achieve the desired results. In a survey of sands used by some Nigerian foundries [5], it was established that most of these foundries are using these sands without knowing their characteristics. The suitability of any particular sand for foundry mould production is determined by the properties and composition it possesses. These properties and composition are of great importance to both foundry engineers and technologists.

Rice husk ash materials, as a waste by-product, are produced during the burning of rice husk after milling of the rice from the husk and present a serious ecological problem associated with their storage and disposal. Research has shown that fly ash and rice husk ash materials have many of the same attributes of foundry sand [6]. These ashes have a very high melting point, can absorb and transmit heat during pouring and have the ability to allow gases to pass through a compacted mass.

Therefore this work is aimed at determining the use of rise husk ash as an additive to River Niger (Itobe deposit) sand for foundry application.

Materials and Method

Materials

The materials that were used in this research are: River Niger natural sand collected from Itobe, located at the Eastern bank of River Niger in Ofu local government area of Kogi state. Rice husk obtained in Egah rice mill located at Egah community (a southern part of Idah) in Idah local government area of Kogi state, moulding box, measurement cylinder, water, and aluminum scraps.

Equipment

The equipment that were used in this research are: moisture tester, permeability machine, strength universal testing machine, speedy mouldability machines, heat treatment furnace, and oven.

Rice Husk Ash Characterization

The rice husk was burnt in open atmosphere, after burning the black ash was packed in the cruicible air tight fire at 1200 °C for 5 h to obtain a whitish color.

Particle Size Analysis

The particle size distributions of 75wt% silica sand and 25wt% rice husk ash were determined using the American Foundry Society (AFS) specification [1, 7]. 100 g each of the dried samples was taken and introduced unto a set of sieves arranged in descending order of fineness (1.60–0.10 mm) and shaken for 15 min which is the recommended shaking time to achieve complete classification of the sand. The weight retained on each sieve was taken and expressed as percentages of the total sample weight. Grain fineness number (GFN) was computed from the percentage weight of the retained sand.

Bulk Density Test

The bulk density of the samples (Db) was determined by measuring the dry weight (D), suspended weight (S) and saturated weight (W) of the test samples. The bulk density was then computed from the relationship shown in Eq. 1 [7, 8].

$$Db = \frac{D}{W - S}\ \text{g/cm}^3 \tag{1}$$

Moisture Content Determination

The moisture content was determined using a speedy moisture tester. A sample of each mixture was weighed on the weighing balance of the tester and then introduced into the flask of the moisture tester. A known weight of calcium carbide was added into the flask for 3 min which is the recommended shaking time and the percentage moisture content of the sample was read directly from the calibrated dial instrument at the top of the flask attached to the machine [1, 9].

Plate 1 Standard test sample for the green sand properties. *Source* Produced by author

Production of Standard Samples for the Determination of Moulding Characteristics of the Sand

Standard test samples for the determination of moulding properties of the sand were prepared by mixing known weight (0–75%) of the silica sand and the percentage of the ash varied from 5, 10, 15, 20 and 25% were used, the mixture was packed into cylinder metal box and then rammed to obtain a cylindrical shape of dimension (6 cm diameter by 6 cm length). The produced samples for green properties are shown in Plate 1. After production, some of the produced test samples were baked at a temperature of 350 °C for 5 h using electric heat treatment furnace, for the purpose of determining the dry strength of the mould.

Permeability Test

The gas permeability was determined using a gas permeability tester. 150 g of the recycled sand mixed with the additives was weighed on a scale and transferred to a specimen sleeve with the base already plugged in the socket. The sand was rammed three times using George Fisher Rammer. The sand sample was then removed from the sleeve and placed on an electric permmeter in an inverted form. The permmeter was switched on and the arrow was allowed to settle. Then the permmeter was adjusted to zero by the control knob. The lever was then moved to check the position. Then the value was read off [10, 11].

Compatibility Test

Compatibility test is carried out to know the way moulding sand will withstand repeated cycles of heating and cooling during casting operations. An empty sand sleeve with the stopper plugged underneath it is placed under the funnel outlet of the compatibility tester's sieve. The sand is sieved until a heap is formed. The heap is then stickled off the sand. The sand is rammed four times and a value X is read off the calibration. The compatibility value is then calculated using Eq. (2) [11, 12].

$$\text{Compatibility} = X\frac{100}{67} \tag{2}$$

Mouldability Test

After ramming, the sand mixture test pieces were placed inside the moulability machine, switched on and allowed to run with the test pieces inside until it stopped. The weight of the sample that fell out of the sieve on the machine was taken as (Wb) and the weight of the initial test pieces as (Wa). Then the mouldability of the test sample was calculated from Eq. (3) [13].

$$\text{Mouldablity} = \frac{Wa - Wb}{Wa} \times 100(\%) \tag{3}$$

Green/Dry Compressive and Shear Strength Determination

The green compressive strengths (GCS) and the green shear strengths (GSS) was determined immediately after ramming, while the dry compression strengths (DCS) and dry shear strength(DSS) were determined after baking the rammed pieces. For the compression strengths, the samples were placed between two parallel plates of a compressible jig, while for the shear strengths the samples were placed between the parallel plates of the shear jig. The samples and the jigs were then placed on the universal sand testing machine in such a manner that the movable jaws clamped the sample to fracture in a slow but continuous movement without shock. The values of the strength were directly read from the calibrated scales attached to the machine [12, 14].

Results and Discussion

The result of the chemical analysis of the clay is shown in Table 1 and the particle size analyses are shown in Tables 2 and 3. The results of the moisture content, bulk density, permeability, compatibility and strengths are shown in Figs. 1, 2, 3, 4, 5, 6 and 7.

Table 1 Chemical composition of the investigated rice husk ash

SiO_2	Al_2O_3	Fe_2O_3	CaO	MgO	K_2O	Na_2O	S	TiO_2
79.18%	8.79%	8.58%	0.12%	0.01%	0.96%	0.12%	0.02%	2.56%

Table 2 The Particle size analysis of the rice husk ash

S/no	Sieve no (mm)	%Wt retained	Multiplier	Product
1	1.60	0	5	0
2	1.00	0.01	10	1
3	0.71	0.14	20	2.8
4	0.63	1.00	30	30.0
5	0.40	2.24	40	89.6
6	0.31	16.15	50	807.5
7	0.20	32.59	70	2281.3
8	0.16	30.77	100	3077
9	0.12	11.05	140	1547
10	0.10	2.02	200	404
11	Pan	1.98	300	594
	Total	97.95		8834.2
	G.F.N			90.19

Table 3 The particle size analysis of river Niger silica sand with rice husk ash

S/no	Sieve no (mm)	%Wt retained	Multiplier	Product
1	1.60	0	5	0
2	1.00	13.5	10	135
3	0.71	24.0	20	480
4	0.63	8.5	30	255
5	0.40	29.5	40	1180
6	0.31	7.5	50	375
7	0.20	6.0	70	420
8	0.16	1.5	100	150
9	0.12	1.2	140	168
10	0.10	0.6	200	120
11	Pan	0.2	300	60
	Total	92.5		3343
	G.F.N			36.14

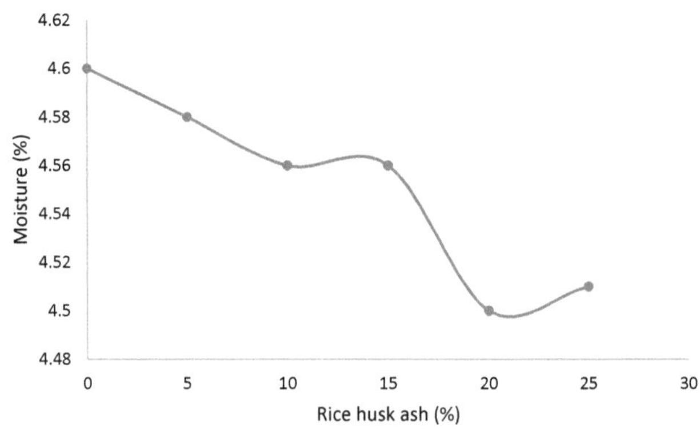

Fig. 1 Variation of % moisture content with % of rice husk ash

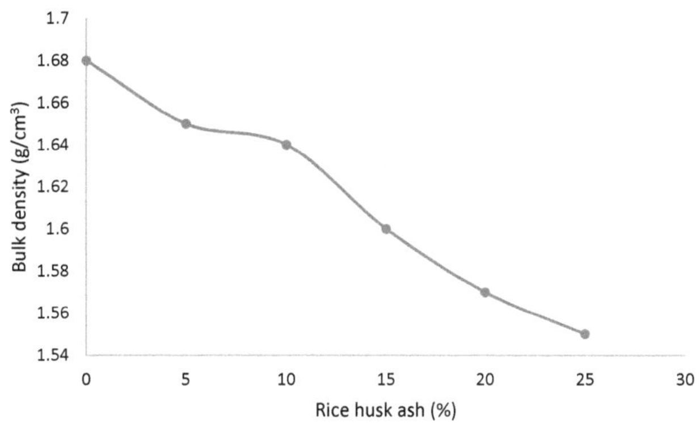

Fig. 2 Variation of bulk density with % of rice husk ash

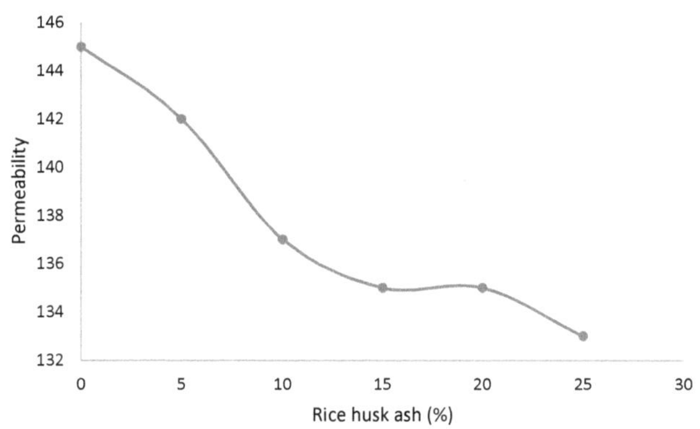

Fig. 3 Variation of permeability with % of rice husk ash

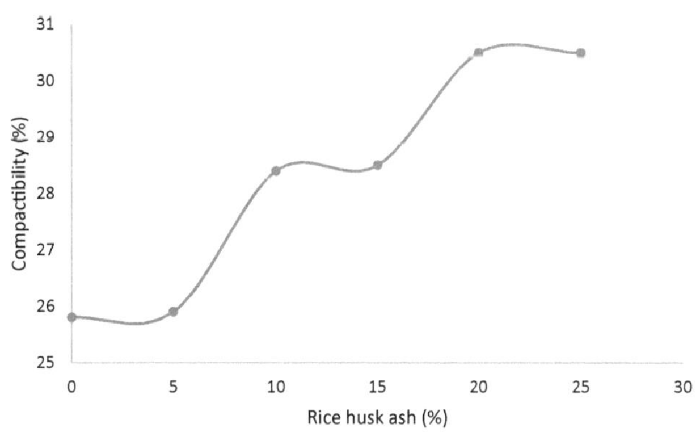

Fig. 4 Variation of compatibility with % of rice husk ash

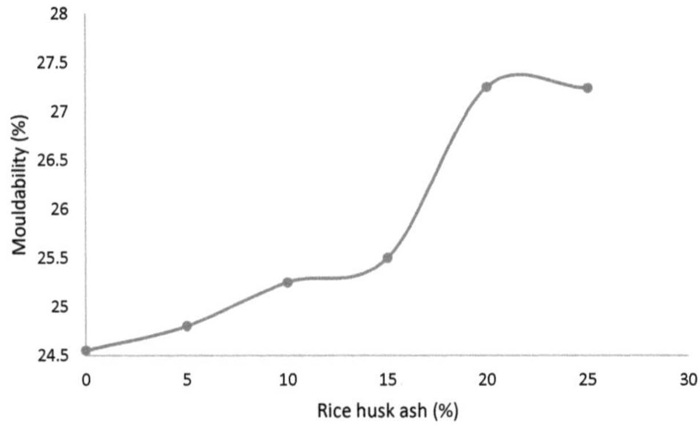

Fig. 5 Variation of mouldability with % rice husk ash

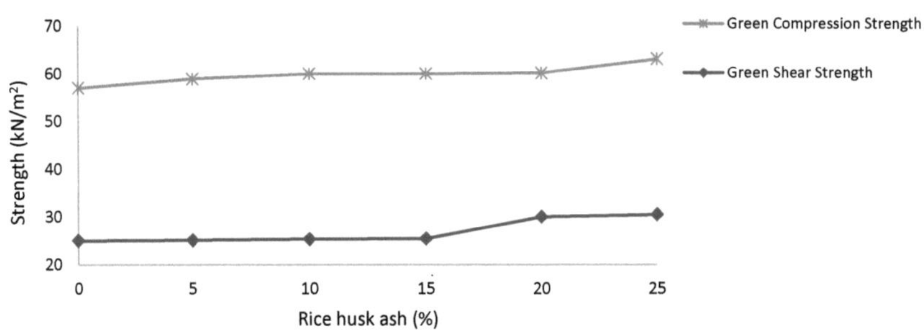

Fig. 6 Variation of green strength with % of rice husk ash

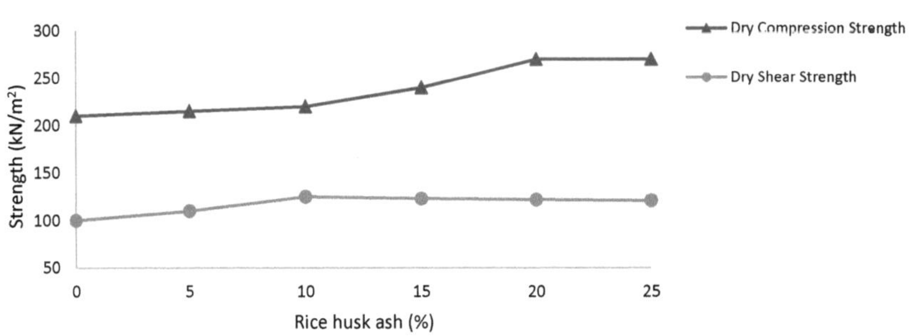

Fig. 7 Variation of dry strength with % of rice husk ash

Plate 2 Surface appearance of aluminum ingots casted from the blended sand

Chemical Analysis

The chemical analysis of the ash shows that it contains 79% SiO_2, 8.79% Al_2O_3, 0.01% MgO and 0.02% S as in Table 1. The values are within the range of acceptable values for typical foundry sand [15, 16].

Sieve Analysis

The sieve analysis of the River Niger silica sand result shows that the sand has 28.96 Green fineness number (G.F.N) which make the sand to be coarse in nature and lower the bounding strength of the sand [1], while the result of the sieve analysis of the ash was 90.19 G.F.N (Table 2). Blending the River Niger silica sand and Rice husk ash in ratio of 3:1 resulted to an increase in the G.F.N of River Niger silica sand from 28.96 to 36.14 G.F.N, this indicates that addition of ash to this sand increased the fineness of the sand (Table 3) This means that casting done with the mixture can result to good surface finish as shown in Plate 2.

Moisture Content

The moisture content slightly decreased with the percentage of ash in the sand mixture. The moisture content decreased from 5.8 to 1.7% at 5 and 25% addition of ash respectively as shown in Fig. 1. The values of the moisture content obtained is an indication that the sand developed plasticity faster with the additions of ash and these values obtained are within the acceptable limit [4].

Bulk Density

The bulk density slightly decreased as the percentage of ash increased in the sand e.g. the bulk density decreased from 1.68 to 1.54 g/cm^3 at 5 and 25% addition of ash (Fig. 2). The slightly decreased of the bulk density is as a result of the bulk density of the ash which was found out to be 1.15 g/cm^3 and that of the silica sand 1.71 g/cm^3.

Permeability

The permeability of the sand mixture decreased to a minimum values of 86 at 25% addition of rice husk ash as shown in Fig. 3, but the permeability values obtained are within the recommended acceptable limit for both ferrous and non-ferrous casting [4] which means that mould produced from these blend will allow escape of gasses freely from the casting hence reducing the chances of getting gas inclusions in the casting.

Mouldability and Compatibility

The mouldability and compatibility values increased as the percentage of rice husk ash in the sand mixture increases as in Figs. 4 and 5, these means that as the percentage of rice husk ash increased the sand develop strong bounding.

Compression Strength and Shear Strength

Figures 6 and 7 shows the compression strength and shear strength variation with rice husk ash, it shows that both the green and dry strength increases as increasing percentage of rice husk ash from 5 to 25% addition respectively. For example, the green compression strength increased from 19 to 42kN/m^2 and shear strength increased from 12 to 31kN/m^2.

The baking strength also increased at the percentage of the rice husk ash increased in the sand mixture, it means that the higher the rice husk ash, the more strength developed with the sand. Figure 7 shows that increased the percentage of rice husk ash beyond 20%wt do not resulted to any further increases in strength of the sand. The increases in compression strength of green sands containing 20%wt rice husk ash indicate its binder properties.

Conclusions

This study revealed that river Niger sand taken from River Niger at Itobe is alumino—silicate with physio—chemical properties that are suitable for non—ferrous alloy casting of low melting point because of its low refractoriness (1093 °C). It responded well to rice husk ash binder that gave good mechanical properties to sand mould specimens. All investigated ash were found capable of being added up to amounts of 20% with their technological properties remaining at sufficient levels which provides an opportunity for obtaining casting of satisfactory quality from sand casting.

References

1. S.A. Yaro, V.S. Aigbodion, Suitability of Rayfield and Rukuba (plateau-state), sand deposits for foundry applications. Afr. J. Nat. Sci. **9**, 167–176 (2006)
2. V.S. Aigbodion, S.B Hassan, S.O. Olajide, O.J Agunloye, A.S. AbdulRahaman, G.E. Okafor. The use of rice husk ash as an aggregate for foundry sand mould production. Paper presented at 25th annual confrence of N.M.S, Akure, Nigeria, pp. 16–22, 2008
3. T.A. Burns. *Foseco Foundry man's Handbook*, 9th edn. (Pergamon Press, London, 1986), pp. 45–134
4. L.O Asquo, S.B Bobo. *Foundry Technology* (ABC s Ltd, Nigeria, 1991), pp. 45–89
5. Raw Material Research and Development Council, *Raw materials sourcing for manufacturing in Nigeria.* (Lagos, Nigeria, 1990)
6. A.O. Apata, Development of Cementing Materials using available raw Materials for Building Construction, Ph.D thesis, Federal University of Technology Akure, 14–35, 2008
7. C.O. Nwajagu. *Foundry theory and Practice* (ABC Publishers Ltd, Nigeria, 1994), pp. 120–229
8. E.O. Atama, A.E. Aye, F.O. Sunday. Possibility of using locally binders and Ochadamu silica for the production of foundry cores. Paper presented at 24th annual conference of N.M.S, Jos, Nigeria, pp. 25–32, 2007
9. M.L. Begeman, B.H. Amstrad, *Manufacturing Process*, 6th edn. (Prentice-hall Ltd, India, 1969), pp. 45–46
10. M.H. Muhammad, M.A Bawa, N.T Akushi, Characterization of Alkaleri and Gombe Foundry sands. Paper presented at 20th annual conference of N.M.S, Ajaokuta Nigeria, pp. 20–25, 2003
11. A.I. Nwankwo, D.P. Seghal, *Survey of Local Sands used by Nigerian Foundries*, (Review of investigation and task on location of new deposits, 1983)
12. V.I. Okezue, Comparative Study of Cudliver and Candlenut Oils as Core Binders in Foundry Application. B.Eng dissertation, Amodu Bello University, Zaria Nigeria pp. 23–35, 2004
13. A.P Ihom, J.S. Jatau, M.H. Muhammed, The assessment of honey as binder for core making in foundry. Paper presented at 23rd annual conference of N.M.S, Abuja, Nigeria, pp. 25–32, 2007
14. O. Aponbiede, Development of Oil-Core for Foundry Application Using Local Vegetable Oil. Ph.D. thesis, Amodu Bello University, Zaria, Nigeria, pp. 24–120, 2000

15. D.A. Aji, N. Igu, M.I. Oseni, A.O. Apata, G. Baba, Characterization of river Niger (Idah Deposit), Ochadamu and Uwowo sands for foundry mould production. J. Emerg. Trends Eng. Appl. Sci. **6**(5), 295–300 (2015)
16. N.E. Idenyi, N.M. Ani. Characterization of Abakaliki and Ugbawka Ricke husks. Paper presented at 23rd annual conference of N.M.S Lagos, pp. 89–94, 2006

Part XV
Advances in Environmental Technologies: Recycling and Sustainability Joint Session: Poster Session

Chemical Reduction of Fe(III) in Nickel Lateritic Wastewater to Recover Metals by Ion Exchange

Amilton Barbosa Botelho Jr, Mónica M. Jiménez Correa, Denise Crocce Romano Espinosa and Jorge Alberto Soares Tenório

Abstract Wastewater from nickel lateritic process still has valuable metals that could be recovered. Using an ion exchange to recovery these metals, the iron becomes a problem due to the fact that it precipitate at pH 2. The aim of this research was to make iron soluble by reducing Fe^{+3} to Fe^{+2}. A synthetic solution was prepared with sulfate ferric. The reducing agent sodium dithionite was used and the potential was varied (initial potential: 700, 600, 500, 450 e 400 mV) using as a reference the Pourbaix diagram for iron created by HydroMedusa software. The pH was ranged between 0.5 and 2.5, the reducing agent's concentration was 0.5 M and the reaction time ranged between 30 and 120 min at room temperature. The ferric iron and total iron concentration were verified using a spectrophotometer with 5-sulfosalicylic acid (SSA) to determine Fe(III) and ammonia to determine Fe(II) and Fe(III) in 500 and 425 nm, respectively. The results indicated that total reduction was at 400 mV in any pH between 0.5 and 2, but less than 40% in pH 2.5 at this potential value.

Keywords Iron · Sodium dithionite · Reduction process

Introduction

The nickel is an important metal in modern world, and it's used in stainless steel alloys, alloys based on nickel, electroplating, among others. Nickel reserves can be found as sulfide and laterite ores. While 60% of world nickel production are from sulfide ores, 70% of world land are laterite. This is due to complexity of mining process from laterites ores. On the other hand, there is a significant increase of laterite ore's nickel mining, because sulfide ores are getting scarcity or are getting technically infeasible to be used [1–3].

A.B. Botelho Jr (✉) · M.M.J. Correa · D.C.R. Espinosa · J.A.S. Tenório
Department of Chemical Engineering, University of São Paulo (USP), São Paulo, Brazil
e-mail: amilton.junior@usp.br

L. Zhang et al. (eds.), *Energy Technology 2017*,
The Minerals, Metals & Materials Series, DOI 10.1007/978-3-319-52192-3_45

Among the methods utilized to extract nickel from these resources, one of then is the leaching, where there is consumption of sulfuric acid [4]. This process bring on considerable quantity of wastewater, so-called liquor, that is deposited in tailings dam, where there are metals wasn't extract [5].

Extracting metals from wastewater can be done via an ion-exchange process, consisting of the interchange of ions between two phases. One ion cross-linked polymer network is in contact with ions in solution of the same charge: ion M^+, in solution, change with cation in solution A^+B^-, and the resign will be with M^+B^- and the solution with A^+ [6].

Amongst the metals in the liquor, iron presents a problem, because it's precipitate in pH 2, presented in Fig. 1—that was constructed with help of the Hydra-Medusa software—, moreover the ferric iron reduces the resin efficiency. Figure 1 presents cation Fe^{+3} and SO_4^{-2}, due to the sulfuric acid used to extract the ore. The liquor has a potential of 700 mV.

A possible solution for this problem could be a chemical reduction of Fe(III) to Fe(II) using a reducing agent. Among the reducing agents available, sodium dithionite (DS) is an option once it's used to remove Fe(III) from mineral called kaolin, (Al_2 Si_2 O_5 $(OH)_4$), whereupon presence of ferric iron is a contamination, due to it affect the brightness of kaolin to use in paper coating [7, 8].

An improved condition to reduce iron would be at pH less or equal than 2, as can see in Fig. 1, because above pH 2 the iron will precipitate mainly at high concentration. DS is more efficient in acidic medium and high temperature; however, in acidic medium there is liberation of SO_2 and H_2S [9], decreasing reducing agent's stability, it could have an elementary sulfur formation [8].

In order for the reduction process to occur, first the DS has to dissociate in aqueous solution (Eq. 1), resulting in bisulfite anion (HSO_3^{-1})—responsible for the reduction process—and thiosulfate anion ($S_2O_3^{-2}$). The bisulfite anion reacts with Fe^{+3} and, through reduction process, results in Fe^{+2}. Equations (2) and (3) present the possible reactions of Fe^{+3}.

$$2Na_2S_2O_4 + H_2O \rightarrow HSO_3^- + S_2O_3^{-2} + 2Na^+ \quad [10] \tag{1}$$

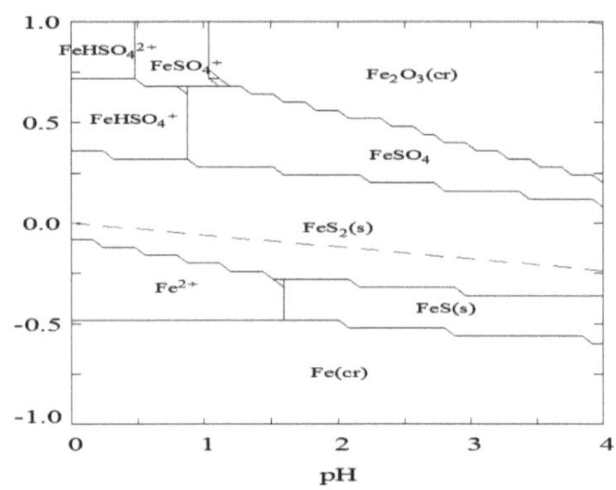

Fig. 1 Pourbaix's diagram constructed using Hydra-Medusa software 25 °C (potential vs. pH)—[Fe] = 334.2 mM

$$S_2O^{2-}_{4(aq.)} + 2Fe_2O_{3(s)} + 2H^+_{(aq.)} \rightarrow 2(SO_4)^{2-}_{(aq.)} + 4Fe^{+2}_{(aq.)} + 2H_2O \quad [7] \quad (2)$$

$$S_2O^{2-}_{4(aq.)} + Fe_2O_{3(s)} + 2H^+_{(aq.)} \rightarrow 2(SO_4)^{2-}_{(aq.)} + 2Fe^{+2}_{(aq.)} + H_2O \quad [7] \quad (3)$$

The aim of this work was to study the chemical reduction of Fe(III) using sodium dithionite (D.S.) at room temperature (25 °C).

Materials and Methods

The Fe^{+3} concentration used in this study was 18,713.5 ppm, based on limonite nickel's iron concentration, and it was prepared using a synthetic solution of sulfate iron octahydrate, in order to simulate iron in the liquor. The influence of potential using an electrode Ag/AgCl (3 M) was investigated starting with initial potential (700 mV, approximately), 600, 500, 450 and 400 mV. Before the chemical reaction, we prepared a 0.5 M DS solution, and added iron solution to reach the desired potential; the pH was varied between 0.5 and 2.5 (using sulfuric acid or sodium hydroxide 2 M for pH adjustment); and the reaction time was studied ranging 0–120 min.

The analyses of Fe^{+3} and Fe_{total} was done using UV/vis spectrophotometry. To perform the analyses, we withdrew an aliquot and diluted it 1:5 with H_2O pH 1.0. Then, a sample of 0.1 mL was removed and added 3 mL of 5-sulfosalicylic acid 10% w/v (SSA) and 97 mL deionized water to quantify the Fe(III) concentration in 500 nm; and added 3 mL of ammonia 25% w/w to quantify Fe_{total} [11, 12].

Results and Discussion

Figure 2 present the results of the reaction at pH 0.5. At 600 mV (E1), there is little ferric iron reduction—values close to 0%. At 400 mV (E4), there is almost all full reduction of Fe(III), and this can be intuited by the system conditions, as pH and environment, (c.f. Fig. 1), but also because of quantity of DS necessary to reach around this environment; there is more bisulfite anion on 400 mV potential

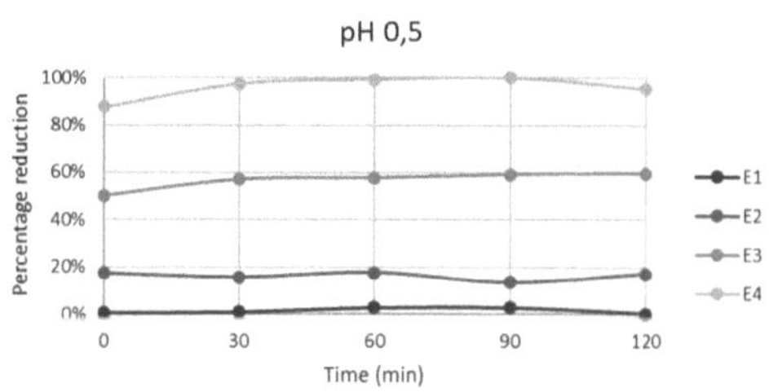

Fig. 2 Results in percent of chemical reduction at pH 0.5, where E1 is 600 mV, E2 is 500 mV, E3 is 450 mV and E4 is 400 mV

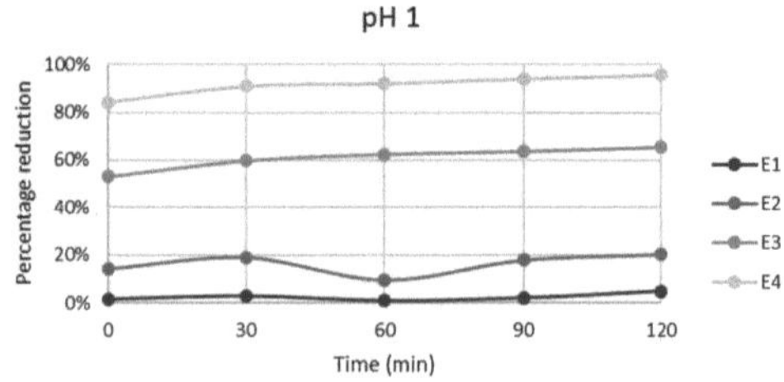

Fig. 3 Results in percent of chemical reduction at pH 1.0, where E1 is 600 mV, E2 is 500 mV, E3 is 450 mV and E4 is 400 mV

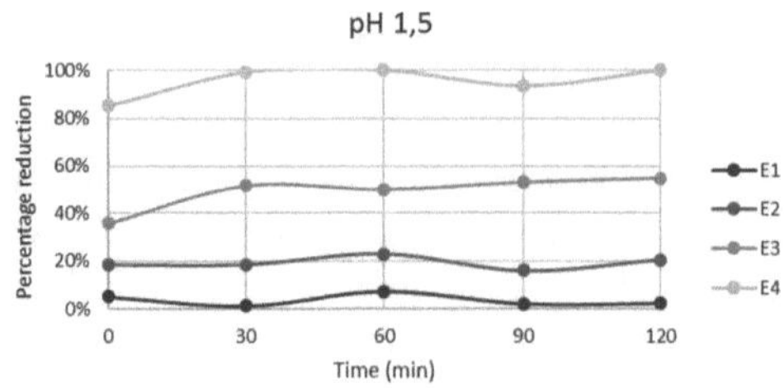

Fig. 4 Results in percent of chemical reduction at pH 1.5, where E1 is 600 mV, E2 is 500 mV, E3 is 450 mV and E4 is 400 mV

available to reduction process than on 600 mV. This can be observed also at 500 and 450 mV.

Figure 3 presents the results of ferric iron chemical reduction at pH 1. These results are consistent with the results at pH 0.5, where in 400 mV (E4) there was almost complete reduction, and in 450 mV (E3) there was almost 60% reduction in both cases. At 500 mV (E2) the percent reduced was below 20%, similar to pH 0.5.

At pH 1.5 (Fig. 4) the percentage of ferric iron reduction is almost 100% on 400 mV, as well as in pH 0.5 and 1.0. However, 450 mV of potential there was no similarity among pH values present before. For pH 1.5, the percentage reduction was almost 50%, while at pH 0.5 and 1 was 60%. For 600 mV (E1) and 500 mV (E2), the reduction values were almost 10 and 20%, respectively. This can be caused by the influence of pH medium.

From Fig. 1, at pH 1.5, the point of 600 mV is close to the border between Fe^{+2} ($FeSO_4$) and Fe^{+3} (Fe_2O_3), and it can harm the ferric iron reduction process. For the pH 2 (Fig. 5) and pH 2.5 (Fig. 6), it can be seen a decrease of reduction in all potential values; especially at pH 2.5, where in 400 mV presented, until then, almost all reduction of Fe(III), and in pH 2.5 there was only 40% of reduction.

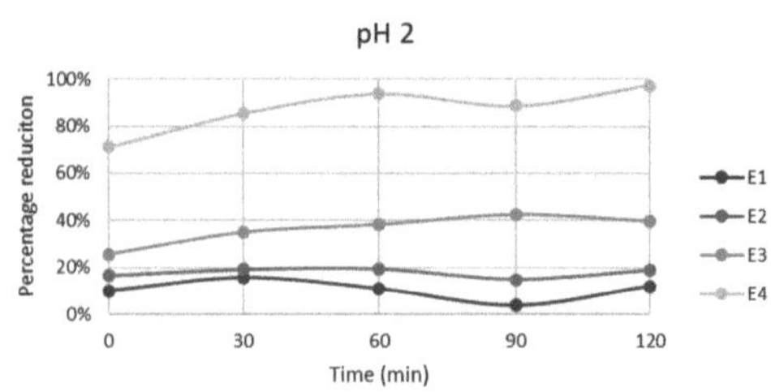

Fig. 5 Results in percent of chemical reduction at pH 2.0, where E1 is 600 mV, E2 is 500 mV, E3 is 450 mV and E4 is 400 mV

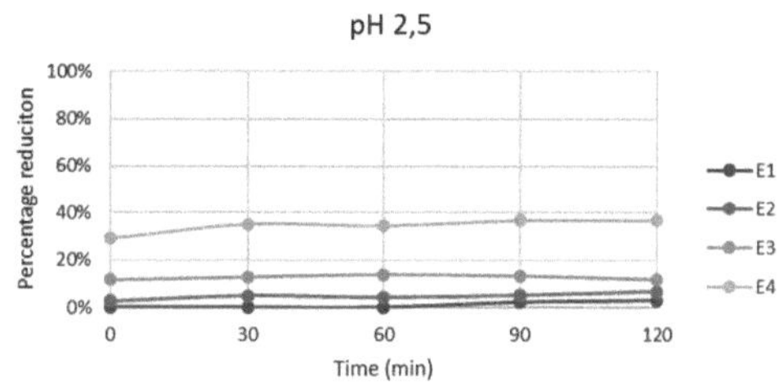

Fig. 6 Results in percent of chemical reduction at pH 2.5, where E1 is 600 mV, E2 is 500 mV, E3 is 450 mV and E4 is 400 mV

This shows the influence of pH medium in ferric iron reduction process, since at pH between 0.5 and 2 can be seen there was almost complete reduction in 400 mV. Meantime, on pH 2.5 there were 40% of reduction on 400 mV (Fig. 6), where shows the influence of pH.

Conclusions

For the potentials 500, 450 and 400 mV, we demonstrated that a more acidic the environment leads to higher ferric iron reduction. In contrast, at 600 mV it was not possible to identify any difference as a function of pH. The percentage of ferric iron reduction between pH 0.5 and 1 did not have differences among the potentials studied, demonstrating it is not necessary decrease the medium pH to 0.5. Among the percentage obtained between pH 1 and 2 to potential of 450 mV there was a decrease of percentage reduction while increase the medium pH; starting in 60% (pH 1), 50% for pH 1.5 and 40% for pH 2. This shows the influence of pH on the reduction process. In spite of the Pourbaix's Diagram showing there will be Fe(II) at the potential studied, probably the kinetic of reduction is slow on 600–450 mV, and fast on 400 mV, supposedly also because the quantity of DS added.

Acknowledgements To the São Paulo Research Foudation (Fundação de Amparo à Pesquisa do Estado de São Paulo - FAPESP), n° 2016/05527-5 and 2012/51871-9, for the financial support through doctorate grant.

References

1. P.C.F. Crowson, Mineral reserves and future minerals availability. Miner. Econ. **24**(1), 1–6 (2011)
2. A.D. Dalvi, W.G. Bacon, R.C. Osborne, in *PDAC 2004 International Convention, no. Figure 2*, The past and the future of nickel laterites, pp. 1–27, 2004
3. G.M. Mudd, Global trends and environmental issues in nickel mining: Sulfides versus laterites. Ore Geol. Rev. **38**(1–2), 9–26 (2010)
4. C.K. Gupta, in *Chemical Metallurgy*, Principles and Practice, 1981
5. A.B. da Luz, J.A. Sampaio, S.C.A. França, *Tratamento de minérios*, vol. 1, (Rio de Janeiro, 2010)

6. S.D. Alexandratos, Ion-exchange resins: a retrospective from industrial and engineering chemistry research. Ind. Eng. Chem. Res. **48**, 388–398 (2009)
7. F. Arruda, N. Gomes, F. Manoel, M. Eloísa, Alvejamento Qímico de caulins brasileiros: Efeito do potencial eletroquímico da polpa e do ajuste de pH. Quim. Nova **34**(2), 262–267 (2011)
8. F.A.N.G. Silva, L.S. Mello, J.A. Sampaio, F.M.S. Garrido, M.E. Medeiros, F.S. Teixeira, A. B. Luz, Alvejamento químico de caulins brasileiros: Efeito do potencial eletroquímico da polpa e do ajuste do ph. Quim. Nova **34**(2), 262–267 (2011)
9. M. Wayman, W.J. Lem, Decomposition of aqueous dithionite. Part II. A reaction mechanism for the decomposition of aqueous sodium dithionite. *Can. J. Chem.* **48**(5), 782 1970
10. V. Cermak, M. Smutek, Mechanism of decomposition of dithionite in aqueous solutions. Collect. Czechoslov. Chem. Commun. **40**(11), 3241–3264 (1975)
11. C. Paipa, M. Mateo, I. Godoy, E. Poblete, M.I. Toral, T. Vargas, Comparative study of alternative methods for the simultaneous determination of Fe^{+3} and Fe^{+2} in leaching solutions and in acid mine drainages. Miner. Eng. **18**(11), 1116–1119 (2005)
12. D.G. Karamanev, L.N. Nikolov, V. Mamatarkova, Rapid simultaneous quantitative determination of ferric and ferrous ions in drainage waters and similar solutions. Miner. Eng. **15**(5), 341–346 (2002)

Chronopotentiometry Applied to the Determination of Copper Transport Properties Through a Cation-Exchange Membrane

K.S. Barros, J.A.S. Tenório and D.C.R. Espinosa

Abstract Over the past several decades, electrodialysis membranes have been employed for the removal of metals from galvanizing industry effluents. However, for its technical viability, it is useful to know some properties of the system to avoid, for example, precipitation of inorganic salts, destruction of the membrane and an increase in the energy consumption. The scope of the present study is to use chronopotentiometry to determine limiting current densities, transition times and resistances of the cationic membrane HDX 100 using a synthesized solution of the galvanizing industry effluent prepared with copper sulfate and sulfuric acid in different concentrations: 0.1, 0.5 and 1.0 g/L of Cu^{+2}. The current-voltage and chronopotentiometric curves obtained have clearly defined regions and it was verified that the Cu^{+2} concentration causes a significant increase in the limiting current densities and in transition times. However, a high decrease in resistance occurred because at the highest concentrations, it is more difficult to reach the zero concentration value at the membrane surface. Besides, the pH increase causes a significant decrease in the limiting current density and in the transport number, but a slight increase in the resistance due to the presence of hydrogen. The results obtained confirm that the chronopotentiometry method is a valuable tool to determine membrane properties.

Keywords Current-voltage · Chronopotentiometry · Limiting current · Membrane resistance · Transition time

K.S. Barros (✉) · J.A.S. Tenório · D.C.R. Espinosa
Department of Chemical Engineering, São Paulo University, Avenida Professor Lineu Prestes, 580, Conjunto Das Químicas, Bloco 18 Cidade, São Paulo, SP CEP: 05508–000, Brazil
e-mail: kayobarros@usp.br

L. Zhang et al. (eds.), *Energy Technology 2017*,
The Minerals, Metals & Materials Series, DOI 10.1007/978-3-319-52192-3_46

Introduction

The treatment of industrial effluents containing metals represents one of the major environmental issues in every industrialized country in the world [1]. In addition, the toxicity of cyanide and its use in some processes as a superficial finish has caused concerns and lead to its substitution to an eco-friendly alternative [2]. In the last decades, ion-exchange membrane processes have been suggested as promising possibilities for the removal and recovery of metals, such as copper, and other inorganic toxic substances which are generated in electroplating processes [3]. Among the different membrane processes, electrodialysis represents one of the most important due to its significant flexibility [4].

However, to ensure good performance for electrodialysis, it is very important to determine some specific properties of the system, as the limiting current density to avoid, for example, precipitation of inorganic salts in the membrane [2]. The resistance of the membrane is also important since it is associated with the difficulty of ion passage through the membrane. Both the limiting current density and the resistance of the membrane can be obtained from the construction of current-voltage curves (CVC). Another important property which has to be determined is the transition time, which corresponds to the moment that the interfacial salt concentration becomes zero and the potential tends to infinity. This property can be determined by the intersection point of two tangents to the regions of slow and fast potential growth of the curves obtained by the chronopotentiometry method [5].

Chronopotentiometry is a powerful dynamic electrochemical technique to obtain information regarding the membrane heterogeneity. Compared with other characterization methods such as impedance spectroscopy and cyclic voltammetry, it allows a direct access to the voltage contributions in different states of the membrane/solution system [3]. CVC can also be constructed using the same chronopotentiometry apparatus and they show three characteristic regions: the first (I) is known as the ohmic region and shows a linear increase in the voltage drop with the imposed current. From the inverse of the slope of this straight line it is possible to determine the resistance of the membrane. The second region (II) is known as the plateau zone and the intersection of its slope with the slope of the first region gives the limiting current value, which is reached when the concentration polarization occurs and causes a scarcity of ions in the depleted layer near the membrane. In the third region (III), or the over-limiting current region, the current intensity increases again with the applied voltage, which is attributed to new mechanisms of ionic mass transport such as the formation of water splitting products or electroconvection occurrence [1, 5]. Figures 1 and 2 show a characteristic current-voltage (U) (CVC) and chronopotentiometric curves under a current density above the limiting current density, respectively.

The main objective of the present work was to study the effect of copper concentration and pH of a synthesized effluent of galvanizing industry on the limiting current, resistance and transition time in a cationic membrane (HDX 100) for the

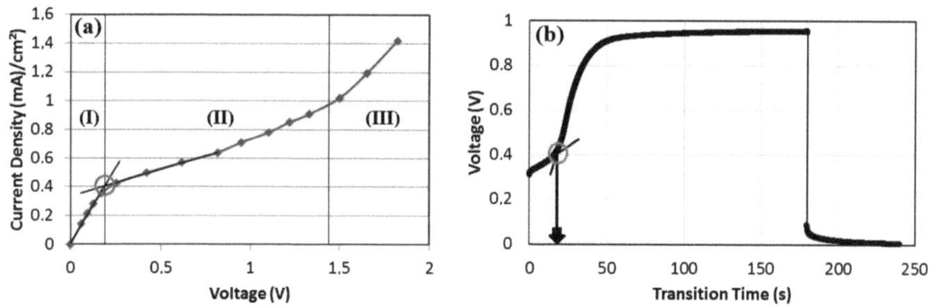

Fig. 1 **a** Typical current-voltage and **b** chronopotentiometric curves of an ion-exchange membrane under a current density above the limiting current density

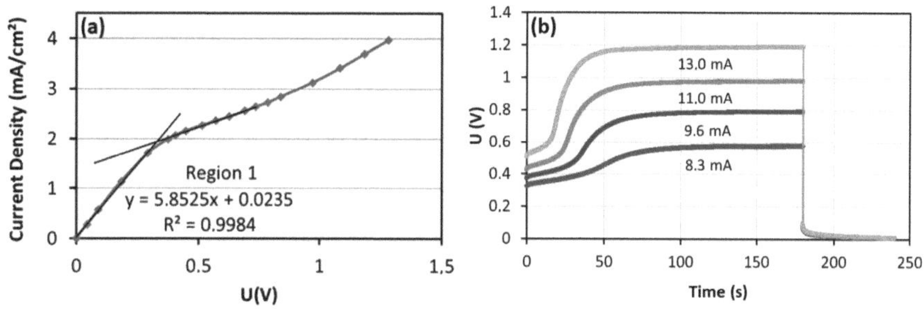

Fig. 2 **a** CVC and **b** chronopotentiometric curves for the solution with 0.5 g/L of Cu^{+2} and pH = 2.6

potential treatment of the effluent by electrodialysis. The technique chosen for this study was to construct current-voltage curves and perform chronopotentiometry by measuring the voltage drop across the cationic membrane as a function of time when an electric current above the limiting current is applied to the system.

Experimental

Solutions

In order to evaluate different conditions of Cu^{+2} concentration and pH of the synthesized effluent of galvanizing industry, the solutions were prepared with $CuSO_4{\cdot}5H_2O$, H_2SO_4, distilled water and KOH solution (0.4%wt) for pH adjustment. All the reagents employed were of analytical grade and the conditions tested are shown in Table 1.

Table 1 Composition of the working solutions

Variable	Cu^{+2} (g/L)	pH	Molar ratio (sulfate:acid)
Effect of Cu^{+2} concentration	0.1	2.6	1:1
	0.5	2.6	1:1
	1.0	2.6	1:1
Effect of pH	0.5	2.0	1:1
	0.5	3.0	1:1
	0.5	4.0	1:1

Ion-Exchange Membrane

The ion-exchange membrane used in this work to evaluate the copper transport was the HDX 100 (provided by Hidrodex®). The anionic HDX 200 was also used to minimize the contribution of the protons generated in the anode on the cation-exchange membrane measurements. The HDX 200 contain quaternary amine groups attached to the membrane matrix and the HDX 100 is charged with sulfonic acid groups and has a similar morphology to that of HDX 200. Both are heterogeneous [6] and have a circular area of 3.52 cm^2. The experiments were accomplished after a membrane equilibration period of 24 h using the same solution used in the experiments. Table 2 presents some of the main properties and characteristics of these membranes.

Electrochemical Cell and Chronopotentiometric Measurements

The chronopotentiometric experiments were accomplished using a three-compartment cell with a volume of 130 mL (each). The cation-exchange

Table 2 Main properties and characteristics of the HDX 100 and 200 membranes

Parameter	HDX 100	HDX 200	Unit
Ion group attached	$-SO_3^-$	$-NR_3^+$	–
Water content	35–50	30–45	%
Ion exchange capacity	≥ 2.0	≥ 1.8	mol kg^{-1} (dry)
Surface resistance (0.1 mol de NaCl)	≤ 20	≤ 20	Ω cm^{-2}
Permselectivity (0.1 mol KCl/0.2 mol KCl)	≥ 90	≥ 89	%
Burst strength	≥ 06	≥ 0.6	MPa
Dimensional change rate	≤ 2	≤ 2	%
Water permeability	≤ 0.1 (<0.2 MPa)	≤ 0.2 (<0.035 MPa)	mL h cm^{-2}

membrane separated the cathodic compartment from the central compartment while the anion-exchange membrane separated the anodic compartment from the central compartment to minimize the contribution of the protons generated in the anode. To measure the potential drop through the cationic membrane two reference electrodes (Ag/AgCl) immersed in Luggin capillaries were used. Moreover, two graphite electrodes were used to impose the current, which was supplied by a potentiostat/galvanostat (AUTOLAB PGSTAT302 N) and a PC data acquisition system was used to register the values obtained. The current-voltage curves were constructed from the steady-state polarization voltage corresponding to an applied current. For the chronopotentiometric curves construction, current densities above the limiting current densities were employed. All experiments were conducted at room temperature without stirring.

Results and Discussion

Evaluation of Cu^{+2} Concentration

The copper concentration (0.1, 0.5 and 1.0 g/L of Cu^{+2}) in the solution in pH = 2.6 was evaluated. Figure 2a shows the CVC obtained for the concentration 0.5 g/L and Fig. 2b shows the chronopotentiometric curves for some applied current densities above the limiting current also for 0.5 g/L concentration. All the results of limiting current densities and resistances for other concentrations are presented in Table 3.

According to the results presented in Table 3, we note that for a constant pH of 2.6, the increase in Cu^{+2} concentration from 0.1 to 1.0 g/L causes an increase in the limiting current density. This was expected since at a higher concentration it is more difficult to reach the zero concentration value at the membrane surface on the diluted compartment when a current density is imposed. This is in accordance with the proportional relationship between these variables in the classical concentration polarization theory expressed by the Eq. (1):

$$I_{lim} = \frac{zAFC_oD}{\delta(\bar{t}_i - t_i)} \quad (1)$$

Table 3 Limiting current density and membrane resistance obtained for solutions of 0.1, 0.5 and 1.0 g/L Cu^{+2} concentrations and pH = 2.6

pH	Cu^{+2} concentration (g/L)	Limiting current density (mA/cm^2)	Membrane resistance ($\Omega\ cm^2$)
2.6	0.1	0.55	422.4
	0.5	1.90	170.9
	1.0	4.00	138.1

Table 4 Transition times obtained for solutions of 0.1, 0.5 and 1.0 g/L Cu^{+2} concentrations and pH = 2.6

Cu^{+2} concentration					
0.1 g/L		0.5 g/L		1.0 g/L	
Current applied (mA)	Transition time (s)	Current applied (mA)	Transition time (s)	Current applied (mA)	Transition time (s)
2.50	51.8	8.30	36.7	17.00	29.2
3.00	31.2	9.00	32.4	18.00	28.1
3.20	26.7	9.30	30.7	19.00	24.2
3.60	19.7	9.60	29.4	20.00	22.5
4.20	14.0	10.00	26.4	21.00	19.7
5.00	9.2	11.00	22.0	22.00	18.8
		12.00	17.9	23.00	17.2
		13.00	15.3	25.00	13.9
		14.00	12.9	27.00	11.6
				29.00	9.7
				34.00	7.2

where δ is the diffusion boundary layer, C_o is the concentration of the counter-ion in the bulk solution, A is the membrane area, D is the electrolyte diffusion coefficient, z is the charge of the counter-ion, F is the Faraday's constant, $\bar{t}_i$ and t_i are the transport number of the counter-ion in the membrane and in the solution, respectively. In an opposite way, the resistance of the membrane strongly decreases as the Cu^{+2} concentration increases, since the presence of more copper ions in solutions facilitates its passage through the membrane and reduces the resistance of current transport.

Transition times for the solutions in 0.1, 0.5 and 1.0 g/L of Cu^{+2} were also determined by the chronopotentiometric curves and the results are shown in Table 4.

As can be observed in Table 4, for all concentrations tested the transition time decreases with the increase in the current applied, as a consequence of the faster ion depletion in the membrane/solution interface, and increases as the Cu^{+2} concentration increases. These results related to transition time (τ), current applied (i) and Cu^{+2} concentration (C_o) are in accordance with Sand equation proposed by [7]:

$$\tau = \frac{\varepsilon^2 \pi D}{4}\left(\frac{zF}{\bar{t}_i - t_i}\right)^2\left(\frac{C_o}{i}\right)^2 \tag{2}$$

Evaluation of PH

The influence of the solution pH was also evaluated and Table 5 shows the results for limiting current density and membrane resistance for pH = 2.0, 3.0 and 4.0 for a concentration of 0.5 g/L Cu^{+2}.

According to Table 5, the decrease in pH increases the limiting current density due to the major presence of H^+ ions in solution, since H^+ and Cu^{+2} compete to pass through the membrane and a higher current density is necessary to reach the zero concentration value at the membrane surface. In relation to the membrane resistance, it is not very sensitive to pH as to the Cu^{+2} concentration, since it decreased to 194.1–169.2 Ω cm^2 when the pH decreased from 4.0 to 2.0 for the same concentration. This behavior can be explained by the major presence of H^+ in lower pH and consequent increase in current passage through the membrane. However, in this case the copper transport number may suffer a negative effect, since the hydrogen molecule is smaller than copper and its passage will have preference.

The transition times for 0.5 g/L Cu^{+2} concentration and different pH were also evaluated and the results are presented in Table 6.

Table 5 Limiting current density and membrane resistance obtained for solutions of 0.5 g/L Cu^{+2} concentration and pH = 2.0, 3.0 and 4.0

Cu^{+2} concentration	pH	Limiting current density (mA/cm^2)	Membrane resistance (Ω cm^2)
0.5 g/L	2.0	2.80	169.2
	3.0	1.50	152.5
	4.0	1.40	194.1

Table 6 Transition times obtained for solutions of 0.5 g/L Cu^{+2} concentration and pH = 2.0, 3.0 and 4.0

pH = 2		pH = 3		pH = 4	
Current applied (mA)	Transition time (s)	Current applied (mA)	Transition time (s)	Current applied (mA)	Transition time (s)
12	46.0	7.0	33.5	6.20	32.7
13	39.2	7.5	30.3	6.50	30.8
14	34.5	8.0	25.3	6.80	28.8
15	30.8	8.5	22.3	7.20	25.4
16	23.0	9.0	19.4	7.60	22.0
17	21.7	9.5	18.1	8.20	20.2
18	16.6	10.0	15.6	8.60	17.6
20	13.4	11.0	13.5	9.20	15.6
22	11.2	12.0	11.1	10.0	12.8
24	9.5	13.0	8.7	11.0	9.5
		14.0	7.9		

As showed in Table 6, the increase in pH causes a decrease the transition time, which emphasizes what was previously discussed: the major presence of hydrogen in lower pH probably causes a decrease in the copper transport number and consequently a reduction in the transition time according to Sand equation (Eq. 2).

Conclusion

Chronopotentiometric and current-voltage curves (CVC) were effectively constructed in this work for the synthesized effluent from the galvanizing industry and the limiting current densities, resistances of the cationic membranes and transition times were determined. We verified that the increase in Cu^{+2} concentration causes a significant increase in the limiting current and in the transition time, but a decrease in the resistance of the membrane. It can be explained by the higher quantity of copper ions in solution and consequent facility to its passage through the membrane. The increase in pH causes a significant decrease in the limiting current density and in the transport number, but a slight increase in the resistance. These results show that both variables highly influence the transport properties of the heterogeneous cationic membrane HDX 100 and confirm the importance of its evaluation. Besides, it is desirable to study the influence of Cu^{+2} concentration and pH solution on the transport number of copper in the membrane ($\bar{t}_J$), since it was suggested that the $\bar{t}_J$ can be negatively affected at lower pH, for example.

Acknowledgements The authors gratefully acknowledge the financial support given by funding agencies CAPES, CNPq and FAPESP (2012/51871-9 and 2014/13351-9).

References

1. M.A.S. Rodrigues, F.D.R. Amado, M.R. Bischoff, C.A. Ferreira, A.M. Bernardes, J.Z. Ferreira, Transport of zinc complexes through an anion exchange membrane. Desalination **227**, 241–252 (2008)
2. I. Herraiz-Cardona, E. Ortega, V. Pérez-Herranz, Evaluation of the Zn^{2+} transport properties through a cation-exchange membrane by chronopotentiometry. J. Colloid Interface Sci. **341**, 380–385 (2010)
3. M. García-Gabaldón, V. Pérez-Herranz, E. Ortega, Evaluation of two ion-exchange membranes for the transport of tin in the presence of hydrochloric acid. J. Membr. Sci. **371**, 65–74 (2011)
4. M. Sadrzadeh, A. Razmi, T. Mohammadi, Separation of monovalent, divalent and trivalent ions from wastewater at various operating conditions using electrodialysis. Desalination **205**, 53–61 (2007)
5. M.C. Martí-Calatayud, M. García-Gabaldón, V. Pérez-Herranz, E. Ortega, Determination of transport properties of Ni(II) through a Nafion cation-exchange membrane in chromic acid. J. Membr. Sci. **379**, 449–458 (2011)

6. M.C. Martí-Calatayud, D.C. Buzzi, M. García-Gabaldón, E. Ortega, A.M. Bernardes, J.A.S. Tenório, V. Pérez-Herranz, Sulfuric acid recovery from acid mine drainage by means of electrodialysis. Desalination **343**, 120–127 (2014)
7. J.-H. Choi, S.-H. Kim, S.-H. Moon, Heterogeneity of ion-exchange membranes: the effects of membrane heterogeneity on transport properties. J. Colloid Interface Sci. **241**, 120–126 (2001)

Effect of Flow Rate on Metals Adsorption of Synthetic Solution Using Chelating Resin Dowex XUS43605 in Column Experiments

Isadora Dias Perez, Mónica M. Jiménez Correa, Flávia P. Cianga Silvas, Jorge A. Soares Tenório and Denise C. Romano Espinosa

Abstract In view of the rising generation of wastewater containing metals, sustainable and clean techniques to recover them are being researched. This study investigated the adsorption of aluminum, cobalt, chromium, copper, iron (III), magnesium, manganese, nickel and zinc presents in a synthetic effluent by chelating ion exchange resin Dowex XUS43605. The resin was tested in columns experiments and was analyzed the flow rate (0.45, 0.66 and 1.11 BV/h), at pH 1.5 and 20 °C. After loading the column, the elution experiments occurred with 1 M of H_2SO_4 and 4 M of NH_4O. The concentration of the metals in solution was obtained by X-ray fluorescence spectrometry energy dispersive (EDX). The results showed that the breakthrough experimental curves demonstrated that the main metals (aluminum, cobalt, manganese and zinc) were immediately broken through just after the initiation of the feeding, while the breakthrough of copper began later. Therefore, the study showed that the chelating resin could be used to treat contaminated effluents with metals.

Keywords Nickel · Copper · Ion exchange · Hydrometallurgy

Introduction

Metals such as nickel, copper and cobalt present in mining effluent have been the target of environmental concern because they have toxicity and can compromise the quality of water resources. In addition, these metals have economic value and in the near future, the mineral deposits of the interested metal will be poor than the current

I.D. Perez (✉) · M.M. Jiménez Correa · J.A. Soares Tenório · D.C. Romano Espinosa
Chemical Engineering Department, Polytechnic School, University of São Paulo, 158 Av. Prof. Luciano Gualberto, Trav. 3, Caixa Postal 61548, São Paulo, SP 05424-970, Brazil
e-mail: dpisadora@gmail.com

F.P. Cianga Silvas
Instituto Tecnológico Vale - Mineração, 31 Av. Juscelino Kubitschek, Bauxita, Ouro Preto, MG 35.400-000, Brazil

L. Zhang et al. (eds.), *Energy Technology 2017*,
The Minerals, Metals & Materials Series, DOI 10.1007/978-3-319-52192-3_47

ones [1]. However, despite the decrease of world reserves, the ore continue to be exploited due to the sophistication of new process, as well as recovery methods and reuse of nickel and other metals from secondary resources [2].

Therefore, the metals recovery from effluents could be separated by hydrometallurgical techniques such as membrane precipitation, reverse osmosis and chelating resins. Among the adsorption techniques available was chosen chelating resin, which is favored by the metal adsorption capacity by ion exchange [3].

In the ion exchange process, the effluent (mobile phase) passes through a stationary bed (stationary phase) formed by ion exchange resin [4]. The ion exchange resins vary according to the physical and chemical characteristics, such as the type of matrix exchange group and its porosity [5].

Thus, it was chosen a chelating resin Dowex XUS43605 for the metals adsorption and due to adsorption capacity and selectivity as well as durability of resin [3].

Materials and Methods

Material

The resin adopted for this research was the chelating resin Dowex XUS43605. Table 1 shows the resin characteristics.

Pretreatment of Resin

The procedure was developed in a glass column with 50 cm height and internal diameter of 1.5 cm. In the pretreatment, the resin was washed with deionized water and eluted with acid solution (6 M HCl) to remove impurities. The procedure was repeated 2 times to ensure that the resin was ready for use.

Table 1 Physical and chemical characteristics of commercial resin Dowex XUS43605 [6]

Manufacturer	Dow chemical company
Functional groups	HPPA
Matrix	Macro styrene
Ionic form	Sulfuric acid
Humidity	40–60%
Size	320 μm
Exchange capacity	35 g/L as Cu

Preparation of Synthetic Solution

Analytical grade reagents were used to prepare the synthetic solution and the concentrations of the metals are 4101.5 ppm of Al [as $Al_2(SO_4)_3.17H_2O$], 78.1 ppm of Co [as $CoSO_4.7H_2O$], 195.2 ppm of Cr [as $Cr_2(SO_4)_3.8H_2O$], 146.9 ppm of Cu [as $CuSO_4.5H_2O$], 18,713.5 ppm of Fe [as $Fe_2(SO_4)_3.8H_2O$], 7774.5 ppm of Mg [as $MgSO_4.7H_2O$], 397.2 ppm of Mn [as $MnSO_4.H_2O$], 2434.2 ppm of Ni [as $NiSO_4.6H_2O$] and 36.7 ppm of Zn [as $ZnSO_4.7H_2O$].

Column Tests

Three tests occurred (Test 1, Test 2 and Test 3) and were conducted in two stages. At first, the loading was done in a column with the synthetic solution. In the second step it was made the elution of metals. For the tests 1, 2 and 3, the flow rate for loading are 0.45 BV/h (bed volumes per hour), 0.66 BV/h, 1.11 BV/h, respectively. The flow rate for the elution of metals using H_2SO_4 are 1.46, 4.45 and 1.33 BV/h. And the flow rate for the elution using NH4OH are 1.77, 2.22 and 1.33 BV/h, respectively for tests 1, 2 and 3.

Loading

The column experiments were conducted preliminarily in the same glass column utilized at pretreatment step, being filled with Dowex XUS43605 until reaching a bed height of 30 cm. The synthetic solution was passed through the column downward and the flow rate was controlled using a peristaltic pump. In order to evaluate the influence of flow rate (0.45, 0.66 and 1.11 BV/h) while the temperature was set at 20 °C and pH 1.5. Thus, the breakthrough curve was studied for each flow rate and each metal. Samples were collected at the end of the column at different time intervals and analyzed by X-ray fluorescence energy dispersive spectroscopy (EDX).

Elution

The elution experiments were done in order to recover the extracted metals and to regenerate the resin. The elution tests occurred after the loading step and they were realized with 1 M H_2SO_4 and 4 M H_4OH. The elution rate was controlled using a peristaltic pump and passed through the column in downflow. To avoid contamination of the resin during the elution experiments, it was used deionized water between each elution procedure.

Results

Three tests were carried out to evaluate the flow rate at pH 1.5 and 20 °C.

Test 1

In the first test, the influence of flow rate 0.45 BV/h on breakthrough curves of the metal is showed in Fig. 1.

Figure 1 shows normalized breakthrough curves for the metals present in the solution during resin loading, where C is the metal ion concentration after their sorption on resin; C0, the metal concentration in the initial solution; and BV, bed volume [4].

According to the curves exhibited for the metal such as Al, Co, Cr, Fe, Mg, Mn, Ni and Zin, the metallic ions were not adsorbed by the resin, so they were moved to the outlet solution. The Fig. 1 shows that the chelating resin has strong selectivity for copper over the other metals. The same behavior was verified by Liebenberg et al. [7]. So the resin affinity sequence is given in the following order: $Cu^{2+} \gg Ni^{2+} > Fe^{3+} > Zn^{2+} > Co^{2+} > Cd^{2+} > Fe^{2+}$ to pH < 2.0 [8]. According to Marston and Rodgers [9], around pH 2.0 the resin has a high affinity for nickel and copper, but not for cobalt.

Analyzing the ratio C/C0 in the early bed volumes (BV), all the metals in solution with the exception of Cu have reached C/C0 = 1, indicating that the displacement of metals to solution, the outlet concentrations become similar to the feeding concentration. This behavior indicates that the metals reached the equilibrium and no more adsorptions occur [10]. When the C/C0 > 1, this means that the metals (Al, Co, Cr, Fe, Mg, Mn, Ni and Zi) were adsorbed by the resin, but later they were displaced from the resin by the ions which have higher affinity, such as copper [11]. The ratio C/C0 < 1 occurs only for copper and indicates no saturation of resin [10].

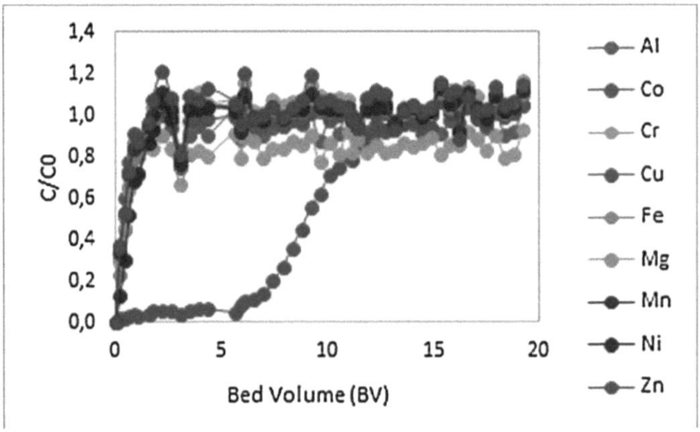

Fig. 1 Test 1—breakthrough curves (pH: 1.5–20 °C–0.45 BV/h)

The Cu adsorption during the experiment was up to 6 BV and after that the outlet concentration of Cu tends to reach the feeding concentration. Due to Cu selectivity 134.3 mg/L was extracted and this corresponds to 95.2% of the total feeding solution. After 6 BV the resin reached the saturation.

If there is an increasing in pH solution, the adsorption of Cu could increase, because the resin has preference for H^+ ions. Thus, at the pH of study (1.5), H^+ ions are in higher concentration than in pH 4.0, e.g., decreasing the load capacity of metals by the resin. At pH < 1.0, a small fraction of the resin functional group will be disassociated. And pH > 4.0, there will be a precipitation of metals such as iron and aluminum [10].

After loading the column, the elution experiments occurred first with 1 M of H_2SO_4 and after with 4 M of NH_4OH. The choice of these chemical reagents is based on studies for transition metals [11]. In addition, the sulfuric acid is effective to elute metals as nickel, iron and cobalt, while the ammonium hydroxide has elution capacity for copper [11]. Nitric acid was not used as eluent, once is an oxidizing agent capable of damaging the properties of the resin [11].

Elution profiles are shown in Fig. 2.

Analyzing the elution profiles in Fig. 2a, it was found that Al and Fe were eluted before the 0.5 BV and at lower concentrations than that contained in the feeding solution. The copper was completely eluted after 0.5 BV, although the elution had a metal concentration about 12 times higher than the concentration in the feeding solution. The same situation was observed for Mendes and Martins [10] however, the metal studied was nickel, and this ion metal was eluted concentrated more than 20 times.

In Fig. 2b it was noted that copper and nickel are desorbed from the resin, but in lower concentrations than in the acid elution. None of the other metals has been completely desorbed in the acid elution and therefore their concentrations are close to zero.

Thus, the combination eluting with H_2SO_4 and NH_4OH, respectively, shows promising results for the elution of metals. The result of the elution was proposed by Diniz et al. [11], when the elution started with NH_4OH, nickel and copper were eluted together to Dowex M-4195 resin. The same behavior not occurred to Dowex

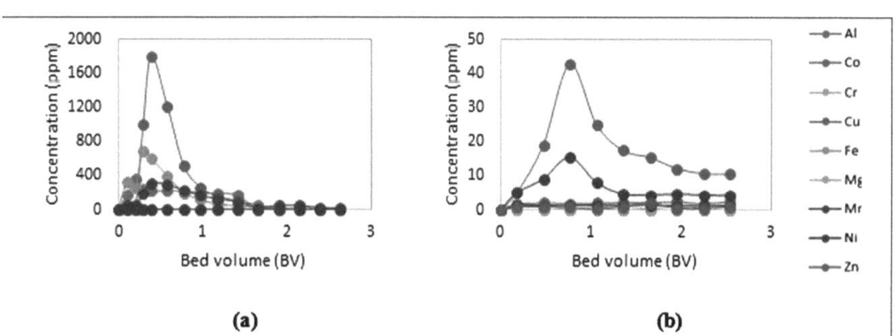

Fig. 2 Test 1—elution curves. Test conditions: **a** elution 1 M H_2SO_4 (1.46 BV/h); **b** elution 4 M NH_4OH (1.77 sBV/h)

XUS43605 resin. For this study, the metals were eluted using 1 M of H_2SO_4, and copper was eluted concentrated more than the nickel. Moreover, nickel is eluted together with copper. Although considerable concentration of impurities was detected, mainly Al and Fe.

Test 2

In test 2, the flow rate was changed from 0.45 to 0.66 BV/h and Fig. 3 shows the result.

The result showed the copper selectivity by the resin again, however, due to the increased flow rate was expected that the breakthrough was before 6 BV (test 1). According to Acheampong et al. [12], when the flow rate decreases it was expected to increase the breakthrough curves. However, the breakthrough curve was obtained at 10 BV. For other metals, the behavior remains the same as the test 1.

The elution profiles are shown in the Fig. 4.

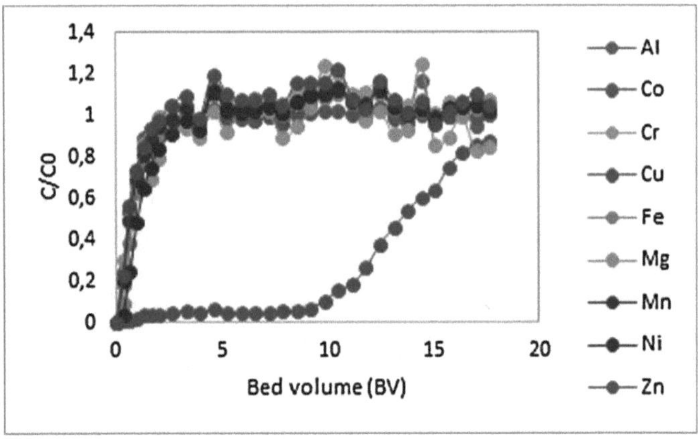

Fig. 3 Test 2—breakthrough curves (pH: 1.5–20 °C–0.66 BV/h)

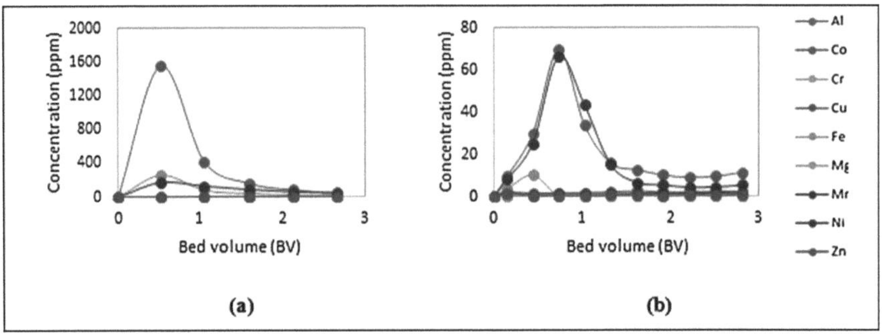

Fig. 4 Test 2—elution curves. Test conditions: **a** elution 1 M H_2SO_4 (4.45 BV/h); **b** elution 4 M NH_4OH (2.22 BV/h)

According to Fig. 4a, copper elution peaked at approximately 0.5 BV even with the increased flow rate, similar to the elution profile of the test 1. However, the eluted copper concentration decreased from 1800 to 1500 ppm. The result expresses the influence of the increase of flow rate in 2.5 times. Nickel and iron had the same elution peak than copper but at concentrations below 270 ppm. It was also observed the elution profiles using 4 M NH_4OH (Fig. 4b).

Comparing the elution using 4 M NH_4OH of test 1 and 2, copper and nickel were desorbed from the resin at the same 0.75 BV. However, in test 2, nickel was eluted in a higher concentration than in the previous study, as well as copper. For test 1, the copper elution was 43 ppm and 15 ppm nickel. However, for test 2, copper was 70 ppm and 66 ppm nickel. It shows that with increasing flow rate was increased concentration of the eluted metals. However, according to Liebenberg et al. [7] was expected that the increasing on the flow rate decreases concentration of the eluted metals. Liebenberg et al. [7], also reported that the copper elution is influenced by the acid concentration, there may be shifting of peaks and increasing peak width. Nevertheless, the concentration of sulfuric acid used remained with 1 M for all tests.

Test 3

In test 3, the loading flow rate was changed to 1.11 BV/h and Fig. 5 shows the results obtained.

Analyzing the Fig. 5 and compared to the test 2, it is noted that the breakthrough was reached before 5.5 BV. The slope of the breakthrough curve increased as the flow rate was changed from 0.66 to 1.11 BV/h, showing that the breakthrough curve becomes steeper with increasing flow rate. Thus, the flow rate 0.66 BV/h (test 2) resulted in a higher residence copper metal in the column and it was found that the increased flow reduces the volume treated even before the bed is saturated, and reducing the metal residence time in the column [12].

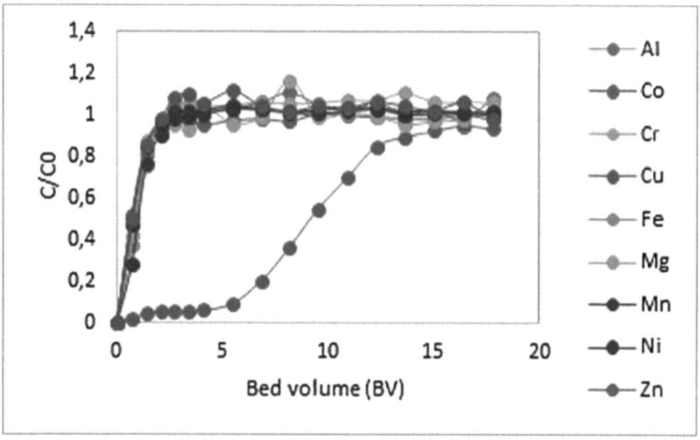

Fig. 5 Test 3—breakthrough curves (pH: 1.5–20 °C–1.11 BV/h)

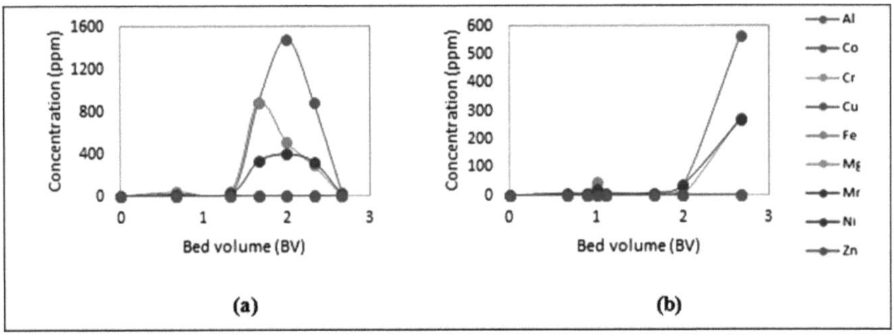

Fig. 6 Test 3—elution curves. Test conditions: **a** elution 1 M H_2SO_4 (1.33 BV/h); **b** elution NH_4OH 4 M (1.33 BV/h)

The elution profiles using H_2SO_4 and NH_4OH are illustrated in Fig. 6.

According to Liebenberg et al. [7], the effect of H_2SO_4 for the nickel elution profile was different than the copper. For the flow rate of 1.33 BV/h, the nickel elution profile was 400 ppm, while the copper was 1500 ppm, approximately. In the opposite of that proposed by Liebenberg et al. [7], with decreasing flow rate, the profiles shifted to the right.

For elution using 4 M NH_4OH, metals such as copper, nickel and iron had their peaks at approximately 2.6 BV and the respective concentrations of metals are 565.5, 270 and 276 ppm. However, the proposed 3 BV were insufficient and interrupted the elution of these metals.

Diniz et al. [11] revealed that the elution process is not simple. Although an effective elution would be crucial for commercial application of resins and the recovery of metals.

Conclusions

The results obtained allowed to conclude that the ion exchange Dowex XUS43605 resin was suitable for copper recovery from synthetic solution similar to effluent produced in nickel mining.

The experimental results confirm the resin selectivity for transition metal depends on the flow rate during the loading step. The copper breakthrough for all test occurred after all the other metals being influenced by the flow rate established for each test. The breakthrough of metals such as aluminum, cobalt, manganese and zinc, occurred before copper. This is determined by comparing the test 2 and 3 and found that for the test 2, having a smaller flow, the breakthrough curve was achieved later. Regeneration of this resin with H_2SO_4 and NH_4OH as eluents agents was very efficient and the regeneration could be reached. Copper was eluted through both acidic and basic elution. And after elution with 1 M H_2SO_4, was concentrated more than 12 times of the feeding solution.

Regeneration of chelating resin Dowex XUS43605 with 1 M H_2SO_4 and 4 M NH_4OH showed by the results the resin regeneration can be achieved. Copper was eluted either by two eluents used H_2SO_4 and NH_4OH. And after the elution with 1 M H_2SO_4, was concentrated more than 12 times of the feeding solution. Therefore, the study shows that the ion exchange resin can be used to treat effluents in particular mining.

Acknowledgements To the Counsel of Technological and Scientific Development (CNPq) for the financial support through doctorate grant.

To the Coordenação de Aperfeiçoamento de Pessoal de Nível Superior (CAPES) for the financial support through master grant.

To the Fundação de Amparo à Pesquisa do Estado de São Paulo (FAPESP) for financial support through the research Project 2012/51871-9.

To the Instituto Tecnológico Vale.

References

1. C.R.M. Butt, D. Cluzel, Nickel laterite ore deposits: weathered serpentinites. Elements **9**(2), 123–128 (2013)
2. P.C.F. Crowson, Mineral reserves and future minerals availability. Miner. Econ. [Internet]. **24** (1):1–6 (2011). Available from: http://link.springer.com/10.1007/s13563-011-0002-9
3. S.M. El-Bahy, Z.M. El-Bahy, Synthesis and characterization of polyamidoxime chelating resin for adsorption of Cu(II), Mn(II) and Ni(II) by batch and column study. J. Environ. Chem. Eng. **4**(1), 276–286 (2016)
4. A. Abrão (ed.), *Operações de Troca Iônica*. Instituto de Pesquisas Energéticas e Nucleares—CNEN/SP. (São Paulo, 2014), 8p
5. J.C. Riani. *Utilização de resinas de troca-ionica em efluentes de galvanoplastia*. Escola Politécnica da Universidade de São Paulo, (2008)
6. Dow Chemical Company. Ion Exchange Resins for Chemical Processing [Internet]. (cited 2016 Apr 5). Available from: http://msdssearch.dow.com/PublishedLiteratureDOWCOM/dh_07c8/0901b803807c8488.pdf?filepath=liquidseps/pdfs/noreg/177-02437.pdf&fromPage=GetDoc
7. C.J. Liebenberg, C. Dorfling, S.M. Bradshaw, G.A. Akdogan, J.J. Eksteen, The recovery of copper from a pregnant sulphuric acid bioleach solution with developmental resin Dow XUS43605. J. S. Afr. Inst. Min. Metall. **113**(5), 389–397 (2013)
8. Zaganiaris EJ. Ion Exchange Resins and Adsorbents in Chemical Processing [Internet]. 2013. Available from: https://books.google.co.uk/books?id=Fdl1AQAAQBAJ. 42p
9. C.R. Marston, M. Rodgers, Process Sep. Copper Nickel Cobalt containing solut. **1**(19), 2010–2012 (2010)
10. F.D. Mendes, A.H. Martins, Selective nickel and cobalt uptake from pressure sulfuric acid leach solutions using column resin sorption. Int. J. Miner. Process. **77**(1), 53–63 (2005)
11. C.V. Diniz, V.S.T. Ciminelli, F.M. Doyle, The use of the chelating resin Dowex M-4195 in the adsorption of selected heavy metal ions from manganese solutions. Hydrometallurgy **78** (3–4), 147–155 (2005)
12. M.A. Acheampong, K. Pakshirajan, A.P. Annachhatre, P.N.L. Lens. Removal of Cu(II) by biosorption onto coconut shell in fixed-bed column systems. J. Ind. Eng. Chem. [Internet]. **19** (3):841–8 (2013). Available from: http://dx.doi.org/10.1016/j.jiec.2012.10.029

Author Index

A
Abed, Abdulkader M., 343
Adams, F.V., 465
Adeleke, A.A., 205
Alvarenga, Rita de Cássia S.S., 291
Amireh, Belal S., 343
Apata, A.O., 465
Aune, Ragnhild E., 267

B
Barros, K.S., 485
Beheshti, Reza, 267
Belliveau, Thomas, 413
Bennett, James, 51, 69
Benouahmane, Zineb, 395
Bojarevics, V., 403
Bones, J.A., 387
Botelho, Amilton Barbosa Jr, 283, 479

C
Cesar, Kléos M.L., 291
Chen, Xiaowen, 87
Choquette, Denis, 413
Cianga Silvas, Flávia P., 495
Correa, Mónica M. Jiménez, 479
Cui, Senlin, 259

D
Dang, Jie, 117
de Paula Santos, Henrique, 291
Ding, Yuchuan, 311
Di Sabatino, M., 387
Djambazov, G., 403
Dorreen, Mark, 15
Dvorak, Petr, 229

E
Ellis, Timothy W., 435
Espelien, Stine, 355
Espinosa, Denise C.R., 239, 277, 283, 479, 485

F
Fontes, Maurício Paulo F., 291
Frydl, Tomas, 229

G
Gabis, Olivier, 413
Gan, Yunhua, 87
Ghodrat, Maryam, 131
Guo, Min, 191

H
Halli, P., 43
Hamad, Bothina M., 343
Hamuyuni, Joseph, 79, 103
He, Wenchao, 117
He, Zhi-jun, 147, 155, 161
Howes, John A., 435
Hsu, Esher, 123
Hu, Meilong, 87
Hu, Y., 387

I
Ikhmayies, Shadia J., 333, 343

L. Zhang et al. (eds.), *Energy Technology 2017*,
The Minerals, Metals & Materials Series, DOI 10.1007/978-3-319-52192-3

J
Jiménez Correa, Mónica M., 495
Jolivet, S., 43
Jung, In-Ho, 259

K
Kaya, Muammer, 445
Klöfverskjöld, A., 43
Kuo, Chen-Ming, 123

L
Lanterne, A., 387
Latostenmaa, P., 43
Li, Haige, 87
Lindberg, Daniel K., 27, 103
Liu, Ji-hui, 155
Li, Xueqin, 117
Li, Yaqiong, 395
Lu, Huimin, 325
Lundström, M., 43, 79
Lv, Xuewei, 117, 301

M
Ma, Naiyang, 173
Marques, Eduardo Antônio G., 291
Matthews, Geoff, 15
Mendes, Beatryz C., 291
Meteleva, Yulia Valery, 343
Morita, Kazuki, 367
Mubarok, M.Z., 247

N
Nagel, James R., 379
Nakano, Anna, 51, 69
Nakano, Jinichiro, 51, 69
Neelameggham, Neale R., 63
Ning, Zhe, 147, 155

O
Okanigbe, D.O., 205
Øvrelid, E., 387

P
Palimaka, Piotr, 181
Pang, Qing-hai, 147, 155, 161
Patel, Pretesh, 15
Perez, Isadora Dias, 495
Pericleous, K., 403
Pietrzyk, Stanislaw, 181
Popoola, A.P.I., 205
Prado, Pedro F.A., 277

Q
Qiu, Jie, 117

R
Rajamani, Raj K., 379
Ramakrishna, Ch., 425
Rashidi, Maria, 131
Romano Espinosa, Denise C., 495
Rosario, Carlos Gonzalo Alvarez, 283
Ruschak, Michael, 413

S
Safarian, Jafar, 355
Samali, Bijan, 131
Schütt, Michael, 3
Selucka, Jana, 229
Sieber, John, 413
Smith, York R., 379
Soares Tenório, Jorge A., 239, 495
Song, Teng-fei, 147
Starkova, Petra, 229
Stepien, Michal, 181
Subramanian, Ganesan, 63
Sukhomlinov, Dmitry, 79

T
Tabeshian, Ali, 267
Taskinen, Pekka, 79
Tenório, Jorge A.S., 277, 283, 479, 485
Tesfaye, Fiseha, 27, 103
Thorsen, F.W., 387
Thriveni, T., 425
Tian, Chen, 161
Tong, Yang, 87
Tranell, Gabriella, 355

U
Utimura, Solange Kazue, 283

V
Vu, Hong, 229

W
Wanderley, Kristine Bruce, 239
Wang, Hui-gang, 191
Wang, Panpan, 325
Wang, Ye, 367
Wei, Ruirui, 301
Whan, Ahn Ji, 425
Wills, Albert, 413
Wilson, B.P., 43

Wolters, Felix, 3
Wong, David S., 15
Wright, Linda, 15
Wyss, Rebecca, 413

X
Xiong, Hao, 311

Y
Yang, Mingrui, 301
Yang, Zhenzhou, 95
Yuan, Shilai, 325
Yudiarto, A., 247

Z
Zhan, Wen-long, 147, 155, 161
Zhang, Jun-hong, 147, 155, 161
Zhang, Lifeng, 395
Zhang, Mei, 191
Zhang, Sen, 161
Zhang, Zuotai, 95
Zhou, Zhiqiang, 311
Zhu, Libin, 311
Zhu, Zizong, 311

Subject Index

A

Acid leaching, 356–358
Aircraft, 259, 260, 266
Alkali metal, 161
Al-Si melts, 326, 330–332
Alumina, 3, 4, 8, 10
Aluminum, 267–271, 273
Aluminum alloy, 260
Aluminium smelting, 15, 16, 21–23
Andalusite ore, 326, 327, 329, 331, 332
Anthropogenic heating, 63, 66
Antimony, 43, 45, 48
Ash chemistry, 29, 32, 33, 36, 37
Assay, 205

B

Basicity, 312, 314, 320
Batteries, 435–439, 442
Bioleaching, 285, 286, 288
Biomass combustion, 27, 28, 33, 35
Boron removal, 367

C

Calcination, 247, 252, 256, 257
Carbonation, 430
Carbothermal reduction, 304, 305, 326, 329, 332
Cement clinker, 95–98, 101
Characterization, 207, 210, 213
Chemical composition, 413
Chemical looping, 69, 70, 76
Chromium removal, 234–237
Chronopotentiometry, 486, 487
Circular economy, 437
Climate change, 63, 64
Climate modeling, 63, 64
CO, 51, 52, 54
CO_2, 161–163, 166, 168
Coke, 161, 162, 164, 165, 168
Combined heat and power generator (CHP), 7
Compatibility and particle size, 465, 467, 469, 470
Compatibility test, 468
Computational modeling, 221
Copper, 267, 268, 270–273, 284, 285, 287, 288, 495, 498–503
Copper extraction, 288
Copper slag, 301–303, 309
Copper smelter dust (CSD), 206, 207, 209, 211, 213–215, 217, 220, 222
Crucible, 387–391, 393, 394
Current-voltage, 487, 489, 492

D

Degradation, 161, 162, 167, 168
Diffusion coefficient, 368–370, 372–374
Dismantling, 445, 453, 454, 459, 460
Dissolution of PDMS, 281
Doré Slag, 43, 44, 46–48
Dust recycling, 182

E

EAF, 131–136, 140, 142
Eco-cost model, 440
Eddy current, 380–382
Effect of leaching conditions on silicon purification, 362
Effect of acid concentration, 362
Effect of leaching time, 363
Effect of temperature, 363
Electrochemical cell and chronopotentiometric measurements, 488
Electro-spraying, 87–89
Electrostatic separation (ES), 452, 458–460
Energy, 103, 104, 106, 107, 110, 112–114
Energy consumption, 10
Energy flexibility, 17, 24
Energy optimization, 3, 8, 10

L. Zhang et al. (eds.), *Energy Technology 2017*,
The Minerals, Metals & Materials Series, DOI 10.1007/978-3-319-52192-3

Energy recovery, 123–125, 127, 128
EnPot, 15–17, 19–24
Environment management, 103, 114
Equilibrium, 334–336, 341
Ethanol, 87, 88, 91
Evaluation of Cu^{+2}concentration, 489
Evaluation of pH, 491
E-waste, 103–113, 446–448, 452–454, 456, 457, 459, 460
Extraction, 217, 219, 220, 222

F
Ferronickel slag, 250
Ferrosilicon alloy, 302, 309
FITs, 123, 124, 127, 128
Foundry sand, 465, 466, 473
FTIR-IR of
  backsheet, 280
  encapsulant, 280

G
Gasification, 51
Gasification reactivity, 153
Granularity, 147, 148, 150, 152, 153, 155, 156, 158, 160
Graphite, 395–400
Gravity separation (GS), 455, 457, 458, 460
Guard bands, 414

H
H_2, 51, 53, 54, 56–58, 60
Heat production, 51, 59
Heat recovery, 119, 120
Heavy metals, 430
Hematite particle surface, 73
Hematite surface, 74
High temperature, 239, 240
H_2O, 51–53, 55, 59
Hydrometallurgical process, 103, 104, 107–109, 111, 112, 114, 207, 217, 222, 496

I
Incineration, 131–133, 135–137, 139–142
In-situ study, 73
Ion exchange, 496, 502, 503
Ion-exchange membrane, 487, 488
Iron, 479–483
Iron and steelmaking, 51
Iron powder, 117, 119–121

L
Laterite ore, 239
Leaching, 43–48, 247, 249, 251–255, 257
  test of, 251
Leaching-SEM study, 358, 360
Life-cycle analysis (LCA), 131, 133–135, 137, 138, 437, 439
Li-ion battery, 270
Limiting current, 486, 487, 489, 491, 492
Lithium, 267–270, 435–439, 442
Livestock waste, 123–128

M
Magnesium
  leaching of, 252
  recovery of, 252, 255
Magnesium addition effect, 359
Magnesium silicide, 357
Magnetic property, 199
Magnetic separation (MS), 458, 460
Measuring the diffusion coefficient of B in the slag, 369
Membrane resistance, 489, 491
Meso-scale combustor, 87
Metal-doped Ni-Zn ferrite, 191, 192, 194, 198–200
Metallic fractions, 455, 458, 460
Metallurgical properties, 312, 313, 317
Metal recovery, 104, 107, 111, 445
Microstructure, 161, 162, 164, 168, 312, 314, 360
Microwave radiation, 147–150, 153, 155, 156
Molten salt, 30, 32
Molten silicon, 369, 374
Monohydrate, 239–241, 243–245
Mouldability, 465, 466, 469, 472, 474
Municipal solid waste, 426

N
Nickel, 495, 498–502
Nonmetallic fractions, 450, 455, 460
Nonmetal recovery, 445

O
Optimized technological process of producing SOG-Si, 331
Overall purification, 363, 365
Oxidation, 69, 70, 74, 75
Oxidative alkali roasting, 229, 230, 233, 237
Oxidative leaching, 229, 230, 232, 236, 237
Oxygen carrier, 69, 70, 76
Oxygen content, 395, 397, 398

P
Paint, 293
Particle powder, 312, 317, 318, 320
PDA, 87–89
Permeability test, 468

Phase diagrams, 334, 341
PM 2.5, 426–428, 430
PM 10, 425, 426, 430
Power grid, 15
Precipitation, 247, 252, 254–257
Printed circuit board (PCB), 446, 447, 449–452, 454, 455, 457, 459, 460
Printed wired board (PWB), 284, 285, 288
PV materials, 380
PV waste, 277
Pyrometallurgy, 103, 104, 106–110, 112–114, 268

Q
Quartz, 333–337, 341, 387, 390, 393

R
Raw materials, 328, 329, 332
Recycling
  aircraft Al, 260
  mechanical, 452, 456
  physical, 452, 459, 460
  process of, 260
Redox gas exposure, 72
Reduction process, 69, 70, 74, 75, 480, 482, 483
Removal of magnesium, 240, 242
Results and discussion, 359
Rice Husk ash, 466, 467, 469, 470, 473–475
Rotary disc atomizer, 117, 120

S
Salt heater, 8
Sandstone, 344–346, 348, 349
Sewage sludge, 95, 96
Silicate-oxide slags, 34
Silicon, 333–335, 340, 341, 345, 349, 355–360, 362–364, 387–390, 393–400
Silicon material, 357
Silicon solar cells, 333
Silicon solidification in Al-Si melts, 330
Simple models, 65
Sintering, 312–314, 316, 317
Slag processing, 27, 30, 173, 174, 177, 178
Slag product, 173–176, 178, 179
Slag refining, 374
Slag utilization, 51, 52, 59
Sodium dithionite, 479–481
Soil-cement, 292, 293, 298
Solar cells, 345, 346, 387, 388
Solar grade silicon, 325, 356
Solidification, 326, 328, 330–332
Solid state reaction, 191, 192, 200
Solutions, 487, 489, 490
Spark-AES, 413, 414
Stabilization, 425
Steelmaking dust utilization, 182–184
Steelmaking slags, 173–179
Strength
  compression, 469, 474, 475
  shear, 465, 469, 474
Sub-molten salts, 229, 230

T
Test methods, 328
TGA of encapsulant, 281
Theory of Eddy current separation, 381
Thermal decomposition, 240, 241
Thermodynamic calculation, 259, 260, 262–266
Thermodynamic equilibrium, 95–97, 99, 101
Thermodynamic evolution, 269
Thermodynamic modeling, 28, 30, 35, 37
Trace element, 95–97, 99–101
Transition time, 486, 490–492
TROF furnace, 48
Tube digester, 3–6
Tumbler strength, 313, 315, 316, 318, 320

U
Ultrasonic, 155–160
Uncertainty, 413–418, 420, 423, 424

V
Vacuum annealing treatment of graphite, 396
Vacuum treatment, 397
Virtual battery, 15
Volume expansion, 59

W
Waste, 291–293
Waste electric and electronic equipment (WEEE), 277, 283
Waste recycling, 124, 128
Waste syringes, 131–133, 135–137, 139–141
Wettability, 396, 397

X
X-ray diffraction, 345
X-ray fluorescence, 345

Z
Zinc recovery, 182
Zn-containing EAFD, 193–195, 199, 200

MIX
Papier aus verantwortungsvollen Quellen
Paper from responsible sources
FSC® C105338

If you have any concerns about our products,
you can contact us on
ProductSafety@springernature.com

In case Publisher is established outside the EU,
the EU authorized representative is:
Springer Nature Customer Service Center GmbH
Europaplatz 3, 69115 Heidelberg, Germany

Printed by Libri Plureos GmbH
in Hamburg, Germany